Computing Patterns in Strings

We work with leading authors to develop the
strongest educational materials in computer science,
bringing cutting-edge thinking and best
learning practice to a global market.

Under a range of well-known imprints, including
Addison-Wesley, we craft high-quality print and
electronic publications which help readers to understand
and apply their content, whether studying or at work.

To find out more about the complete range of our
publishing, please visit us on the World Wide Web at:
www.pearsoneduc.com

Computing Patterns in Strings

Bill Smyth

McMaster University
Curtin University of Technology

Harlow, England • London • New York • Boston • San Francisco • Toronto
Sydney • Tokyo • Singapore • Hong Kong • Seoul • Taipei • New Delhi
Cape Town • Madrid • Mexico City • Amsterdam • Munich • Paris • Milan

Pearson Education Limited
Edinburgh Gate
Harlow
Essex CM20 2JE
England

and Associated Companies throughout the world

Visit us on the World Wide Web at:
www.pearsoneduc.com

First published 2003

ISBN 0 201 39839 7

British Library Cataloguing-in-Publication Data
A catalog record for this book is available from the British Library

Library of Congress Cataloging-in-Publication Data
A catalog record for this book is available from the Library of Congress

10 9 8 7 6 5 4 3 2 1
07 06 05 04 03

Typeset by 68
Printed and bound in Great Britain by Biddles Ltd, Guildford and King's Lynn

Contents

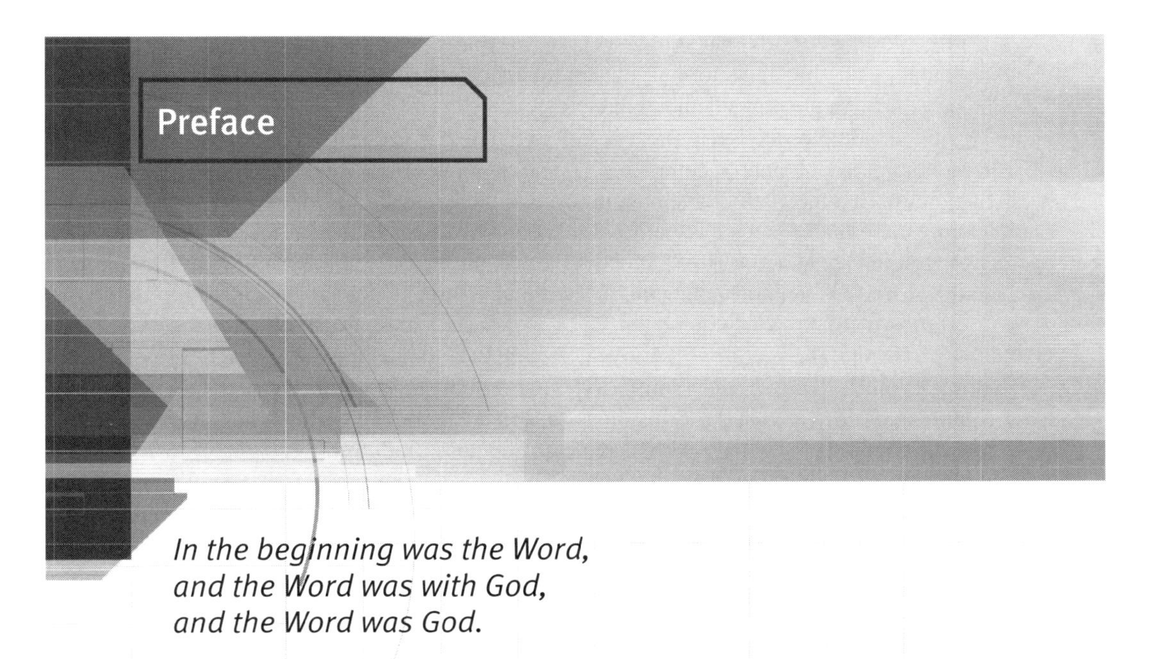

Preface

In the beginning was the Word,
and the Word was with God,
and the Word was God.

— *John 1:1*

The computation of patterns in strings is a fundamental requirement in many areas of science and information processing. The operation of a text editor, the lexical analysis of a computer program, the functioning of a finite automaton, the retrieval of information from a database — these are all activities which may require that patterns be located and/or computed. In other areas of science, the algorithms that compute patterns have applications in such diverse fields as data compression, cryptology, speech recognition, computer vision, computational geometry, and molecular biology. And computing patterns in strings is not a topic whose importance lies only in its current practical applications: it is a branch of combinatorics that includes many simply-stated problems which often turn out to have solutions — and often more solutions than one — of great subtlety and elegance.

It is surprising therefore that academic Departments of Mathematics or Computer Science do not generally include in their undergraduate or graduate curricula courses which provide an introduction to this interesting, important, and heavily researched topic. It is perhaps even more surprising that so few texts have been written with the purpose of putting together in a uniform way some of the basic results and algorithms that have appeared over the past quarter-century. I know of five books [St94, CR94, G97, SM97, CHL01] and three fairly long survey articles [BY89a, A90, Nav01] whose subject matter overlaps significantly with that of this volume. Of these, the survey articles, [St94], and [CR94] are written more as summaries of research than as texts for students; while [G97] and [SM97] focus heavily on the (important) applications in molecular biology. The final monograph [CHL01] does indeed function as both a monograph and a textbook on string algorithms, and is moreover both clearly and elegantly written; unfortunately, it is currently available only in French.

The purpose of this book then is to begin to fill a gap: to provide a general introduction to algorithms for computing patterns in strings that is useful to experienced researchers in this and other fields, but that also is accessible to senior undergraduate and graduate students. Let us linger a moment over three of the words used in the preceding sentence: "accessible", "algorithms", and "patterns".

An overriding objective in this book is to make the material *accessible* to students who have completed or nearly completed a mathematics or computer science undergraduate curriculum that has included some emphasis on discrete structures and the algorithms that operate upon them.

A first consequence of this objective is that the mathematical background required to read this book is general rather than specific to strings. It would certainly be provided by the standard IEEE/ACM courses in Discrete Mathematics, Data Structures, and Analysis of Algorithms. The reader will know what stacks, queues, linked lists and arrays are, for example, and will have some familiarity with the analysis of algorithms and the "asymptotic complexity" notation used for this purpose, some experience with mathematical assertions and the methods used to establish their correctness, and some knowledge of important algorithms on graphs and trees. In addition, the assumption is made that the reader is familiar with some computer programming language, and has the ability to read and understand algorithms expressed in such a language.

A second consequence is that no claim is made to completeness: my objective is to lure the student and the reader into a fascinating field, not to write an encyclopædia of algorithms that compute patterns in strings. In particular, I have been selective in two main ways: I focus on results that are (I believe) important and that moreover can be explained with reasonable economy and simplicity. Inevitably, this means that the exposition of some interesting and valuable material is omitted. However, I hope that, both by providing references to much of this material and by stimulating interest, I will encourage readers to investigate the literature for themselves.

The underlying subject matter of this book is a mathematical object called by most computer scientists a "string" (or, in Europe and by most mathematicians, a "word"). But the focus of this book is on *algorithms* — that is, on precise methods or procedures for doing something — and it is thus more properly thought of as a text in computer science rather than in mathematics. This book will therefore take quite a different approach from that of the classic monograph [L83], and its descendants [L97, L02], that elucidates mathematical properties above all. We shall rather be interested primarily in algorithms that find various kinds of patterns in strings; for the most part, only as a byproduct of that focus will interest be displayed in the mathematical properties of the strings themselves. This does not mean that results will not be proved rigorously, only that the selection of those results will generally depend on their relevance to the behaviour of some algorithm. A final remark here: in the exposition I confine myself to sequential algorithms on strings in one dimension, making no reference to the extensive literature on the corresponding parallel (especially PRAM) algorithms or to the growing literature on multi-dimensional (especially two-dimensional) strings.

Another focus of this book will be on *patterns*. That is, the algorithms we discuss will virtually all be devoted to finding some sort of a pattern in a string. I say "some sort" of pattern, because three main kinds will be distinguished — specific, generic, and intrinsic — that provide this book with three of its four main divisions.

A *specific pattern* is one that can be specified by listing characters in their required order; for example, if we were searching for the pattern $\boldsymbol{u} = abaab$ in the string $\boldsymbol{x} = abaababaabaab$ we would find it (three times), but we would not find the pattern $\boldsymbol{u} = ababab$.

(Sometimes the pattern that we are looking for can contain "don't-care" symbols, and sometimes the match that we seek need only be "approximate" in some well-defined sense, but these are refinements to be dealt with later.)

By contrast, a *generic pattern* is one that is described only by structural information of some kind, not by a specific statement of the characters in it. For example, we might ask for all the "repetitions" in $\boldsymbol{x}$ — that is, for all cases in which two or more adjacent substrings of $\boldsymbol{x}$ are identical. (The response in this case would be a list of repetitions including $(aba)(aba)$, $(abaab)(abaab)$, aa (three separate times), and several others — see if you can find them all.)

I call the final kind of pattern that we search for an *intrinsic pattern* — one that requires no characterization, one that is inherent in the string itself. Here I discuss various patterns that in one way or another expose the periodic structure of a given string $\boldsymbol{x}$; for example, normal form, suffix tree, Lyndon decomposition, s-factorization. These intrinsic patterns are used so frequently in algorithms that compute specific or generic patterns as to be almost ubiquitous. Collectively, they form the basis for the efficient processing of strings. The variety of these intrinsic patterns is remarkable: the normal form of our example string $\boldsymbol{x}$ is $(abaababa)(abaab)$ while its Lyndon decomposition is $(ab)(aabab)(aab)(aab)$ and its s-factorization is $(a)(b)(a)(aba)(baaba)(ab)$ — all of these patterns are computationally useful.

The organization of this book is as follows. Part I gives basic information about both strings and algorithms on strings. It provides the terminology, notation and essential properties of strings that will be used throughout; in addition, it describes the main kinds of algorithms to be presented and illustrates some of them on certain famous strings; these strings are also "infamous" as examples of worst case behaviour for many string algorithms. It is particularly Chapter 2 that provides a kind of key to the rest of the book: it explains precisely which problems are to be solved and directs the reader to the later sections that present the algorithms that solve them. Thus the book may be used fairly easily by the reader who has selective interests. Part I also discusses qualities of "good" algorithms on strings, and raises the interesting question of how implementation of these algorithms should in practice be validated. Parts II–IV deal with algorithms for computing intrinsic, specific and generic patterns, respectively, as described above.

Altogether there are 13 chapters distributed over the four parts. As indicated in the table of contents, these are broken down into sections, each of which ends with a collection of exercises. Where appropriate, chapters include a summary of the topics/algorithms covered and a discussion of related results, additional topics and open problems.

A note about the exercises, of which there are some 500 or so: they are an integral part of the book, used for four main purposes:

- to make sure the reader has understood;
- to clarify, or to put into a different context, principles or ideas that have arisen in the text — in a phrase, to make *connections*;
- to handle extensions or modifications of algorithms or mathematical results that the reader should be aware of, but should not need to have explained to him in detail;
- to deal with details (of algorithms, of proofs) that would otherwise unnecessarily clutter the presentation.

Wherever possible, by explaining in the text only what really needs to be explained, I have tried through the exercises to involve the reader in the development or analysis of the algorithms presented. What I myself have discovered by taking this approach is that for the most part the algorithms, and the improvements to algorithms, depend on a very simple new idea, an insight that is not complicated but that has somehow previously escaped other researchers. Once that idea is captured, what remains consists mainly of technicalities — tedious and convoluted perhaps, but still a direct consequence of the main idea. This observation seems to be true of string algorithms: I wonder what fields of study it is not true of?

A note also about the dreaded index: if on page p of the text I have cited a work authored or co-authored by person P, then the index entry for P should include p.

I am very sensible of the fact that sometimes I can be hasty (as Treebeard [T55] would say), consequently error-prone. I have therefore spent a great deal of time reworking this book in order to correct errors, rectify oversights, or smooth over inelegancies; nevertheless I cannot imagine that there will not be many defects to be found. I will be maintaining a website

`http://www.cs.curtin.edu.au/~smyth/patterns.shtml`

to record corrections and suggestions for improvement, and I would be grateful if readers would contact me at

`smyth@computing.edu.au` or `smyth@mcmaster.ca`

with their comments.

The material in this book is at least sufficient to cover two one-semester (12–14 week) courses for graduate or advanced undergraduate students. Indeed, I hope that it is more than sufficient: I hope that it is also suitable. The initial chapters have already been presented several times to graduate computer science students in the Departments of Computer Science & Systems and of Computing & Software at McMaster University, Hamilton, Ontario, Canada; also to graduate students in the Department of Computer Science, University of Debrecen, Hungary. These students have contributed materially to the book's development.

I wish particularly to express my deep gratitude to the School of Computing, Curtin University, Perth, Western Australia, and to its past and present Heads of School, Dennis Moore, Terry Caelli, Svetha Venkatesh and Geoff West, for generous support and encouragement, both intellectual and practical, over a period of several years. Most of this book has been written during my sheltered visits to Curtin. I am grateful also to Professor Pethő Attila, Chair of the Department of Computer Science at Debrecen, for his interest and support: it was in Debrecen late in 2001 that the last bits of LaTeX were finally keyed in. It is a pleasure to express my debt to my friends and colleagues, Leila Baghdadi, Jerry Chapple, Franya Franěk, Costas Iliopoulos, Thierry Lecroq, Yin Li, Dennis Moore, Pat Ryan, Jamie Simpson and Xiangdong Xiao for their valuable contributions. And many thanks also to two anonymous referees whose constructive comments have contributed materially to the final form of the book. Finally, kudos to Jocelyn Smyth for her entertaining selection and careful verification of "string" and "word" quotations!

W. F. S.

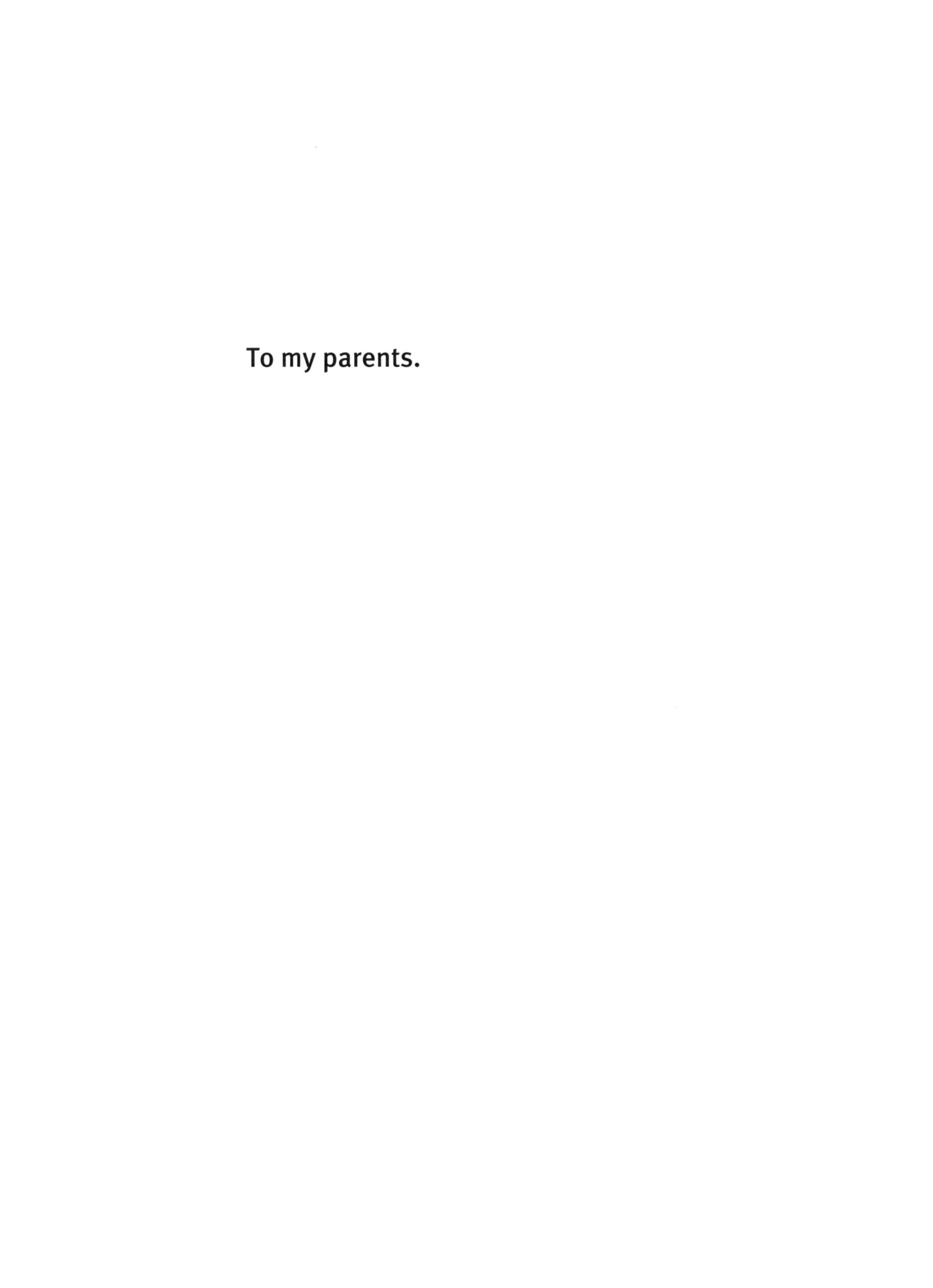

To my parents.

Part I

Strings and Algorithms

Part I

A word in time saves nine.

— Anonymous

Chapter 1

Properties of Strings

1.1 Strings of Pearls

Words form the thread on which we string our experiences.

— Aldous HUXLEY (1894–1963), *Brave New World*

Consider a string of pearls. Imagine that the string is laid out on the table before you, so that one end is on the left and the other on the right. Suppose that there are n pearls in the string, and that each pearl has a tiny label pasted on it. Suppose further that the labels are integers in the range $1..n$ and that they satisfy the following rules:

- the label on the leftmost pearl is 1;
- for every integer $i = 1, 2, \ldots, n-1$, the pearl to the right of the pearl labelled i has label $i+1$.

These rules seem to satisfy our intuitive idea of what makes a string of pearls a "string": the pearls all lie on a single well-defined path, and the path can be traversed from one end to the other by moving from the current pearl to an adjacent one. Reflecting on the rules, however, we realize that we need not be so specific. First of all, of course, we do not really need to speak of "pearls": we can speak more generally of (undefined) ***elements***. But a second, more fundamental, observation is that the labels do not need to be chosen in any specific order, and they do not need to be integers: they could be colours, for example, or letters of the alphabet. What really matters is that

(0) every element has a label that is unique;

(1) every element with some label x (except at most one, called the ***leftmost***) has a unique determinable ***predecessor*** labelled $p(x)$;

(2) every element with some label x (except at most one, called the ***rightmost***) has a unique determinable ***successor*** labelled $s(x)$;

(3) whenever an element with label x is not leftmost, $x = s(p(x))$;

(4) whenever an element with label x is not rightmost, $x = p(s(x))$;

(5) for any two distinct elements with labels x and y, there exists a positive integer k such that either $x = s^k(y)$ or $x = p^k(y)$.

These rules capture the essential idea of ***concatenation***: each element has either a unique predecessor or a unique successor, and, except at the extremes, actually has both. Furthermore, by following a finite sequence of either successors or predecessors, we can reach any element with label y from any other element with label x. Fortified with these observations, then, we boldly state:

Definition 1.1.1 *A* ***string*** *is a collection of elements that satisfies rules 0–5.*

A critical feature of this definition, not as yet discussed, is the condition, included in rules 1 and 2, that there be *at most* one leftmost or rightmost element. For consider what happens when the clasp is fastened on the original string of pearls, forming a necklace. Now there is no longer either a "leftmost" or a "rightmost" element — but that turns out not to be a problem, since rules 0–5 continue to apply. According to our definition, a necklace is also a string!

Furthermore, suppose that the number of pearls in the original string were infinite: beginning at the lefthand edge of the table but stretching away without end toward a forever unseen edge at the right. We see that this infinite string also is covered by the definition: it has a leftmost element, but no rightmost one. And, perhaps most surprising of all, we see that even a string which extends to infinity in both directions is covered by rules 0–5: in this case there is again neither a leftmost nor a rightmost element.

In this book we will at various times become interested in all of these different kinds of strings. To prevent confusion, therefore, we adopt the following conventions. A string with a finite number of elements including both a leftmost and a rightmost element will be called a ***linear string***. A string with a finite but nonzero number of elements and neither a leftmost nor a rightmost element will be called a ***necklace*** (in the literature also called a ***circular string***). A string with an infinite number of elements, of which one is leftmost, will be called an ***infinite string***; while a string with an infinite number of elements, of which none is either leftmost or rightmost, will be called an ***infinite necklace***. When the meaning is clear from the context, we will just use the word "string" to refer to any object satisfying Definition 1.1.1. In practice it will be easy to distinguish between linear strings and necklaces, because necklaces will normally be defined in terms of a corresponding linear string x and written

$C(\boldsymbol{x})$ — we think of $C(\boldsymbol{x})$ as being the necklace formed from $\boldsymbol{x}$ by making its leftmost element the successor of its rightmost element.

Exercises 1.1

1. Explain why concatenation rule 3 is required. Give an example of a mathematical object which satisfies rules 0–2 but not rule 3.
2. Can rule 4 for concatenation be derived from rules 0–3? Explain your answer.
3. Explain why concatenation rule 5 is required. Characterize the mathematical objects that satisfy rules 0–4 but not rule 5.
4. Does the infinite set $\{a, a^2, \ldots.\}$ of strings contain an infinite string?
5. According to Definition 1.1.1, can a string consist of a single element a? Could such a string be a necklace?
6. Is Definition 1.1.1 satisfied by a string with *no* elements in it (a so-called ***empty*** string)? Does the above definition of a linear string include the empty string? What about the definition of a necklace?
7. In view of the preceding exercise, how many different kinds of string are included in the classification of strings given in this section?
8. Our classification of strings omits the following cases:

   (a) a string with an infinity of elements including a rightmost one but no leftmost one;

   (b) a string with a finite number of elements including either a leftmost one or a rightmost one, but not both.

   Comment on these omissions.
9. Is there any way to *prove* that Definition 1.1.1 defines a string?

1.2 Linear Strings

He who has been bitten by a snake
fears a piece of string.

— Persian proverb

From the discussion of the preceding section, it becomes clear that the idea of a string, though a simple one, is also very general. A string might be

- a word in the English language, whose elements are the upper and lower case English letters together with apostrophe (') and hyphen (-);
- a text file, whose elements are the ASCII characters;
- a book written in Chinese, whose elements are Chinese ideographs;
- a computer program, whose elements are certain "separators" (space, semicolon, colon, and so on) together with the "words" between separators;
- a DNA sequence, perhaps three *billion* elements long, containing only the letters C, G, A and T, standing for the nucleotides cytosine, guanine, adenine and thymine, respectively;
- a stream of bits beamed from a space vehicle;
- a list of the lengths of the sides of a convex polygon, whose values are drawn from the real numbers.

All of these examples are instances of what we have called in Section 1.1 a "linear string". Indeed, most of this book will deal with linear strings, and so in this section we introduce notation and terminology useful for talking about them. Much, but not all, of this terminology will also apply to necklaces, infinite strings, and the empty string.

The examples make clear that an important feature of any string is the nature of its elements: bits, members of $\{C, G, A, T\}$, real numbers, as the case may be. It is in fact customary to describe a string by identifying a set of which every element in the string is a member. This set is called an ***alphabet***, and so naturally its members are referred to as ***letters*** — though, as we have seen, the term "letter" must be interpreted much more broadly than is usual in English. We say then that a string is defined ***on*** its alphabet. Of course, if a string $\boldsymbol{x}$ is defined on an alphabet A, then $\boldsymbol{x}$ is also defined on any superset of A, so an alphabet for $\boldsymbol{x}$ is not unique. A minimum alphabet for $\boldsymbol{x}$ is just the set of all the distinct elements that actually occur in $\boldsymbol{x}$. Sometimes it is convenient to define the alphabet of a string as the minimum one ("bits"), sometimes as a set that is far from minimum ("real numbers").

Throughout this book A will denote an alphabet and $\alpha = |A|$ its cardinality. In the common cases that α is 2, 3 or 4, we say that the alphabet A is ***binary***, ***ternary*** or ***quaternary***, respectively; as we shall see, there are many interesting strings on a binary alphabet, and a quaternary alphabet is of particular importance because of applications to the analysis of DNA sequences.

In general, apart from the distinctness of the elements of A that follows from the set property, we place no other restriction upon them: the elements of the alphabet may be finite (even zero!) in number, countably infinite (for example, the integers), or uncountably infinite (for example, the reals). And the elements of the alphabet may be totally ordered (so that a comparison of any distinct pair of them yields the result "less" or "greater"), unordered, or somewhere in between ("partially ordered"). For many of the algorithms discussed in this book, it will be sufficient to use an unordered alphabet; as discussed in Chapter 4, the nature of the alphabet on which an algorithm operates is very important to the selection of test data for the algorithm as well as to its computational efficiency.

For a given alphabet A, let A^+ be the set of all possible nonempty finite concatenations of the letters of A. Thus, for example, if $A = \{a\}$, then $A^+ = \{a, a^2, \ldots.\}$, where we write a^2 for aa, a^3 for aaa, etc.; and if $A = \{0, 1\}$, then A^+ consists of all distinct nonempty finite sequences of bits, and so may be thought of as including all the nonnegative integers. As suggested in Exercise 1.1.6, it is convenient also to introduce the idea of the ***empty string***, usually written ε, which we use to define the sets $A' = A \cup \{\varepsilon\}$ and $A^* = A^+ \cup \{\varepsilon\}$.

This terminology allows us to express another definition of linear strings equivalent to that given in the previous section:

> **Definition 1.2.1** *An element of A^+ is called a* ***linear string*** *on alphabet A. An element of A^* is called a* ***finite string*** *on A.*

Thus A^+ is the set of all linear strings on a given alphabet A. Note that the empty string is *not* a linear string: after all, it has neither a leftmost nor a rightmost element!

Throughout this book strings will consistently be denoted by boldface letters, almost always lower case: $\boldsymbol{p}$, $\boldsymbol{t}$, $\boldsymbol{x}$, and so on. We will implement the rules 0–5 of Section 1.1 by treating strings as one-dimensional arrays; alternatively, we might have used a linked-list or some other representation, but arrays are a simple and natural model, as we have seen with the string of pearls. Thus for any string, say $\boldsymbol{x}$, containing $n \geq 0$ letters drawn from an alphabet A, the implicit declaration will be

$$\boldsymbol{x} : \textbf{array } [1..n] \textbf{ of } A.$$

In this case we will say that the string has ***length*** $n = |\boldsymbol{x}|$ and ***positions*** $1, 2, \ldots, n$. For any integer $i \in 1..n$, the letter in position i is $\boldsymbol{x}[i]$, so that we may write

$$\boldsymbol{x} = \boldsymbol{x}[1]\boldsymbol{x}[2] \cdots \boldsymbol{x}[n],$$

which we recognize from Definition 1.1.1 as a concatenation of n strings of length 1. In fact, in this formulation the position i plays the role of the label introduced in Definition 1.1.1.

Note that the array model works also for the empty string $\boldsymbol{x} = \varepsilon$ which corresponds to an empty array and has length 0.

Digression 1.2.1 We have said that arrays are a "simple and natural" representation of strings, a statement that obscures a significant computational issue. Certainly an array is a simple data structure, but whether or not it is natural depends on assumptions about the mechanisms by which the elements of the array are accessed. As we have seen, strings are defined in terms of concatenation, and so it would seem to be "natural" to access elements using the successor (next) and predecessor (previous) operations introduced in Section 1.1. These are mechanisms compatible with a linked-list representation, where access to an element at position i from the "current" position j would at least require time proportional to $|j - i|$. However, an arbitrary element in a computer array can normally be accessed in constant time simply by specifying its location i, quite independent of the array position j most recently visited. An array representation of strings is thus more powerful than a linked-list representation, and so arguably not suitable, since it could justify execution time estimates for algorithms lower than those attainable using list processing.

In practice, algorithms on strings almost always begin either at the left (position 1) or at the right (position n), and inspect adjacent (successor or predecessor) positions one by one. On the other hand, the output of string algorithms often specifies string positions, the implicit assumption being that the user can access these positions in constant time. One way to reconcile these different models of string access is to suppose that strings are initially available as linked-lists — so that their elements are accessible only one-by-one, either left-to-right or right-to-left — but copied as they are input into an array. This copying, if it were necessary, would require only $\Theta(n)$ time and $\Theta(n)$ additional space, and so would not affect the asymptotic complexity of any of the algorithms considered in this book. We therefore adopt the rather odd convention that strings are processed as linked-lists on input, but may in some cases be regarded as arrays on output. We promise to alert the reader if ever we deviate from this convention. □

Equality between strings is defined in an obvious way. A string $\boldsymbol{x}$ of length n and a string $\boldsymbol{y}$ of length m are said to be ***equal*** (written $\boldsymbol{x} = \boldsymbol{y}$) if and only if $n = m$ and $\boldsymbol{x}[i] = \boldsymbol{y}[i]$ for every $i = 1, \ldots, n$. Thus the empty string ε is equal only to itself. Note further that, by this definition, prepending or appending ε to a given string $\boldsymbol{x}$ does not change its value; thus we may if we please write $\boldsymbol{x} = \varepsilon\boldsymbol{x}\varepsilon$.

Corresponding to any pair of integers i and j that satisfy $1 \le i \le j \le n$, we may define a ***substring*** $\boldsymbol{x}[i..j]$ of $\boldsymbol{x}$ as follows:

$$\boldsymbol{x}[i..j] = \boldsymbol{x}[i]\boldsymbol{x}[i+1]\cdots\boldsymbol{x}[j].$$

We say that $\boldsymbol{x}[i..j]$ ***occurs*** at position i of $\boldsymbol{x}$ and that it has ***length*** $j-i+1$. If $j-i+1 < n$, then $\boldsymbol{x}[i..j]$ is called a ***proper substring*** of $\boldsymbol{x}$. Two noteworthy kinds of substring are $\boldsymbol{x} = \boldsymbol{x}[1..n]$ of length n, and $\boldsymbol{x}[i] = \boldsymbol{x}[i..i]$ of length 1. For every pair of integers i and j such that $i > j$, we adopt the convention that $\boldsymbol{x}[i..j] = \varepsilon$, a substring of length zero. As we have already seen, since $\boldsymbol{x} = \varepsilon\boldsymbol{x}\varepsilon$, ε may be regarded as a substring of any element of A^*.

Let $k \in 1..n$ be an integer, and consider positions $i_k = 1, 2, \ldots, n-k+1$ of $\boldsymbol{x}$. Each of these $n-k+1$ positions represents the starting position of a substring $\boldsymbol{x}[i_k..i_k+k-1]$ of length k. Thus every string of length n has $n-k+1$ (not necessarily distinct) substrings of length k.

If $\boldsymbol{u}$ is a substring (respectively, proper substring) of $\boldsymbol{x}$, then $\boldsymbol{x}$ is said to be a ***superstring*** (respectively, ***proper superstring***) of $\boldsymbol{u}$. Of course substrings and superstrings are also strings.

If A is ordered, we may use the substring notation to define a corresponding induced order on the elements of A^* called ***lexicographic order*** — that is, dictionary order. More precisely, suppose we are given two strings $\boldsymbol{x} = \boldsymbol{x}[1..n]$ and $\boldsymbol{y} = \boldsymbol{y}[1..m]$, where $n \ge 0$, $m \ge 0$. We say that $\boldsymbol{x} < \boldsymbol{y}$ ($\boldsymbol{x}$ is lexicographically less than $\boldsymbol{y}$) if and only if one of the following (mutually exclusive) conditions holds:

- $n < m$ and $\boldsymbol{x}[1..n] = \boldsymbol{y}[1..n]$ (as we shall see shortly, this is the case in which $\boldsymbol{x}$ is a "proper prefix" of $\boldsymbol{y}$);
- $\boldsymbol{x}[1..i-1] = \boldsymbol{y}[1..i-1]$ and $\boldsymbol{x}[i] < \boldsymbol{y}[i]$ for some integer $i \in 1..\min\{n, m\}$ (this is the case in which there is a first position i in which $\boldsymbol{x}$ and $\boldsymbol{y}$ differ).

Then, for example, using the order of the English alphabet:

- $ab < abc$ (because $i = 2 = n < 3 = m$);
- $\varepsilon < a$ (because $i = 0 = n < 1 = m$);
- $ab < aca$ (because $i = 2$ and $b < c$).

Observe that this definition is valid also in cases where one or both of n, m is infinite; that is, also for infinite strings. Based on this definition, the other order relations $(\leq, >, \geq)$ are defined in the usual way: $\boldsymbol{x} \leq \boldsymbol{y}$ if and only if $\boldsymbol{x} = \boldsymbol{y}$ or $\boldsymbol{x} < \boldsymbol{y}$, $\boldsymbol{x} > \boldsymbol{y}$ if and only if $\boldsymbol{y} < \boldsymbol{x}$, $\boldsymbol{x} \geq \boldsymbol{y}$ if and only if $\boldsymbol{y} \leq \boldsymbol{x}$.

Writing $\boldsymbol{x} = \boldsymbol{u_1 u_2} \cdots \boldsymbol{u_k}$ where the $\boldsymbol{u_i}$ are nonempty substrings, $i \in 1..k$, is called a ***factorization*** or ***decomposition*** of $\boldsymbol{x}$ into ***factors*** $\boldsymbol{u_i}$ (see Section 1.4 and Chapter 6). Thus a factor is just a nonempty substring.

There are two special kinds of substring $\boldsymbol{x}[i..j]$ which are of particular importance, and to which we give special names. For any integer $j \in 0..n$, we say that $\boldsymbol{x}[1..j]$ is a ***prefix*** of $\boldsymbol{x}$, sometimes written pref($\boldsymbol{x}$); if in fact $j < n$, we say that $\boldsymbol{x}[1..j]$ is a ***proper prefix*** of $\boldsymbol{x}$. Similarly, for any integer $i \in 1..n+1$, we say that $\boldsymbol{x}[i..n]$ is a ***suffix*** of $\boldsymbol{x}$, written suff($\boldsymbol{x}$), and a ***proper suffix*** if $i > 1$. Note that, in accordance with the identity $\boldsymbol{x} = \varepsilon \boldsymbol{x} \varepsilon$, these definitions allow us to include the empty string ε as both a proper prefix and a proper suffix of $\boldsymbol{x}$. Thus, for example, the string

$$\boldsymbol{f} = abaababaabaab$$

has prefixes

$$\varepsilon, a, ab, aba, \ldots, \boldsymbol{f} = abaababaabaab$$

and suffixes

$$\varepsilon, b, ab, aab, \ldots, \boldsymbol{f}.$$

The proper prefixes and suffixes of $\boldsymbol{f}$ are obtained simply by omitting $\boldsymbol{f}$ itself from these lists.

A concept whose value is not immediately apparent, but which we will find to be useful in many different contexts, is that of a "border".

Definition 1.2.2 *A **border** $\boldsymbol{b}$ of $\boldsymbol{x}$ is any proper prefix of $\boldsymbol{x}$ that equals a suffix of $\boldsymbol{x}$.*

We see that, according to this definition, $\boldsymbol{x}$ always has an empty border $\boldsymbol{b} = \varepsilon$ of length $\beta = 0$, but that $\boldsymbol{x}$ itself is not a border of $\boldsymbol{x}$. In general, we use the symbol β to denote the length $|\boldsymbol{b}|$ of $\boldsymbol{b}$. Often we will be particularly interested in the ***longest border***, denoted by $\boldsymbol{b}^*$ with length $\beta^* = |\boldsymbol{b}^*|$, where $0 \leq \beta^* \leq n-1$. The string $\boldsymbol{f}$ introduced above has two nonempty borders: ab and $abaab$. The string

$$\boldsymbol{g} = abaabaab$$

has exactly the same two borders, but observe that in this case the longest one, $abaab$, overlaps with itself. Similarly, the string a^n has borders $a, a^2, \ldots, a^{n-1}$, of which, for $i = \lceil (n+1)/2 \rceil$, the borders $a^i, a^{i+1}, \ldots, a^{n-1}$ overlap. As we shall discover presently, overlapping borders are in fact characteristic of strings that contain repetitive substrings: in the above example, observe that $\boldsymbol{g}$ can be written in the form $(aba)^2 ab = (aba)(aba)ab$, thus representing it as a string in which two occurrences of aba are followed by a prefix of aba.

We now apply the idea of a border to generalize this observation and to derive what we call a "normal form" for a given nonempty string $\boldsymbol{x}$ of length n. Suppose that a border $\boldsymbol{b}$ and its length β have been computed. (We shall see in Section 1.3 how to compute *every* border of $\boldsymbol{x}$ in $\Theta(n)$ time.) By Definition 1.2.2 it must be true that

$$\boldsymbol{x}[1..\beta] = \boldsymbol{x}[n-\beta+1..n], \tag{1.1}$$

from which we see that the quantity

$$p = n - \beta \geq 1$$

measures the displacement between positions of $\boldsymbol{x}$ that are required to be equal. (Observe that the larger the value of β, the smaller the value of p.) Thus, for every integer $i \in 1..\beta$, it must be true that

$$\boldsymbol{x}[i] = \boldsymbol{x}[i+p]. \tag{1.2}$$

In particular, if $\beta = 0$ ($p = n$), we see that (1.1) and (1.2) are trivially true; while if $2\beta \geq n$ ($2p \leq n$), then $\boldsymbol{x}$ must contain at least two equal adjacent substrings $\boldsymbol{x}[1..p]$ and $\boldsymbol{x}[p+1..2p]$. More precisely, we see that $\boldsymbol{x}$ consists of $\lfloor n/p \rfloor$ identical substrings, each of length p, followed by a possibly empty suffix of length $n \bmod p$. Setting $r = n/p$ and letting $\boldsymbol{u} = \boldsymbol{x}[1..p]$, we see that the values p and r which we have derived from β permit us to express any string $\boldsymbol{x}$ in the form

$$\boldsymbol{x} = \boldsymbol{u}^{\lfloor r \rfloor}\boldsymbol{u}', \tag{1.3}$$

where $\boldsymbol{u}' = \boldsymbol{x}\big[1..n - \lfloor r \rfloor p\big]$ is a proper prefix (possibly empty) of u. Alternatively, we can separate r into its integral and fractional parts by writing $r = \lfloor r \rfloor + k/p$ for some integer $k \in 0..p-1$. Then, interpreting $\boldsymbol{u}^{k/p} = \boldsymbol{u}[1..p]^{k/p}$ to mean simply $\boldsymbol{u}[1..k]$, we find that (1.3) can be rewritten in the compact form

$$\boldsymbol{x} = \boldsymbol{u}^{r}. \tag{1.4}$$

We call p a ***period*** and r an ***exponent*** of $\boldsymbol{x}$. The prefix $\boldsymbol{u} = \boldsymbol{x}[1..p]$ we call a ***generator*** of $\boldsymbol{x}$. Note that since every string $\boldsymbol{x}$ has an empty border $\boldsymbol{b} = \varepsilon$, it therefore has a ***trivial period*** $p = n$, a ***trivial exponent*** $r = 1$, and a ***trivial generator*** $\boldsymbol{x}$.

Looking over the previous paragraph, we see that what has essentially been done is to compute a period $p = p(\beta)$ and a corresponding exponent $r = r(\beta)$ as functions of β. It is clear that p is monotone decreasing and r monotone increasing in β. Therefore with the choice $\beta = \beta^*$, p achieves its minimum value p^* and r its maximum r^*, the ***minimum period*** and the ***maximum exponent*** respectively. Generally, the values p^* and r^* will be the ones we are most interested in, and so, when there is no ambiguity, we will simply refer to p^* as ***the period*** and r^* as ***the exponent***. Similarly, we refer to $\boldsymbol{u} = \boldsymbol{x}[1..p^*]$ as ***the generator***.

Definition 1.2.3 *Let p^* be the minimum period of $\boldsymbol{x} = \boldsymbol{x}[1..n]$, and let $r^* = n/p^*$, $u = x[1..p^*]$. Then the decomposition*

$$\boldsymbol{x} = \boldsymbol{u}^{r^*} \tag{1.5}$$

*is called the **normal form** of $\boldsymbol{x}$.*

The normal form (1.5) leads to a useful and important taxonomy, or classification system, of strings:

- if $r^* = 1$, we say that $\boldsymbol{x}$ is ***primitive***; otherwise, $\boldsymbol{x}$ is ***periodic***;
- if $r^* \geq 2$, we say that $\boldsymbol{x}$ is ***strongly periodic***; if $1 < r^* < 2$, $\boldsymbol{x}$ is said to be ***weakly periodic***;
- if $r^* \geq 2$ is an integer, we say that $\boldsymbol{x}$ is a ***repetition*** (or, equivalently, that $\boldsymbol{x}$ is ***repetitive***); in the special cases that $r^* = 2$ or 3, $\boldsymbol{x}$ is called a ***square*** or a ***cube***, respectively.

Thus we see that $\boldsymbol{x}$ must be either primitive or periodic, and, if it is periodic, then one of strongly periodic or weakly periodic; further, if $\boldsymbol{x}$ is repetitive, then it must also be strongly periodic. Observe that $r^* \geq 2$ if and only if $\boldsymbol{x}$ has a border of length $\beta \geq n/2$. Here are some examples of these definitions:

- $\boldsymbol{x} = aaabaabab$ is primitive ($p^* = n$);
- $\boldsymbol{f} = abaababaabaab = (abaababa)(abaab)$ is weakly periodic with period $p^* = 8$, exponent $r^* = 13/8$ and generator $abaababa$;
- $\boldsymbol{g} = abaabaab = (aba)^2ab$ is strongly periodic with period $p^* = 3$, exponent $r^* = 8/3$ and generator aba;
- $\boldsymbol{x} = (ab)^4$ is repetitive with period $p^* = 2$, exponent $r^* = 4$ and generator ab;
- $\boldsymbol{x} = (abcabd)^2$ is a square with period $p^* = 6$ and generator $abcabd$.

We remark that the normal form (1.5) is actually a kind of "intrinsic" pattern, in the sense of Part II of this book: every string $\boldsymbol{x}$ has the pattern called a "normal form" that can be used to assign it its place in a taxonomy of strings. We remark further that the simple ideas introduced in this section (e.g. border, primitive, period) will recur again and again in our discussions of various string algorithms.

Exercises 1.2

1. Try to reconcile the definitions of linear string given in Sections 1.1 and 1.2; that is, to prove that a linear string according to one definition is necessarily a linear string according to the other.

2. Suppose that an alphabet A contains exactly α letters. Given some nonnegative integer n, how many elements of A^* have length n? How many have length *at most* n?
3. It was remarked above that for $A = \{0, 1\}$, A^+ may be thought of as including all the nonnegative integers. How many times is each integer included?
4. Based on the definition of equality in strings, is it true that $\varepsilon = \varepsilon^2$? Justify your answer.
5. Using the definitions of equality and lexicographic order, prove that for arbitrary strings $\boldsymbol{x}$ and $\boldsymbol{y}$ on an ordered alphabet, $\boldsymbol{x} = \boldsymbol{y}$ if and only if $\boldsymbol{x} \not< \boldsymbol{y}$ and $\boldsymbol{y} \not< \boldsymbol{x}$. In particular, demonstrate that this result holds in the case $\boldsymbol{x} = \boldsymbol{y} = \varepsilon$, and thus show that ε is the unique lexicographically least element of A^*.
6. Based on the usual ordering of the English lower-case letters, arrange the following strings in increasing lexicographical order:

$$abbac, abbba, abb, abc, a, \varepsilon^2, ab, aba, \varepsilon b.$$

7. Prove that the operator $<$ as we have defined it satisfies transitivity; that is, that

$$\boldsymbol{x} < \boldsymbol{y} \quad \text{and} \quad \boldsymbol{y} < \boldsymbol{z} \;\Rightarrow\; \boldsymbol{x} < \boldsymbol{z}.$$

8. Give an independent definition of the order relation $>$, then use it together with the definition of $<$ given in the text to show that $\boldsymbol{x} > \boldsymbol{y}$ if and only if $\boldsymbol{y} < \boldsymbol{x}$.
9. Given a string $\boldsymbol{x}$ of length n, find the length of the following substrings of $\boldsymbol{x}$, and state the conditions on i and k for which your answer is valid:

   (a) $\boldsymbol{x}[i..i+k-1]$;
   (b) $\boldsymbol{x}[i-k+1..i]$;
   (c) $\boldsymbol{x}[i+1..k-1]$;
   (d) $\varepsilon\boldsymbol{x}[1..k]$.

10. What is the maximum number of distinct substrings that there can possibly be in a string $\boldsymbol{x}$ of length n? Give an example of a string that attains this maximum. Then characterize the set of strings of length n that attains it.
11. In the preceding exercise there is an unstated assumption that the alphabet size should be regarded as unbounded. Vályi Sándor suggests that the question becomes more interesting (and much more difficult) if the size α of the alphabet is finite and fixed. What progress can you make with this apparently unsolved research problem?

    Hint: Perhaps a good place to start is with the following question: for given positive integers $\alpha = |A|$ and k, what is the longest string on alphabet A that contains no substring of length k more than once?
12. Describe an algorithm that computes all the distinct substrings in $\boldsymbol{x}$. Establish your algorithm's correctness and asymptotic complexity (try to achieve $O(n^2)$).

13. If $\boldsymbol{y}$ is a nonempty string of length m and k is a positive integer, determine $|\boldsymbol{x}|$ as a function of m and k in each of the following cases:

(a) $\boldsymbol{x} = \boldsymbol{y}^k$;

(b) $\boldsymbol{x} = \boldsymbol{y}^{|\boldsymbol{y}|}$;

(c) $\boldsymbol{x} = \boldsymbol{y}^{|\boldsymbol{y}|^k}$.

14. What is the length of the string

$$\boldsymbol{x} = (ab)^n a(ab)^{n-1}a \cdots (ab)^2 a(ab)a?$$

15. Are the ideas of prefix, suffix and border defined for the empty string ε?

16. Determine the longest border, the period, and the exponent of each of the following strings, and hence classify each one as primitive, strongly periodic, weakly periodic, or repetitive:

(a) $abaababaabaababaababa$;

(b) $abcabacabcbacbcacb$;

(c) $abcabdabcabdabcabd$;

(d) $a(ab)^3a(ab)^3aa$.

17. A string $\boldsymbol{x}$ of length n is said to be a ***palindrome*** if it reads backwards the same as forwards; more precisely, if for every $i = 1, 2, \ldots, \lfloor n/2 \rfloor$, $\boldsymbol{x}[i] = \boldsymbol{x}[n-i+1]$. Jamie Simpson believes every border of a palindrome is itself a palindrome. Prove him right or wrong.

Remark: To inspect some nontrivial palindromes, you could consult the following URLs:

`www.growndodo.com/wordplay/palindromes/dogseesada.html`

`complex.gmu.edu/neural/personnel/ernie/witty/palindromes.html`

18. Show that period and exponent are "well-defined"; in other words, whenever $\boldsymbol{x} = \boldsymbol{u}^k\boldsymbol{u}'$ for some string $\boldsymbol{u}$, some positive integer k, and a proper prefix $\boldsymbol{u}'$ of $\boldsymbol{u}$, that there exists a unique corresponding border.

19. Show that if $\boldsymbol{u}^{r^*}$ is the normal form of $\boldsymbol{x}$, then $\boldsymbol{u}$ is not a repetition. (This fact becomes important in Section 2.3 when we consider the encoding of the repetitions in a string.)

20. Consider $(ab)^{3/2} = aba$. What is $((ab)^{3/2})^2$? Is it $(aba)^2$? Or is it $(ab)^{(3/2)2} = (ab)^3$?

21. Show that no string has two distinct primitive generators.

22. Can you find a way to compute the number of strings on $\{a, b\}$ that are of length n and primitive?

Hint: Observe that for odd n, an arbitrary string $\boldsymbol{x}[1..n]$ can be formed from strings $\boldsymbol{x}\big[1..(n-1)/2\big]$ and $\boldsymbol{x}\big[(n+1)/2..n\big]$, and for even n from strings $\boldsymbol{x}[1..n/2]$ and $\boldsymbol{x}[n/2+1..n]$. If on due reflection you still have trouble with this exercise, consult [GO81a, GO81b].

23. The taxonomy of strings given above is expressed in terms of values r^*. Re-express it in terms of values of the longest border β^*.

24. Observe that what we have called the "normal form" of $\boldsymbol{x}$ should really more properly be called the "left" normal form. That is, instead of writing $\boldsymbol{x}$ in the form $\boldsymbol{u}^r\boldsymbol{u}'$, where $\boldsymbol{u}'$ is a prefix of $\boldsymbol{u}$, we might just as well have written $\boldsymbol{x} = \boldsymbol{v}'\boldsymbol{v}^s$, where $\boldsymbol{v}'$ is a suffix of $\boldsymbol{v}$.

(a) Derive the right normal form of $\boldsymbol{x}$.

(b) Show that, in the taxonomy of strings, the classification of $\boldsymbol{x}$ according to right normal form is the same as it is according to left normal form.

25. Can a necklace be a substring of a linear string? Can a linear string be a substring of a necklace?

26. Draw a labelled tree which represents the taxonomy of strings introduced in this section.

1.3 Periodicity

Be not careless in deeds, nor confused in words,
nor rambling in thought.

— Marcus Aurelius ANTONINUS (121–180), *Meditations VIII*

As we shall see in subsequent chapters, the maximum border of a string $\boldsymbol{x}$ turns out to be a very useful quantity in numerous contexts, not only as a means of classifying strings. Therefore, in this section we take the time to show in detail how to compute the maximum border; indeed, we show how to compute the maximum border of every nonempty prefix of $\boldsymbol{x}$, and to do so in $\Theta(n)$ time. We then go on to prove an important lemma that describes a fundamental periodicity property of strings.

Of course there is an obvious way to compute the length β^* of the maximum border $\boldsymbol{b}^*$ of $\boldsymbol{x}$. Begin by setting $\beta^* \leftarrow 0$. Then compare $\boldsymbol{x}[1]$ with $\boldsymbol{x}[n]$; if equal, set $\beta^* \leftarrow 1$. Next compare $\boldsymbol{x}[1..2]$ with $\boldsymbol{x}[n-1..n]$; if equal, set $\beta^* \leftarrow 2$. Proceed in this way, adding one to the length of the prefixes and suffixes compared at each stage, until finally $\boldsymbol{x}[1..n-1]$ is compared with $\boldsymbol{x}[2,n]$, with β^* set to $n-1$ in case of equality. The final value of β^* is then the length of the longest border $\boldsymbol{b}^*$ of $\boldsymbol{x}$. We do not present this algorithm formally because it is inefficient: we leave as Exercise 1.3.1 the exact determination of its asymptotic complexity.

Imagine that successive values of β^* are stored in an array (or string!) $\boldsymbol{\beta}[1..n]$; then for $i = 1, 2, \ldots, n$, $\boldsymbol{\beta}[i]$ gives the length of the longest border of $\boldsymbol{x}[1..i]$. We call $\boldsymbol{\beta}[1..n]$ the ***border array*** of $\boldsymbol{x}[1..n]$. Then $\boldsymbol{\beta}[n]$ is the length of the longest border of $\boldsymbol{x}$, the quantity required to compute the normal form of $\boldsymbol{x}$. We make the following observations:

- $\boldsymbol{\beta}[1] = 0$ (since ε is the longest border of $\boldsymbol{x}[1..1]$);
- for $2 \le i \le n$, if $\boldsymbol{x}[1..i]$ has a border of length $k > 0$, then $\boldsymbol{x}[1..i-1]$ has a border of length $k-1$; thus, in particular, for $1 \le i \le n-1$, $\boldsymbol{\beta}[i+1] \le \boldsymbol{\beta}[i] + 1$;
- for $1 \le i \le n-1$, $\boldsymbol{\beta}[i+1] = \boldsymbol{\beta}[i] + 1$ if and only if $\boldsymbol{x}[i+1] = \boldsymbol{x}\left[\boldsymbol{\beta}[i]+1\right]$ (since $\boldsymbol{\beta}[i]+1$ is the position of $\boldsymbol{x}$ immediately to the right of the prefix $\boldsymbol{x}\left[1..\boldsymbol{\beta}[i]\right]$ which is the longest border of $\boldsymbol{x}[1..i]$);
- if $\boldsymbol{b}$ is a border of $\boldsymbol{x}$, and $\boldsymbol{b}'$ is a border of $\boldsymbol{b}$, then $\boldsymbol{b}'$ is a border of $\boldsymbol{x}$.

These observations, particularly the third and fourth, suggest that it may be possible to compute $\boldsymbol{\beta}[i+1]$ from $\boldsymbol{\beta}[1], \boldsymbol{\beta}[2], \ldots, \boldsymbol{\beta}[i]$. The chief difficulty evidently arises when $\boldsymbol{\beta}[i] > 0$ but $\boldsymbol{x}[i+1] \ne \boldsymbol{x}\left[\boldsymbol{\beta}[i]+1\right]$: in this case, it is necessary to look at the *second-longest* border of $\boldsymbol{x}[1..i]$. If this border is empty, then $\boldsymbol{\beta}[i+1] = 1$ (if $\boldsymbol{x}[1] = \boldsymbol{x}[i+1]$) or 0 (otherwise). If this border is not empty, then, denoting its length by $\boldsymbol{\beta}^2[i] \ge 1$, we will need to compare $\boldsymbol{x}[i+1]$ with $\boldsymbol{x}\left[\boldsymbol{\beta}^2[i]+1\right]$: if equal, then $\boldsymbol{\beta}[i+1] \leftarrow \boldsymbol{\beta}^2[i]+1$; if not, then we go on to consider the third-longest border, of length $\boldsymbol{\beta}^3[i]$, of $\boldsymbol{x}[1..i]$. And so on, until finally for some k, $\boldsymbol{\beta}^k[i] = 0$.

The argument made here will be easier to follow in terms of an example. Suppose that $\boldsymbol{x} = abaababa*$, where the $*$ indicates that the 9th letter is not as yet known. The border array of $\boldsymbol{x}$ is then 00112323?, where the ? indicates that the 9th entry remains to be computed. Observe that, for $i = 8$, $\boldsymbol{\beta}[i] = 3$. If in fact it turns out that

$$\boldsymbol{x}[9] = \boldsymbol{x}[i+1] = \boldsymbol{x}\left[\boldsymbol{\beta}[i]+1\right] = \boldsymbol{x}[4] = a,$$

then we immediately have $\boldsymbol{\beta}[9] = 4$. If not, however, then we must consider the second border of $\boldsymbol{x}[1..8]$; that is,

$$\boldsymbol{\beta}^2[8] = \boldsymbol{\beta}\left[\boldsymbol{\beta}[8]\right] = \boldsymbol{\beta}[3] = 1.$$

Thus if

$$\boldsymbol{x}[8+1] = \boldsymbol{x}\left[\boldsymbol{\beta}^2[8]+1\right] = \boldsymbol{x}[2] = b,$$

we have $\boldsymbol{\beta}[9] = 2$. But if $\boldsymbol{x}[9]$ is neither a nor b, and instead turns out to be a new letter c, we conclude that $\boldsymbol{\beta}[9] = 0$.

In general, consider the quantities $\boldsymbol{\beta}^j[i]$, $j = 1, 2, \ldots, k$, representing the jth-longest borders of $\boldsymbol{x}[1..i]$. (Here we take $\boldsymbol{\beta}^1[i] = \boldsymbol{\beta}[i]$ and $\boldsymbol{\beta}^k[i] = 0$.) Since for every $j = 2, 3, \ldots, k$, $\boldsymbol{\beta}^j[i] < \boldsymbol{\beta}^{j-1}[i]$, it follows that

$$\boldsymbol{x}\left[1..\boldsymbol{\beta}^j[i]\right]$$

is a border of $\boldsymbol{x}\left[1..\boldsymbol{\beta}^{j-1}[i]\right]$. That is, the jth-longest border of $\boldsymbol{x}[1..i]$ is a border of the $(j-1)$th-longest border of $\boldsymbol{x}[1..i]$, and, in particular, it is the *longest* border of the $(j-1)$th-longest border of $\boldsymbol{x}[1..i]$! In symbols,

$$\boldsymbol{\beta}^j[i] = \boldsymbol{\beta}\left[\boldsymbol{\beta}^{j-1}[i]\right]. \tag{1.6}$$

If $\boldsymbol{x}[i+1] \neq \boldsymbol{x}\left[\boldsymbol{\beta}^{j-1}[i]+1\right]$, then we can determine the next position to be tested against $\boldsymbol{x}[i+1]$ by computing $\boldsymbol{\beta}^{j}[i]$ according to (1.6), so that the next test would compare $\boldsymbol{x}[i+1]$ against $\boldsymbol{x}\left[\boldsymbol{\beta}^{j}[i]+1\right]$.

It will be useful to summarize these observations in a lemma:

Lemma 1.3.1 *For some integer $n \geq 1$, let $\boldsymbol{x} = \boldsymbol{x}[1..n]$ be a string with border array $\boldsymbol{\beta} = \boldsymbol{\beta}[1..n]$. Let k be the least integer such that $\boldsymbol{\beta}^{k}[n] = 0$. Then*

(a) for every integer $j \in 1..k$, $\boldsymbol{x}[1..n]$ has a border $\boldsymbol{x}\left[1..\boldsymbol{\beta}^{j}[n]\right]$;
(b) for any choice of letter λ, the only possible borders of $\boldsymbol{x}[1..n+1] = \boldsymbol{x}[1..n]\lambda$ are those whose lengths are members of the descending sequence

$$\left\langle \boldsymbol{\beta}[n]+1, \boldsymbol{\beta}^{2}[n]+1, \ldots, \boldsymbol{\beta}^{k}[n]+1, 0\right\rangle. \quad (1.7)$$

□

Equation (1.6) leads to an efficient algorithm for the computation of the array $\boldsymbol{\beta}[1..n]$, which contains the lengths of the longest borders of each of the prefixes $\boldsymbol{x}[1..i]$, $i = 1, 2, \ldots, n$, of the given string $\boldsymbol{x}$. This is presented as Algorithm 1.3.1. As we shall see in Chapter 7, all of these lengths, not only $\boldsymbol{\beta}[n]$, are potentially useful when $\boldsymbol{x}$ is a specific pattern to be matched against some other string: they are used to determine how far to shift the pattern along the other string in case of a mismatch or "failure". For this reason the computation of $\boldsymbol{\beta}[1..n]$ has sometimes been called the ***failure function*** algorithm [AHU74].

Algorithm 1.3.1 (Border Array)

– *Compute the border array $\boldsymbol{\beta}$ of $\boldsymbol{x}[1..n]$*

```
β[1] ← 0
for i ← 1 to n−1 do
  b ← β[i]
  while b > 0 and x[i + 1] ≠ x[b + 1] do
    b ← β[b]
  if x[i + 1] = x[b + 1] then
    β[i + 1] ← b + 1
  else
    β[i + 1] ← 0
```

Theorem 1.3.2 *Algorithm 1.3.1 correctly computes $\boldsymbol{\beta}[1..n]$.*

Proof First consider the **while** loop: this loop handles the computation (1.6), replacing $b = \boldsymbol{\beta}^{j-1}[i]$ by $b = \boldsymbol{\beta}^{j}[i]$. Thus b is monotone decreasing and so the **while** loop must terminate. Since the **for** loop terminates when $i = n$, it follows that Algorithm 1.3.1 terminates after $n - 1$ steps.

Let us consider the **while** loop further. Exit from this loop occurs if either $b = 0$ or $\boldsymbol{x}[i+1] = \boldsymbol{x}[b+1]$, or both. If $\boldsymbol{x}[i+1] = \boldsymbol{x}[b+1]$, we should set $\boldsymbol{\beta}[i+1] \leftarrow b+1$ regardless of whether or not $b = 0$; otherwise, it is clear that we must set $\boldsymbol{\beta}[i+1] \leftarrow 0$. Thus the **if** structure in the algorithm deals correctly with the result of the **while** loop: it selects the greatest possible length from the sequence (1.7). We conclude that Algorithm 1.3.1 does indeed correctly compute $\boldsymbol{\beta}[i+1]$ for every $i \in 0..n-1$. □

Now we consider the time required by the algorithm. All the steps within the **for** loop, except possibly the **while** loop, require only constant time. Then Algorithm 1.3.1 requires $\Theta(n)$ time plus the total time used within the **while** loop. To estimate this total time, consider the values that b assumes during the execution of the algorithm: b is initially zero, and can be increased in value only by one, and only by the assignment

$$\boldsymbol{\beta}[i+1] \leftarrow b+1$$

at the end of an iteration for i, followed by the assignment

$$b \leftarrow \boldsymbol{\beta}[i]$$

at the beginning of the iteration for $i+1$. Thus b can be incremented by one at most $n-2$ times, and each such incrementation uses up one iteration of the **for** loop. At the same time, the only way that b can be decreased is in the **while** loop: since the number of decrements (always by at least one) cannot be greater than the number of increments (always by exactly one), it follows that the assignment statement $b \leftarrow \boldsymbol{\beta}[b]$ within this loop can be executed no more than $n-2$ times in total. Hence

Theorem 1.3.3 *Algorithm 1.3.1 requires $\Theta(n)$ time and constant additional space for its execution.* □

Digression 1.3.1 This is the first time of many in this book that we have occasion to discuss the asymptotic space and time complexity of an algorithm. There are a couple of important points to be made about the assumptions that underlie the analysis leading to Theorem 1.3.3:

- Consider the integer values i and b of Algorithm 1.3.1, as well as those stored in the border array $\boldsymbol{\beta}$: each of these values may be as large as $n-1$, and so for each one $\lceil \log_2 n \rceil$ bits need to be reserved for storage. Thus, strictly speaking, the storage requirement, therefore the processing requirement, for an element of $\boldsymbol{\beta}$ is actually

$$\log_2 n/w \in O(\log n),$$

where w is the word length of the computer. This is not constant at all!

Based on this analysis, the space required for i and b is $O(\log n)$, while that for β is $O(n \log n)$, and the overall processing time becomes $O(n \log n)$ rather than $\Theta(n)$. We resolve this difficulty here, and throughout this book, by making the assumption that the size of the problem considered is not gigantic, that it is subject to some reasonable bound. Thus, if n is the problem size, we are supposing that there always exists a (small) constant k such that

$$\log_2 n/w \leq k.$$

Indeed, in any practical context, we will not go far wrong if we assume that $k = 1$: in a modern computer, $w \geq 32$, and so $\log_2 n/w \leq 1$ provided

$$n \leq 2^{32} \approx 4.3 \times 10^9.$$

If we are dissatisfied with this limitation, we may instead luxuriate in the supposition that $k = 2$, hence that

$$n \leq 2^{64} \approx 1.8 \times 10^{19},$$

certainly large enough to deal with any normal person's string processing requirements.

In fact, my pocket calculator tells me that just to scan 1.8×10^{19} computer words (string elements) at a rate of one billion per second will require 570 years — it is not we, but rather our distant descendants, who will finally get to the end of such a string!

That being said, I am still not unsympathetic to the purist who multiplies occurrences of n in the time and space complexities quoted in this book by a $\log n$ factor. In some sense the factor should "really" be there: it is only in deference to ordinary practical considerations that it is omitted.

■ The second point to be made relates to the sets O, Ω and Θ used throughout this book to characterize the asymptotic time and space requirements of algorithms. We suppose that the reader is familiar with the definitions of these sets; if not, the seminal reference is [K76], and [RS92] may also be found useful. But the story does not end there. When $O(n)$, for example, is used to describe a property of an algorithm, an intellectual leap is made that requires justification.

If we say

$$\text{execution time} \in O(n),$$

we treat the execution time of the algorithm as if it were a function of the size n of the problem instance, when of course it is no such thing — it is instead a function of the problem instance itself. As a rule there will be a large number, in fact usually an infinite number, of problem instances corresponding to any given problem size n.

Nevertheless, the approach adopted in this book will be to treat both execution space and execution time *as if* they were functions of problem size. Thus, for example, to say that

$$\text{execution time} \in \Theta(n) \tag{1.8}$$

will mean that, over all problem instances whose size exceeds some fixed value n_0, the algorithm requires time at least $k_1 n$ and at most $k_2 n$, where k_1 and k_2 are also fixed. By contrast, the statement

$$\text{execution space} \in O(n)$$

makes the considerably weaker assertion that, again over all problem instances of size greater than n_0, the algorithm requires space at most $k_2 n$. Thus, in this case, it is possible that there may exist problem instances whose space requirement is proportional to a quantity, perhaps $\log n$ or $\sqrt{n}$ or $n/\log n$ or even 1, that is asymptotically less than n; in fact, it is even possible that *all* problem instances have space requirement proportional to such a quantity.

Throughout this book, we will use Θ to describe algorithmic properties only when the strong condition illustrated in (1.8) is satisfied. Usually, when we use $O(f(n))$ (respectively, $\Omega(f(n))$), we will mean that there do exist collections of problem instances for which the property is in fact exactly of order $f(n)$, but that also there exist other collections of problem instances for which the property is asymptotically less than (respectively, greater than) order $f(n)$.

□

We have seen that Algorithm 1.3.1 requires $\Theta(n)$ time for its execution over all problem instances; since any algorithm that computed the border array would need to access each of n positions at least once, we are therefore justified in describing Algorithm 1.3.1 as ***asymptotically optimal***: no algorithm could compute $\boldsymbol{\beta}[1..n]$ in less than $\Theta(n)$ time and constant space for any problem instance of size n. As discussed in Section 4.1, Algorithm 1.3.1 is also in a certain sense an ***on-line*** algorithm, yielding at each step a result for the current position i as the given string is processed from left to right. Thus, without backtracking, the algorithm yields a result, not only for the given string, but also for every prefix of it. However, in a stricter sense, Algorithm 1.3.1 would not be described as on-line: since any position $i' < i$ in $\boldsymbol{x}$ may need to be visited in order to assist in the calculation for i, the entire string needs to be available for processing at every step of the calculation.

As an example of a border calculation, suppose that the string

$$\begin{array}{cccccccccccccc} & 1 & 2 & 3 & 4 & 5 & 6 & 7 & 8 & 9 & 10 & 11 & 12 & 13 \\ \boldsymbol{f} = & a & b & a & a & b & a & b & a & a & b & a & a & b, \end{array}$$

introduced in Section 1.2, is given. Then the border array $\boldsymbol{\beta}[1..n]$ of $\boldsymbol{f}$ is

$$0011232345645.$$

Observe how, in a string such as $\boldsymbol{f}$ with many repetitive substrings, the values in the border array may fall back and then rise again. For example, $\boldsymbol{\beta}[6..8] = 323$ and $\boldsymbol{\beta}[11..13] = 645$.

The way in which the border array $\boldsymbol{\beta}$ of a given string $\boldsymbol{x}$ is calculated makes it clear that *all* of the borders of $\boldsymbol{x}$ (indeed, of every prefix of $\boldsymbol{x}$) can be computed from $\boldsymbol{\beta}$. This follows from the observation stated in (1.6) that the second-longest border of $\boldsymbol{x}[1..i]$ is the longest border of $\boldsymbol{x}\big[1..\beta[i]\big]$. We state this important result as a theorem:

Theorem 1.3.4 *The border array $\boldsymbol{\beta}$ of any string $\boldsymbol{x}$ contains all the information required to compute all borders, periods and exponents of any nonempty prefix of $\boldsymbol{x}$.* □

Digression 1.3.2 The border array is the first of several string data structures that report in linear space information that may in fact be supralinear. To see this, consider the example string $\boldsymbol{f}$. The borders reported by $\boldsymbol{\beta}[1..n]$ in this case are as follows:

i	*border*
13	5, 2, 0
12	4, 1, 0
11	6, 3, 1, 0
10	5, 2, 0
⋮	⋮

The point being made here is that the user must agree to accept the array $\boldsymbol{\beta}$ as a compact representation of the information provided perhaps more conveniently or more explicitly in the above table. If indeed he or she were to insist that we produce the table, then as much as $\Theta(n \log n)$ processing time could be required, and so the border array computation would no longer be linear. □

We conclude this section with a result that identifies an important arithmetic relationship between distinct periods of a string. This result is a consequence of, *inter alia*, the properties of a border array.

Theorem 1.3.5 ("The Periodicity Lemma" [FW65, LS62]) *Let p and q be two periods of $\boldsymbol{x} = \boldsymbol{x}[1..n]$, and let $d = \gcd(p, q)$. If $p + q \leq n + d$, then d is also a period of $\boldsymbol{x}$.*

Proof The proof of this theorem is interesting. It depends on reducing the conditions as they are stated based on parameters (d, p, q, n) to an equivalent set of conditions based on parameters $(d, p, q - p, n - p)$. This reduction mimics the iterative reduction contained in the Euclidean algorithm [K73a] for the calculation of $\gcd(p, q)$, and in fact the validity of the proof depends upon the correctness of the Euclidean algorithm. Recall that for $q > p$ this algorithm computes d based on the reduction

$$d = \gcd(p, q) = \gcd(p, q - p), \tag{1.9}$$

terminating after a finite number of steps with $d = \gcd(d, d)$.

Let $\mathrm{H}(d, p, q, n)$ denote the hypothesis of the theorem as it is stated; that is, that p and q are periods of $\boldsymbol{x}[1..n]$ with $d = \gcd(p, q)$ and $p + q \leq n + d$. Without loss of generality we suppose that $q > p$. (If $q = p$ the result holds trivially.) Hence $\mathrm{H}(d, p, q - p, n - p)$

denotes the hypothesis that p and $q-p$ are periods of $\boldsymbol{x}[1..n-p]$ with $d = \gcd(p, q-p)$ and $q \le n-p+d$. We shall show that

$$\mathrm{H}(d,p,q,n) \Rightarrow \mathrm{H}(d,p,q-p,n-p),$$

thus allowing us to replace the hypothesis related to $\boldsymbol{x}[1..n]$ by an analogous hypothesis related to the shorter string $\boldsymbol{x}[1..n-p]$. We imagine this reduction being carried out a finite number of times until a hypothesis $\mathrm{H}(d,d,d,n')$ is reached for which it is trivially true that d is a period of $\boldsymbol{x}[1..n']$. We then show that d is a period of $\boldsymbol{x}[1..n-p]$ if and only if d is a period of $\boldsymbol{x}$; since d is a period of $\boldsymbol{x}[1..n']$, it therefore follows that d is a period of $\boldsymbol{x}$.

We now suppose that $\mathrm{H}(d,p,q,n)$ holds. Let $\beta_1 = n-p$ and $\beta_2 = n-q$ be the lengths of the borders $\boldsymbol{b_1}$ and $\boldsymbol{b_2}$ corresponding to p and q, respectively. Then as we have seen in Lemma 1.3.1, $\boldsymbol{b_2}$ is necessarily a border of $\boldsymbol{b_1}$; in other words, $\boldsymbol{x}[1..\beta_1] = \boldsymbol{x}[1..n-p]$ has border $\boldsymbol{x}[1..\beta_2]$, hence period

$$\beta_1 - \beta_2 = n-p-n+q = q-p.$$

Furthermore, note that since $d = \gcd(p,q)$, it must therefore be true that $d \le q-p$. Then by hypothesis

$$p \le q-d \le n-p, \tag{1.10}$$

so that, since $\boldsymbol{x}$ has period p, $\boldsymbol{x}[1..n-p]$ must also have period p. Thus we have so far shown that $\boldsymbol{x}[1..n-p]$ has periods p and $q-p$. By (1.9) we know that $d = \gcd(p, q-p)$, and since $p+q \le n+d$ if and only if

$$p + (q-p) \le (n-p) + d,$$

we see that $\mathrm{H}(d,p,q,n)$ implies $\mathrm{H}(d,p,q-p,n-p)$.

It remains to show that d is a period of $\boldsymbol{x}$ if and only if it is a period of $\boldsymbol{x}[1..n-p]$. Clearly if d is a period of $\boldsymbol{x}$, it is also a period of $\boldsymbol{x}[1..n-p]$. To establish the converse, observe that if d is a period of $\boldsymbol{x}[1..n-p]$, it is also by (1.10) a period of $\boldsymbol{x}[1..p]$; and since d must divide p, it follows that

$$\boldsymbol{x}[1..p] = \boldsymbol{x}[1..d]^{p/d},$$

a repetition unless in fact $d = p$. Thus, recalling that $\boldsymbol{x}[1..n-p]$ is a border of $\boldsymbol{x}$, we find that d is a period of

$$\boldsymbol{x}[1..d]^{p/d}\boldsymbol{x}[1..n-p] = \boldsymbol{x}[1..p]\boldsymbol{x}[p+1..n] = \boldsymbol{x},$$

as required.

Thus the reduced string $\boldsymbol{x}[1..n-p]$ has d as a period if and only if d is a period of the original string $\boldsymbol{x}$. It remains to remark that, by the correctness of the Euclidean algorithm, we arrive after a finite number of such reductions at the reduced hypothesis $\mathrm{H}(d,d,d,n')$, for some integer $n' \in d..n-p$, where d is trivially a period of $\boldsymbol{x}[1..n']$, hence of $\boldsymbol{x}$. □

In Section 1.4 we shall see one of the many applications of this result. As an example of the Periodicity Lemma, consider the string

$$\boldsymbol{x} = abaaabaaabaaabaaab$$

of length $n = 18$ and periods $q = 12$, $p = 8$: since $d = \gcd(p, q) = 4$ and $p + q = 20 < n + d = 22$, we conclude that $d = 4$ is also a period of $\boldsymbol{x}$. To see that sometimes the Periodicity Lemma holds even though the conditions for it are not satisfied, consider the related example

$$\boldsymbol{y} = abaaabaaabaaab$$

of length $n = 14$, also with periods $q = 12$ and $p = 8$: even though in this case $p + q = 20 > 18 = n + d$, nevertheless $d = 4$ is a period of $\boldsymbol{y}$. However, to see that the condition $p + q \leq n + d$ is required in Theorem 1.3.5, consider the example

$$\boldsymbol{z} = abaaba$$

of length $n = 6$ with periods $q = 5$ and $p = 3$: in this case $d = \gcd(3, 5) = 1$ is *not* a period of $\boldsymbol{z}$, and $p + q = 8 = n + d + 1$.

The Periodicity Lemma has recently been extended to apply to strings with three periods [CMR99].

Exercises 1.3

1. Determine the asymptotic complexity of the "obvious" algorithm for finding the longest border of a given string $\boldsymbol{x}$. Give an example of a string of arbitrary length n which gives rise to the algorithm's worst-case behaviour. In the worst case, exactly how many letter comparisons are required between the individual elements of $\boldsymbol{x}$?
2. Since our objective is to compute the longest border of $\boldsymbol{x}$, we might well have thought a little more about the problem in order to come up with a somewhat more efficient algorithm that tested possible borders in *descending* rather than ascending order of length. Present such an algorithm and determine its asymptotic complexity. In which cases is its time requirement the same as that of the obvious algorithm? In which cases does it execute in time linear in the string length?
3. If $\boldsymbol{\beta}$ is thought of as a string, on what alphabet is it defined?
4. Prove the statement made in the text that if $\boldsymbol{b}$ is a border of $\boldsymbol{x}$, and $\boldsymbol{b}'$ is a border of $\boldsymbol{b}$, then $\boldsymbol{b}'$ is a border of $\boldsymbol{x}$.
5. Prove the following lemma (due to Jiandong Jiang):

   If $\boldsymbol{x} = \boldsymbol{b}\boldsymbol{x}'\boldsymbol{b}$, where $\boldsymbol{b}$ is primitive, $\boldsymbol{x}'$ is either empty or primitive, and $\boldsymbol{x}' \neq \boldsymbol{b}$, then $\boldsymbol{b}$ is the only nonempty border of $\boldsymbol{x}$.

6. We have regarded Lemma 1.3.1 as proved even though it has not been clearly demonstrated that there exists no border of $\boldsymbol{x}[1..n]$ of length other than $\boldsymbol{\beta}^j[n]$, $j = 1, 2, \ldots, k$. Complete the proof.

7. Software engineers seek to apply formal methods to establish the correctness of software. In particular, for loops they define the following:

- A ***loop invariant*** is a Boolean condition satisfied at the beginning and the end of each pass through the loop.
- A ***loop variant*** is an expression which assumes positive integer values if and only if the loop is still being executed, and which decreases at every pass through the loop.

Find invariants and variants for the two loops in Algorithm 1.3.1.

8. Apply Algorithm 1.3.1 to the following strings and compute the border array $\boldsymbol{\beta}[1..n]$ for each:

(a) $abcdefg$;
(b) a^n;
(c) $(ab)^n$;
(d) $abaababaabaab$;
(e) $(aba)^3a(aba)^2a(aba)^3$;
(f) a^4ba^5b.

9. The **if** statement in Algorithm 1.3.1 is a little annoying, because it duplicates a letter comparison already performed in the **while** statement. Making use of an additional position $\boldsymbol{\beta}[0]$ in the border array, can you find a way to eliminate the **if** statement and so streamline the border array algorithm?

10. Exercise 1.2.24 points out that "normal form" should perhaps more properly be called "left normal form". Similarly, since the border array specifies the longest border of every prefix of $\boldsymbol{x}$, it might better be called the "left border array"; then the ***right border array*** would specify the longest border of every suffix of $\boldsymbol{x}$.
Write a linear-time algorithm that computes the right border array.

Remark: This is not just an academic exercise designed to torture the student. The right border array is important to one of the most famous string algorithms of all, the Boyer-Moore algorithm discussed in Section 7.2.

11. Let $\boldsymbol{\rho}$ be the right border array of a string $\boldsymbol{x}$ of length n, and let $\boldsymbol{\beta}_T$ be the border array of the transposed string

$$\boldsymbol{x}_T = \boldsymbol{x}[n]\boldsymbol{x}[n-1]\cdots\boldsymbol{x}[1].$$

Prove that, for every $i \in 1..n$, $\boldsymbol{\rho}[i] = \boldsymbol{\beta}_T[n-i+1]$.

12. Making use of Lemma 1.3.1, design an algorithm that determines whether or not a given string $\boldsymbol{y} = \boldsymbol{y}[1..n]$ of nonnegative integers is a feasible border array for some string on some alphabet. Your algorithm should require $\Theta(n)$ time for its execution; prove that in fact it does. And, of course, prove that your algorithm is correct.

Caution: This difficult exercise is better viewed as a research problem — in fact, it is closely related to ideas developed in Sections 4.2 and 4.3, and it has already led to two published papers [FGLR02, DLL02]. Despite all this effort, there is still probably more to do!

13. Let $\boldsymbol{x}'$ be a string formed by permuting the letters of the alphabet found in a given string $\boldsymbol{x}$. (For example, $\boldsymbol{x} = abcab$ might be transformed into $\boldsymbol{x}' = cabca$ under the permutation $a \to c$, $b \to a$, $c \to b$.) Show that the border array computed for $\boldsymbol{x}'$ is equal to the border array of $\boldsymbol{x}$. Hence show that if the alphabet contains more than one letter, no string on that alphabet is uniquely determined by its border array. (We return to this important observation in Section 4.3.)

14. We have claimed that Algorithm 1.3.1 is asymptotically optimal. Does this claim remain valid if we regard the purpose of Algorithm 1.3.1 to be the computation of $\boldsymbol{\beta}[n]$ only?

15. In the discussion leading up to Theorem 1.3.3, the rather vague statement was made that "the number of decrements cannot exceed ... the number of increments". To make this idea more precise, consider functions

$$x(v) = v + 1 \text{ and } y(v) = v - 1$$

defined on any real value v. Let $\boldsymbol{s}$ be any string of $j \geq 0$ x's and $k \geq 0$ y's representing successive function evaluations. Then $|\boldsymbol{s}| = j + k$ and $\boldsymbol{s}(v)$ is the value resulting from these evaluations. For example, for $\boldsymbol{s} = xyx$, $|\boldsymbol{s}| = 3$ and $\boldsymbol{s}(0) = 1$.

Prove that $\boldsymbol{s}(v) - v = j - k$ and show how this fact relates to Theorem 1.3.3.

16. State and prove a lemma similar to that of the preceding exercise in the more general case that

$$x(v) = v + r \text{ and } y(v) = v - s$$

for positive real numbers r and s.

17. The author became quite excited when he thought that he had invented a brand new linear-time border array algorithm. This is an outline of it:

> Begin with $\boldsymbol{\beta}[1] \leftarrow 0$ and compare $\boldsymbol{x}[1]$ with every $\boldsymbol{x}[j]$, $2 \leq j \leq n$, recording $\boldsymbol{\beta}[j] \leftarrow 1$ if equal, otherwise $\boldsymbol{\beta}[j] \leftarrow 0$. At the same time, create a linked-list L of positions j such that $\boldsymbol{\beta}[j] = 1$.
>
> For every $i = 2, 3, \ldots$, exit if L is empty. Otherwise, for every $j < n$ in L, compare $\boldsymbol{x}[i]$ with $\boldsymbol{x}[j+1]$; if equal, set $\boldsymbol{\beta}[j+1] \leftarrow i$ and replace j in L with $j+1$; if not equal, or if $j = n$, delete j from L.

In the style of Algorithm 1.3.1, write down an implementation of this algorithm, establish whether or not it is correct, and then determine its time complexity. Try to characterize cases in which the algorithm will execute efficiently ($\Theta(n)$ time) or inefficiently.

18. Can you find a way to modify the border array algorithm presented in the preceding exercise so as to reduce its worst case time complexity?

19. Show that if the hypothesis of the Periodicity Lemma is satisfied for $p < q$, then $p \leq n/2$.

1.4 Necklaces

Poetry teaches the enormous force of a few words.

— Ralph Waldo EMERSON (1803–1882), *preface to Parnassus*

Necklaces become important surprisingly often in string algorithms, even when they do not appear at first to have any connection with the problem being solved. This section discusses some of the main properties of necklaces.

As mentioned in Section 1.1, a necklace is normally defined in terms of a linear string $\boldsymbol{x}$. Thus we use $C(\boldsymbol{x})$ to denote the necklace formed from $\boldsymbol{x}$ by concatenating $\boldsymbol{x}[1]$ at the right of $\boldsymbol{x}[n]$ — thus turning $\boldsymbol{x}$ into a string that has neither a leftmost nor a rightmost position. For example, corresponding to $\boldsymbol{f} = abaababa$, $C(\boldsymbol{f})$ might be represented as follows:

$$\begin{array}{ccccc} & & a & & \\ & a & & b & \\ b & & & & a \\ & a & & a & \\ & & b & & \end{array}$$

To make it possible for an algorithm to process a necklace, it is often convenient to think of $C(\boldsymbol{x})$ in terms of the "rotations" of $\boldsymbol{x}$. Given a string $\boldsymbol{x} = \boldsymbol{x}[1..n]$, the jth ***rotation*** (or ***cyclic shift***) of $\boldsymbol{x}$ is defined to be the string

$$R_j(\boldsymbol{x}) = \boldsymbol{x}[j+1..n]\boldsymbol{x}[1..j]$$

for every integer $j \in 0..n-1$. Thus $R_0(\boldsymbol{x}) = \boldsymbol{x}$ and the eight rotations of the string $\boldsymbol{f}$ defined above are as follows:

$$\begin{aligned} R_0(\boldsymbol{f}) &= abaababa \\ R_1(\boldsymbol{f}) &= baababaa \\ R_2(\boldsymbol{f}) &= aababaab \\ &\vdots \\ R_7(\boldsymbol{f}) &= aabaabab \end{aligned}$$

Note that even though $aabaa$, for example, is not a substring of $\boldsymbol{f}$, it is however a substring of $R_7(\boldsymbol{f})$, hence of $C(\boldsymbol{f})$. We see then that when a necklace $\boldsymbol{x}$ is processed, it is important to keep available all of its rotations. In practice, this is accomplished by storing $\boldsymbol{x}^2$, which contains every rotation $R_j(\boldsymbol{x})$ as a substring.

The preceding discussion enables us to speak meaningfully of a ***substring*** of $C(\boldsymbol{x})$; that is, a substring of any of the rotations of $\boldsymbol{x}$ or, equivalently, a substring of $\boldsymbol{x}^2$ of length at most n. However, since necklaces have neither a leftmost nor a rightmost letter, the special substrings introduced in Section 1.2 — prefix, suffix and border — can be defined only with respect to the individual rotations, not the necklace as a whole. As a consequence, the idea of a period, which is derived from the border, cannot be applied to the entire necklace either. Indeed, as the next two theorems show, different rotations of $C(\boldsymbol{x})$ may have quite different properties when $\boldsymbol{x}$ is not a repetition.

Theorem 1.4.1 *Let $\boldsymbol{x}$ be a string of length $n \geq 3$ on a binary alphabet ($\alpha = 2$).*

(a) *If $\boldsymbol{x}$ is a repetition, then so is every rotation of $\boldsymbol{x}$;*

(b) *otherwise, there exists at least one periodic rotation of $\boldsymbol{x}$.*

Proof To prove (a), let $\boldsymbol{x} = \boldsymbol{u}^r$ for integers $p < n$ and $r > 1$ such that $n = pr$ and $\boldsymbol{u} = \boldsymbol{x}[1..p]$. Then

$$R_1(\boldsymbol{x}) = \boldsymbol{x}[2..n]\boldsymbol{x}[1] = \big(\boldsymbol{u}[2..p]\boldsymbol{u}[1]\big)^r,$$

and the result follows by induction.

To prove (b), we try to construct $\boldsymbol{x}$ in such a way that every rotation of $\boldsymbol{x}$ is primitive. Then $\boldsymbol{x}[1] \neq \boldsymbol{x}[n]$, since otherwise $\boldsymbol{x}$ itself would be periodic. Assume therefore without loss of generality that $\boldsymbol{x}[1] = a$, $\boldsymbol{x}[n] = b$. But then it must be true that $\boldsymbol{x}[2] = b$, for otherwise $R_1(\boldsymbol{x})[1] = R_1(\boldsymbol{x})[n] = a$, so that $R_1(\boldsymbol{x})$ would be periodic. Similar reasoning leads to the conclusion that $\boldsymbol{x}[n-1] = a$. But for $n = 3$ this implies that $\boldsymbol{x}[2] \neq \boldsymbol{x}[3-1]$, an impossibility, while for $n > 3$ we find that $\boldsymbol{x}$ has border ab and so is periodic. □

As we shall discover in Section 3.3, it is possible to construct arbitrarily long repetition-free strings on an alphabet of three letters. If we generate one of these strings of length $n-1$ and then append to it a single letter that does not already appear in the string, we shall have constructed, as we see in Exercise 1.4.3, a string of length n on four letters whose every rotation is primitive. We see then that Theorem 1.4.1(b) cannot be generalized to alphabets of size 4 or greater. On the other hand, we discover in Exercise 1.4.2 that Theorem 1.4.1(a) is true in general.

A second useful result connects rotations, repetitions and periodicity in another way. It tells us that the rotations of $\boldsymbol{x}$ are all distinct except when $\boldsymbol{x}$ is a repetition. The proof depends on the Periodicity Lemma.

Theorem 1.4.2 *Let $\boldsymbol{x}$ be an arbitrary string of length n and minimum period p^*, and let $j \in 1..n-1$ be an integer. Then $R_j(\boldsymbol{x}) = \boldsymbol{x}$ if and only if $\boldsymbol{x}$ is a repetition and p^* divides j.*

Proof To prove sufficiency, it suffices to observe that if $\boldsymbol{x} = \boldsymbol{x}[1..p^*]^{r^*}$, where $r^* = n/p^*$ is an integer, then for every $j = p^*, 2p^*, \ldots, (r^*-1)p^*$, $R_j(\boldsymbol{x}) = \boldsymbol{x}$.

To prove necessity, suppose that $R_j(\boldsymbol{x}) = \boldsymbol{x}$. Then $\boldsymbol{x}[i+j] = \boldsymbol{x}[i]$ for every integer $i \in 1..n-j$, so that $\boldsymbol{x}$ has border $\boldsymbol{x}[1..n-j]$ and period j. Also, $\boldsymbol{x}[n-j+i] = \boldsymbol{x}[i]$ for every integer $i \in 1..j$, so that $\boldsymbol{x}$ has border $\boldsymbol{x}[1..j]$ and period $n-j$. We now invoke Theorem 1.3.5 parameterized as $\mathrm{H}(d, j, n-j, n)$, where $d = \gcd(j, n-j)$. Since $j + (n-j) < n + d$, the theorem tells us that d is a period of $\boldsymbol{x}$, so that

$$\begin{aligned}\boldsymbol{x} &= \boldsymbol{x}[1..n-j]\boldsymbol{x}[1..j] \\ &= \boldsymbol{x}[1..d]^{(n-j)/d}\boldsymbol{x}[1..d]^{j/d} \\ &= \boldsymbol{x}[1..d]^{n/d}.\end{aligned}$$

Thus $\boldsymbol{x}$ is a repetition. To see that p^* divides d, hence j, we invoke Theorem 1.3.5 again, now parameterized as $\mathrm{H}(d', p^*, d, n)$, where $d' = \gcd(p^*, d)$. Since $p^* \le d \le n/2$, it follows that $p^* + d < n + d'$, so that d' is also a period of $\boldsymbol{x}$. But this means that $d' = p^*$, and so p^* divides d, as required. □

In a sense Theorem 1.4.2 is discouraging because it makes clear that, apart from the special case of repetitions, we might need to consider all of the rotations of given strings $\boldsymbol{x}$ and $\boldsymbol{y}$ in order to be able to decide whether or not $C(\boldsymbol{x}) = C(\boldsymbol{y})$. This problem arises in a surprising way in computer graphics and computational geometry: a convex polygon may be represented by two strings, one that gives, in clockwise order of occurrence, the length of each side, another that gives, in the same order, the angle that each side makes with the next side. In order to determine, then, whether two polygons are congruent, it is necessary to determine whether or not their length strings and angle strings are equal. Let $\boldsymbol{x}$ be a string for one polygon, $\boldsymbol{y}$ the corresponding string for the other. Since one polygon could be differently oriented on the screen than the other, the first value $\boldsymbol{x}[1]$ occurring in $\boldsymbol{x}$ might not be the first in $\boldsymbol{y}$. In other words, as shown in Figure 1.4, the two polygons could be congruent even though $\boldsymbol{x} \ne \boldsymbol{y}$: it is still possible that $C(\boldsymbol{x}) = C(\boldsymbol{y})$.

A natural approach to this problem is to define a unique "canonical form" for a necklace so that $C(\boldsymbol{x})$ and $C(\boldsymbol{y})$ could both be expressed in that form prior to comparison. In practice, for a given string $\boldsymbol{x}$, this would amount to identifying a particular rotation $R_j(\boldsymbol{x}), 0 \le j \le n-1$, that is efficiently computable in a deterministic manner. For strings on an ordered alphabet, one solution is provided by

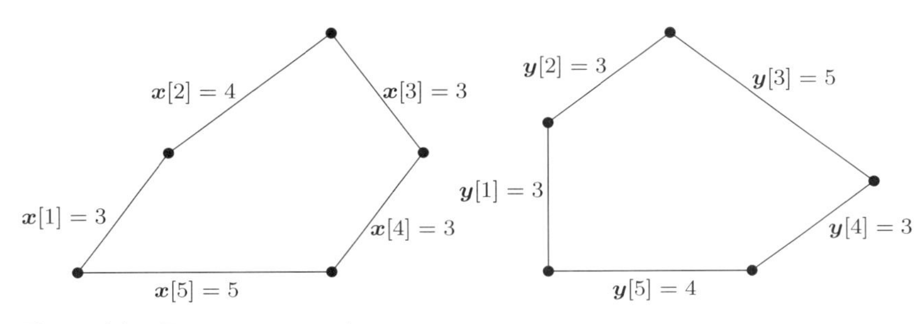

Figure 1.1. Two congruent polygons

Definition 1.4.3 *The* ***canonical form*** *of a necklace $C(\boldsymbol{x})$ on an ordered alphabet is a lexicographically least rotation $R_j(\boldsymbol{x})$ of $\boldsymbol{x}$; that is, $R_j(\boldsymbol{x}) \leq R_i(\boldsymbol{x})$ for every $i \in 0..n-1$.*

The canonical form of a necklace is very closely related to another important idea:

Definition 1.4.4 *A nonempty string $\boldsymbol{w}$ on an ordered alphabet is said to be a* ***Lyndon word*** *if and only if $\boldsymbol{w} < R_j(\boldsymbol{w})$ for every integer $j \in 1..n-1$.*

Thus, for example, every single letter must be a Lyndon word while by Theorem 1.4.2 no repetition can be. Further, we see that while ab is a Lyndon word, ba is not; similarly, aab is a Lyndon word, but neither aba nor baa is. More generally, a further consequence of Theorem 1.4.2 is that whenever a string $\boldsymbol{x}$ is *not* a repetition, the rotations of $\boldsymbol{x}$ are all distinct, and so there exists one and only one rotation of $\boldsymbol{x}$ that is a Lyndon word. Thus, when $\boldsymbol{x}$ is not a repetition, we see that there exists a Lyndon word $\boldsymbol{w} = R_j(\boldsymbol{x})$ that is exactly the canonical form of the necklace $C(\boldsymbol{x})$. Another way of stating this is that the set of all Lyndon words on a given ordered alphabet A is exactly the set of all canonical forms on A that are not repetitions.

We now prove a sequence of three lemmas (adapted from [D83]) that give fundamental and important properties of Lyndon words. These lemmas will in turn be used to establish an elegant "Lyndon decomposition" theorem.

Lemma 1.4.5 *Every Lyndon word is primitive.*

Proof We need to show that no Lyndon word has a nonempty border. Suppose on the contrary that there exists a Lyndon word $\boldsymbol{w}$ such that $\boldsymbol{w} = \boldsymbol{u}\boldsymbol{v}' = \boldsymbol{v}''\boldsymbol{u}$ for nonempty strings $\boldsymbol{u}$, $\boldsymbol{v}'$ and $\boldsymbol{v}''$. Therefore $|\boldsymbol{v}'| = |\boldsymbol{v}''|$ and, by the definition of Lyndon word,

$$\boldsymbol{u}\boldsymbol{v}' = \boldsymbol{w} < \boldsymbol{u}\boldsymbol{v}'' \quad \text{and} \quad \boldsymbol{v}''\boldsymbol{u} = \boldsymbol{w} < \boldsymbol{v}'\boldsymbol{u}.$$

We conclude that $\boldsymbol{v}' < \boldsymbol{v}''$ and $\boldsymbol{v}'' < \boldsymbol{v}'$, an impossibility. □

Lemma 1.4.6 *A string $\boldsymbol{w}$ is a Lyndon word if and only if for every nonempty proper suffix $\boldsymbol{v}$ of $\boldsymbol{w}$, $\boldsymbol{w} < \boldsymbol{v}$.*

Proof To prove sufficiency, observe that if $\boldsymbol{w} = \boldsymbol{u}\boldsymbol{v} < \boldsymbol{v}$ for nonempty strings $\boldsymbol{u}$ and $\boldsymbol{v}$, then certainly $\boldsymbol{u}\boldsymbol{v} < \boldsymbol{v}\boldsymbol{u}$.

To prove necessity, suppose that w is a Lyndon word and that $w = uv$, where u and v are nonempty. Then $uv < vu$. Now we consider the possible orderings of uv and v. Clearly $uv \neq v$ since $|uv| > |v|$. Suppose then that $uv > v$. Two cases arise: either v is a prefix of uv or it is not. But in the first case v would be a border of w, contrary to Lemma 1.4.5, and so this is impossible. In the second case, it must be true that $w[1..|v|] > v$, which implies that $w > vu$, contrary to the Lyndon word property, and so this also is impossible. We conclude that $uv < v$, as required. □

Lemma 1.4.7 *Suppose that w_1 and w_2 are Lyndon words. Then w_1w_2 is a Lyndon word if and only if $w_1 < w_2$.*

Proof Necessity is an immediate consequence of Lemma 1.4.6: if w_1w_2 is a Lyndon word, then $w_1 < w_1w_2 < w_2$.

To prove sufficiency, let v be a nonempty proper suffix of w_1w_2 and suppose that $w_1 < w_2$. Then one of the following cases arises:

- $v = w_1'w_2$ where w_1' is a nonempty proper suffix of w_1. Then since w_1 is a Lyndon word, $w_1 < w_1'$ by Lemma 1.4.6. Since $|w_1| > |w_1'|$, it follows that $w_1w_2 < w_1'w_2 = v$.
- $v = w_2$. If w_1 is not a prefix of w_2, it follows, since $w_1 < w_2$, that $w_1w_2 < w_2 = v$. On the other hand, if w_1 is a prefix of w_2, we may write $w_2 = w_1w_2''$ for some proper suffix w_2'' of w_2. Then, since w_2 is a Lyndon word, we can apply Lemma 1.4.6 again to conclude that $w_2 < w_2''$, hence that $w_1w_2 < w_1w_2'' = w_2 = v$.
- $v = w_2'$ where w_2' is a proper suffix of w_2. Since w_2 is a Lyndon word, Lemma 1.4.6 tells us that $w_2 < w_2'$. Since w_2 is not a prefix of w_2', and since by hypothesis $w_1 < w_2$, it follows that w_1 is not a prefix of w_2' either. Therefore $w_1w_2 < w_2 < w_2' = v$, as required.

In each of these cases we have shown that $w_1w_2 < v$ where v is a suffix of w_1w_2. Thus by Lemma 1.4.6, w_1w_2 is a Lyndon word. This completes the proof. □

In order to use these results, we first need

Definition 1.4.8 *Let x be a nonempty string on an ordered alphabet A. The decomposition $x = w_1w_2 \cdots w_k$ is called a **Lyndon decomposition** if and only if each w_i, $i = 1, 2, \ldots, k$, is a Lyndon word and*

$$w_1 \geq w_2 \geq \cdots \geq w_k.$$

Now we can state and prove a remarkable theorem (due to [CFL58]):

Theorem 1.4.9 *For every nonempty string $\boldsymbol{x}$ on an ordered alphabet, there exists a unique Lyndon decomposition.*

Proof We show first that there exists at least one Lyndon decomposition of $\boldsymbol{x}$. Begin with an initial decomposition

$$\boldsymbol{x} = \boldsymbol{w}_1^{(0)} \boldsymbol{w}_2^{(0)} \cdots \boldsymbol{w}_n^{(0)},$$

where $\boldsymbol{w}_i^{(0)} = \boldsymbol{x}[i]$ for every $i \in 1..n$. As we have seen, this is a decomposition into Lyndon words, but it is not necessarily a Lyndon decomposition, since possibly $\boldsymbol{w}_j^{(0)} < \boldsymbol{w}_{j+1}^{(0)}$ for one or more integers j. For every such j, Lemma 1.4.7 tells us that $\boldsymbol{w}_j^{(0)} \boldsymbol{w}_{j+1}^{(0)}$ is a Lyndon word. Thus, by repeated application of Lemma 1.4.7 we achieve a second decomposition

$$\boldsymbol{x} = \boldsymbol{w}_1^{(1)} \boldsymbol{w}_2^{(1)} \cdots \boldsymbol{w}_{n_1}^{(1)},$$

where each $\boldsymbol{w}_i^{(1)}$ is a Lyndon word and $n_1 \leq n$. If $n_1 < n$, we continue the process, applying Lemma 1.4.7 at each stage, until after a finite number s of stages, a Lyndon decomposition

$$\boldsymbol{x} = \boldsymbol{w}_1^{(s)} \boldsymbol{w}_2^{(s)} \cdots \boldsymbol{w}_{n_s}^{(s)}$$

is achieved, with $\boldsymbol{w}_1^{(s)} \geq \boldsymbol{w}_2^{(s)} \geq \cdots \geq \boldsymbol{w}_{n_s}^{(s)}$.

Now suppose that the Lyndon decomposition is not unique. Then there are two distinct Lyndon decompositions

$$\boldsymbol{x} = \boldsymbol{w}_1 \boldsymbol{w}_2 \cdots \boldsymbol{w}_k \quad \text{and} \quad \boldsymbol{x} = \boldsymbol{w}_1' \boldsymbol{w}_2' \cdots \boldsymbol{w}_{k'}',$$

and we may suppose without loss of generality that there exists some least positive integer $i < \min\{k, k'\}$ such that $\boldsymbol{w}_i' = \boldsymbol{w}_i \boldsymbol{v}$ for some nonempty $\boldsymbol{v}$ and such that $\boldsymbol{w}_j' = \boldsymbol{w}_j$ for every $j \in 1..i-1$. Then there exists a unique integer $r \geq 1$ and a nonempty string $\boldsymbol{v}^*$ such that

$$\boldsymbol{w}_i \boldsymbol{w}_{i+1} \cdots \boldsymbol{w}_{i+r-1} \boldsymbol{v}^* = \boldsymbol{w}_i \boldsymbol{v}$$

and

$$|\boldsymbol{w}_i \boldsymbol{w}_{i+1} \cdots \boldsymbol{w}_{i+r}| \geq |\boldsymbol{w}_i \boldsymbol{v}|.$$

Let $\boldsymbol{w}^* = \boldsymbol{w}_i \boldsymbol{w}_{i+1} \cdots \boldsymbol{w}_{i+r-1}$, so that $\boldsymbol{w}_i' = \boldsymbol{w}^* \boldsymbol{v}^*$. Since $\boldsymbol{w}_i'$ is a Lyndon word, Lemma 1.4.6 tells us that $\boldsymbol{w}_i' < \boldsymbol{v}^*$; further, since $\boldsymbol{w}^*$ is a proper prefix of $\boldsymbol{w}_i'$, it follows that $\boldsymbol{w}^* < \boldsymbol{v}^*$.

Now since $\boldsymbol{v}^*$ is a prefix of $\boldsymbol{w}_{i+r}$, we conclude that $\boldsymbol{v}^* \leq \boldsymbol{w}_{i+r}$, and since $\boldsymbol{w}_i$ is a prefix of $\boldsymbol{w}^*$, we find that $\boldsymbol{w}_i \leq \boldsymbol{w}^*$. Hence

$$\boldsymbol{w}_i \leq \boldsymbol{w}^* < \boldsymbol{v}^* \leq \boldsymbol{w}_{i+r},$$

contradicting the assumption that $\boldsymbol{x} = \boldsymbol{w}_1 \boldsymbol{w}_2 \cdots \boldsymbol{w}_k$ is a Lyndon decomposition. □

Finally, we return to the remark made above that whenever $\boldsymbol{x}$ is not a repetition, the canonical form of $C(\boldsymbol{x})$ is exactly the rotation of $\boldsymbol{x}$ that is a Lyndon word. Suppose we have available an algorithm — for example, one of those described in Section 6.1 — that computes the Lyndon decomposition of a given string $\boldsymbol{x}$. We express this decomposition in the compressed form

$$\boldsymbol{x} = \boldsymbol{w_1}^{q_1}\boldsymbol{w_2}^{q_2}\cdots\boldsymbol{w_k}^{q_k}, \tag{1.11}$$

where now $\boldsymbol{w_1} > \boldsymbol{w_2} > \cdots > \boldsymbol{w_k}$ and the q_i, $i = 1, 2, \ldots, k$, are positive integers. As made clear in Exercise 1.4.14, we may assume that our algorithm outputs all q_i copies of $\boldsymbol{w_i}$ at the same time.

Our task is to identify a position MSP$(\boldsymbol{x}) = i^*$ in $\boldsymbol{x}$, called a ***minimum starting point***, such that $R_{i^*-1}(\boldsymbol{x})$ is a minimum over all rotations of $\boldsymbol{x}$. Of course, as we have seen, the position i^* is unique and $R_{i^*-1}(\boldsymbol{x})$ is a Lyndon word if and only if $\boldsymbol{x}$ is not a repetition. We note first some special cases:

- If $k = 1$, then MSP$(\boldsymbol{x}) = 1$.
- If $k = 2$, then MSP$(\boldsymbol{x}) = q_1|\boldsymbol{w_1}| + 1$.
- If $\boldsymbol{w_k}^{q_k}$ is not a prefix of $\boldsymbol{w_{k-1}}^{q_{k-1}}$, then MSP$(\boldsymbol{x}) = n - q_k|\boldsymbol{w_k}| + 1$.
- If $\boldsymbol{x}$ is a repetition u^m, $m > 1$, then MSP$(\boldsymbol{x}) = MSP(\boldsymbol{u})$.

We now consider the application of the algorithm to the string $\boldsymbol{x^2}$. If $\boldsymbol{x}$ is not a repetition, the algorithm will yield a decomposition $\boldsymbol{v_1}R_{i^*-1}(\boldsymbol{x})\boldsymbol{v_2}$, where $\boldsymbol{v_1}$ and $\boldsymbol{v_2}$ are decompositions of $\boldsymbol{x}[1..i^*-1]$ and $\boldsymbol{x}[i^*..n]$, respectively, and $R_{i^*-1}(\boldsymbol{x})$ is a Lyndon word. Thus, in this case, we need only look for a Lyndon word of length n in order to identify MSP$(\boldsymbol{x}) = |\boldsymbol{v_1}| + 1$. In the case that $\boldsymbol{x} = \boldsymbol{u^m}$ for some integer $m > 1$, we need only look for a block of identical Lyndon words $\boldsymbol{w_i}^{q_i}$ such that $q_i|\boldsymbol{w_i}| = n$, again identifying MSP$(\boldsymbol{x}) = |\boldsymbol{v_1}| + 1$. Thus any algorithm that computes the Lyndon decomposition can be used, with minor modification, to compute the canonical form. However, as shown in Exercise 1.4.16, it is not straightforward to determine, based only on the Lyndon decomposition of $\boldsymbol{x}$, which factor in the decomposition identifies MSP$(\boldsymbol{x})$. We return briefly to this question in Section 6.2.

We shall also see in Section 6.2 that Lyndon decomposition further provides a basis for solving other problems, notably those of efficiently finding the lexicographically least and the lexicographically greatest suffixes of every prefix of a string.

Exercises 1.4

1. Show that in fact, to preserve all information about $C(\boldsymbol{x})$, it suffices to store $\boldsymbol{x}^{(2n-1)/n}$.
2. For every nonempty string $\boldsymbol{x} = \boldsymbol{u}^k$, $k \geq 1$, and every integer $j \in 0..|\boldsymbol{u}|-1$, prove by induction that $R_j(\boldsymbol{x}) = (R_j(\boldsymbol{u}))^k$. (This makes explicit the "induction" mentioned in the proof of Theorem 1.4.1(a).) Hence show that Theorem 1.4.1(a) is true for any alphabet.

3. Suppose that x is a repetition-free string of length $n-1$ on $A = \{a, b, c\}$. Prove the claim made in the text that every rotation of xd is primitive.

4. The discerning reader will not have failed to observe that no mention was made in the text of the existence of primitive strings on three letters whose rotations are all primitive. Perhaps the reason for this omission will become clear from a consideration of the following computer-generated results: no such strings exist of lengths $n = 5, 7, 9, 10, 14, 17$, but they do exist for all other values of $n \in 1..19$. So here is an interesting research problem, proposed by Jamie Simpson: for exactly which values of n do there exist primitive strings on three letters whose rotations are all primitive?

   Bulletin: After spending a great deal of time on this problem, your author became convinced that for every $n > 17$, these strings exist. He became even more convinced that their existence was very hard to prove. However, very recently James Currie [C02] found a nice approach that establishes this result.

5. Without using the Periodicity Lemma, show that if two nonempty strings $\boldsymbol{x}$ and $\boldsymbol{y}$ satisfy $\boldsymbol{xy} = \boldsymbol{yx}$, then there exist positive integers k_1, k_2 and a string $\boldsymbol{u}$ such that $\boldsymbol{x} = \boldsymbol{u}^{k_1}$ and $\boldsymbol{y} = \boldsymbol{u}^{k_2}$.

   Hence provide an alternative proof of the fact (Theorem 1.4.2) that if $\boldsymbol{x} = R_j(\boldsymbol{x})$ for some $j \in 1..n-1$, then $\boldsymbol{x}$ is a repetition.

6. The discerning reader identified above may also have noticed that in the proof of Lemma 1.4.6 there is a "missing lemma"; that is, a more general result could have been established earlier in order to make the proof of the lemma a little more transparent. This result is as follows:

   Let $\boldsymbol{x}$ and $\boldsymbol{y}$ be strings on an ordered alphabet A. If $\boldsymbol{x} > \boldsymbol{y}$ and $\boldsymbol{y}$ is not a prefix of $\boldsymbol{x}$, then for all strings $\boldsymbol{z}, \boldsymbol{z}' \in A^*$, $\boldsymbol{xz} > \boldsymbol{yz}'$.

   Can you prove the missing lemma?

7. Similarly in the proof of Lemma 1.4.7, there are two other missing lemmas that it might have been useful to introduce:

   If $\boldsymbol{x} < \boldsymbol{y}$ and $\boldsymbol{x}$ is not a prefix of $\boldsymbol{y}$, then for all strings $\boldsymbol{z}, \boldsymbol{z}' \in A^*$, $\boldsymbol{xz} < \boldsymbol{yz}'$.
   If $\boldsymbol{x} < \boldsymbol{y}$ and $|\boldsymbol{x}| \geq |\boldsymbol{y}|$, then for all strings $\boldsymbol{z}, \boldsymbol{z}' \in A^*$, $\boldsymbol{xz} < \boldsymbol{yz}'$.

   Prove these missing lemmas.

8. In a similar vein, show that if for strings $\boldsymbol{x}, \boldsymbol{y}, \boldsymbol{x}', \boldsymbol{y}' \in A^*$, $\boldsymbol{xy} < \boldsymbol{x}'\boldsymbol{y}'$ and $\boldsymbol{x}$ and $\boldsymbol{x}'$ are not prefixes of each other, then $\boldsymbol{x} < \boldsymbol{x}'$.

9. The first part of the proof of Theorem 1.4.9 suggests an algorithm for determining the Lyndon decomposition of a given string $\boldsymbol{x}$. Specify this algorithm and determine an upper bound on its asymptotic complexity.

10. Find the Lyndon decomposition of the following strings:

    (a) $xyzabc$;

    (b) $abcxyz$;

    (c) $abaababa$;

(d) $(baa)^n$;

(e) $abacabadabacabae$;

(f) $abacabadabacabad$;

(g) $abacabadabacabab$;

(h) $abacabadabacabaa$.

11. For each of the strings listed in the preceding exercise, find the canonical form of the corresponding necklace.

12. Give an example of

(a) a primitive string that contains a repetition;
(b) a repetition-free string that is not primitive.

13. Let us define a necklace $C(\boldsymbol{x})$ to be ***repetition-free*** if and only if every rotation of $\boldsymbol{x}$ is repetition-free. Then prove the following two results:

The canonical form of $C(\boldsymbol{x})$ is primitive if and only if $\boldsymbol{x}$ is not a repetition.
$C(\boldsymbol{x})$ is repetition-free if and only if every rotation of $\boldsymbol{x}$ is primitive.

14. Suppose that a Lyndon decomposition algorithm processes $\boldsymbol{x}$ from left to right and has at a certain point identified a prefix of Lyndon words followed by a repetition:

$$\boldsymbol{x} = \boldsymbol{w_1}^{q_1} \boldsymbol{w_2}^{q_2} \cdots \boldsymbol{w_j}^{q_j} \boldsymbol{u}^{\boldsymbol{m}} \lambda \cdots,$$

for integers $j \geq 0$, $m \geq 2$ and some letter λ. Describe the processing required in each of the following three cases:

- $\lambda < \boldsymbol{u}[1]$;
- $\lambda = \boldsymbol{u}[1]$;
- $\lambda > \boldsymbol{u}[1]$.

Hence show that all identical Lyndon words are recognized by the algorithm at the same time.

15. Prove the correctness of the four "special cases" given in the text for MSP($\boldsymbol{x}$).

16. It is tempting to imagine (as the author at one time did) that MSP$(\boldsymbol{x}) = i_k$, where we let i_j be the starting point of $\boldsymbol{w_j}^{q_j}$ in the compressed Lyndon decomposition (1.11), $j = 1, 2, \ldots, k$. Show by example that it can happen that MSP$(\boldsymbol{x}) = i_{k-1}$ or even that MSP$(\boldsymbol{x}) = i_{k-2}$.

Chapter 2

Patterns? What Patterns?

How long is a piece of string?

— *Aussie rhetorical question*

By this time the reader may have started to think that the title of this book is rather misleading: there has been some mention of "patterns", but very little compared to the coverage of strings and their properties in Chapter 1. To redress the balance, we present in this chapter an overview of the patterns for whose computation we shall, in Parts II–IV of this book, provide detailed algorithms. Thus this chapter provides a compact statement of the main problems to be solved.

Throughout this chapter, as indeed throughout the book, $\boldsymbol{x} = \boldsymbol{x}[1..n]$ is an arbitrary string of length n.

2.1 Intrinsic Patterns (Part II)

Free speech is intended to protect the controversial and even outrageous word; and not just comforting platitudes too mundane to need protection.

— General Colin POWELL (1937–)

As mentioned in the Preface, intrinsic patterns are those that are in some sense inherent in a string — while not every string will contain a specific pattern bb, for example, nor a given generic pattern such as a repetition, on the other hand intrinsic patterns are those that are found in all circumstances and to some extent characterize the string. Thus, like the "normal form" of Section 1.2, these patterns can be thought of as providing standard descriptions of each string that are independent of other considerations. The interesting thing is that there are so many useful and quite different such "standard descriptions".

The normal form has already been defined in Section 1.2 and its calculation as a byproduct of the border array computation described in Section 1.3. It provides a useful and compact means of describing the periodic structure of a string.

Problem 2.1 **(Compute Normal Form)** *Express $\boldsymbol{x}$ in the form $\boldsymbol{u}^r$ where $\boldsymbol{u}$ is a prefix of $\boldsymbol{x}$ chosen to maximize the value of r (Section 1.3).*

Although not identified as such, the border array itself is a kind of intrinsic pattern associated with a given string. The key result here is Theorem 1.3.4 which establishes the central importance of the border array in computing the borders, periods, exponents and normal forms for the prefixes of $\boldsymbol{x}$.

Problem 2.2 **(Compute Border Array)** *Compute the longest border of every prefix of $\boldsymbol{x}$ as an array (string) $\boldsymbol{\beta} = \boldsymbol{\beta}[1..n]$ (Section 1.3).*

In Chapter 5 we study two kinds of tree: the border tree and the suffix tree.

The border tree is a structure that is already implicit in the border array, as we shall see below; one reason for making this structure explicit is that it is useful for dealing with problems on "covers", a concept defined below (Section 2.3). In fact, we shall find in Chapter 13 that every string has an intrinsic pattern called a "cover array" whose structure mimics that of the border array. The border tree contains exactly $n+1$ nodes labelled by distinct integers in the range $0..n$. The parent-child relationship in the tree is simply defined: node i is a child of node $\boldsymbol{\beta}[i]$. The proof that the structure resulting from these statements is in fact a rooted tree is left as Exercise 1.4.3. Section 5.1 gives an example of the construction of a border tree.

Problem 2.3 **(Compute Border Tree)** *Represent the border array of $\boldsymbol{x}$ as a rooted tree (Section 5.1).*

The three problems already discussed in this section are of course closely related, but the next problem is quite different, and gives a new perspective on the structures inherent in strings. To introduce the idea of a suffix tree, we will first review a common data structure called a "trie" [K73b, pp. 481–499] for re*trie*val.

Suppose a set $X = \{\boldsymbol{x}_1, \boldsymbol{x}_2, \ldots, \boldsymbol{x}_m\}$ of pairwise distinct strings is given. Then a ***trie*** on X is a search tree containing exactly $m+1$ terminal nodes, one for each $\boldsymbol{x}_i$, $i = 1, 2, \ldots, m$, plus one for ε. The edges of a trie are labelled with the letters that occur in the strings of X plus a special end-of-string ***sentinel*** conventionally denoted by \$. Accordingly the edges from the root of the trie to the terminal node for a string $\boldsymbol{x}_i$ spell out $\boldsymbol{x}_i$ followed by \$. More generally, we associate with *every* downward path in a trie, from node N_1 to N_2, say, the string $\boldsymbol{u}$ spelled out by the path. To avoid notational complexities and terminological excesses, we shall simply say that the node N_2 "is" the string $\boldsymbol{u}$, meaning that if N_1 is chosen to be the root, then the path from N_1 to N_2 spells out $\boldsymbol{u}$: of course all such strings $\boldsymbol{u}$ are prefixes of at least one string $\boldsymbol{x}_i\$$. When $N_2 = N_1$, so that the path is empty, $\boldsymbol{u} = \varepsilon$; thus in particular the root of the trie is the empty string.

The use of \$ ensures that the $m+1$ terminal nodes are in fact leaf nodes of the structure, as illustrated for $X = \{ab, abc\}$ in Figure 2.1(a): without the use of \$, the terminal node labelled 1 would be internal, reflecting the fact that ab is a prefix of abc. This figure also displays another characteristic property of these data structures: common prefixes (such as ab or ε) of elements of X appear only once. Thus the nodes in a trie are constrained to be pairwise distinct.

Next to the trie, we find in Figure 2.1(b) a matching ***Patricia trie*** (or ***compacted trie***) [M68], constructed from a trie by eliminating all internal nodes of degree 2 (those with a parent and just a single child), thus forming edges that spell out a substring rather than just a single letter.

In each of these figures, an integer label $i \le n$ of a terminal node identifies the element $\boldsymbol{x}_i\$$ represented at that node. The trees are searched for a given string $\boldsymbol{u}$ by starting at the root and following the edge whose label begins with the next letter of $\boldsymbol{u}$: if at any point there is no such edge, or a nonterminal node is reached that is not a prefix of $\boldsymbol{u}$, or a terminal node is reached that is not $\boldsymbol{u}\$$, then $\boldsymbol{u} \notin X$; otherwise, $\boldsymbol{u} \in X$. It is left as Exercise 2.1.3 to show that the first letters of the downward edges from any node of $T_{\boldsymbol{x}}$ are pairwise distinct.

Note that both tries and Patricia tries can conveniently be used to search, not only for any string in X, but also for any prefix of any string in X. But it should be remarked that the search for a prefix in a Patricia trie is in a sense more complicated than the search for a prefix in an ordinary trie. In a trie a prefix $\boldsymbol{u}$ is sure to be a node of the tree, whereas in a Patricia trie, it must in some cases (for example, the prefix a or abc in Figure 2.1(b)) be recognized somehow as occurring "on an edge" rather than only at a node. Thus a Patricia trie (and therefore, as we shall see in Section 5.2.1, a suffix tree) must incorporate techniques for the scanning of edges as well as the selection of nodes.

Succinctly put, a ***suffix tree*** for $\boldsymbol{x}$ (sometimes called a ***subword tree*** [A85] for $\boldsymbol{x}$) is just a Patricia trie on the set X of suffixes of $\boldsymbol{x}$ (including ε, the empty suffix). We shall denote the suffix tree of $\boldsymbol{x}$ by $T_{\boldsymbol{x}}$. Note that $T_{\boldsymbol{x}}$ contains exactly $n+1$ terminal nodes and, as shown

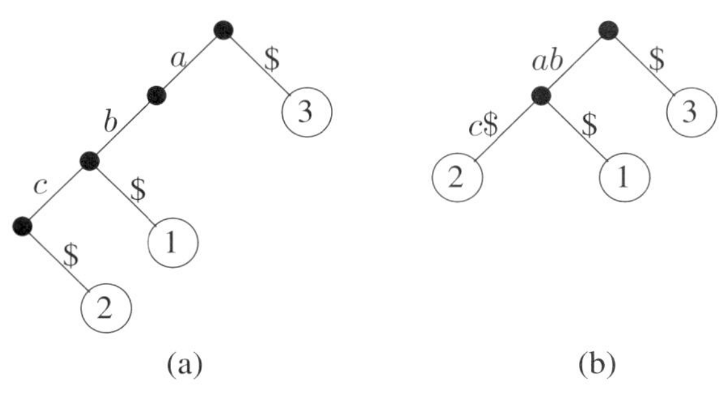

Figure 2.1. Tries on $X = \{ab, abc\}$

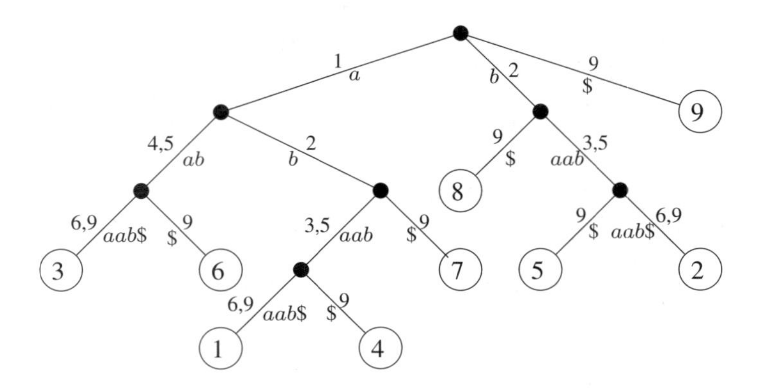

Figure 2.2. The suffix tree $T_{\boldsymbol{g}}$ for $\boldsymbol{g} = abaabaab$

in Exercise 2.1.4, at most n internal nodes called ***branch nodes***. Thus, there are at most $2n+1$ nodes and at most $2n$ edges in $T_{\boldsymbol{x}}$. The storage required for each edge label can be reduced to a constant by replacing the substring on each edge by two integers specifying a position of the substring in $\boldsymbol{x}$. (But recall Digression 1.2: this replacement assumes that the substring can be accessed in constant time from its position, hence that the input string is now stored as an array.) It follows then that the suffix tree is a "desirable" data structure in the sense that it requires only $\Theta(n)$ storage (though see Digression 2.1 below). Figure 2.2 shows $T_{\boldsymbol{g}}$ for

$$\begin{array}{cccccccccc} & & 1 & 2 & 3 & 4 & 5 & 6 & 7 & 8 \\ \boldsymbol{g} & = & a & b & a & a & b & a & a & b. \end{array}$$

Note that now the labels of the terminal nodes identify the positions in the string at which each suffix begins, with label $n+1$ denoting the empty suffix.

The suffix tree for $\boldsymbol{x}$, introduced by Wiener [W73], has become a very important intrinsic pattern. Given a string $\boldsymbol{u}$, $T_{\boldsymbol{x}}$ can be used in a straightforward way to determine whether or not $\boldsymbol{u}$ is a suffix of $\boldsymbol{x}$, but of course this can be determined, probably with greater efficiency, directly from $\boldsymbol{x}$. More importantly, since every substring of $\boldsymbol{x}$ is a prefix of some suffix of $\boldsymbol{x}$, $T_{\boldsymbol{x}}$ can also be used to determine whether or not $\boldsymbol{u}$ is a substring of $\boldsymbol{x}$. Further, it turns out that, with some massaging, $T_{\boldsymbol{x}}$ can also be used to locate the first occurrence of $\boldsymbol{u}$ in $\boldsymbol{x}$, and in fact to locate *all* occurrences of $\boldsymbol{u}$ in $\boldsymbol{x}$. Thus it has applications also to computing repetitions in strings [AP83, GS98, SG98]. Since a suffix tree can be formed from a set of strings just by concatenating the strings of the set, all of these capabilities can easily be extended from a single string $\boldsymbol{x}$ to a collection X of strings. Efficient algorithms for the construction of suffix trees are described in Section 5.2, where also more insight into applications is provided.

Finally, in Chapter 11, we discover that the suffix tree has a natural interpretation as a finite automaton (FA), and that searching the suffix tree for a match with a string $\boldsymbol{p}$ is equivalent to executing an FA on input $\boldsymbol{p}$.

Problem 2.4 **(Compute Suffix Tree)** *Construct a suffix tree $T_{\boldsymbol{x}}$ for $\boldsymbol{x}$ (Section 5.2).*

Digression 2.1.1 A word about the efficiency of building and searching a suffix tree.

In Section 5.2 we shall study three algorithms for building suffix trees, two that require an ordered alphabet (Section 4.1) — that is, an alphabet in which for every pair of distinct letters λ, μ, it can be decided in constant time whether or not $\lambda > \mu$. These two algorithms both require $O(n \log \alpha)$ time, where α is alphabet size; thus, since possibly $\alpha \in \Theta(n)$, the time requirement in the worst case is $\Omega(n \log n)$. On an ordered alphabet, this lower bound on the time complexity of suffix tree construction is least possible.

However, the third algorithm in Section 5.2 is a recent breakthrough that executes in $\Theta(n)$ time on an indexed alphabet (Section 4.1) — that is, an alphabet that can be treated as if it consists of the integers $\{1, 2, \ldots, \alpha\}$ for some $\alpha \in O(n)$. An indexed alphabet is very common in practice: any binary alphabet can be treated as indexed, as well as any subset of the ASCII characters, including the English alphabet, DNA and RNA alphabets, and many others.

In fact, even on an alphabet that is ordered but not necessarily indexed, there are cases in which suffix tree construction is unexpectedly efficient. One of the ordered alphabet suffix tree construction algorithms discussed in Section 5.2 is ***on-line***; that is, the suffix tree $T_{\boldsymbol{x}\lambda}$ for any letter λ can be computed in $O(\log n)$ time from $T_{\boldsymbol{x}}$. This means that when new strings are formed by ***right extension*** — by adding a suffix — the suffix tree of each new string can be computed incrementally with very little time investment. Similarly, a suffix tree of an increasing *collection* of strings can be efficiently maintained using this algorithm.

Complementary to suffix tree construction is suffix tree search. At each nonterminal node of $T_{\boldsymbol{x}}$, it needs to be decided, based on the current letter $\boldsymbol{u}[i]$ of $\boldsymbol{u}$, which downward edge (if any) in the tree should be followed. This decision requires testing $\boldsymbol{u}[i]$ against the current letter on at least one (and possibly all) of these edges.

If the alphabet A is ordered, these tests can be performed in $O(\log \alpha)$ time for each $\boldsymbol{u}[i]$ by implementing some standard data structure at each node (a search tree, for example, or an ordered array); alternatively, for small alphabets it will be at least as efficient to implement $T_{\boldsymbol{x}}$ as an equivalent binary tree (see Exercise 2.1.7). Thus, overall, we must suppose that using $T_{\boldsymbol{x}}$, $\Omega(|\boldsymbol{u}| \log \alpha)$ time can be required to search for $\boldsymbol{u}$ in $\boldsymbol{x}$. If $|\boldsymbol{u}|$ is small with respect to $n = |\boldsymbol{x}|$, and if the alphabet size is not too large, this search time can compare very favourably with the $O(n)$ time normally required to search $\boldsymbol{x}$.

Searching a suffix tree becomes still more attractive in the common case of an indexed alphabet, where the $\log \alpha$ factor can disappear altogether, since the correct downward edge can be computed in constant time using a table-lookup procedure. On the other hand, if α is large, this time-efficient solution can impose an unacceptable $\Theta(\alpha)$ space requirement at each node of $T_{\boldsymbol{x}}$. When the alphabet is ordered, it may sometimes be worthwhile to use hash table techniques to select the correct edge, again sacrificing space in order to achieve near-constant retrieval time.

Overall, the suffix tree is an extremely effective and widely-used tool for searching strings. Its Achilles' heel is its high space requirement, especially in the case of large ordered alphabets, though even with small alphabets the penalty can be severe — see the discussion in Subsection 5.2.5. Generally, the use of suffix trees for searching is most attractive when $\boldsymbol{x}$ is static — that is, subject to infrequent change — the alphabet is small and fixed, and $\boldsymbol{u}$ is relatively short. These requirements are often satisfied when DNA sequences are searched: the alphabet is fixed (C, G, A, T), so also is $\boldsymbol{x}$, and $|\boldsymbol{u}| \ll n$. $\square$

In an effort to develop more convenience or efficiency of implementation, various other interesting suffix structures have been proposed. In Section 5.3 we discuss two of these: DAWGs (**D**irected **A**cyclic **W**ord **G**raphs) and suffix arrays.

Problem 2.5 (Compute Suffix Structures) *Construct other suffix structures (Section 5.3).*

Intrinsic patterns of particular interest are those that provide a decomposition or factorization of a given string $\boldsymbol{x}$. In Chapter 6 we first consider Lyndon decomposition, already discussed in some detail in Section 1.4, where its connection to the computation of a canonical form for a necklace was explained. Solutions to this problem also provide a basis for computing the lexicographically least and greatest suffixes of every prefix of $\boldsymbol{x}$. Efficient algorithms for Lyndon decomposition and related problems are described in Sections 6.1 and 6.2.

Problem 2.6 (Compute Lyndon Decomposition) *Compute the Lyndon decomposition of $\boldsymbol{x}$ (Section 6.1).*

Another important (and quite different) form of decomposition is the so-called s-factorization [LZ76] that was first employed as a tool for string compression [ZL77], then later applied [C84, M89] in a surprising way as the basis of efficient algorithms for computing the repetitions in a string. In brief, the ***s*-factorization** of a given string $\boldsymbol{x}$ is a decomposition $\boldsymbol{x} = \boldsymbol{w}_1\boldsymbol{w}_2\cdots\boldsymbol{w}_k$ where every $w_j, j = 1, 2, \ldots, k$, is either a single letter that does not occur in $\boldsymbol{w}_1\boldsymbol{w}_2\cdots\boldsymbol{w}_{j-1}$ or else the longest substring that occurs both as suffix and proper substring of $\boldsymbol{w}_1\boldsymbol{w}_2\cdots\boldsymbol{w}_j$. We define the s-factorization more formally and describe its calculation in Section 6.3, leaving until Section 12.2 the explanation of its application to computing repetitions.

Problem 2.7 (Compute s-Factorization) *Compute the s-factorization of $\boldsymbol{x}$ (Section 6.3).*

Exercises 2.1

1. Show that a "border tree" is a tree (acyclic and connected) rooted at node 0.
2. Give examples of a Patricia trie and a suffix tree whose root nodes have but one child.
3. Show that the labels of the downward edges from any node of a Patricia trie have no common nonempty prefix.

4. Suppose that in a given tree with k terminal nodes, every internal node has exactly two children. Show that this tree contains exactly $k-1$ internal nodes. Hence show that a suffix tree $T_{\boldsymbol{x}}$ contains at most n internal nodes.
5. Generalize the preceding result to show that a Patricia trie for a set X of k strings requires storage for at most $O(k)$ nodes and edges. Show by example that this result does not hold for ordinary tries.
6. Draw the suffix tree for $\boldsymbol{x} = ababbabbaba$.
7. Suppose that a suffix tree $T_{\boldsymbol{x}}$ has been constructed for a certain string $\boldsymbol{x}$ on an ordered alphabet A, where $\alpha = |A|$ is a known fixed integer. For each of the following cases, design a data structure for the nodes that will permit efficient selection of the correct downward edge during a search of the tree. (You may well come up with more than one solution in each case, depending for example on whether α is "small" or "large".)

   (a) $A = \{1, 2, \ldots, \alpha\}$;
   (b) A consists of ordered but not necessarily consecutive letters $\lambda_1 < \lambda_2 < \cdots < \lambda_\alpha$.

   For each of the data structures you devise, justify your decisions, taking into account the time *and* the space requirements, and the trade-offs between them. (If you have never implemented a tree as a binary tree before, perhaps it is time you did!)
8. Design your own algorithm to construct a suffix tree for a given string $\boldsymbol{x}$ on a fixed alphabet A. You will find you need to consider the cases that arise in the preceding exercise. Try to make your algorithm as efficient as possible — not worse than $O(n^2)$!
9. Exercise 1.2.12 asked you to design an algorithm, hopefully requiring $O(n^2)$ time, to compute all the distinct substrings in a given string $\boldsymbol{x}$. On the assumption that the suffix tree $T_{\boldsymbol{x}}$ of $\boldsymbol{x}$ is given, show that the distinct substrings in $\boldsymbol{x}$ can actually be computed in $\Theta(n)$ time.

   Hint: Observe that the distinct substrings that begin at position i are exactly the strings $\boldsymbol{x}[i..j]$, $j \leq n$, that have as prefix the minimum-length distinct substring that begins at i. Thus these strings are determined simply by reporting this mimimum-length distinct prefix.

2.2 Specific Patterns (Part III)

Words — so innocent and powerless as they are,
as standing in a dictionary, how potent for good and evil
they become in the hands of one who knows how to combine them.

— Nathaniel HAWTHORNE (1804–1864)

The patterns dealt with in this section are the ones computed by the "classical" pattern-matching algorithms of computer science: given a string (sometimes called the ***text***), identify one or all (if any) of the occurrences of a second specified string (called the ***pattern***) within it. An algorithm to solve this problem has been executed thousands of times (to me it feels like millions!) by the Unix editor that I am using to prepare the LaTeX files for this book: searching for snippets of my own prose is the main mechanism I use to find my way around in these files. Around the world such an algorithm is probably executed in total hundreds of millions of times a day by ordinary people using their PCs to edit files. Other applications may not be so ordinary: a molecular biologist might have accessed a DNA segment 5,000,000 ***base-pairs*** (letters) long from an international database of genomic information in order to locate the occurrences of a particular shorter segment only 300 base-pairs long within it. (Perhaps the shorter segment is one of the characteristic patterns marking the beginning or end of a gene.) Again, the search for the shorter segment is an application of classical pattern-matching as described above.

Algorithm 2.2.1 (Easy)

– *Find all occurrences of* $\boldsymbol{p}$ *in* $\boldsymbol{x}$

for $i \leftarrow 1$ **to** $n-m+1$ **do**
  $j \leftarrow$ *match*(i, m)
  if $j = m+1$ **then output** i

But the reader may be wondering what all the fuss is about. With a few strokes of the pen, he or she has just written an algorithm, similar to Algorithm 2.2.1, that takes care of the problem: for each position i in the text at which a match *could* occur, it executes a trivial routine *match*(i, m) that compares $\boldsymbol{p}[1..m]$ with $\boldsymbol{x}[i..i+m-1]$ position-by-position from left to right, returning either $m+1$ (if $\boldsymbol{p} = \boldsymbol{x}[i..i+m-1]$) or else the least integer $j \in 1..m$ such that $\boldsymbol{p}[1..j] \neq \boldsymbol{x}[i..i+j-1]$. Of course, if $j = m+1$, an occurrence of $\boldsymbol{p}$ in $\boldsymbol{x}$ has been located at position i. Easy! Algorithm 2.2.1 is perfectly correct, a statement whose proof is left to Exercise 2.2.2. The trouble with it is however that it may be inefficient; for example, in the case

$$\boldsymbol{x} = a^n;\ \boldsymbol{p} = a^{m-1}b$$

the routine *match* must perform m character comparisons for every $i \in 1..n-m+1$, hence $m(n-m+1)$ comparisons altogether. Thus Algorithm 2.2.1 requires $O(mn)$ time for its execution, perhaps a rather serious matter when $n = 5{,}000{,}000$ and $m = 300$! Although on average we would expect Algorithm 2.2.1 not to require so much time, still we would like to avoid using an algorithm whose worst-case behaviour is so time-consuming and, as far as we know, unpredictable. As we have seen, using suffix trees allows us to find each occurrence of $\boldsymbol{p}$ in $O(m \log \alpha)$ time — at least once the suffix tree itself is computed. And as we shall find in Chapters 7 and 8, there are many algorithms that guarantee that *all* occurrences of $\boldsymbol{p}$ can be located in total time $O(m+n)$, a big improvement on Easy. In Chapter 7 we discuss the four "basic" pattern-matching algorithms: Knuth-Morris-Pratt [KMP77], Boyer-Moore [BM77], Karp-Rabin [KR87], and Dömölki-(Baeza-Yates)-Gonnet [D68, BYG92]. Then in Chapter 8 we consider a very few of the dozens, if not hundreds, of more sophisticated algorithms that have been proposed in the search either for optimal behaviour in theory or for optimal behaviour in practice — the two are not necessarily the same.

Problem 2.8 (Compute All Patterns) *With guaranteed efficiency, compute all occurrences of a nonempty pattern* $\boldsymbol{p}$ *in a given string* $\boldsymbol{x}$ *(Chapters 7 and 8).*

For many string-searching problems, the model presented so far is not adequate. In a DNA or protein sequence, a particular pattern may be *almost* duplicated at various locations, but not exactly — in such a circumstance one wants to treat the given pattern $\boldsymbol{p}$ as "approximate", so that matches with $\boldsymbol{p}$ will be recognized even though some of them may not be exact. For example, given $\boldsymbol{p} = CGAT$, one might want to accept interchange in either the first two or the last two letters as being close enough; thus one would recognize $\boldsymbol{p} \approx GCAT$ and $\boldsymbol{p} \approx CGTA$, so that in

$$\boldsymbol{x} = TCTAGG\underline{CGAT}TCG\underline{GCAT}ATTCG\underline{CGTA}GCTCTA, \tag{2.1}$$

the positions of the underlined strings would all be returned.

The fundamental idea on which the recognition of approximate patterns is commonly based is "distance" between two strings. In general, the ***distance*** between two strings $\boldsymbol{p_1}$ and $\boldsymbol{p_2}$ is the weighted number of standard edit operations required to transform $\boldsymbol{p_1} \to \boldsymbol{p_2}$. The edit operations normally considered are the following:

- ***insertion*** — for example, inserting one G into $GCAT$ to form $GCGAT$;
- ***deletion*** — for example, deleting one G from $GCGAT$ to form $CGAT$;
- ***substitution*** — for example, substituting G for C in $GCAT$ to form $GGAT$.

Of course these edit operations are defined for nonempty characters only. They give rise to several different kinds of distance, some of which we now define:

(a) If two strings $\boldsymbol{x_1}$ and $\boldsymbol{x_2}$ have the same length n, the ***Hamming distance*** $d_H(\boldsymbol{x_1}, \boldsymbol{x_2})$ [H82] is defined to be the minimum number of *substitutions* required to transform $\boldsymbol{x_1}$ into $\boldsymbol{x_2}$. Thus $d_H(GCAT, CGAT) = 2$. Observe that a substitution for $\boldsymbol{x_1}[i]$ can be regarded as deletion of $\boldsymbol{x_1}[i]$ followed by insertion of the substitute letter immediately after $\boldsymbol{x_1}[i-1]$.

(b) For arbitrary strings $\boldsymbol{x_1}$ and $\boldsymbol{x_2}$, the ***Levenshtein distance*** $d_L(\boldsymbol{x_1}, \boldsymbol{x_2})$ [L66] is defined to be the minimum number of *deletions* and *insertions* required to transform $\boldsymbol{x_1}$ into $\boldsymbol{x_2}$. Thus $d_L(GCAT, CGAT) = 2$ while

$$d_L(GCGAT, CGAT) = d_L(CAT, CGAT) = 1.$$

To see that when $\boldsymbol{x_1}$ and $\boldsymbol{x_2}$ have the same length, it is *not* necessarily true that $d_L(\boldsymbol{x_1}, \boldsymbol{x_2}) = d_H(\boldsymbol{x_1}, \boldsymbol{x_2})$, consider the example $\boldsymbol{x_1} = CGA$, $\boldsymbol{x_2} = AGT$: $d_H(CGA, AGT) = 2$ while $d_L(CGA, AGT) = 4$. Observe that the sequence of insertions and deletions required to achieve the minimum may not be obvious: $CGATA$ can be transformed into $ATACG$ by four deletions (of CG and TA) followed by four insertions (of AT and CG), but the same transformation can also be effected by only two deletions (of prefix CG) followed by only two insertions (of suffix CG).

(c) For arbitrary strings $\boldsymbol{x_1}$ and $\boldsymbol{x_2}$, the ***edit distance*** $d_E(\boldsymbol{x_1}, \boldsymbol{x_2})$ [L66] is defined to be the minimum number of *deletions*, *insertions* and *substitutions* required to transform $\boldsymbol{x_1}$ into $\boldsymbol{x_2}$. Thus the calculation of d_E permits substitution for a single letter to take place at a cost of only one operation, whereas for d_L substitution requires two operations, one deletion and one insertion. To see better how d_E relates to d_H and d_L, consider $\boldsymbol{x_1} = CGACG$, $\boldsymbol{x_2} = GTCGA$. $d_H(\boldsymbol{x_1}, \boldsymbol{x_2}) = 5$, since only substitutions are allowed and corresponding positions in $\boldsymbol{x_1}$ and $\boldsymbol{x_2}$ are all distinct: $d_L(\boldsymbol{x_1}, \boldsymbol{x_2}) = 4$, since the prefix C can be deleted, a suffix A inserted, and then the interior A transformed to T by a deletion and an insertion; however, $d_E(\boldsymbol{x_1}, \boldsymbol{x_2}) = 3$, since the transformation $A \rightarrow T$ costs only one substitution, rather than a deletion and an insertion.

The reader is warned that the terms "Levenshtein distance" and "edit distance" are not used consistently in the literature: sometimes what we have called edit distance is called Levenshtein distance, and *vice versa*. I do however undertake to use them consistently within the confines of this book!

(d) The preceding definitions have been based on the assignment of unit values to edit operations. However, at least for the substitution operation, unit values are not always appropriate. It may be that substitution of one letter for another (for example, $C \rightarrow G$) is more likely or more frequent than some other substitution (say $C \rightarrow T$). To reflect this distinction, it is common, in applications to molecular biology, to assign variable weights to substitutions using a so-called ***scoring matrix*** W; that is, an $\alpha \times \alpha$ matrix of nonnegative real numbers in which $W[i, j]$ gives the weight to be applied when the jth letter in the alphabet A is substituted for the ith letter. For arbitrary strings $\boldsymbol{x_1}$ and $\boldsymbol{x_2}$, then, the ***weighted distance*** $d_W(\boldsymbol{x_1}, \boldsymbol{x_2})$ is defined to be the minimum sum of the number of insertions, the number of deletions, and the substitution weights required to transform $\boldsymbol{x_1}$ into $\boldsymbol{x_2}$, using the scoring matrix W. For example, suppose that for $A = \{C, G, A, T\}$, W is specified as follows:

	C	G	A	T
C	0.0	0.4	0.9	1.0
G	0.4	0.0	1.0	0.9
A	0.9	1.0	0.0	0.5
T	1.0	0.9	0.5	0.0

Then, using the example from (c), we find that

$$d_W(CGACG, GTCGA) = 2.5,$$

since now the transformation $A \rightarrow T$ costs only 0.5. We shall study a more sophisticated version of the scoring matrix in Section 13.3.

It is important to remark that any distance function d defined on any domain D is normally required to satisfy the conditions of a ***metric***; that is, for every $u, v, w \in D$:

$$d(u, v) \geq 0; \qquad \textbf{(Nonnegativity)} \qquad (2.2)$$

$$d(u, v) = 0 \iff u = v; \qquad \textbf{(Uniqueness)} \qquad (2.3)$$

$$d(u, v) = d(v, u); \qquad \textbf{(Symmetry)} \qquad (2.4)$$

$$d(u, w) \leq d(u, v) + d(v, w). \qquad \textbf{(Triangle Inequality)} \qquad (2.5)$$

We leave as Exercise 2.2.1 the demonstration that d_H, d_L and d_E are in fact metrics, as well as the specification of conditions that make d_W a metric. Here we note that in some applications it can happen that the matrix W is not symmetric, or that there is a requirement for other distance functions, such as those that permit weighting of insertions and deletions as well. In fact, returning to the example (2.1) above, we see that none of the distance functions defined here captures the idea that *interchanges* between certain adjacent (or even nonadjacent) letters ($C \leftrightarrow G$, $A \leftrightarrow T$) should be considered instead of, or in addition to, insertions, deletions and substitutions. Such interchanges are thought to occur sometimes in DNA as part of the evolutionary process. There is no single "correct" idea of distance, and the results obtained by algorithms, as well as their efficiency, depend very much on the often rather arbitrary definition of the distance function that is used.

The first and fundamental problem related to distance is of course

> **Problem 2.9 (Compute Distance)** *Given a distance function d and two arbitrary strings $\boldsymbol{x_1}$ and $\boldsymbol{x_2}$, compute $d(\boldsymbol{x_1}, \boldsymbol{x_2})$ (Chapter 9).*

In this context, there is another problem of interest. Given a string $\boldsymbol{x}$, consider the string

$$\boldsymbol{x'} = \boldsymbol{x}[i_1]\boldsymbol{x}[i_2]\cdots\boldsymbol{x}[i_k],$$

where $1 \leq i_1 < i_2 < \cdots < i_k \leq n$ and $1 \leq k \leq n$. Then $\boldsymbol{x'}$ is called a ***subsequence*** of $\boldsymbol{x}$. For completeness, we also define ε to be a subsequence of every string. We see then that every substring of $\boldsymbol{x}$ is also a subsequence of $\boldsymbol{x}$. On the other hand, $\boldsymbol{x'} = cold$ is a subsequence, but not a substring, of $\boldsymbol{x} = scrolled$. Given two strings $\boldsymbol{x_1}$ and $\boldsymbol{x_2}$, we may look for a ***longest common subsequence*** $\text{LCS}(\boldsymbol{x_1}, \boldsymbol{x_2})$ — we might find, for example, that

$$cold = \text{LCS}(scrolled, could).$$

It turns out that this problem, which also arises in molecular biology, is intimately connected to Problem 2.9, especially for Levenshtein distance d_L. In fact, as established in Exercise 2.2.9, any algorithm that determines a minimum sequence of deletes and inserts to transform a string $\boldsymbol{x_1}$ into another string $\boldsymbol{x_2}$ effectively determines an $\text{LCS}(\boldsymbol{x_1}, \boldsymbol{x_2})$: the deletes can all be performed first, and the string that results from the deletes is an LCS. Thus algorithms to solve the LCS problem can be thought of as variants of algorithms that compute Levenshtein distance. We discover in Exercise 2.2.15 that $\text{LCS}(\boldsymbol{x_1}, \boldsymbol{x_2})$ is not necessarily unique.

Problem 2.10 **(Compute LCS)** *Compute a longest string $\boldsymbol{x}'$ that is a subsequence of two given strings $\boldsymbol{x}_1$ and $\boldsymbol{x}_2$ (Chapter 9).*

Although the problem of computing $\text{LCS}(\boldsymbol{x}_1, \boldsymbol{x}_2, \ldots, \boldsymbol{x}_k)$, $k > 2$, is NP-Complete [K72], on the other hand for $k = 2$ and strings of length n, it has been shown to require $\Omega(n^2)$ time if only equality comparisons are permitted on letters of the alphabet [AHU76], $\Omega(n \log n)$ time if the alphabet is ordered [H78]. For a discussion of alphabet types, see Section 4.1.

Building on solutions to Problem 2.9, the approximate pattern-matching problem is usually stated in terms of a maximum distance k between patterns:

Problem 2.11 **(Compute All Approximate Patterns)** *Given a string $\boldsymbol{x}$, a nonempty pattern $\boldsymbol{p}$, a distance function d, and a number $k \geq 0$, compute all substrings $\boldsymbol{x}'$ of $\boldsymbol{x}$ such that $d(\boldsymbol{p}, \boldsymbol{x}') \leq k$ (Chapter 10).*

There is however another important way to capture the idea of approximate equality between strings: grouping together strings that conform to some "pattern", especially one based on regular expressions. Before discussing these patterns in their full generality, we give a somewhat informal description of the use of ***wild-card symbols*** in patterns to represent matches with any letter or letters of the alphabet. The use of these symbols is convenient for searches of English-language text and so for example has been implemented in various contexts within the Unix operating system (`grep`, for example). The wild-card symbols foreshadow the "metacharacters" that occur in regular expressions. There are two wild-card symbols commonly used:

$\bullet$ matches any single letter of A;
$*$ matches any string of A^*.

For example, given

$$\boldsymbol{x} = aaabaabbaabbba,$$

the pattern $\boldsymbol{p}_1 = a \bullet a$ matches aaa and aba, while the pattern $\boldsymbol{p}_2 = a * a$ matches aa, aaa, aba, $abba$ $abbba$, and several other substrings including $\boldsymbol{x}$ itself. This example illustrates a major difficulty that arises in wild-card pattern-matching, one that precludes using any straightforward extensions of the basic pattern-matching algorithms of Chapter 7: the transitivity of matching is lost. As we see in the example, the two strings matched by $a \bullet a$ do not match each other!

Problem 2.12 **(Compute All Wild-Card Matches)** *Compute all the substrings of a given string $\boldsymbol{x}$ that match a nonempty pattern $\boldsymbol{p}$ possibly containing wild-card symbols $\bullet$ and/or $*$ (Section 10.4).*

We turn now to patterns that are regular expressions. Matching these patterns against substrings of $\boldsymbol{x}$ partakes somewhat of both approximate and multiple pattern-matching: as we shall see, many different patterns can be specified by a regular expression, and these patterns can often be thought of as "approximations" of it. We begin by introducing ***metacharacters***; that is, characters in a pattern $\boldsymbol{p}$ that are *not* letters of A and that have a special meaning. Without saying so explicitly, we have already made use of metacharacters

$$(\;)^{<\text{integer expression}>}$$

as a convenience, so that certain strings, such as $ababababaabaabacccc$, can be more compactly and appropriately expressed in a form such as $(ab)^2(aba)^3c^4$. And of course we see now that the wild-card symbols just introduced are also metacharacters, but ones that go beyond mere convenience: they provide a notation that permits classes of strings that were not heretofore expressible to be specified by a single pattern. We now go on to define the more powerful or expressive metacharacters that give rise to regular expressions:

(M1) For any pair of patterns $\boldsymbol{p_1}$ and $\boldsymbol{p_2}$, $\boldsymbol{p_1}|\boldsymbol{p_2}$ means any one of $\{\boldsymbol{p_1}, \boldsymbol{p_2}\}$; thus $\boldsymbol{p_1}|\boldsymbol{p_2}$ matches both $\boldsymbol{p_1}$ and $\boldsymbol{p_2}$. For example,

- $a^3|b^3$ matches both a^3 and b^3;
- $C|G|A$ matches any one of $\{C, G, A\}$;
- $\varepsilon|a|b|ab$ matches any one of $\{\varepsilon, a, b, ab\}$.

We see that for a finite alphabet A, the metacharacter $|$ is a generalization of the wild-card symbol $\bullet$: it can be used to ensure a match with any single letter of A, but it also extends to other cases not expressible using $\bullet$.

In the Unix operating system, and in query languages for many database systems, expressions such as $a|b$ are often represented by the notation $[ab]$. By a further extension, when the alphabet A is ordered, expressions such as $[a - zA - Z]$ can be used in place of $a|b|\cdots|z|A|B|\cdots|Z$. These are syntactic considerations only and have no effect on the expressiveness of the pattern.

(M2) For any pattern $\boldsymbol{p}$, $\boldsymbol{p}^*$ means zero or more (but finitely many) concatenated occurrences of $\boldsymbol{p}$. For example,

- a^* matches any element of $\{a^i\}$, where i ranges over all nonnegative integers and we define $a^0 \equiv \varepsilon$;
- $(a|b)^*$ matches any element of A^* where $A = \{a, b\}$;
- $(a|b)^*(a|b)(a|b)^*$ matches any element of A^+ where $A = \{a, b\}$;
- $a^*|b^*$ matches any one of $\{a^i, b^i\}$, where i ranges over all nonnegative integers;
- a^*b^* matches any string a^ib^j, where i and j are nonnegative integers;
- $(ab)^*$ matches any string $(ab)^i$, where i is a nonnegative integer.

We see from the second of these examples that, for a finite alphabet A, the metacharacters $|$ and * can be combined to represent the wild-card symbol $*$.

However, the metacharacter * alone can be used to represent patterns not representable using any combination of wild-card symbols; for example, the patterns a^* and $(ab)^*$.

(M3) For any pattern $\boldsymbol{p}$, $\tilde{}\boldsymbol{p}$ matches any string *not* matched by $\boldsymbol{p}$. For example, on the alphabet $A = \{a, b\}$,

- $\tilde{}(a^*)$ matches $(a|b)^*b(a|b)^*$ — that is, any string that contains b;
- $\tilde{}(a|b)^*$ matches no string whatever, not even the empty string;
- $\tilde{}((a|b)^*(a|b)(a|b)^*)$ matches ε;
- $\tilde{}(a^*b^*)$ matches any string in which some b precedes some a — that is, $a^*bb^*a(a|b)^*$;
- $\tilde{}(a^*|b^*)$ matches any string that contains both a and b — that is,

$$((a|b)^*a(a|b)^*b(a|b)^*) \mid ((a|b)^*b(a|b)^*a(a|b)^*); \tag{2.6}$$

- $\tilde{}(ab)^*$ matches any nonempty string that is not a power of ab — that is, $(ab)^*(a|(aa|b)(a|b)^*)$.

In these examples, it is no accident that all the patterns that include the metacharacter $\tilde{}$ can be re-expressed so as to exclude it: it can be shown, though we will not do so, that this property holds in general. Thus $\tilde{}$ is another "metacharacter of convenience"; its use does not extend the classes of strings that can be expressed by patterns using $|$ and *.

It should be pointed out that, in the preceding exposition of metacharacters, there is a slight abuse of notation: the symbol $\boldsymbol{p}$ used for a string is also used for a pattern representing a *set* of strings. I hope that this abuse has not confused the reader: the motive has been to avoid additional formalism and additional notation.

We are now in a position to state

Definition 2.2.1 *A **regular expression** is a pattern that may include the metacharacters $|$ and * as described in (M1) and (M2). A **regular set** is the set of strings represented by a regular expression.*

The associated problem of course is

Problem 2.13 **(Compute All Regular Expression Matches)** *In a given string $\boldsymbol{x}$ compute all substrings that match a given regular expression $\boldsymbol{p}$ (Section 11.1).*

Matching patterns approximately is one important extension to pattern-matching; matching a set of patterns, rather than a single one, is another. In Section 11.2 we deal with two forms of this problem: exact and approximate. To find all exact occurrences of a set $P = \{\boldsymbol{p_1}, \boldsymbol{p_2}, \ldots, \boldsymbol{p_r}\}$ of patterns in a text string $\boldsymbol{x} = \boldsymbol{x}[1..n]$ is actually a special case of regular expression matching, since we are really matching against

$$\boldsymbol{p} = \boldsymbol{p_1}|\boldsymbol{p_2}|\cdots|\boldsymbol{p_r}.$$

However, as we shall see, the special case can be handled in $\Theta(m+n)$ time, where $m = |\boldsymbol{p_1}| + |\boldsymbol{p_2}| + \cdots + |\boldsymbol{p_r}|$. Also in Section 11.2, we present two algorithms for the more difficult problem of computing all k-approximate occurrences of multiple patterns in $\boldsymbol{x}$.

Problem 2.14 **(Compute All Multiple Patterns)** *Compute all (exact or approximate) occurrences of all patterns* $P = \{\boldsymbol{p_1}, \boldsymbol{p_2}, \ldots, \boldsymbol{p_m}\}$ *in a given string* $\boldsymbol{x}$ *(Section 11.2).*

Exercises 2.2

1. Write the "trivial" routine *match* and prove its correctness.
2. Prove that Algorithm 2.2.1 is correct for every pattern $\boldsymbol{p}$. Be sure that your proof takes all cases into account, including $n = 0$, $m = 0$, $m = n$, $m > n$. Does it make sense to search an empty string? To search for an empty pattern?
3. Suppose that Algorithm 2.2.1 is applied to text strings $\boldsymbol{x}$ and patterns $\boldsymbol{p}$ (of fixed lengths n and m, respectively) that are somehow "randomly" generated on an alphabet of size $\alpha \geq 2$. Show that the expected number of letter comparisons is approximately

$$(n-m+1)\left(\frac{\alpha}{\alpha-1}\right)\left(1-\left(\frac{1}{\alpha}\right)^m\right).$$

4. For strings $\boldsymbol{x_1}$ and $\boldsymbol{x_2}$ of the same length n, show that

   (a) $d_L(\boldsymbol{x_1}, \boldsymbol{x_2}) \leq 2d_H(\boldsymbol{x_1}, \boldsymbol{x_2}) \leq 2n$;
   (b) $d_H(\boldsymbol{x_1}, \boldsymbol{x_2}) - d_L(\boldsymbol{x_1}, \boldsymbol{x_2}) \leq n-2, n \geq 2$;
   (c) $d_E(\boldsymbol{x_1}, \boldsymbol{x_2}) \leq \min\{d_H(\boldsymbol{x_1}, \boldsymbol{x_2}), d_L(\boldsymbol{x_1}, \boldsymbol{x_2})\}$.

   Give examples of strings for which equality holds in each of (a)–(c).
5. Successively substituting d_H, d_L and d_E for d, compute the following:

   (a) $d(\boldsymbol{x}, \varepsilon)$;
   (b) $d((ab)^3, (ba)^3)$;
   (c) $d(ababbba, aabbaab)$.

6. In the discussion of d_L, it is claimed that $d_L(CGACG, GTCGA) = 4$, based on two specified deletions and two insertions. Find another set of four insertions and deletions that transform $CGACG \rightarrow GTCGA$.

7. For any string $\boldsymbol{x}[1..n]$, show that for every $j \in 0..n-1$,

$$d_H(\boldsymbol{x}, R_j(\boldsymbol{x})) \neq 1,$$

where R_j is the jth rotation of $\boldsymbol{x}$ (Section 1.4).

8. For edit or Levenshtein distance d, find nonempty strings $\boldsymbol{x_1}$, $\boldsymbol{x_2}$ and $\boldsymbol{x_3}$ such that $|\boldsymbol{x_1}| = |\boldsymbol{x_2}|$ and

$$d(\boldsymbol{x_1}, \boldsymbol{x_2}) > d(\boldsymbol{x_1}, \boldsymbol{x_2 x_3}).$$

9. Suppose there exist three strings $\boldsymbol{x_1}$, $\boldsymbol{x_2}$, $\boldsymbol{x_3}$ such that $\boldsymbol{x_1}$ equals a subsequence of $\boldsymbol{x_2 x_3}$ and

$$|\boldsymbol{x_2 x_3}| - |\boldsymbol{x_1}| < d(\boldsymbol{x_1}, \boldsymbol{x_2}),$$

where d denotes either edit or Levenshtein distance. Weilin Lu believes that in this case $d(\boldsymbol{x_1}, \boldsymbol{x_2}) > d(\boldsymbol{x_1}, \boldsymbol{x_2 x_3})$. Is Weilin's belief justified?

10. Let us define ***deletion distance*** $d_D(\boldsymbol{x_1}, \boldsymbol{x_2})$ between two given strings $\boldsymbol{x_1}$ and $\boldsymbol{x_2}$ to be the minimum number of deletions required to transform $\boldsymbol{x_1}$ and $\boldsymbol{x_2}$ into the same string. (For example, $d_D(ab, bac) = 3$ since three deletions are required to transform both ab and bac into a — or, alternatively, into b — while more than three deletions would be required to transform ab and bac into any other string.) Show that

$$d_D(\boldsymbol{x_1}, \boldsymbol{x_2}) = |\boldsymbol{x_1}| + |\boldsymbol{x_2}| - 2|\text{LCS}(\boldsymbol{x_1}, \boldsymbol{x_2})| = d_L(\boldsymbol{x_1}, \boldsymbol{x_2}).$$

11. Let us define the ***asymmetric edit distance*** $d_A(\boldsymbol{x_1}, \boldsymbol{x_2})$ between two given strings $\boldsymbol{x_1}$ and $\boldsymbol{x_2}$ to be the minimum number of deletions in $\boldsymbol{x_2}$ plus the minimum number of substitutions/deletions in $\boldsymbol{x_1}$ required to transform $\boldsymbol{x_1}$ and $\boldsymbol{x_2}$ into the same string. Show that $d_A(\boldsymbol{x_1}, \boldsymbol{x_2}) = d_E(\boldsymbol{x_1}, \boldsymbol{x_2})$.

12. Show that the distance functions d_H, d_L and d_E are metrics according to (2.2)–(2.5) on any domain A^* of strings.

13. Show that d_W is a metric if and only if W is symmetric with zero diagonal and positive off-diagonal elements.

14. Show that Levenshtein distance d_L is independent of the order in which the edit operations (delete, insert) are performed. Interpreting substitution as a delete together with an insert, show that the same result holds for edit distance d_E as well as for weighted distance d_W. Are any of the axioms of a metric required for your proof?

15. Let $\boldsymbol{x_1}$ and $\boldsymbol{x_2}$ be strings such that neither one is a subsequence of the other. Show that $\boldsymbol{x} = \text{LCS}(\boldsymbol{x_1}, \boldsymbol{x_2})$ if and only if $d_L(\boldsymbol{x_1}, \boldsymbol{x}) + d_L(\boldsymbol{x_2}, \boldsymbol{x}) = d_L(\boldsymbol{x_1}, \boldsymbol{x_2})$.

16. Show by example that $\mathrm{LCS}(\boldsymbol{x_1}, \boldsymbol{x_2})$ is not unique. Show further that for any positive integer k, strings $\boldsymbol{x_1}$ and $\boldsymbol{x_2}$ of length $2k$ can be found such that two distinct LCSs of $\boldsymbol{x_1}$ and $\boldsymbol{x_2}$ differ in k positions.

17. Easier than the LCS problem is the LCF problem — given two strings $\boldsymbol{x_1}$ and $\boldsymbol{x_2}$, to compute $\mathrm{LCF}(\boldsymbol{x_1}, \boldsymbol{x_2})$, the ***longest common factor*** (or substring). Design an algorithm that uses suffix trees to compute the LCF of two given strings. Determine your algorithm's complexity.

18. On the alphabet $A = \{a, b\}$, characterize the strings represented by $\boldsymbol{p} = a \bullet b * a$.

19. Write the following pattern expressions in a simpler form:

(a) $(a|b)^*(a|b)(a|b)^*$;

(b) $(a|b)^*b(a|b)^*$;

(c) $(a^*|b^*)^*$.

20. Equation (2.6) gives two options for strings that contain both a and b. Thomas doubts that both of these options are necessary. To relieve his agony, specify a string that is included in the lefthand option but not the righthand one (and *vice versa*). Then show that (2.6) is equivalent to $((a^*ab)|(b^*ba))\ (a|b)^*$. Can this expression be simplified by replacing ab and ba by a and b, respectively?

2.3 Generic Patterns (Part IV)

And once sent out a word takes wing —
irrevocably.

— HORACE (65–8 BC), *Epistles*

In this section we state problems that require the computation of generic patterns in strings; that is, patterns specified in terms of their internal structure rather than in terms of any specific substrings or classes of substrings described by them. For the most part, the structure of these patterns will be some kind of relaxation of the idea of periodicity introduced in Section 1.2. Thus, generic patterns may roughly be thought of as those that in some sense display "approximate periodicity".

We begin with patterns that in fact display *exact* periodicity — repetitions, defined in Section 1.2. Repetitions have been of interest to mathematicians since early in the 20th century, when Axel Thue [T06] considered the construction of infinite repetition-free strings on an alphabet of three letters. In Sections 3.2 and 3.3 we consider some of these "Thue strings". A problem more in the spirit of computer science is the computation of all the repetitions in a given string, apparently first considered by Main and Lorentz [ML79] in the late 1970s. As mentioned earlier, this problem and its extensions have applications to molecular biology and to data compression. In designing an algorithm for this computation, a critical question that arises is the maximum number of repetitions that can possibly occur in a string: this number determines the maximum size of the output, and so establishes a lower bound on the complexity of the algorithm. Consider the string

$$\boldsymbol{x} = aaaaaa.$$

This string contains five occurrences of the square a^2, three of $(aa)^2$, and one of $(aaa)^2$ — thus altogether nine distinct occurrences of squares. In general, it is not difficult to show that the string a^n contains $\lfloor n^2/4 \rfloor$ occurrences of squares, from which we naturally conclude that the computation of all the squares in a given string $\boldsymbol{x}$ of length n must require $\Omega(n^2)$ time.

But our "natural" conclusion is wrong! If we inspect the squares enumerated for a^6, we find that they can all be deduced from the observation that $\boldsymbol{x}[1] = a$ is repeated six times. Thus in this case it suffices to output a "Crochemore triple" [C81]

$$(i, p^*, r^*) = (1, 1, 6)$$

indicating that the substring of length $p^* = 1$ occurring at position $i = 1$ is repeated $r^* = 6$ times. Here p^* is the period and r^* the exponent of $\boldsymbol{x}$ as defined in Definition 1.2.3, and so the triples (i, p^*, r^*) provide a breakdown into normal form of each maximal repetition within a given string $\boldsymbol{x}$. As a further example, consider $\boldsymbol{x} = a^2b^3a^2b^3a$: here the repetitions returned should be

$$(1,1,2),\ (3,1,3),\ (6,1,2),\ (8,1,3),\ (1,5,2),\ (2,5,2).$$

We see in general that, since the normal form provides a complete description of any repetition, it is necessary only to output the normal form of each maximal repetition found within the given string $\boldsymbol{x}$. Recall from Exercise 1.2.19 that in the normal form $\boldsymbol{u}^{r^*}$ of any string, the generator $\boldsymbol{u}$ itself cannot be a repetition.

We say that the triples (i, p^*, r^*) provide an ***encoding*** of the output. We shall show in Section 3.4 that, in terms of this encoding, Fibostrings $\boldsymbol{f_n}$ with $f_n \equiv |\boldsymbol{f_n}|$ contain $\Theta(f_n \log f_n)$ maximal repetitions, for each of which one triple must be output. Thus, at least for this encoding, $\Omega(n \log n)$ gives a lower bound on the complexity of any algorithm that finds all repetitions in a given string $\boldsymbol{x}$ of length n. We shall see in Chapter 12 that in fact this lower bound is attained by algorithms solving

Problem 2.15 **(Compute All Repetitions)** *Compute all the repetitions in a given string $\boldsymbol{x}$ in $O(n \log n)$ time (Section 12.1).*

Digression 2.3.1 The encoding of repetitions described above has the same curious property that the border array has (see Digression 1.3.2): it allows us, by agreement with the user, to output more information in less space. The border array holds what could be thought of as $\Theta(n \log n)$ information in $\Theta(n)$ space; here repetitions that may number as many as $\Theta(n^2)$ are compressed, effectively using the normal form, into only $O(n \log n)$ triples.

In the case of the border array, we cannot hope to reduce the storage requirement below $\Theta(n)$, but for repetitions, a legitimate question arises: could we find an encoding that would allow us to output all the repetitions in something less than $\Theta(n \log n)$ space — say $\Theta(n \log \log n)$ or even $\Theta(n)$? As noted above, there is an $\Omega(n \log n)$ lower bound on the time required to compute all the repetitions in $\boldsymbol{x}$, but there is no information-theoretic lower bound, other than the trivial one of $\Omega(n)$, on the space required to output those repetitions. In fact, as we shall see in Section 3.4, the repetitions in a Fibostring $\boldsymbol{f_n}$ can be described in $\Theta(f_n)$ space — if, for strings with many repetitions, those repetitions can be output using only linear space, why should it not be possible to achieve linear space also for strings with fewer repetitions? And if linear space is achievable, why not also linear time?

We return to this interesting question in Chapter 12; meanwhile, the definitions given below will help to make it clearer. □

Problem 2.15 is the "classical" problem related to repeating substrings that was studied for the most part prior to 1990. It has since been realized, however, that the idea of a repetition can be modified or extended in various useful ways. For instance, nonadjacent repeating substrings may also be of great interest in various contexts. One example is in DNA sequence analysis, where particular nonadjacent subsequences may signal special conditions, such as the beginning or the end of a gene. Another example is in the storage, retrieval and analysis of musical texts [CIR98], where there may be a requirement to recognize repeated motifs or melodies within the same score or across several selected scores.

We begin our discussion of these "extended" repetitions with the following:

Definition 2.3.1 *A **repeat** in a string $\boldsymbol{x}$ is a tuple*

$$M_{\boldsymbol{x},\boldsymbol{u}} = (p;\ i_1, i_2, \ldots, i_r),\ r \geq 2,$$

where $i_1 < i_2 < \cdots < i_r$ and

$$\boldsymbol{u} = \boldsymbol{x}[i_1..i_1+p-1] = \boldsymbol{x}[i_2..i_2+p-1] = \cdots = \boldsymbol{x}[i_r..i_r+p-1].$$

*$\boldsymbol{u}$ is said to be a **repeating substring** of $\boldsymbol{x}$ and the **generator** of $M_{\boldsymbol{x},\boldsymbol{u}}$. As for repetitions, we call $p = |\boldsymbol{u}|$ the **period** of the repeat and r its **exponent**. If the tuple includes all the occurrences of $\boldsymbol{u}$ in $\boldsymbol{x}$, then $M_{\boldsymbol{x},\boldsymbol{u}}$ is said to be **complete** and is written $M^*_{\boldsymbol{x},\boldsymbol{u}}$.*

In its most general form, then, our task may be defined to be the computation of $M^*_{\boldsymbol{x},\boldsymbol{u}}$ for every repeating $\boldsymbol{u}$. As we shall see below, this task may be simplified by introducing the idea of "nonextendibility". First however we need to associate the new idea of a repeat with the familiar idea of a repetition.

A repeat is clearly a generalization of a repetition, with the same associated terminology (generator, period, exponent) also generalized. To make the relationship precise, let us define a ***gap*** to be the difference

$$g_j = i_{j+1} - i_j, \; 1 \leq j \leq r-1,$$

between consecutive elements of a repeat $M_{\boldsymbol{x},\boldsymbol{u}}$. We can then classify repeats according to their gaps:

- $M_{\boldsymbol{x},\boldsymbol{u}}$ is said to be a ***repetition*** (or ***tandem repeat***) iff every gap $g_j = p$; in this case, of course, $M_{\boldsymbol{x},\boldsymbol{u}}$ may be written in the abbreviated form

$$(i,p,r) \equiv (p;\; i, i+p, \ldots, i+(r-1)p).$$

- $M_{\boldsymbol{x},\boldsymbol{u}}$ is said to be ***split*** iff every gap $g_j > p$.
- $M_{\boldsymbol{x},\boldsymbol{u}}$ is said to be an ***overlap*** iff every gap $g_j < p$.
- $M_{\boldsymbol{x},\boldsymbol{u}}$ is said to be a ***cover*** iff every $g_j \leq p$; in this case $M_{\boldsymbol{x},\boldsymbol{u}}$ is said to ***cover*** the string $x[i_1..i_r+p-1]$, which is also said to be ***coverable*** or $\boldsymbol{u}$***-coverable***.

If $M_{\boldsymbol{x},\boldsymbol{u}}$ does not fall into any of the above categories, it is said to be ***mixed***. To make these ideas clear, consider

$$\begin{array}{cccccccccccccccccccccc} & 1 & 2 & 3 & 4 & 5 & 6 & 7 & 8 & 9 & 10 & 11 & 12 & 13 & 14 & 15 & 16 & 17 & 18 & 19 & 20 & 21 \\ \boldsymbol{x} = & a & b & a & a & b & a & b & a & a & b & a & a & b & a & b & a & a & b & a & b & a \end{array}$$

and $\boldsymbol{u} = aba$. Then a variety of possible repeats $M_{\boldsymbol{x},\boldsymbol{u}}$ may be defined:

- $(3;\; 1,4)$ and $(3;\; 6,9,12)$ are tandem;
- $(3;\; 1,6,12,17)$ is split;
- $(3;\; 4,6)$ is an overlap;
- $(3;\; 1,4,6,12)$ is mixed;
- last but not least, the complete repeat

$$M^*_{\boldsymbol{x},\boldsymbol{u}} = (3;\; 1,4,6,9,12,14,17,19)$$

  is actually a cover of $\boldsymbol{x}$!

Observe that by computing $M^*_{\boldsymbol{x},\boldsymbol{u}}$ we effectively compute all the repetitions of $\boldsymbol{u}$, all the overlaps, and all the maximum-length $\boldsymbol{u}$-coverable substrings: a single scan of $M^*_{\boldsymbol{x},\boldsymbol{u}}$ suffices to output all of this information. In the above example, such a scan would reveal that

- $(3;\; 1,4)$, $(3;\; 6,9,12)$ and $(3;\; 14,17)$ are the repetitions of $\boldsymbol{u} = aba$ in $\boldsymbol{x}$;
- $(3;\; 4,6)$, $(3;\; 12,14)$ and $(3;\; 17,19)$ are the overlaps;
- there is exactly one maximum-length substring coverable by $\boldsymbol{u}$ and that is $\boldsymbol{x}$ itself.

In our example string, let us look now at the complete repeat of b:

$$M^*_{\boldsymbol{x},b} = (1;\; 2,5,7,10,13,15,18,20).$$

Inspection of this repeat reveals an interesting fact: every occurrence of b is followed by a, and so reporting the repeat

$$M^*_{\boldsymbol{x},ba} = (2;\ 2,5,7,10,13,15,18,20)$$

implicitly reports $M^*_{\boldsymbol{x},b}$ also! That is, we know from $M^*_{\boldsymbol{x},ba}$ that

$$\boldsymbol{x}[2..3] = \boldsymbol{x}[5..6] = \cdots = \boldsymbol{x}[20..21],$$

so that

$$\boldsymbol{x}[2] = \boldsymbol{x}[5] = \cdots = \boldsymbol{x}[20].$$

But inspection of $M^*_{\boldsymbol{x},ba}$ reveals further that every occurrence of ba is preceded by a, so that reporting

$$M^*_{\boldsymbol{x},aba} = (3;\ 1,4,6,9,12,14,17,19)$$

implicitly reports both $M^*_{\boldsymbol{x},b}$ and $M^*_{\boldsymbol{x},ba}$.

The reader will have realized that here again, as with borders and repetitions (see Digressions 1.3.2 and 2.3.1), we may be in a position to encode the outputs of repeats in such a way as to reduce both the space and time requirements for their computation. In order to describe this encoding precisely, we need further definitions:

Definition 2.3.2 *A repeat $M_{\boldsymbol{x},\boldsymbol{u}} = (p;\ i_1, i_2, \ldots, i_r)$ is called **left-extendible** (respectively, **right-extendible**) if and only if*

$$(p;\ i_1-1, i_2-1, \ldots, i_r-1)\ \ (\textit{respectively, } (p;\ i_1+1, i_2+1, \ldots, i_r+1))$$

*is a repeat. If $M_{\boldsymbol{x},\boldsymbol{u}}$ is neither left-extendible nor right-extendible, it is said to be **nonextendible**. We abbreviate these terms as **LE**, **RE** and **NE** respectively.*

It is interesting to consider the special case of Definition 2.3.2 in which the repeat $M_{\boldsymbol{x},\boldsymbol{u}}$ is actually a repetition or tandem repeat $(\boldsymbol{u}[1..p])^r$, $r \geq 2$. Suppose for example that

$$\boldsymbol{x} = \cdots a(aba)^r a \cdots,$$

with generator $\boldsymbol{u} = aba$ of period $p = 3$. We observe that the repetition $(aba)^r$ is LE because $(aab)^r$ is a repeat of aab of period $p = 3$, also RE because $(baa)^r$ is a repeat of baa. As this example illustrates, in the tandem case, to say that $M_{\boldsymbol{x},\boldsymbol{u}}$ is LE implies that $\boldsymbol{u}^r$ is preceded in $\boldsymbol{x}$ by a letter $\lambda = \boldsymbol{u}[p]$; thus the repetition $M_{\boldsymbol{x},\boldsymbol{u}}$ is an extension of another repetition $M_{\boldsymbol{x},\boldsymbol{u}'}$ of the same period p, where $\boldsymbol{u}' = \lambda\boldsymbol{u}[1..p-1]$. Similarly, if $M_{\boldsymbol{x},\boldsymbol{u}}$ is RE, it extends to another repetition $M_{\boldsymbol{x},\boldsymbol{u}''}$, where $\boldsymbol{u}'' = \boldsymbol{u}[2..p]\boldsymbol{u}[1]$. From Section 1.4 we recognize $\boldsymbol{u}'$ and $\boldsymbol{u}''$ as *rotations* of $\boldsymbol{u}$, and so we realize that repetitions that are extendible determine repetitions of certain rotations of $\boldsymbol{u}$. In terms of the Crochemore encoding, this means that if the repetition (i, p^*, r^*) is LE, then $(i-1, p^*, r^*)$ is also a repetition, while if (i, p^*, r^*) is RE, then $(i+1, p^*, r^*)$ is also a repetition. We are thus able to elucidate an important new idea:

Definition 2.3.3 *Given a string $\boldsymbol{x}$, we say that a 4-tuple (i, p^*, r^*, t) is a **run** in $\boldsymbol{x}$ iff*

(a) *$t \in 0..p^*-1$ is an integer; and*
(b) *(i, p^*, r^*) is a non-LE (**NLE**) repetition in $\boldsymbol{x}$; and*
(c) *$(i+t, p^*, r^*)$ is a non-RE (**NRE**) repetition in $\boldsymbol{x}$.*

*We call t the **tail** of the run.*

The run seems to have been first introduced by Main in [M89], where it was called, less concisely but more evocatively, a ***maximal periodicity***. The term ***maximal repetition*** is also used [KK00].

It was mentioned in Digression 2.3.1 that the repetitions in a Fibostring could actually be reported in linear time; we shall see in Section 3.4 that it is exactly the idea of a run that permits this efficiency to be achieved. In a more general context, we shall discover in Section 12.2 that the number of runs in *any* string is in fact linear in the string length: this opens up the prospect of computing all the runs in at least some strings in less time than the $O(n \log n)$ time required to compute repetitions. Thus we have

Problem 2.16 **(Compute All Runs)** *Efficiently compute all the runs in a given string $\boldsymbol{x}$ (Section 12.2).*

In addition to runs, the above definitions make it possible to state problems related both to covers and to repeats in general, as we now explain.

It has been seen above that the generalization of repetition to repeat allows us to define not only split repetitions but also maximum-length coverable substrings. Thus we may immediately state a generalization of Problem 2.15 that replaces repetitions (gap $= p$) by coverable substrings (gap $\leq p$):

Problem 2.17 **(Compute All Coverable Substrings)** *Compute all the NE coverable substrings of a given string $\boldsymbol{x}$.*

There are three known algorithms [AE93, BP00, IM01] that solve a problem very similar to this one: they actually compute all the NRE coverable substrings of $\boldsymbol{x}$. Unfortunately these algorithms are all rather complicated and technical, and so we do not describe any of them in detail in this book. However, in Section 13.2, where a closely related problem is considered (Problem 2.19), we discuss these solutions and speculate on future improvements to them.

The idea of a "cover" also gives rise to an interesting problem that may have application to data compression: the computation of the covers of a string. In this context, the cover of a string may be thought of as a generalization of the generator of a repetition. For this reason, a cover of a string is sometimes called its ***quasiperiod***, and a string that has a cover is said to be ***quasiperiodic***.

Several algorithms were proposed in the 1990s to compute quasiperiods; in Chapter 13 we present one of them [LS02] which computes a new intrinsic pattern called the "cover array" closely analogous to the border array discussed in Section 1.3. Like the border array, the cover array is computable in linear time; and just as the border array determines all the periods of every prefix of a string $\boldsymbol{x}$, so the cover array determines all the quasiperiods of every prefix of $\boldsymbol{x}$.

> **Problem 2.18 (Compute All Covers)** *Compute all the covers of a given string $\boldsymbol{x}$ (Section 13.1).*

Since it is clearly only necessary to report those repeats that are NE, we can state a further generalization of Problem 2.15 (the "repeats" problem in its most general form):

> **Problem 2.19 (Compute All Repeats)** *Compute all the nonextendible repeats in a given string $\boldsymbol{x}$ (Section 13.2).*

As we shall see in Section 13.2, this general problem can also be solved in $\Theta(n \log n)$ time, and so its solution can also provide a basis for solving both Problem 2.15 and Problem 2.17.

The All-Repeats algorithm (Section 13.2) is important in computational biology, but even more important is an algorithm that computes approximate repeats of substrings that differ by at most a distance k, where "distance" may take any of the forms discussed earlier (Section 2.2).

> **Problem 2.20 (Compute Approximate Repeats)** *Given a distance function d and an integer $k \geq 0$, compute all the k-approximate repeats in a given string $\boldsymbol{x}$ (Section 13.3).*

Despite its importance, this "approximate repeats" problem has been relatively little studied, certainly in comparison to approximate pattern-matching (Section 2.2). A major reason for this no doubt is the difficulty of the problem — even the best algorithms fail to achieve $O(n^2)$ execution time. The difficulty becomes apparent as soon as one begins to think seriously about how to define the problem, as discussed below.

We mentioned earlier the nontransitivity of matching when wild-card characters are involved (Problem 2.12). The same phenomenon arises when we try to define "approximate equality" of two substrings $\boldsymbol{u_1}$ and $\boldsymbol{u_2}$ of a given string $\boldsymbol{x}$. Using Hamming distance, for example, we would find that

$$d_H(ab, aa) = d_H(ab, bb) = 1,$$

so that it would be reasonable to regard either aa or bb as an "approximate repeat" of ab under the condition that the maximum distance between the repeating substrings should be $k = 1$. However, because $d_H(aa, bb) = 2$, we cannot regard any set of repeats containing both aa and bb as satisfying this condition. So right at the beginning we are faced with the awkward choice of either reporting approximate repeats only in pairs, thus multiplying the number of outputs required and presenting a fragmented report to the user, or alternatively reporting together only those approximate repeats that pairwise satisfy the stated condition, thus complicating the processing while still perhaps presenting the output in a fragmented form.

Also, the concept of nonextendibility, important for reducing the output of exact repeats, applies in a curious way to approximate repeats. As above, consider the substrings ab and aa of a given string $\boldsymbol{x}$, approximately equal for $k = 1$. Suppose that aa always occurs as a suffix of baa in $\boldsymbol{x}$ while ab occurs in two contexts, bab and aab — as found in Fibostrings, for example. Since, using the first of these contexts, $d_H(bab, baa) = 1$, we see that in this case the approximate repeat $\{ab, aa\}$ is in some sense left-extendible, whereas, in the second context, when ab is preceded by a, it is not left-extendible because $d_H(aab, baa) = 2 > k$. We thus find ourselves in the position of wishing to report some occurrences of ab but not others in the approximate repeat $\{ab, aa\}$.

Perhaps the most awkward problem results from the fact that, except when Hamming distance is being used, a string $\boldsymbol{u_1}$ can be an approximate repeat of another string $\boldsymbol{u_2}$ even though $|\boldsymbol{u_1}| \neq |\boldsymbol{u_2}|$. For example, since $d_L(ab, abc) = 1$, ab and abc would be approximate repeats for any $k \geq 1$. Thus the whole concept of period disappears, or at least changes radically, whenever approximate repeats are determined using anything other than Hamming distance. More generally, the differences among the various definitions of distance have such diverse consequences that it would seem that algorithms for approximate repetition should be specialized according to the kind of distance used. Observe, for example, that if "approximate equality" between $\boldsymbol{u_1}$ and $\boldsymbol{u_2}$ is determined by $d(\boldsymbol{u_1}, \boldsymbol{u_2}) \leq k$, then for Hamming distance all strings of length at most k are trivially approximately equal, whereas for Levenshtein distance, $\boldsymbol{u_1}$ and $\boldsymbol{u_2}$ will be approximately equal whenever $|\boldsymbol{u_1}| + |\boldsymbol{u_2}| \leq k$.

We return to this discussion in Section 13.4. For now we content ourselves with stating our final problem, admittedly in a somewhat imprecise form:

Problem 2.21 **(Compute Approximate Period)** *Given a distance function* d *and an integer* $k \geq 0$*, compute a* k*-approximate period of a given string* $\boldsymbol{x}$ *(Section 13.4).*

Exercises 2.3

1. Prove that for every integer $n \geq 0$, the string a^n contains $\lfloor n^2/4 \rfloor$ distinct occurrences of squares.
2. Find all the maximal repetitions in the Fibostring

$$\boldsymbol{f}_6 = abaababaabaab$$

and encode them as triples (i, p^*, r^*). Count the number of triples you have found. Approximately how many maximal repetitions would you expect to find in $\boldsymbol{f}_7 = \boldsymbol{f}_6(abaababa)$?

3. A string $\boldsymbol{x}$ is said to be a ***weak*** repetition iff $\boldsymbol{x} = \boldsymbol{u_1 u_2} \cdots \boldsymbol{u_k}$ for some integer $k \geq 2$, where for every $i \in 2..k$, the letters of $\boldsymbol{u_i}$ are a permutation of the letters of $\boldsymbol{u_1}$.

   (a) Show that every repetition in the ordinary sense is also a weak repetition.

   (b) Find all the *weak* repetitions in $\boldsymbol{f_6}$.

4. Find all the covers, if any, of $\boldsymbol{f_6}$ and $\boldsymbol{f_7}$.
5. Provide a definition that captures what is meant by a ***cover*** $\boldsymbol{u}$ of a necklace $C(\boldsymbol{x})$, hence ***coverable*** and ***u-coverable*** for necklaces. Is the set of covers of $C(\boldsymbol{x})$ identical to the set of covers of all the rotations of $\boldsymbol{x}$?
6. Characterize the covers of $C((abc)^n)$.
7. For $k = 2$–4 determine the minimum number of k-strings required on a chip U to ensure that every necklace $C(\boldsymbol{x})$ on the alphabet $A = \{a, b\}$ can be covered by strings from U.
8. What is the maximum number of distinct k-strings that could possibly be required in a minimum-cardinality k-cover of a string $\boldsymbol{x}$?
9. Using Definition 2.3.2, prove that

$$M_{\boldsymbol{x},\boldsymbol{u}} = (p;\ i_1, i_2, \ldots, i_r)$$

is a LE tandem repeat if and only if

$$M_{_{-1}\boldsymbol{x},\boldsymbol{u}'} = (p;\ i_1 - 1, i_2 - 1, \ldots, i_r - 1)$$

is a RE tandem repeat.

10. In order that Definition 2.3.3 should be a "good" definition, it is clearly desirable that for every integer $t' \in 1..t-1$, $(i + t', p^*, r^*)$ should be a repetition that is both LE and RE. Show that this fact follows from the definition.

Chapter 3

Strings Famous and Infamous

3.1 Avoidance Problems and Morphisms

Harp not on that string.

— William SHAKESPEARE (1564–1616), *Richard III*

In this chapter we consider three special strings that have been much studied because of their intriguing properties. These strings have two main features:

- they can be categorized in terms of the nature of their repetitions;
- they can be generated by substitutions usually called "morphisms".

These strings are of interest to us as extremal cases for some of our algorithms, especially those that compute generic patterns. Before discussing each of them in detail, we first provide an appropriate framework.

A string $\boldsymbol{x}$ is said to ***avoid*** a certain (specific or generic) pattern $\boldsymbol{p}$ if and only if there are no occurrences of $\boldsymbol{p}$ in $\boldsymbol{x}$. We may also say that $\boldsymbol{x}$ is ***p-free***. For example, $\boldsymbol{f}_5 = abaababa$ avoids $\boldsymbol{p} = bb$, and $\boldsymbol{x} = abacabca$ is square-free. The classical avoidance problem has generally been formulated in terms of repetitions in infinite strings:

Problem 3.1 **(Construct (α, r)-Avoidance)** *Construct an infinite string (or infinite necklace) on an alphabet of size α that avoids repetitions of exponent r (but contains repetitions of exponents $2, 3, \ldots, r-1$).*

Observe that some of these problems are trivial. For example, for $\alpha = 1$, and for every choice of $r > 1$, there is no string on a single letter of length r or more that avoids repetitions of exponent r. Thus there is no construction for the $(1, r)$-problem. Exercise 3.1.1 shows that there is no construction for the $(2, 2)$-problem either; that is, there is no square-free infinite string on an alphabet of two letters. However, as we shall soon discover, there may nevertheless be infinite strings on $\{a, b\}$ that avoid cubes or fourth powers; in other words, there are constructions for the $(2, 3)$- and $(2, 4)$-problems.

Axel Thue [T06] was the first to study avoidance problems — specifically, the $(2, 3)$- and $(3, 2)$-problems. In this chapter we shall look at Thue's celebrated constructions for both of these problems: in Section 3.2 we consider cube-free strings on a binary alphabet and then in Section 3.3 square-free strings on a ternary alphabet. Finally, in Section 3.4, we discuss Fibostrings, as we call them, derived from Fibonacci numbers and examples of constructions for the $(2, 4)$-problem: fourth-power-free strings on a binary alphabet that nevertheless contain squares and cubes. All of these strings are justly famous, but they may also be thought of as infamous: as we shall see, both the Thue strings $(2, 3)$ and the Fibostrings contain "many" repetitions in a well-defined sense; in fact the Fibostrings contain the maximum number of repetitions (asymptotically) that may be contained in any string. Thus these strings represent challenging cases for algorithms that compute repetitions. On the other hand, the Thue strings $(3, 2)$ contain no repetitions at all, and so oblige these algorithms to execute in vain, to produce no output. We shall see that Fibostrings are also worst cases for algorithms that compute other generic patterns in strings.

An interesting generalization of the avoidance problem that we call the ***weak*** (α, r)-***avoidance problem*** was first proposed by Erdős [E61]: if possible, construct infinite strings on an alphabet of size α that avoid weak repetitions of exponent r. Since as we have seen (Exercise 2.3.3), every repetition is also a weak repetition, it follows that if there is no construction for a particular avoidance problem, then there is no construction for the corresponding weak avoidance problem either. Of course the converse is not necessarily true.

In particular, Erdős asked what is the least value of α for which there exists an infinite string on α letters that avoids weak (Abelian) squares. In 1970 Pleasants [P70] published a construction for the weak $(5, 2)$-problem, and recently, with the assistance of a computer, Keränen [K92] found a construction for the weak $(4, 2)$-problem. Exercise 3.1.3 shows that Keränen's result is best possible.

We turn now to the second of the two main features mentioned at the beginning of this section: a ***morphism***, or substitution rule, is a mapping h that assigns a specific string $h(\lambda) \in A^+$ to each letter $\lambda \in A$. The mapping h is extended to any string $\boldsymbol{x} = \boldsymbol{x}[1..n] \in A^+$ by the identity

$$h(\boldsymbol{x}) = h(\boldsymbol{x}[1])\, h(\boldsymbol{x}[2]) \cdots h(\boldsymbol{x}[n]). \tag{3.1}$$

For completeness, we extend the mapping further to A^* by supposing that for every morphism h,

$$h(\varepsilon) = \varepsilon.$$

Thus any morphism h can be ***iterated*** from any ***initial string*** $\boldsymbol{x}_0$ to generate a ***sequence of iterates*** $h^*(\boldsymbol{x}_0)$. We use the notation

$$h^*(\boldsymbol{x_0}) = \{h^0(\boldsymbol{x_0}), h^1(\boldsymbol{x_0}), h^2(\boldsymbol{x_0}), \ldots.\},$$

where $h^0(\boldsymbol{x_0}) \equiv \boldsymbol{x_0}$ and $h^k(\boldsymbol{x_0}) = h(h^{k-1}(\boldsymbol{x_0}))$ for every integer $k \geq 1$.

For example, the morphism defined by

$$h(\lambda) = \lambda, \quad \forall \lambda \in A,$$

is called the ***identity morphism***, yielding the identity sequence

$$h^*(\boldsymbol{x}) = \{\boldsymbol{x}, \boldsymbol{x}, \ldots.\}$$

for every $\boldsymbol{x} \in A^*$. On $A = \{a, b\}$, another simple morphism is

$$h(a) = a, \quad h(b) = ab$$

generating

$$\begin{aligned} h^*(a) &= \{a, a, \ldots.\}; \\ h^*(b) &= \{b, ab, a^2b, \ldots, a^kb, \ldots.\}; \\ h^*(ab) &= \{ab, a^2b, \ldots, a^kb, \ldots.\}; \\ h^*(ba) &= \{ba, aba, a^2ba, \ldots, a^kba, \ldots.\}; \end{aligned}$$

and so on.

The infinite sequences of strings studied in the next three sections of this chapter are generated by the following morphisms:

$$\text{Thue Strings } (2,3) : h(a) = ab, \quad h(b) = ba; \tag{3.2}$$

$$\begin{aligned} \text{Thue Strings } (3,2) : h(a) &= abcab, \\ h(b) &= acabcb, \\ h(c) &= acbcacb; \end{aligned} \tag{3.3}$$

$$\text{Fibostrings } (2,4) : h(a) = ab, \quad h(b) = a. \tag{3.4}$$

We shall see that each of these morphisms in fact possesses an important additional property that we now describe. Let $h^*(A)$ be the set of strings contained in all the sequences $h^*(\lambda)$, $\lambda \in A$; that is,

$$h^*(A) = \bigcup_{\lambda \in A} h^*(\lambda). \tag{3.5}$$

We say that the morphism h is a ***code*** if and only if, for every string $\boldsymbol{x} \in h^*(A)$, there is at most one string $\boldsymbol{y} \in A^*$ such that $h(\boldsymbol{y}) = \boldsymbol{x}$. Of course, for every string $\boldsymbol{x} \in h^*(A) - A$, it follows from (3.5) that there will be *at least* one such string $\boldsymbol{y}$; it is only for $\boldsymbol{x} \in A$ that no such $\boldsymbol{y}$ may exist. When h is a code and $h(\boldsymbol{y}) = \boldsymbol{x}$, we call $\boldsymbol{y}$ the ***antecedent*** of $\boldsymbol{x}$ and write $\boldsymbol{y} = h^{-1}(\boldsymbol{x})$, where h^{-1} is the inverse mapping that, if the reader will allow a slight abuse of language, we will call the ***inverse morphism***.

Thus, informally, a code is a morphism with a well-defined inverse. Note however that the definition of the inverse is confined to $h^*(A) - A$: there may exist strings in $A^* - h^*(A)$ that have no antecedent or that even have multiple antecedents. For example, the morphism

$$h : a \to a,\ \ b \to ab,\ \ c \to abc$$

on $\{a, b, c\}$ is a code even though numerous simple strings such as ba and c^2 have no inverse under h. Similarly, the morphism

$$h : a \to a,\ \ b \to ab,\ \ c \to bab$$

is also a code even though the string $abab$, which is not an element of $h^*(A)$, has two distinct antecedents ac and b^2.

The fact that their morphisms are codes is critical for proving the basic properties of the Thue strings $(3, 2)$ and the Fibostrings.

Exercises 3.1

1. Prove that there is no construction for the $(2, 2)$-avoidance problem.
2. For every nonnegative integer k, construct a square-free string of length $n = 2^k$ on an alphabet of $k + 1$ letters.
3. Prove that there is no construction for the weak $(3, 2)$-avoidance problem, hence that Keränen's result is best possible.
4. (a) Consider the morphism on the alphabet $A = \{a, b, c, d\}$ defined by

$$h(a) = b,\ \ h(b) = c,\ \ h(c) = d,\ \ h(d) = \varepsilon.$$

   For every $\boldsymbol{x} \in A$, what are $h^4(\boldsymbol{x})$ and $h^5(\boldsymbol{x})$?

   (b) Change the morphism of (a) so that $h(d) = a$. Now what is $h^4(\boldsymbol{x})$?
5. Use (3.1) to show that if $\boldsymbol{x} = \boldsymbol{uv}$, then

$$h(\boldsymbol{x}) = h(\boldsymbol{u})h(\boldsymbol{v}).$$

6. Suppose that a morphism h and an initial string $\boldsymbol{x}_0$ are given. We say that $\boldsymbol{x} \equiv \boldsymbol{y}$ if and only if both $\boldsymbol{x}$ and $\boldsymbol{y}$ are elements of the sequence $h^*(\boldsymbol{x}_0)$. Is the relation "$\equiv$" an equivalence relation? Justify your answer.
7. Suppose an alphabet A and a morphism h on A are given. Is it possible that any iteration sequence $h^*(\boldsymbol{x}_0)$ terminates after a finite number of terms?
8. Is an inverse morphism a morphism?
9. Determine whether or not the following morphisms are codes. Justify your answers.

   (a) $h : a \to aba,\ \ b \to ab,\ \ c \to aab;$

   (b) $h : a \to a,\ \ b \to ab,\ \ c \to ab;$

   (c) $h : a \to aba,\ \ b \to ab.$

3.2 Thue Strings (2, 3)

A word to the wise ain't necessary —
it's the stupid ones that need the advice.

— Bill COSBY (1937–)

In this section we study Thue strings $\boldsymbol{t_n}$ determined by the morphism h of (3.2):

$$h(a) = ab, \quad h(b) = ba.$$

Setting

$$h^*(\boldsymbol{t_0}) = \{\boldsymbol{t_0}, \boldsymbol{t_1}, \boldsymbol{t_2}, \ldots.\},$$

where $\boldsymbol{t_n} = h(\boldsymbol{t_{n-1}})$ for every integer $n \geq 1$, and letting $\boldsymbol{t_0} = a$, we find

$$\begin{aligned} \boldsymbol{t_1} &= ab; \\ \boldsymbol{t_2} &= abba; \\ \boldsymbol{t_3} &= abbabaab; \\ \boldsymbol{t_4} &= abbabaabbaababba; \\ &\vdots \end{aligned}$$

Clearly $|\boldsymbol{t_n}| = 2^n$. Observe that, at least in the first few terms, $\boldsymbol{t_n}$ has the curious property that the first 2^{n-1} letters are the "conjugates" of the next 2^{n-1} letters; that is,

$$\boldsymbol{t_n}\left[i + 2^{n-1}\right] = a \quad \Leftrightarrow \quad \boldsymbol{t_n}[i] = b, \ \forall\, i \in 1..2^{n-1}. \tag{3.6}$$

More generally, introducing the notation $\overline{a} = b$ and $\overline{b} = a$ to express conjugacy, we see that we can re-express (3.2) as

$$h : \lambda \to \lambda\overline{\lambda} \tag{3.7}$$

for any letter $\lambda \in \{a, b\}$. This notation is easily extended to arbitrary strings of $\{a, b\}^*$: given a string $\boldsymbol{x}$ of length n, we say that $\boldsymbol{y} = \overline{\boldsymbol{x}}$ (in words, $\boldsymbol{y}$ is a ***conjugate*** of $\boldsymbol{x}$) if and only if $|\boldsymbol{y}| = n$ and

$$\boldsymbol{y}[i] = \overline{\boldsymbol{x}[i]}$$

for every $i \in 1..n$. Note that, according to this definition, every string $\boldsymbol{x}$ has a unique conjugate; further that only the empty string $\boldsymbol{\varepsilon}$ is self-conjugate. Now let $h(\boldsymbol{x})$ be the string formed by applying the substitution rule (3.7) to $\boldsymbol{x}$. Then

$$h(\boldsymbol{x}) = \boldsymbol{x}[1]\overline{\boldsymbol{x}[1]}\boldsymbol{x}[2]\overline{\boldsymbol{x}[2]} \cdots \boldsymbol{x}[n]\overline{\boldsymbol{x}[n]}$$

and, using the fact that $\lambda = \overline{\overline{\lambda}}$, we can compute the conjugate of $h(\boldsymbol{x})$:

$$\begin{aligned}\overline{h(\boldsymbol{x})} &= \overline{\boldsymbol{x}[1]}\boldsymbol{x}[1]\overline{\boldsymbol{x}[2]}\boldsymbol{x}[2]\cdots\overline{\boldsymbol{x}[n]}\boldsymbol{x}[n] \\ &= h\left(\overline{\boldsymbol{x}[1]\boldsymbol{x}[2]}\cdots\overline{\boldsymbol{x}[n]}\right) \\ &= h(\overline{\boldsymbol{x}}). \end{aligned} \tag{3.8}$$

We are now in a position to formally state and prove the conjugacy relationship (3.6):

Lemma 3.2.1 *For every integer $n \geq 1$, $\boldsymbol{t_n} = \boldsymbol{t_{n-1}}\overline{\boldsymbol{t_{n-1}}}$.*

Proof We know that $\boldsymbol{t_1} = ab = a\overline{a} = \boldsymbol{t_0}\overline{\boldsymbol{t_0}}$, and so the result is true for $n = 1$. Then suppose the lemma to be true for some integer $n' = n-1 \geq 1$. We see that

$$\begin{aligned}\boldsymbol{t_n} &= h(\boldsymbol{t_{n-1}}) \\ &= h\left(\boldsymbol{t_{n-2}}\overline{\boldsymbol{t_{n-2}}}\right), \text{ by the inductive hypothesis} \\ &= h(\boldsymbol{t_{n-2}})h\left(\overline{\boldsymbol{t_{n-2}}}\right), \text{ by (3.1)} \\ &= \boldsymbol{t_{n-1}}\overline{\boldsymbol{t_{n-1}}}, \text{ by (3.8)}\end{aligned}$$

and the result follows by induction. □

From a structural point of view, this lemma tells us that the substitution rule for the strings $\boldsymbol{t_n}$ is ***prefix-preserving***; that is, $\boldsymbol{t_n}$ has $\boldsymbol{t_{n-1}}$ as prefix. As we shall see in Section 3.4, this is one of several properties shared both by Thue strings $\boldsymbol{t_n}$ and by Fibostrings $\boldsymbol{f_n}$. In the case of Thue strings, prefix-preservation implies that $\boldsymbol{t_n}$ has the prefixes $\boldsymbol{t_i}\overline{\boldsymbol{t_i}}$, $i = 0, 1, \ldots, n-1$, while $\overline{\boldsymbol{t_n}}$ has prefixes $\overline{\boldsymbol{t_i}}\boldsymbol{t_i}$. We draw two immediate conclusions:

Lemma 3.2.2 *(a) $\boldsymbol{t_n}$ and $\overline{\boldsymbol{t_n}}$ have no common nonempty prefix;*
(b) neither $\boldsymbol{t_n}$ nor $\overline{\boldsymbol{t_n}}$ has a square prefix. □

We leave as an exercise the proof of the lemma conjugate to this result, in which "prefix" is replaced by "suffix".

We use these results to characterize the overlaps in $\boldsymbol{t_\infty}$ between occurrences of $\boldsymbol{t_n}$ and occurrences of either $\boldsymbol{t_n}$ or $\overline{\boldsymbol{t_n}}$. The main lemma involves our old friend, the border: we find that every border of a Thue string is itself a Thue string. As we shall see (Lemma 3.4.4), a very similar result holds also for Fibostrings.

Lemma 3.2.3 *For every integer $n \geq 2$, the only borders of $\boldsymbol{t_n}$ are $\boldsymbol{t_i}$, where*

$$i = n-2, n-4, \ldots, n \bmod 2.$$

Proof It suffices to show that the longest border of $\boldsymbol{t_n}$ is $\boldsymbol{t_{n-2}}$. Since

$$\begin{aligned}\boldsymbol{t_n} &= \boldsymbol{t_{n-1}}\overline{\boldsymbol{t_{n-1}}} &(3.9)\\ &= \boldsymbol{t_{n-2}}\overline{\boldsymbol{t_{n-2}}\boldsymbol{t_{n-2}}}\boldsymbol{t_{n-2}}, &(3.10)\end{aligned}$$

it is clear that t_{n-2} is one of the borders of t_n; we must show that there is no longer border.

Suppose that b is a border of t_n such that $|b| > 2^{n-2}$. If it were also true that $|b| \geq 2^{n-1}$, it would follow from (1.1) that t_n was strongly periodic, hence prefixed by a square, in contradiction to Lemma 3.2.2(b). Thus $|b| < 2^{n-1}$.

Since both b and t_{n-2} are borders of t_n, t_{n-2} must be a border of b. Since $|b| < 2^{n-1}$, b is therefore strongly periodic, hence has a square prefix. But b is a prefix of t_n, which as we have just seen can have no square prefix. We conclude that b does not exist. □

Of course we conclude also from Lemma 3.2.3 the conjugate result that for $n \geq 2$, $\overline{t_n}$ has borders $\overline{t_i}$ only, where $i = n-2, n-4, \ldots, n \bmod 2$. Observe further that since $t_{n+1} = t_n\overline{t_n}$ (respectively, $\overline{t_{n+1}} = \overline{t_n}t_n$), Lemma 3.2.3 is equivalent to the assertion that the only prefixes of t_n (respectively, $\overline{t_n}$) that are also suffixes of $\overline{t_n}$ (respectively, t_n) are t_i (respectively, $\overline{t_i}$), $i = n-1, n-3, \ldots, (n-1) \bmod 2$. Lemma 3.2.3 thus also completely specifies the possible overlaps between occurrences of t_n and $\overline{t_n}$.

We embark now on a demonstration that t_n leads to a construction for the $(2,3)$-avoidance problem; that is, even though every t_n, $n \geq 2$, contains at least one square, no t_n contains a cube. In fact, we prove a considerably stronger result: that every square in t_n is ***nonextendible*** in the sense defined in Section 2.3 — for every position i such that for some $p \geq 1$,

$$t_n[i..i+2p-1] = \left(t_n[i..i+p-1]\right)^2,$$

neither $t_n[i-1..i+2p-2]$ nor $t_n[i+1..i+2p]$ is a square.

Lemma 3.2.4 *Suppose that u^2 is a square in t_n for some $n \geq 2$. If $|u|$ is odd, then u is one of $\{a, b, aba, bab\}$.*

Proof Let $k = |u| \geq 1$. By (3.7), u^2 must take one of two forms:

(a) $u^2 = \lambda_1\overline{\lambda_1}\lambda_2\overline{\lambda_2}\cdots\lambda_k\overline{\lambda_k}$, where $v = \lambda_1\lambda_2\cdots\lambda_k$ is a substring of t_{n-1};

(b) $u^2 = \overline{\lambda_0}\lambda_1\overline{\lambda_1}\cdots\lambda_{k-1}\overline{\lambda_{k-1}}\lambda_k$, where $v = \lambda_0\lambda_1\cdots\lambda_k$ is a substring of t_{n-1}.

In case (a), since k is odd,

$$\begin{aligned} u &= \lambda_1\overline{\lambda_1}\cdots\overline{\lambda_{\lfloor k/2\rfloor}}\lambda_{\lfloor k/2\rfloor+1} \\ &= \overline{\lambda_{\lfloor k/2\rfloor+1}}\lambda_{\lfloor k/2\rfloor+2}\cdots\lambda_k\overline{\lambda_k}. \end{aligned}$$

Without loss of generality, we may let $\lambda_1 = a$, so that $\overline{\lambda_1} = b$. Then

$$\begin{aligned} \overline{\lambda_{\lfloor k/2\rfloor+1}} &= \lambda_1 = a, & \lambda_{\lfloor k/2\rfloor+2} &= \overline{\lambda_1} = b, \\ \lambda_2 &= \overline{\lambda_{\lfloor k/2\rfloor+2}} = b, & \lambda_{\lfloor k/2\rfloor+3} &= \overline{\lambda_2} = a, \end{aligned}$$

and so on, so that u consists of alternating a's and b's. Since k is odd, u must therefore begin and end with a; on the other hand, since $\overline{\lambda_{\lfloor k/2\rfloor+1}} = a$, it must be true that $u[k] = \lambda_{\lfloor k/2\rfloor+1} = b$, a contradiction. We conclude that case (a) is impossible.

Suppose then that (b) holds, so that

$$\begin{aligned}\boldsymbol{u} &= \overline{\lambda_0}\lambda_1\overline{\lambda_1}\cdots\lambda_{\lfloor k/2\rfloor}\overline{\lambda_{\lfloor k/2\rfloor}}\\ &= \lambda_{\lfloor k/2\rfloor+1}\overline{\lambda_{\lfloor k/2\rfloor+1}}\cdots\lambda_{k-1}\overline{\lambda_{k-1}}\lambda_k.\end{aligned}$$

Again without loss of generality, choose $\lambda_k = a$. Then

$$\begin{aligned}\overline{\lambda_{\lfloor k/2\rfloor}} &= \lambda_k = a, &\qquad \overline{\lambda_{k-1}} &= \lambda_{\lfloor k/2\rfloor} = b,\\ \overline{\lambda_{\lfloor k/2\rfloor-1}} &= \lambda_{k-1} = a, & \overline{\lambda_{k-2}} &= \lambda_{\lfloor k/2\rfloor-1} = b,\end{aligned}$$

and so on, so that in this case also $\boldsymbol{u}$ consists of alternating a's and b's. For $k = 1$, $\boldsymbol{u} = a$ or b, while for $k = 3$, $\boldsymbol{u} = aba$ or bab. However, for $k \geq 5$, $\boldsymbol{u}$ has prefix $ababa$ or $babab$; furthermore, since $\boldsymbol{u}$ begins with $\overline{\lambda_0}$, it is therefore preceded in $\boldsymbol{t_n}$ by λ_0. Thus one of the substrings

$$ababab,\ bababa$$

exists in $\boldsymbol{t_n}$, so that one of the forbidden substrings $\lambda_0\lambda_1\lambda_2 = aaa$ or bbb must occur in $\boldsymbol{t_{n-1}}$. Therefore $k < 5$, and the proof is complete. □

Of course it is clear that the squares aa and bb actually occur in $\boldsymbol{t_n}$ and $\overline{\boldsymbol{t_n}}$ for every $n \geq 2$, but to find the first examples of $(aba)^2$ and $(bab)^2$, we need to look at

$$\boldsymbol{t_5} = abbabaabbaa\underline{babbab}aababbaabbabaab.$$

Here we discover $(bab)^2$ centred at position $2^4 - 1 = 15$ (and as by symmetry one would expect, also its conjugate $(aba)^2$, centred at position 19). Then, by the prefix-preserving property (Lemma 3.2.1), we know that all four of the distinct squares identified by Lemma 3.2.4 must occur in every $\boldsymbol{t_n}$, $n \geq 5$.

It is also clear that whenever an "odd" square $\boldsymbol{u^2}$ occurs in $\boldsymbol{t_n}$, corresponding "even" squares $(h^r(\boldsymbol{u}))^2$ occur in $\boldsymbol{t_{n+r}}$, for every positive integer r. We now show that these rth iterates of odd squares are in fact the *only* even squares in $\boldsymbol{t_\infty}$.

Consider therefore a square $\boldsymbol{u^2}$ in $\boldsymbol{t_n}$ for which $k = |\boldsymbol{u}| \geq 2$ is even. The same possibilities exist as for odd squares:

(a) $\boldsymbol{u^2} = \lambda_1\overline{\lambda_1}\lambda_2\overline{\lambda_2}\cdots\lambda_k\overline{\lambda_k}$, where $\boldsymbol{v} = \lambda_1\lambda_2\cdots\lambda_k$ is a substring of $\boldsymbol{t_{n-1}}$;

(b) $\boldsymbol{u^2} = \overline{\lambda_0}\lambda_1\overline{\lambda_1}\cdots\lambda_{k-1}\overline{\lambda_{k-1}}\lambda_k$, where $\boldsymbol{v} = \lambda_0\lambda_1\cdots\lambda_k$ is a substring of $\boldsymbol{t_{n-1}}$.

In case (a),

$$\begin{aligned}\boldsymbol{u} &= \lambda_1\overline{\lambda_1}\cdots\lambda_{k/2}\overline{\lambda_{k/2}}\\ &= \lambda_{k/2+1}\overline{\lambda_{k/2+1}}\cdots\lambda_k\overline{\lambda_k},\end{aligned}$$

so that

$$(\lambda_1\lambda_2\cdots\lambda_{k/2})(\lambda_{k/2+1}\lambda_{k/2+2}\cdots\lambda_k)$$

is a square also in $\boldsymbol{t_{n-1}}$. Similarly in case (b),

$$u = \overline{\lambda_0}\lambda_1\overline{\lambda_1}\cdots\lambda_{k/2-1}\overline{\lambda_{k/2-1}}\lambda_{k/2}$$
$$= \overline{\lambda_{k/2}}\lambda_{k/2+1}\overline{\lambda_{k/2+1}}\cdots\lambda_{k-1}\overline{\lambda_{k-1}}\lambda_k,$$

so that

$$(\lambda_1\lambda_2\cdots\lambda_{k/2})(\lambda_{k/2+1}\lambda_{k/2+2}\cdots\lambda_k) \quad (3.11)$$

is a square in $\boldsymbol{t_{n-1}}$. Thus every even square of length $2k$ in $\boldsymbol{t_n}$ is derived from a square of length k in $\boldsymbol{t_{n-1}}$. Of course the square of length k in $\boldsymbol{t_{n-1}}$ must either also be even (and so derived from a square in $\boldsymbol{t_{n-2}}$) or else be one of the four special odd squares identified by Lemma 3.2.4. Hence we have our result:

Lemma 3.2.5 *Every square in $\boldsymbol{t_\infty}$ takes the form $(h^r(\boldsymbol{u}))^2$, where r is a nonnegative integer and $\boldsymbol{u} \in \{a, b, aba, bab\}$.* □

Now consider an odd square $\boldsymbol{u^2} = a^2$ in $\boldsymbol{t_n}$ of length $2k = 2$. By Lemma 3.2.2(b) $\boldsymbol{u^2}$ is neither a prefix nor a suffix of $\boldsymbol{t_n}$ and so must be embedded in a substring $\boldsymbol{v} = b\boldsymbol{uu}b$ of length $2k + 2$ that has the following properties:

$$\boldsymbol{v}[1] \neq \boldsymbol{u}[k], \quad \boldsymbol{v}[2k+2] \neq \boldsymbol{u}[1]. \quad (3.12)$$

Observe that these properties ensure that the square $\boldsymbol{u^2}$ is nonextendible. In fact, it is not hard to prove that the same properties hold for all squares of length $2k$ that are derived from the application of h^r to $baab$, so that all the squares $h^r(aa)$ are also nonextendible. This fact is proved in Exercise 3.2.7, and then in Exercise 3.2.8 it is shown that all squares $h^r(abaaba)$ are nonextendible as well. We have therefore the main result of this section:

Theorem 3.2.6 *Every square in every Thue string $\boldsymbol{t_n}$ is nonextendible.* □

There is a small step yet to take. The set $h^*(\boldsymbol{t_0})$ of strings with which we began this section is an infinite set, containing every finite $\boldsymbol{t_n}$, but not containing the limiting string $\boldsymbol{t_\infty}$, the ***infinite Thue string***. We observe however that if $\boldsymbol{t_\infty}$ contains a cube, then the cube must be finite, and so must be a substring of some finite Thue string, a circumstance ruled out by Theorem 3.2.6. Thus

Theorem 3.2.7 *The infinite Thue string $\boldsymbol{t_\infty}$ solves the $(2, 3)$-avoidance problem.* □

A briefer and perhaps more elegant proof of this result is given in [B63], but it does not however provide as much information about the structure of $\boldsymbol{t_\infty}$ — we have preferred therefore to take a longer route.

We conclude this section by considering a question important for computation: the number of squares in $\boldsymbol{t_n}$, or at least of those that take the form $h^r(a)$ or $h^r(b)$. Let us agree to call squares of this form ***regular***, since each such square is constructed by concatenating a suffix of $\boldsymbol{t_n}$ (respectively, $\overline{\boldsymbol{t_n}}$) with a prefix of $\overline{\boldsymbol{t_n}}$ (respectively, $\boldsymbol{t_n}$). We saw earlier, as a consequence of Lemma 3.2.3, that new regular squares $\overline{\boldsymbol{t_i}}, i = n-2, n-4, \ldots, n \bmod 2$, are formed at the centre of $\boldsymbol{t_n} = \boldsymbol{t_{n-1}}\overline{\boldsymbol{t_{n-1}}}$. These squares are therefore $\lfloor n/2 \rfloor$ in number, a fact expressed by the recurrence relation

$$q(n) - 2q(n-1) = \lfloor n/2 \rfloor, \; n \geq 2, \tag{3.13}$$

where $q(n)$ is the total number of regular squares in $\boldsymbol{t_n}$. Then, assuming n odd, we may solve (3.13) in a standard way:

$$\begin{aligned} q(n) - 2q(n-1) &= (n-1)/2 \\ 2q(n-1) - 4q(n-2) &= 2(n-1)/2 \\ &\vdots \\ 2^{n-3}q(3) - 2^{n-2}q(2) &= 2^{n-3}(2/2) \\ 2^{n-2}q(2) - 2^{n-1}q(1) &= 2^{n-2}(2/2). \end{aligned}$$

Observing that $q(1) = 0$, summing these equations, and then massaging the result a little, we obtain

$$\begin{aligned} q(n) &= \frac{3}{2}\left[(n-1) + (n-3)2^2 + (n-5)2^4 + \cdots + (2)2^{n-3}\right] \\ &= 3 \sum_{i=0}^{(n-3)/2} (n-2i-1)2^{2i-1}, \end{aligned}$$

a summation in which every term takes a value that is at most half that of the term immediately to its right. It follows that

$$3 \cdot 2^{n-3} \leq q(n) \leq 3 \cdot 2^{n-2},$$

hence that $q(n) \in \Theta(2^n)$. We draw the same conclusion also when n is even, and so we have

Theorem 3.2.8 *The number $q(n)$ of regular squares in $\boldsymbol{t_n}$ satisfies*

$$q(n) \in \Theta(|\boldsymbol{t_n}|). \quad \square$$

By some mysterious process, Jamie Simpson has divined what the exact value of $q(n)$ is; his formula is proved in Exercise 3.2.10.

In Section 3.4 we shall see that Fibostrings $\boldsymbol{f_n}$ of length f_n actually give rise to $\Theta(f_n \log f_n)$ repetitions, and so in an asymptotic sense contain "more" repetitions than Thue strings $\boldsymbol{t_n}$ do.

Exercises 3.2

1. Suppose the morphism (3.7) is applied to the initial string $\boldsymbol{t'_0} = b$, yielding the ***reverse Thue strings*** $\boldsymbol{t'_n}$, $n = 0, 1, \ldots$. Show that $\boldsymbol{t'_n} = \overline{\boldsymbol{t_n}}$.
2. Recall that a string $\boldsymbol{x}$ is called a ***palindrome*** if it reads backwards the same as forwards. Show that for every even integer n, $\boldsymbol{t_n}$ is a palindrome.
3. Equation (3.7) and Lemma 3.2.1 both give rise to iterative algorithms for computing $\boldsymbol{t_n}$. Specify each of these algorithms and comment on their relative efficiency.
4. Prove the "conjugate" of Lemma 3.2.2.
5. In Lemma 3.2.3 why did we consider only borders $\boldsymbol{b}$ of length greater than $|\boldsymbol{t_{n-2}}|$? Supply the missing arguments that complete the proof.
6. A dubious Thomasina is worried about the proof of Lemma 3.2.5. She remarks that not only is (3.11) a square in $\boldsymbol{t_{n-1}}$, but so also is

$$(\lambda_0\lambda_1 \cdots \lambda_{k/2-1})(\lambda_{k/2}\lambda_{k/2+1} \cdots \lambda_{k-1}),$$

so that (3.11) is left-extendible, in contradiction to Theorem 3.2.6. Can you alleviate Thomasina's state of dubiety?

7. Show that (3.12) holds for all strings

$$\boldsymbol{v} = h^r(baab) = h^r(b)h^r(a^2)h^r(b),$$

where $\boldsymbol{u} = h^r(a)$, $|\boldsymbol{u}| = k$, $r \geq 0$. Hence show that all squares $\boldsymbol{u^2}$ are nonextendible.

8. Show that all squares $h^r((aba)^2)$ in $\boldsymbol{t_n}$ are nonextendible.
9. Prove that if $\boldsymbol{u^2}$ is a square in $\boldsymbol{t_n}$, then $|\boldsymbol{u}| = 2^r$ or $3 \cdot 2^r$ for some integer $r \geq 0$.
10. Prove by induction Jamie Simpson's formula for $q(n)$, $n \geq 0$:

$$q(n) = 2^{n+1}/3 - n/2 - e_n,$$

where $e_n = 2/3$ if n is even, $5/6$ if n is odd.

11. Can you go even further and compute the number of *all* squares, regular or not, in $\boldsymbol{t_n}$?
12. Show that the morphism (3.2) is a code.
13. [B63] claims that $\boldsymbol{t_\infty}$ can be generated in the following way. First write down all the nonnegative integers in binary form:

$$0, 1, 10, 11, 100, 101, 110, \ldots.$$

Then add up the digits in each of these integers *modulo* 2:

$$0, 1, 1, 0, 1, 0, 0, \ldots.$$

Remove the commas, replace 0 by a and 1 by b, and hey presto! you have $\boldsymbol{t}_\infty$. Prove or disprove this claim.

3.3 Thue Strings (3, 2)

Words without actions are the assassins of idealism.

— Herbert HOOVER (1874–1964)

In this section we study Thue strings $\boldsymbol{\tau}_n$ that are in a sense the opposite of the Thue strings $\boldsymbol{t}_n$ studied in the preceding section: the strings $\boldsymbol{\tau}_n$ will be found to contain no repetitions at all. These strings are generated by the morphism h of (3.3) as follows:

$$h(a) = abcab, \;\; h(b) = acabcb, \;\; h(c) = acbcacb.$$

Setting

$$h^*(\boldsymbol{\tau}_0) = \{\boldsymbol{\tau}_0, \boldsymbol{\tau}_1, \boldsymbol{\tau}_2, \ldots.\},$$

where $\boldsymbol{\tau}_n = h(\boldsymbol{\tau}_{n-1})$ for every integer $n \geq 1$, and letting $\boldsymbol{\tau}_0 = a$, we find

$$\begin{aligned} \boldsymbol{\tau}_1 &= abcab; \\ \boldsymbol{\tau}_2 &= (abcab)(acabcb)(acbcacb)(abcab)(acabcb); \\ &\vdots \end{aligned}$$

Each of the three components $abcab$, $acabcb$ and $acbcacb$ of the morphism h is called a ***block***. Observe that we can write each block $\boldsymbol{B}$ in the form $\boldsymbol{IT}$, where both $\boldsymbol{I}$ and $\boldsymbol{T}$ have prefix a and a suffix that is a string on $\{b, c\}$. We will call $\boldsymbol{I}$ and $\boldsymbol{T}$ ***initiator*** and ***terminator*** respectively. Observe further that all initiators and terminators are distinct:

$$\{abc, ac, acbc; \;\; ab, abcb, acb\}.$$

In fact, the set of all initiators and terminators is exactly the set of all strings with prefix a and a square-free suffix of length 1–3 on $\{b, c\}$.

The distinctness of the initiators/terminators means that each initiator occurring in any string of $h^*(\boldsymbol{\tau}_0)$ is recognizable and determines its following terminator; similarly, each terminator in $h^*(\boldsymbol{\tau}_0)$ is recognizable and determines its preceding initiator. Also, every string τ_n, $n > 0$, of $h^*(\tau_0)$ is a concatenation of blocks; that is, a concatenation of alternating initiators and terminators. Thus, for every block $\boldsymbol{B}$ in τ_n, its antecedent $h^{-1}(\boldsymbol{B})$ under the inverse morphism h^{-1} is well-defined.

Lemma 3.3.1 *The Thue string τ_n is square-free if and only if τ_{n-1} is square-free.*

Proof Clearly τ_{n-1} is square-free if τ_n is square-free, so only the converse needs to be proved. Suppose then that there exists a square $\boldsymbol{u} = \boldsymbol{u_L u_R}$ in τ_n for some $n \geq 1$.

If $\boldsymbol{u_R}$ has an initiator $\boldsymbol{I_i}$ as a prefix, for some $i \in 1..3$, then so does $\boldsymbol{u_L}$, which (since no terminator can match any initiator) must therefore take the form

$$\boldsymbol{u_L} = (\boldsymbol{I_i T_i})(\boldsymbol{I_{j_1} T_{j_1}}) \cdots (\boldsymbol{I_{j_k} T_{j_k}})$$

for some $k \geq 0$. Thus both $\boldsymbol{u_L}$ and $\boldsymbol{u_R}$ are composed of an integral number of blocks, and so $h^{-1}(\boldsymbol{u_L})$ is well-defined, and so $h^{-1}(\boldsymbol{u_L u_R})$ is a square in τ_{n-1}.

Similarly, if $\boldsymbol{u_R}$ has a terminator $\boldsymbol{T_i}$ as a prefix, we conclude that

$$\boldsymbol{I_i u_L} = (\boldsymbol{I_i T_i})(\boldsymbol{I_{j_1} T_{j_1}}) \cdots (\boldsymbol{I_{j_k} T_{j_k}})\boldsymbol{I_i} = \boldsymbol{v_L I_i}$$

for some $k \geq 0$, while

$$\boldsymbol{v_R} = \boldsymbol{T_i}(\boldsymbol{I_{j_1} T_{j_1}}) \cdots (\boldsymbol{I_{j_k} T_{j_k}})$$

is a prefix of $\boldsymbol{u_R}$. Thus $\boldsymbol{v^2} = \boldsymbol{v_L I_i v_R}$ is a square in which $\boldsymbol{v} = \boldsymbol{v_L}$ is composed of an integral number of blocks, so that again $h^{-1}(\boldsymbol{v})$ is well-defined and is therefore the generator of a square in τ_{n-1}.

Now suppose that $\boldsymbol{u_R}$ does not have a as a prefix. Since no substring that excludes a can be a square, there must be at least one a in both $\boldsymbol{u_L}$ and $\boldsymbol{u_R}$. Thus $\boldsymbol{u_R}[1] = \boldsymbol{u_L}[1]$ occurs either in some $\boldsymbol{I_i}$ or in some $\boldsymbol{T_i}$. Arguments analogous to those just used above then lead to the conclusion that τ_n contains a square $\boldsymbol{v^2}$ such that $h^{-1}(\boldsymbol{v^2})$ is well-defined and denotes a square in τ_{n-1}.

We have thus shown in every possible case that existence of a square in τ_n implies the existence of a square in τ_{n-1}, as required. □

The immediate corollary of Lemma 3.3.1 is

Theorem 3.3.2 *If τ_0 is square-free, so is every string of $h^*(\tau_0)$.* □

To complete this analysis, we argue as in Section 3.2 that if the infinite Thue string τ_∞ contains a square, so also does some finite Thue string, in contradiction to Theorem 3.3.2. Hence:

Theorem 3.3.3 *The infinite Thue string τ_∞ solves the $(3,2)$-avoidance problem.* □

The morphism (3.3) was the first one discovered that generates square-free strings on three letters [BEM79], but it is not the only such morphism. A simpler one is the following [B63, C91]:

$$\begin{aligned} h(a) &= acb, \\ h(b) &= c, \\ h(c) &= ab. \end{aligned} \tag{3.14}$$

There is a very pretty demonstration that this morphism does in fact yield infinite square-free strings. Given the Thue string

$$\boldsymbol{t}_\infty = abbabaabbaababba \cdot \cdots,$$

consider a new infinite string $\boldsymbol{u}_\infty$ on $\{0,1,2\}$ formed by counting the number of b's between consecutive a's in $\boldsymbol{t}_\infty$:

$$\boldsymbol{u}_\infty = 2102012 \cdot \cdots.$$

It is not hard to see that, if a square were to exist in $\boldsymbol{u}_\infty$, an extendible square would have to exist in $\boldsymbol{t}_\infty$, in contradiction to Theorem 3.2.6. Thus $\boldsymbol{u}_\infty$ must be square-free. Finally, observe that $\boldsymbol{u}_\infty$ is derived from the morphism

$$\begin{aligned} h(2) &= 210, \\ h(0) &= 1, \\ h(1) &= 20. \end{aligned} \tag{3.15}$$

With the substitution $\{0,1,2\} \to \{b,c,a\}$, this becomes the morphism (3.14). Exercise 3.3.5 gives the details of this demonstration.

Another morphism that satisfies the $(3,2)$-avoidance problem is the following [L57], in which we observe that $h(b)$ is obtained from $h(a)$, and $h(c)$ from $h(b)$, by a cyclic permutation of the letters of the alphabet:

$$\begin{aligned} h(a) &= abcbacbcabcba, \\ h(b) &= bcacbacabcacb, \\ h(c) &= cabacbabcabac. \end{aligned} \tag{3.16}$$

Square-free morphisms on larger alphabets can also be of interest. In the Tower of Hanoi puzzle, three pegs numbered 1–3 and n disks of distinct diameters $d \in 1..n$ are given. Initially, the disks are piled on peg 1 in decreasing order of diameter (smallest on top), and the object is to move the disks one-by-one to one of the other pegs so that the disks are again piled in the same order. A move is legal if and only if the moved disk is not placed on top of a smaller disk. There are therefore six possible kinds of move: from pegs 1 to 2, 2 to 3, 3 to 1, and the inverses of these moves. We code these moves using the letters a, b, c and $\bar{a}$, $\bar{b}$, $\bar{c}$, respectively. It turns out that a minimum sequence of moves to solve the puzzle is represented by the following morphism on $\{a, b, c, \bar{a}, \bar{b}, \bar{c}\}$:

$$\begin{aligned} &h(a) = a\bar{c},\ h(b) = c\bar{b},\ h(c) = b\bar{a}; \\ &h(\bar{a}) = ac,\ h(\bar{b}) = cb,\ h(\bar{c}) = ba. \end{aligned} \tag{3.17}$$

This morphism solves the $(6, 2)$-avoidance problem, and so it follows that in the minimum sequence of moves that yields a solution to the Tower of Hanoi puzzle, there are no repeated subsequences of moves. For further details, consult [AARS94].

Exercises 3.3

1. Show that the morphism (3.3) is a code.
2. Specify an algorithm that implements the morphism (3.3) a given number of times on a given initial letter $\boldsymbol{\tau}_0$.
3. Complete the details of the proof of Lemma 3.3.1.
4. Show that the six blocks that we have denoted $\boldsymbol{I_i}, \boldsymbol{T_i}, i = 1, 2, 3$, can be arranged in ordered pairs in 120 distinct ways. Of these, how many define a square-free morphism?
5. In order to convince yourself that the morphism (3.14) does in fact generate square-free strings, prove the following:

   (a) If $\boldsymbol{u}_\infty$ contains a square, then $\boldsymbol{t}_\infty$ contains a substring $\boldsymbol{x}^2\lambda$ with a longest border of length 1.

   (b) The morphism (3.15) generates $\boldsymbol{u}_\infty$ as the limit string of $h^*(0)$.

6. Show further that the morphism (3.14) generates strings of lengths $3 \cdot 2^i$, $i \geq 0$, whose rotations are all square-free. Then look back at the "research problem" given in Exercise 1.4.4.
7. Herendi Tamás remarks that nevertheless the morphism (3.14) does not have the same nice property that the Thue morphism does; that is, (3.14) does not map *every* square-free string into a square-free string. Give an example to support Tamás's observation.
8. Prove that the morphism (3.16) also solves the $(3, 2)$-avoidance problem.
9. Prove that the morphism (3.17) solves the $(6, 2)$-avoidance problem (or consult [AARS94]!).

3.4 Fibostrings (2, 4)

When ideas fail, words come in very handy.

— Johann Wolfgang von GOETHE (1749–1832), *Faust I*

We conclude this chapter by studying Fibonacci strings (Fibostrings) generated by the morphism h of (3.4):

$$h(a) = ab, \quad h(b) = a.$$

As in the two preceding sections, we set

$$h^*(\boldsymbol{f_0}) = \{\boldsymbol{f_0}, \boldsymbol{f_1}, \boldsymbol{f_2}, \ldots\},$$

where now $\boldsymbol{f_n} = h(\boldsymbol{f_{n-1}})$ for every integer $n \geq 1$ and, by convention, $\boldsymbol{f_0} = b$. We shall see that the lengths $f_n = |\,\boldsymbol{f_n}|$ of Fibostrings are Fibonacci numbers, known and renowned since the late Middle Ages. It is only a slight exaggeration to say that Fibonacci numbers are ubiquitous in combinatorial work; indeed, they are of such interest and importance that an entire journal, *The Fibonacci Quarterly*, is devoted to the study of their properties and those of other iteratively-defined sequences of numbers.

From (3.4) we compute the first few Fibostrings as follows:

$$\begin{aligned}
\boldsymbol{f_0} &= b;\\
\boldsymbol{f_1} &= a;\\
\boldsymbol{f_2} &= ab;\\
\boldsymbol{f_3} &= aba; \qquad (3.18)\\
\boldsymbol{f_4} &= abaab;\\
\boldsymbol{f_5} &= abaababa;\\
\boldsymbol{f_6} &= abaababaabaab;\\
&\ \vdots \qquad (3.19)
\end{aligned}$$

with lengths $f_0 = 1$, $f_1 = 1$, $f_2 = 2$, $f_3 = 3$, $f_4 = 5$, $f_5 = 8$, $f_6 = 13$, respectively. More generally, in order to be able to relate Fibostrings to the usual definition of the Fibonacci numbers, we state the following basic result:

Lemma 3.4.1 *The Fibostrings $\boldsymbol{f_n}$ satisfy the recurrence*

$$\boldsymbol{f_0} = b; \ \ \boldsymbol{f_1} = a; \ \ \boldsymbol{f_n} = \boldsymbol{f_{n-1}}\boldsymbol{f_{n-2}}, \ \ \forall\, n \geq 2. \qquad (3.20)$$

Proof Easily proved by induction (see Exercise 3.4.2). □

It is then immediate that the lengths of the $\boldsymbol{f_n}$ satisfy the rule

$$f_n = f_{n-1} + f_{n-2}, \ \forall\, n \geq 2, \tag{3.21}$$

which is the defining relation for the Fibonacci numbers [K73a]. Observe also from (3.20) that, like the Thue strings $(2, 3)$, the Fibostrings $\boldsymbol{f_n}$, $n > 0$, are prefix-preserving. Thus it makes sense to define the ***infinite Fibostring*** $\boldsymbol{f_\infty}$ to be the string that has every $\boldsymbol{f_n}$, $n > 0$, as a prefix.

We saw in Section 3.2 that the length of the Thue strings $(2, 3)$, which doubled at each application of h, was 2^n. Because the length of a Fibostring "almost" doubles at each step, it is natural to conjecture that its length f_n is exponential in n. If so, then there has to exist some real number $\phi > 1$ such that

$$f_n \in \Omega(\phi^n).$$

To prove this, observe first from (3.21) that if

$$f_{n-1} \geq \phi^{n-2} \quad \text{and} \quad f_{n-2} \geq \phi^{n-3},$$

then it necessarily follows that

$$f_n \geq \phi^{n-2} + \phi^{n-3} = \phi^{n-3}(\phi + 1). \tag{3.22}$$

Thus, if ϕ were chosen to satisfy the relation

$$\phi^2 = \phi + 1, \tag{3.23}$$

then the desired result, that $f_n \geq \phi^{n-1}$, would be an immediate consequence of the corresponding results for f_{n-1} and f_{n-2}. But (3.23) is just a simple quadratic equation in ϕ, easily solved to yield (choosing the positive square root)

$$\phi = \left(1 + \sqrt{5}\right)/2 \approx 1.618034, \tag{3.24}$$

an interesting number in its own right, often called "the golden mean" [K73a]. It remains then to check, using (3.24), that $f_0 = 1 \geq \phi^{-1} = (\sqrt{5} - 1)/2 \approx 0.618034$ and that $f_1 = 1 \geq \phi^0 = 1$. From (3.22) and (3.23), it follows that $f_2 \geq \phi^1$, hence that $f_3 \geq \phi^2$, and so on; thus, by induction, we have the general result that

$$f_n \geq \phi^{n-1}, \tag{3.25}$$

where ϕ is given by (3.24). It is shown in Exercise 3.4.4 that $f_n \leq \phi^n$.

Having completed the basic preliminaries for Fibostrings, we turn now to consider their repetitions. We shall establish two facts that at first seem contradictory:

- In terms of the (i, p^*, r^*) encoding introduced in connection with Problem 2.15, we shall show that the number of repetitions in a Fibostring $\boldsymbol{f_n}$ is $\Theta(f_n \log f_n)$. Thus Fibostrings constitute a worst case for algorithms that compute all the repetitions in a string.

- Nevertheless, we shall also show that the repetitions in Fibostrings can be *implicitly* reported in $\Theta(n)$ time by making use of an extended (i, p^*, r^*, t) encoding. This encoding implements "runs" of repetitions, an idea already introduced more formally in Section 2.3. Informally, a run is a repetition $\boldsymbol{u}[1..p^*]^{r^*}$ followed by a possibly empty proper prefix $\boldsymbol{u}[1..t]$ of $\boldsymbol{u}[1..p^*]$.

The demonstration of these properties of Fibostrings will take us on a rather roundabout journey that will however visit some marvellous spots. In particular, as mentioned earlier, we shall see that Fibostrings provide a solution to the $(2,4)$-avoidance problem: they contain cubes but not fourth powers. We will also find that not only can we count the repetitions in Fibostrings, we can even characterize them: what they look like, and where they are.

We begin by generalizing the idea of a Fibostring in an interesting way. Recall that the morphism $h : A \rightarrow A^*$ of (3.4) operates on two given letters a, b as follows:

$$h(a) = ab, \;\; h(b) = a.$$

We extend h to $h : A^* \rightarrow A^*$ so that it operates in the same manner on any two given strings $\boldsymbol{x}$ and $\boldsymbol{y}$:

$$h(\boldsymbol{x}) = \boldsymbol{xy}, \;\; h(\boldsymbol{y}) = \boldsymbol{x}. \tag{3.26}$$

Thus the Fibonacci morphism (3.4) is seen as the restriction of (3.26) to the special case $\boldsymbol{x} = a$, $\boldsymbol{y} = b$. Analogous to (3.18), we can compute

$$\begin{aligned} h^2(\boldsymbol{y}) &= \boldsymbol{xy}; \\ h^3(\boldsymbol{y}) &= \boldsymbol{xyx}; \\ h^4(\boldsymbol{y}) &= \boldsymbol{xyxxy}; \\ &\vdots \end{aligned}$$

Then the nth ***generalized Fibostring*** is defined to be

$$\boldsymbol{f_n}(\boldsymbol{x}, \boldsymbol{y}) = h^n(\boldsymbol{y}), \tag{3.27}$$

where $h^n(\boldsymbol{y})$ is the $(n+1)$th entry in the sequence

$$h^*(\boldsymbol{y}) = \{\boldsymbol{y}, h(\boldsymbol{y}), h^2(\boldsymbol{y}), \ldots\}$$

determined by h. Of course the first few strings are:

$$\begin{aligned} \boldsymbol{f_0}(\boldsymbol{x}, \boldsymbol{y}) &= \boldsymbol{y}; \\ \boldsymbol{f_1}(\boldsymbol{x}, \boldsymbol{y}) &= \boldsymbol{x}; \\ \boldsymbol{f_2}(\boldsymbol{x}, \boldsymbol{y}) &= \boldsymbol{xy}; \\ \boldsymbol{f_3}(\boldsymbol{x}, \boldsymbol{y}) &= \boldsymbol{xyx}; \\ \boldsymbol{f_4}(\boldsymbol{x}, \boldsymbol{y}) &= \boldsymbol{xyxxy}; \\ \boldsymbol{f_5}(\boldsymbol{x}, \boldsymbol{y}) &= \boldsymbol{xyxxyxyx}; \\ \boldsymbol{f_6}(\boldsymbol{x}, \boldsymbol{y}) &= \boldsymbol{xyxxyxyxxyxxy}; \\ &\vdots \end{aligned}$$

and, in general,

$$\boldsymbol{f_n(x, y)} = \boldsymbol{f_{n-1}(x, y) f_{n-2}(x, y)}, \ n \geq 2. \tag{3.28}$$

By the same analogy, the ***infinite generalized Fibostring*** $\boldsymbol{f_\infty(x, y)}$ is the string that contains every $\boldsymbol{f_n(x, y)}$, $n > 0$, as a prefix.

Since $h^n(\boldsymbol{y}) = h^{n-k}(h^k(\boldsymbol{y}))$, it follows from (3.27) that

$$\boldsymbol{f_n(x, y)} = \boldsymbol{f_{n-k}} \left(h^k(\boldsymbol{x}), h^k(\boldsymbol{y}) \right).$$

Then, setting $\boldsymbol{x} = a$ and $\boldsymbol{y} = b$, and noting that $h^k(\boldsymbol{x}) = h^{k+1}(\boldsymbol{y})$, we obtain a rather startling result for ordinary Fibostrings:

Lemma 3.4.2 *For all integers k and n such that $0 \leq k \leq n-1$,*

$$\boldsymbol{f_n} = \boldsymbol{f_{n-k}(f_{k+1}, f_k)}. \ \square$$

This result enables us to express any Fibostring entirely in terms of selected Fibostrings $\boldsymbol{f_{k+1}}$ and $\boldsymbol{f_k}$; for example, we can write $\boldsymbol{f_7}$ in terms of $\boldsymbol{f_3}$ and $\boldsymbol{f_2}$:

$$\boldsymbol{f_7} = \boldsymbol{f_5(f_3, f_2)} = (aba)(ab)(aba)(aba)(ab)(aba)(ab)(aba).$$

Of course, Lemma 3.4.2 extends also to the infinite case:

$$\boldsymbol{f_\infty} = \boldsymbol{f_\infty(f_{n+1}, f_n)} = \boldsymbol{f_{n+1}\, f_n\, f_{n+1}\, f_{n+1}\, f_n\, f_{n+1}\, f_n\, f_{n+1}} \cdots . \tag{3.29}$$

To this equation we apply two easy results proved in Exercise 3.4.11:

Lemma 3.4.3 *For every integer $n \geq 2$, $\boldsymbol{f_n^2} = \boldsymbol{f_{n+1} f_{n-2}}$.* $\square$

Lemma 3.4.4 *For every integer $n \geq 3$, $\boldsymbol{f_n}$ has borders $\boldsymbol{f_i}$,*

$$i = n-2, n-4, \ldots, 2-(n \bmod 2). \ \square$$

It follows from these lemmas and from the prefix-preserving property of Fibostrings that, for $n \geq 3$, every occurrence of $\boldsymbol{f_{n+1}}$ in $\boldsymbol{f_\infty}$ gives rise to a square $\boldsymbol{f_n^2}$. Observe moreover that every occurrence of $\boldsymbol{f_n}$ is followed either by $\boldsymbol{f_{n+1} f_n}$ or by $\boldsymbol{f_{n+1}^2} = \boldsymbol{f_{n+1} f_n f_{n-1}}$. Repeated application of Lemma 3.4.1 tells us therefore that

$$\boldsymbol{f_n\, f_{n+1}\, f_n} = \boldsymbol{f_n^3\, f_{n-3}\, f_{n-2}}, \tag{3.30}$$

so that every occurrence of $\boldsymbol{f_n}$ in the expansion (3.29) actually gives rise to the cube $\boldsymbol{f_n^3}$. Thus, for every $n \geq 3$, both $\boldsymbol{f_n}$ and $\boldsymbol{f_n^2}$ actually *cover* $\boldsymbol{f_\infty}$ in the sense of Problem 2.18. Since $\boldsymbol{f_n}$ covers $\boldsymbol{f_\infty}$, it follows that any other occurrence of $\boldsymbol{f_n}$ in $\boldsymbol{f_\infty}$ that does not form part of one of the repetitions identified in (3.29) must in fact be identical to some rotation

$R_j(\boldsymbol{f_n})$, $j > 0$. But Theorem 1.4.2 tells us that this is impossible unless $\boldsymbol{f_n}$ is a repetition; since of course it is not, we conclude that all the occurrences of $\boldsymbol{f_n}$ (and therefore of $\boldsymbol{f_n^2}$ and $\boldsymbol{f_n^3}$) in $\boldsymbol{f_\infty}$ have been identified for $n \geq 3$.

Having found a way to identify the occurrences of $\boldsymbol{f_n^3}$, we can therefore infer the existence of squares of every rotation $R_j(\boldsymbol{f_n})$, $j > 0$, as shown in Exercise 3.4.12. Further, we now find it easy to observe that no fourth powers of $\boldsymbol{f_n}$ can occur in $\boldsymbol{f_\infty}$. For consider the expansion, derived from Lemma 3.4.1:

$$\boldsymbol{f_n^4} = \boldsymbol{f_n^3 f_{n-2} f_{n-3} f_{n-2}}.$$

Comparing this expression with (3.30), we find that $\boldsymbol{f_n^4}$ can exist if and only if $\boldsymbol{f_{n-3} f_{n-2}} = \boldsymbol{f_{n-2} f_{n-3}}$. It is shown in Exercise 3.4.13 that this equality cannot hold, and so we conclude that $\boldsymbol{f_n^4}$ does not exist.

There is still a fly in the ointment: how do we know that the *only* squares in $\boldsymbol{f_\infty}$ are squares of Fibostrings or their rotations? To answer this question, it is convenient to extend somewhat the idea of an inverse morphism introduced in Section 3.1. There the inverse was defined only on the strings of $h^*(A) - A$. In the case of Fibostrings, however, we may define h^{-1} on shorter substrings by observing that ab occurs in $\boldsymbol{f_n}$ only as a result of the mapping $h : a \to ab$, while aa occurs only because the leftmost a resulted from the mapping $h : b \to a$. We thus convince ourselves that an inverse morphism for Fibostrings may be expressed as follows:

$$h^{-1} : ab \to a, \;\; a \not\to b, \tag{3.31}$$

where we mean by the notation $\not\to$ that $a \to b$ if and only if a is not followed on the right by b; that is, either a is followed by another a or else a is the rightmost letter of $\boldsymbol{f_n}$. It is clear from (3.4) that bb cannot occur in a Fibostring; hence (3.31) really defines an algorithm that proceeds from left to right across $\boldsymbol{f_n}$, inspecting each occurrence of a and performing an inverse mapping h^{-1} based on the letter occurring to the right of the a. Since this algorithm recaptures the antecedent of every $\boldsymbol{f_n}$, $n > 0$, we have thus shown that the morphism (3.4) for Fibostrings is a code.

Observing that the substring a^3 cannot occur in a Fibostring (see Exercise 3.4.9), we are now ready to get rid of that fly:

Theorem 3.4.5 *A square $\boldsymbol{u^2}$ occurs in $\boldsymbol{f_\infty}$ if and only if $\boldsymbol{u}$ is a rotation of $\boldsymbol{f_n}$ for some integer $n \geq 0$.*

Proof We have already seen that for every rotation $\boldsymbol{u} = R_j(\boldsymbol{f_n})$ of $\boldsymbol{f_n}$, where $j \in 0..f_n - 1$, $\boldsymbol{u^2}$ necessarily occurs in $\boldsymbol{f_\infty}$. This proves sufficiency.

To prove necessity, suppose that $\boldsymbol{u}$ is not any such rotation but that nevertheless $\boldsymbol{u^2}$ occurs in $\boldsymbol{f_\infty}$. Suppose further that $\boldsymbol{u}$ is the shortest string with this property, and let n be the least integer such that $\boldsymbol{u}$ is a proper substring of $\boldsymbol{f_n}$. We shall show that $h^{-1}(\boldsymbol{f_n}) = \boldsymbol{f_{n-1}}$ contains a square $\boldsymbol{v^2}$, where $|\boldsymbol{v}| < |\boldsymbol{u}|$ and $\boldsymbol{v}$ is not a rotation of any Fibostring, a contradiction that proves the theorem. There are four possible cases:

- $\boldsymbol{u} = b \cdots b$ Impossible because then $\boldsymbol{u^2}$ contains bb.
- $\boldsymbol{u} = a \cdots b$ Impossible because then $\boldsymbol{u} = a \cdots ab$ so that $\boldsymbol{v} = h^{-1}(\boldsymbol{u})$ is well-defined with $|\boldsymbol{v}| < |\boldsymbol{u}|$.
- $\boldsymbol{u} = a \cdots a$ Since $\boldsymbol{u^2} = a \cdots aa \cdots a$ exists and a^3 is impossible, it follows that $\boldsymbol{u^2} = ab \cdots baab \cdots ba$; then, since bb is impossible, $\boldsymbol{u} = aba \cdots aba$ for which $\boldsymbol{v} = h^{-1}(\boldsymbol{u})$ is well-defined with $|\boldsymbol{v}| < |\boldsymbol{u}|$.
- $\boldsymbol{u} = b \cdots a$ Since bb cannot exist, $\boldsymbol{u}$ must be preceded by a; thus the string

$$ab \cdots \lambda ab \cdots \lambda a = (ab \cdots \lambda)^2 a$$

exists for some $\lambda \in \{a, b\}$. This case then reduces either to $a \cdots a$ or to $a \cdots b$.

Observe that whenever $\boldsymbol{v}$ is well-defined, it cannot be a Fibostring, since if it were, $\boldsymbol{u} = h(\boldsymbol{v})$ would also be a Fibostring, in contradiction to the stated hypothesis. □

This theorem tells us immediately that no rotation of a Fibostring can occur as a fourth power, since

$$(R_j(\boldsymbol{f_n}))^4 = ((R_j(\boldsymbol{f_n}))^2)^2$$

and, as we saw earlier, a square such as $(R_j(\boldsymbol{f_n}))^2$ cannot be a Fibostring. Observing that $\boldsymbol{f_7}$ contains $(aba)^3$, we therefore have

Theorem 3.4.6 *For $n \geq 7$, the Fibostrings $\boldsymbol{f_n}$ all contain cubes, but every Fibostring avoids fourth powers.* □

We argue as in Sections 3.2 and 3.3 that if $\boldsymbol{f_\infty}$ contains a fourth power, then so does some $\boldsymbol{f_n}$, an impossibility by Theorem 3.4.6. Hence

Theorem 3.4.7 *The infinite Fibostring $\boldsymbol{f_\infty}$ solves the $(2,4)$-avoidance problem.* □

We now have sufficient information at our disposal to estimate the number of repetitions in $\boldsymbol{f_n}$. To establish a rough lower bound on this quantity, we shall count, for every integer $k \in 3..n-4$, the number of repetitions associated with the occurrences of $\boldsymbol{f_k}$ in $\boldsymbol{f_n}$. Merely for ease of counting, we thus ignore in our calculation the squares a^2 and $(ab)^2$, but we shall see that this makes no difference in an asymptotic sense.

In Exercise 3.4.15(a) it is shown that there are f_{n-2} occurrences of b in $\boldsymbol{f_n}$; it follows then from Lemma 3.4.2 that there are f_{n-k-2} occurrences of $\boldsymbol{f_k}$ in $\boldsymbol{f_n}$. Since possibly one of these occurrences is a suffix of $\boldsymbol{f_n}$, there are at least $f_{n-k-2} - 1$ occurrences of $\boldsymbol{f_k}$ that give rise to cubes $\boldsymbol{f_k^3}$. Now observe that each cube in turn gives rise to at least f_k distinct squares formed from every rotation of $\boldsymbol{f_k}$:

$$(R_j(\boldsymbol{f_k}))^2,\ j = 0, 1, \ldots, f_k - 1.$$

Thus the total number of repetitions associated with the $f_{n-k-2} - 1$ occurrences of $\boldsymbol{f_k^3}$ in $\boldsymbol{f_n}$ is at least $f_k(f_{n-k-2} - 1)$.

Since as we have seen $f_n \in \Theta(\phi^n)$, it follows that

$$f_k f_{n-k-2} \in \Theta(\phi^k \phi^{n-k-2}),$$

hence that

$$f_k(f_{n-k-2} - 1) \in \Theta(f_{n-2}), \tag{3.32}$$

independent of k. Since (3.32) holds for every $k \in 3..n-4$, we conclude that there are at least $\Theta((n-6)f_{n-2})$ repetitions in $\boldsymbol{f_n}$. Since moreover $n \approx \log_\phi f_n$ and $f_{n-2} \approx f_n/\phi^2$, we have

Theorem 3.4.8 *The number $q(n)$ of repetitions in $\boldsymbol{f_n}$ satisfies*

$$q(n) \in \Omega(f_n \log f_n). \quad \square$$

As mentioned earlier, since there are algorithms (Section 12.1) that compute all repetitions in a string of length n in $\Theta(n \log n)$ time, the lower bound of Theorem 3.4.8 is in fact also an upper bound, and represents, in an asymptotic sense, the largest number of repetitions that can be contained in any string.

Having gone to a great deal of trouble to establish Theorem 3.4.8, we now propose to convince the reader, as announced at the beginning of this section, that "really" the number of repetitions in $\boldsymbol{f_n}$ is linear — that is, of order $\Theta(f_n)$! We do this not to confuse, but to make clear a point raised earlier in Digressions 1.3.2 and 2.3.1: the encoding of information is critical to the efficiency of string algorithms.

Refer back to the preceding development. Clearly what has implicitly been used to describe repetitions is the (i, p^*, r^*) encoding introduced in Section 2.3: for every position i at which a square or cube of $R_j(\boldsymbol{f_k})$ occurs, one repetition — $(i, f_k, 2)$ or $(i, f_k, 3)$ — is counted. But in this development we have also seen that every cube $\boldsymbol{f_k^3}$ implies an additional $f_k - 1$ repetitions that really do not need to be explicitly output. Consider, for example, the repetition $\boldsymbol{f_4^3} = (abaab)^3$ at position 9 of

$$f_8 = abaababa\underline{abaababaababaaba}ababaabaab.$$

This cube and all of the repetitions of its rotations can be expressed by a device mentioned above and defined formally in Section 2.3; that is, by the ***run***

$$(i, p^*, r^*, t) = (9, 5, 3, 1),$$

where the final element (or ***tail***) $t = 1$ identifies the total number of rotations of $\boldsymbol{f_5}$, excluding $\boldsymbol{f_5}$ itself, that are cubes. Put another way, t is the maximum length of the substring on the right of $(abaab)^3$ that is a prefix of $abaab$: in the language of Definition 2.3.2, the run (i, p^*, r^*, t) is not *right-extendible*. Thus $(9, 5, 3, 1)$ stands for the cubes

$$(abaab)^3 \quad \text{and} \quad (baaba)^3$$

together with the squares

$$(abaab)^2, \quad (ababa)^2 \quad \text{and} \quad (babaa)^2.$$

Provided the user agrees, then, we can in general output 4-tuples (i, p^*, r^*, t) to describe the runs of cubes and squares in $\boldsymbol{f_n}$. This means that for every occurrence of a repetition of $\boldsymbol{f_k}$ in $\boldsymbol{f_n}$, $k = 1, 2, \ldots, n-2$, we need to output only a single 4-tuple. As we have seen, for each k there are f_{n-k-2} occurrences of $\boldsymbol{f_k}$, so that altogether at most

$$q'(n) = \sum_{k=1}^{n-2} f_{n-k-2} = \sum_{k=0}^{n-3} f_k$$

tuples need to be output. It is easily proved by induction that $\sum_{k=0}^{n-3} f_k < f_{n-1}$. Hence

Theorem 3.4.9 *The number $q'(n)$ of runs of repetitions in $\boldsymbol{f_n}$ satisfies*

$$q'(n) < f_{n-1}. \quad \square$$

We are faced then with an intriguing situation. As Theorem 3.4.8 shows, Fibostrings have a large number of repetitions. Despite this fact, with a little tolerance on the part of the user those repetitions can be expressed in terms of runs whose number is linear in the string length, as shown by Theorem 3.4.9. We could be pardoned for making the following wild surmise: is it possible that the repetitions in *every* string $\boldsymbol{x}$ can be reported as a linear collection of runs? We return to this subject in Chapter 12.

One further point should be made: *both* of the encodings (i, p^*, r^*) and (i, p^*, r^*, t) identify repetitions by specifying a location i in the string, and thus, as discussed in Digression 1.2.1, implicitly require the string to be represented as an array.

We conclude this section with a further application of (3.26), this time to compute precisely the locations of squares $\boldsymbol{f_n^2}$ in $\boldsymbol{f_\infty}$ for every integer $n \geq 3$. We shall see that this computation leads directly to a linear-time algorithm to compute all the repetitions in a Fibostring. In (3.26) let

$$\boldsymbol{x} = f_{n+1}, \quad \boldsymbol{y} = f_n$$

for any nonnegative integer n, where now we express $\boldsymbol{x}$ and $\boldsymbol{y}$ in the alphabet of the Fibonacci numbers. Then we may consider

$$\boldsymbol{f_\infty}(f_{n+1}, f_n),$$

so that, for example, for $n = 3$,

$$\boldsymbol{f}_\infty(f_4, f_3) = 535535355355353553 \cdots\cdot. \tag{3.33}$$

Now, for all nonnegative integers i and n, let $\Sigma_{i,n}$ be the sum of the first i values in $\boldsymbol{f}_\infty(f_{n+1}, f_n)$. Then $\Sigma_{0,n} = 0$ for all n and, for example, in (3.33),

$$\Sigma_{1,3} = 5,\ \Sigma_{2,3} = 8,\ \Sigma_{5,3} = 21,\ \Sigma_{9,3} = 46.$$

Based on this notation, it is possible to specify positions in $\boldsymbol{f}_\infty$ where $\boldsymbol{f}_{n+1}$ and $\boldsymbol{f}_n^2$ occur; as we have already seen, these positions are the *only* ones at which powers of $\boldsymbol{f}_n$ can possibly occur.

Theorem 3.4.10 *For every integer* $i \geq 0$,

(a) $\boldsymbol{f}_{n+1} = \boldsymbol{f}_\infty[\Sigma_{i,n}+1..\Sigma_{i,n}+f_{n+1}],\ n \geq 2;$
(b) $\boldsymbol{f}_n^2 = \boldsymbol{f}_\infty[\Sigma_{i,n}+1..\Sigma_{i,n}+2f_n],\ n \geq 3.$

Proof Since $\boldsymbol{f}_\infty$ avoids b^2, it follows that every occurrence of $\boldsymbol{f}_n$ in (3.29) is followed by an occurrence of $\boldsymbol{f}_{n+1} = \boldsymbol{f}_{n-1}\boldsymbol{f}_{n-2}\boldsymbol{f}_{n-1}$, hence that $\boldsymbol{f}_{n+1}$ occurs at every occurrence of $\boldsymbol{f}_n$ in $\boldsymbol{f}_\infty$. Then (a) is an immediate consequence of Lemma 3.4.2 and (3.29): it merely says that $\boldsymbol{f}_{n+1}$ occurs as a prefix of $\boldsymbol{f}_\infty$ and then at displacements of f_{n+1} or f_n, depending on whether the current term in (3.29) is $\boldsymbol{f}_{n+1}$ or $\boldsymbol{f}_n$, respectively. By Lemma 3.4.3 and the fact that, for $n \geq 3$, $\boldsymbol{f}_{n-2}$ is a prefix of both $\boldsymbol{f}_{n+1}$ and $\boldsymbol{f}_n$, we see that these positions also mark occurrences of $\boldsymbol{f}_n^2$, and so (b) follows. □

The objection may be raised to Theorem 3.4.10(b) that it is restricted to the cases $n \geq 3$ and so does not locate squares $\boldsymbol{f}_1^2 = a^2$ and $\boldsymbol{f}_2^2 = (ab)^2$. However, as the next result shows, these squares are implicitly included:

Lemma 3.4.11 *For* $n = 1, 2$, $\boldsymbol{f}_n^2$ *occurs in* $\boldsymbol{f}_\infty$ *only as a substring of* $\boldsymbol{f}_{n+2}^2$.

Proof See Exercise 3.4.16. □

In view of this result, Theorem 3.4.10 effectively allows us to locate all occurrences of squares $\boldsymbol{f}_n^2$ in $\boldsymbol{f}_\infty$ for every $n \geq 1$. As we have seen, occurrences of $\boldsymbol{f}_n$ in the expansion (3.29) mark occurrences of cubes $\boldsymbol{f}_n^3$ and hence of squares or cubes of rotations $R_j(\boldsymbol{f}_n)$, $j \in 0..n-1$, $n \geq 3$. Using the (i, p^*, r^*, t) encoding of runs described above, it thus becomes possible in principle to compute all the repetitions in $\boldsymbol{f}_n$, and to do so in linear time; an algorithm that implements this principle is given in [IMS97]. In [FS99] formulæ are given both for the exact number of squares and for the exact number of distinct squares in a Fibostring; in [KK00] the surprising result is proved that the number of runs (Definition 2.3.3) in a Fibostring is one less than the number of distinct squares.

Digression 3.4.1 Fibostrings have wonderful properties, but perhaps even more wonderful is the fact that most of the important ones can be generalized to Sturmian strings, a class of which Fibostrings are but one among an infinite number of members.

A ***Sturmian string*** is an infinite string in which, for every integer $k \geq 0$, there are exactly $k + 1$ distinct strings of length k. Thus the alphabet for Sturmian strings is binary. A finite prefix of a Sturmian string is also called ***Sturmian***. The value $k + 1$ is often called the ***complexity*** of the string. An infinite string $\boldsymbol{x}$ which for some integer k has complexity k can only take the form $\boldsymbol{x}[1..k]^{\infty}$, so Sturmian strings can be thought of as the strings of least complexity that are also interesting.

Indeed, Sturmian strings are *very* interesting. Because they have low complexity — that is, few distinct substrings — it is natural to suppose that they have many repetitions. This turns out to be true [FKS00]: finite Sturmian strings of length n all contain $\Theta(n \log n)$ repetitions that can however be reported in $\Theta(n)$ runs and computed in $\Theta(n)$ time, just like Fibostrings. Furthermore, for every integer $r \geq 4$, there exists a Sturmian string that solves the $(2, r)$-avoidance problem, as Fibostrings solve the $(2, 4)$-avoidance problem.

Also like Fibostrings, Sturmian strings can be defined in terms of morphisms that can be used to establish the properties outlined above [FKS00] as well as to recognize in linear time whether or not a given string is in fact Sturmian [BF84]. It has recently been shown that all of these properties extend to a further generalization of Sturmian strings called "two-pattern strings" [FJLS02].

Sturmian strings have many different definitions and many other fascinating properties; as a result, they have been much studied [A94, B96, BS93, L02, LM94, Mi89, MS93, S76]. In addition, other classes of strings of low complexity have been identified and studied [Mi89a, AR91, R94]. □

Exercises 3.4

1. We have already used Fibostrings as examples in the two previous chapters. Can you find where?
2. For $n \geq 2$, give an alternative definition of $\boldsymbol{f_n}$ using the ideas of "suffix" and "prefix". Then go on to prove Lemma 3.4.2.
3. Based on (3.20), specify a recursive algorithm which computes the first 20 Fibostrings along with their lengths. Then eliminate the recursion and rewrite your algorithm as an iterative one. Which algorithm is easier to understand?
4. Apply the same inductive method used to establish (3.25) to show that $f_n \leq \phi^n$. Hence prove that $f_n \in \Theta(\phi^n)$.
5. More precisely, prove that

$$f_n = \left(\phi^{n+1} - \hat{\phi}^{n+1}\right) / \sqrt{5},$$

where $\hat{\phi} = 1 - \phi$. Hence show further that $f_n = [\phi^{n+1}/\sqrt{5}]$, where $[x]$ denotes the integer nearest to x.

6. A ***reverse Fibostring*** is defined by

$$\boldsymbol{f}_0' = b;\ \boldsymbol{f}_1' = a;\ \boldsymbol{f}_n' = \boldsymbol{f}_{n-2}'\, \boldsymbol{f}_{n-1}',\ \forall\, n \geq 2.$$

Show that the morphism

$$h' : a \to ba,\ b \to a$$

generates reverse Fibostrings from an initial string $\boldsymbol{f}_0' = b$.

7. Prove that $h(\boldsymbol{f}_\infty) = \boldsymbol{f}_\infty$.
8. Show that $\boldsymbol{f}_n$ avoids proper prefixes b, prefixes aa, and suffixes aa.
9. Show that $\boldsymbol{f}_n$ avoids b^2, a^3, $(ab)^3$ and $(ba)^3$.
10. Specify an algorithm that implements (3.31) on $\boldsymbol{f}_n$.
11. (a) Prove Lemma 3.4.3.

    (b) Prove Lemma 3.4.4. Can you prove more, as in Lemma 3.2.3, that the borders specified are in fact the *only* borders?

    (c) Use Lemmas 3.4.3 and 3.4.4 to prove the assertions made in the text, that every occurrence of $\boldsymbol{f}_{n+1}$ (respectively, $\boldsymbol{f}_n$) in $\boldsymbol{f}_\infty$ gives rise to $\boldsymbol{f}_n^2$ (respectively, $\boldsymbol{f}_n^3$).
12. Show that every occurrence of $\boldsymbol{f}_n^3$ in $\boldsymbol{f}_\infty$ gives rise to squares $(R_j(\boldsymbol{f}_n))^2$ for every integer $j \in 1..f_n - 1$.
13. Prove that $\boldsymbol{f}_{n-1}\,\boldsymbol{f}_n \neq \boldsymbol{f}_n\,\boldsymbol{f}_{n-1}$.
14. In the text it was rather casually stated that "of course" no Fibostring $\boldsymbol{f}_n$ is a repetition. Prove or disprove this statement.
15. (a) Prove that $\boldsymbol{f}_n$ contains f_{n-1} occurrences of a and f_{n-2} occurrences of b.

    (b) Prove that f_n is even if and only if $n = 3k+2$ for some nonnegative integer k.

    (c) Use (a) and (b) to provide an alternate proof of the preceding exercise.
16. Prove Lemma 3.4.11.
17. The development leading up to Theorem 3.4.8 can obviously be tightened up considerably so as to yield a more precise lower bound on the number of repetitions in $\boldsymbol{f}_n$. Can you tighten it up enough to compute an *exact* number of repetitions in $\boldsymbol{f}_n$? (Using Lemma 3.4.11 might help!)
18. Similarly, can you improve on Theorem 3.4.9 to come up with an exact number of runs of repetitions in $\boldsymbol{f}_n$?
19. In order to specify the locations of all the repetitions in $\boldsymbol{f}_\infty$, it turns out to be convenient to be able to determine in constant time the value of $\boldsymbol{f}_\infty[i]$ for any given integer $i > 0$. Can you find a formula to compute this value?
20. Show how to construct a repetition-free string of length 2^k on an alphabet of $k+1$ letters.

21. Show that a "Sturmian morphism"

$$h' : a \to aab, \;\; b \to ab$$

generates strings $\boldsymbol{f'_n} = R_{n-1}(\boldsymbol{f_n})$, n odd, from an initial string $\boldsymbol{f'_1} = a$.

22. Show that if for some integer $k > 0$ a string $\boldsymbol{x}$ has complexity k, then $\boldsymbol{x} = \boldsymbol{x}[1..k]^r$ for some integer $r \geq 2$.

Chapter 4

Good Algorithms and Good Test Data

Take but degree away, untune that string,
And hark! what discord follows; each thing meets
In mere occupancy.

— William SHAKESPEARE (1564–1616), *Troilus and Cressida*

In a book focussed on algorithms, it is perhaps not inappropriate, before beginning a discussion of specific algorithms, to deal with some of their more general features. This is the role of Section 4.1. One of the features discussed in this section is correctness, a quality that depends heavily on the availability of satisfactory test data. The special strings introduced in Chapter 3 certainly constitute excellent test data for several string algorithms, but more general kinds of test data are also necessary. In Sections 4.2 and 4.3 we describe approaches to the generation of test data sets designed to provide testing for particular algorithms that is both efficient and effective. Some of these generation algorithms have been implemented in an experimental software package [L96].

4.1 Good Algorithms

. . . do not say a little in many words,
but a great deal in a few.

— PYTHAGORAS (582–507 BC)

In this section we discuss briefly six features of algorithms on strings that one way or another affect their usefulness in practice:

- correctness;
- speed and storage (time and space complexity);
- freedom from backtracking (on-line);
- alphabet-independence;
- dependence on preprocessing;
- encoding of the output.

Correctness

There is no doubt that correctness is a necessary feature of algorithms: an incorrect algorithm is useless. Worse still, an incorrect algorithm implemented in computer code can be dangerous if, for example, it controls the brakes of the family car, or the flight of a commercial passenger plane, or the decision to launch an ICBM. Ensuring the correctness of software — indeed, even determining exactly what is meant by "correctness" — is the great unresolved problem of "software engineering". For the mathematical algorithms dealt with in this book, correctness is much easier to establish than it is for many other types of algorithm, but this does not mean that it is easy. Some years ago, a research assistant of mine implemented four published algorithms, including one of my own, in a C program to compute the canonical form of a necklace (Definition 1.4.3) in four different ways: we discovered that three of the algorithms, including my own, were incorrect!

There are two main stages in establishing the correctness of an algorithm. The first stage employs mathematical reasoning to "prove" correctness; the second stage executes a hopefully correct implementation of the algorithm on test data for which the correct output is hopefully known. Both of these stages are notoriously unreliable: as we have just seen, mathematical reasoning applied to algorithms is not to be trusted, and execution against test data cannot establish correctness because the test data is never exhaustive. Nevertheless, it is better to use unreliable tools than no tools at all. In Sections 4.2 and 4.3 we show how, for many string algorithms, the generation of test data can be approached so as to provide comprehensive testing over a wide range of string lengths at greatly reduced processing cost.

Time and Space Complexity

Of course time and space complexity are major considerations in the evaluation of any algorithm, and of course an algorithm whose worst-case execution time on a string of length n was $\Theta(n)$ would generally be preferred to one whose worst case was $O(n^2)$. But the matter is not clear cut. For example, Algorithm Easy, the simple-minded algorithm introduced in Section 2.2 to compute all occurrences of a specific pattern of length m in a string of length n, is not, most of the time, such a bad algorithm. It is only in pathological cases virtually unknown in practice that its $O(mn)$ worst-case time bound is achieved: most of the time it executes in time $O(m+n)$ and in fact, precisely because it is simple, probably executes faster on average than the KMP Algorithm (Section 7.1) that elegantly avoids all backtracking by never having to retest positions in the string to the left of the current position.

As another example, use of suffix trees (Section 2.1) is inhibited by the fact that, even though they can be computed in $\Theta(n \log \alpha)$ time for an alphabet of cardinality α, and require only $\Theta(n\alpha)$ space, the constants of proportionality in these asymptotic bounds can be prohibitively large, especially for larger alphabets. Similarly, extremely elaborate pattern-matching algorithms that reduce the number of letter comparisons to close to a theoretical minimum [GG92, CH92] are unlikely to be used in practice either because of their intellectual (rather than time) complexity or because of "housekeeping" processes that again increase the constant of proportionality.

On the whole, time complexity is a more decisive criterion than space complexity for algorithm selection, reflecting the fact that, currently, unit storage costs are much lower than unit processing costs. Even so, as we have seen, favourable time complexity can sometimes be outweighed by other considerations.

Freedom from Backtracking

Since strings are naturally processed from left to right, it is a particularly attractive feature of a string algorithm that it be ***on-line*** — that is, having performed calculations based on position i of the input string, the algorithm does not ever have to visit position i again. In other words, an on-line algorithm never has to ***backtrack***.

There are also weaker definitions of an on-line algorithm in common use. For example, if there exists a positive constant k such that at most k positions in the range $i-k+1..k$ may need to be revisited during the processing of position i, then the algorithm may be said to be on-line with ***window*** k. An even weaker definition allows the algorithm to be described as on-line if the value corresponding to position i never needs to be recalculated, even though positions $i' < i$ may need to be revisited in order to make the calculation possible.

Since it may be true for each i that $i = n$ — that i is in fact the rightmost position of the input string $\boldsymbol{x}$ — it follows that (according to any of the above definitions) an on-line algorithm on a string effectively solves the given problem for each prefix $\boldsymbol{x}[1..i]$ on the way to solving it for $\boldsymbol{x} = \boldsymbol{x}[1..n]$. Thus in a sense an on-line algorithm actually solves n problems for the price of one.

We have seen above that the KMP Algorithm is on-line. We remark that the border array calculation (Algorithm 1.3.1) is also on-line (in the weakest sense of the term): the border array $\boldsymbol{\beta}[1..i]$ for $\boldsymbol{x}[1..i]$ is computed when $\boldsymbol{x}[i]$ is visited for the first time and is never thereafter changed. We shall see in Section 13.1 that a refinement of the border array called the "cover array" can also be computed using an on-line algorithm.

Independence from the Alphabet

As we learned in Section 1.2, a string can be defined on a set of elements of unrestricted generality. Thus the most desirable algorithms are those that execute efficiently on strings drawn from any arbitrary alphabet — in other words, those that are alphabet-independent. An example of such an algorithm is the border array calculation (Algorithm 1.3.1): it is an asymptotically optimal linear-time algorithm that depends only on the computer's ability to decide in constant time whether any two letters of the alphabet are equal or unequal — no other property of the alphabet is required.

On the other hand, it often occurs that the time complexity of an algorithm can be reduced if the alphabet has special properties. If in fact alphabets with these properties occur frequently in practice, then it will be worthwhile to design special versions of algorithms that take advantage of them. To make this idea more precise, consider the following simple problem:

Problem 4.1 **(Count Occurrences)** *Count the number of occurrences of every letter in a given string* $\boldsymbol{x}$ *of length* n.

We consider this problem in the context of three different kinds of alphabet A:

- ***general (unordered) alphabet***: $\forall\ \lambda, \mu \in A$, it is decidable in constant time whether $\lambda = \mu$ (for example, A is a set of Chinese ideographs);
- ***ordered alphabet***: an alphabet in which $\forall\ \lambda, \mu \in A$, it is decidable in constant time whether $\lambda < \mu$ for some order relation $<$ (for example, A is a set of real numbers or a set of English-language words);
- ***indexed alphabet***: it is possible to declare an array T such that, $\forall\ \lambda \in A$, $T[\lambda]$ is accessible in constant time (for example, A is a subset of the ASCII characters).

Observe that an indexed alphabet A must have finite cardinality $\alpha = |A|$. Then on an indexed alphabet, Problem 4.1 can be solved in a straightforward manner: first initialize an integer array $T[1..\alpha]$ to zero, then for each position $i \in 1..n$ in $\boldsymbol{x}$, increment $T[\boldsymbol{x}[i]]$ by one. By keeping an auxiliary list of the distinct letters encountered, the nonzero entries in T can then be output in time proportional to the number of distinct letters in $\boldsymbol{x}$. Thus the time requirement in this case is $\Theta(\alpha + n)$.

In the case of an ordered alphabet, a balanced search tree (such as an AVL tree or a 2–3 tree) can be constructed from the letters found as $\boldsymbol{x}$ is processed from left to right: each node of the tree is identified by a pair (λ, c), where λ is a letter found in $\boldsymbol{x}$ and c is a counter that is set to one when the node is created and thereafter incremented by one for every occurrence of λ. The time requirement in this case for the solution of Problem 4.1 is $O(n \log n)$. Note that if the cardinality of the alphabet α is large with respect to n, this algorithm may execute more quickly than the one on an indexed alphabet.

For an unordered alphabet, since the only mechanism for identifying letters is an equality test, it may be that $\Theta(n)$ tests need to be performed at each of $\Theta(n)$ steps. Thus the solution to Problem 4.1 in this case requires $O(n^2)$ time. Observe however that this solution would nevertheless be the simplest and the fastest if, for example, the alphabet were binary. It should also be remarked in this context that, in the digital computer model of computation, letters in an unordered alphabet must be encoded as bit strings to which, in practice, the usual order relations apply; thus, in practice, alphabets can almost always be regarded as ordered.

We see then that for this (not atypical) problem, the choice of an appropriate algorithm depends very heavily on the strings we expect to process as well as on the characteristics of the alphabet. Indeed, as discussed in the next subsection, the matter does not end there.

Dependence on Preprocessing

We have seen in Section 2.1 that the computation of suffix trees can be thought of as a form of preprocessing of the string $\boldsymbol{x}$: the tree is then used to facilitate multiple subsequent searches of $\boldsymbol{x}$, each requiring time proportional to the length of the pattern being searched for. Similarly, in the KMP algorithm, the border array of the pattern is initially computed in a preprocessing phase, then used during the search to avoid backtracking. Thus it seems that the ability of an algorithm to make use of preprocessing can be a valuable feature. As we now demonstrate, this point may also be made in the context of Problem 4.1 introduced in the preceding subsection.

Suppose that the alphabet A is fixed — that is, specified in advance — and finite, and that the occurrence count is required for $m \geq 1$ strings on A. To simplify the discussion, we may suppose without loss of generality that each of the m strings is of length n. Then the alphabet may be preprocessed in time $\Theta(\alpha)$ into a hash table T so that individual letters can be accessed in constant or near-constant time. The hash table entry $T[\lambda]$ for each letter $\lambda \in A$ would contain a string counter $j \in 0..m$, identifying the string in which λ most recently occurred; initially j would be set to zero for each λ. $T[\lambda]$ would also contain an occurrence counter c. Once the preprocessing phase was complete, each of the m strings could be processed. For each position i in the jth string $\boldsymbol{x_j}$, $T[\boldsymbol{x_j}[i]]$ would be accessed: if the string counter were equal to j, then c would be incremented by one; otherwise, c would be set to one and the string counter set to j. Thus the average time requirement in this case, ***amortized*** over the m strings, would be $\Theta(\alpha/m + n)$ per string, no matter what alphabet was used. Clearly this algorithm that makes use of preprocessing would in many cases outperform the other algorithms discussed above — always provided the alphabet can be specified in advance.

Observe that all of the algorithms described above that count occurrences of letters in $\boldsymbol{x}$ are actually on-line algorithms.

Encoding of the Output

We have already seen in Section 2.3 the critical role played by output encoding in Problem 2.15, the computation of the repetitions in a given string, effectively reducing the size of the output, and thus the complexity of the algorithms, from $O(n^2)$ to $O(n \log n)$. Thus, for example, as pointed out in connection with Problem 2.15, by reporting the repetition a^6 in what is effectively the normal form $(i, 1, 6)$, repetition algorithms assume that the occurrences of the various component repetitions a^2, a^3, a^4 are *implicitly* reported.

In general, as dicussed also in Digressions 1.3.2 and 2.3, it is not clear what criteria should be applied in order to determine whether an output encoding is acceptable or not. Consider, for example, the border array $\boldsymbol{\beta}$ of a given string $\boldsymbol{x}$. Usually, the border array is accepted as a suitable encoding of the borders of $\boldsymbol{x}$. But in order to actually list the borders of $\boldsymbol{x}$, it is necessary to compute $\boldsymbol{\beta}[n], \boldsymbol{\beta^2}[n], \ldots$, as described in Lemma 1.3.1. Recall from Lemma 3.2.3 that Thue strings $(2, 3)$, like Fibostrings, have a number of borders logarithmic in their lengths. Thus, in general, it may require $\Theta(\log n)$ time or, as far as we know, even more time to list the borders of $\boldsymbol{x}$ given the border array $\boldsymbol{\beta}$. A doubting Thomas might well then argue that in providing the border array, we have not really completed the solution of the problem, since a logarithmic amount of work may remain to be done. Perhaps the main reason that the border array is accepted as a solution is a practical one: usually the longest border is the focus of interest, and shorter borders are considered only after the longest one has been dealt with.

Exercises 4.1

1. Specify the four algorithms given in the text that compute the number of occurrences of each letter in a given string $\boldsymbol{x}$.

2. Discuss the time complexity of these four algorithms; identify cases in which each algorithm would be preferred to the others.
3. Show that the letters in an indexed alphabet A of size α can be mapped one-one in linear time onto the natural numbers $N_\alpha = \{1, 2, \ldots, \alpha\}$. Hence conclude that, without loss of generality, an indexed alphabet may be assumed to be the set N_α.
4. It has been stated in the text, but not proved, that Thue strings $(2, 3)$ have $\Theta(\log n)$ borders, while Fibostrings have $\Omega(\log n)$ borders, where n is the length of the string. Using Lemmas 3.2.3 and 3.4.4, prove these statements.

4.2 Distinct Patterns

Originality is not seen in single words or even sentences.

— Isaac Bashevis SINGER (1904–1991)
New York Times Magazine, 12 March 1978

What makes one string "distinct" from another? This is a fundamental question for the generation of data designed to test string algorithms: one would like to include as many "distinct" strings as possible, while at the same time avoiding duplication of strings that are not "distinct".

The conventional approach — and the one implicitly adopted in Section 1.2 — is to treat two strings as distinct if and only if they are not position-by-position equal. Thus, according to this definition, on a finite alphabet A of cardinality α, there are exactly α^n distinct strings of length n, and all of these strings should be entered in order to provide exhaustive testing over strings of length n on A.

This approach is not necessarily the only one or the best one, however, for many algorithms. Consider the simple problem stated in Section 4.1 of counting the occurrences of each letter in a given string $\boldsymbol{x}$. On unordered alphabets, the algorithms that solve this problem treat the letters of $\boldsymbol{x}$ individually, with at most tests of equality/inequality between any pair of them. Thus, for these algorithms, the processing performed on the input string $\boldsymbol{x} = aab$ will be identical to the processing performed on $\boldsymbol{y} = ddc$: it may be important that the test data should include one of these strings, but both are not required.

This example suggests the idea of ***pattern-equivalence***, where given strings $\boldsymbol{x}$ and $\boldsymbol{y}$ are considered to be equivalent if and only if each corresponds to the same "pattern". To make the idea precise, we state the following:

Definition 4.2.1 *Two strings $\boldsymbol{x}$ and $\boldsymbol{y}$ are said to be **p-equivalent** if and only if*

(a) $|\boldsymbol{x}| = |\boldsymbol{y}| = n$ *for some integer* $n \geq 0$;

(b) *for all integers i and j satisfying $1 \leq i \leq j \leq n$,*

$$\boldsymbol{x}[i] = \boldsymbol{x}[j] \Leftrightarrow \boldsymbol{y}[i] = \boldsymbol{y}[j].$$

*If $\boldsymbol{x}$ and $\boldsymbol{y}$ are not p-equivalent, they are said to be **p-distinct**.*

In this section, following [MSM99], we explore the properties of p-distinct strings; in particular, we show both how to count and how to compute all p-distinct strings of length n, and we show that these strings are fewer in number, by an exponential factor, than the α^n strings that are "distinct" in the usual sense. We show further that these strings can be computed in constant time per string.

In the next section we will introduce the idea of border-equivalence (or b-equivalence), an even coarser equivalence relation than p-equivalence. This will lead us to define b-distinct strings, which we show are fewer than p-distinct strings by a further exponential factor. Border-distinct strings also can be generated in constant time per string.

Thus, for the algorithms for which they are suitable, p-distinct and b-distinct strings provide test data that is both effective (exhaustive testing) and efficient (rapid generation of exponentially fewer strings).

The first important fact about p-equivalence is that it is an equivalence relation between two strings and therefore separates the strings of length n into equivalence classes of p-distinct strings (see Exercise 4.2.2). It is convenient to identify a unique representative of each of these classes. To do so, we introduce a countably infinite ***standard alphabet***

$$\Lambda = \{\lambda_1, \lambda_2, \ldots, \lambda_\alpha, \ldots\}, \tag{4.1}$$

with subalphabets $\Lambda_\alpha = \{\lambda_1, \lambda_2, \ldots, \lambda_\alpha\}$ for every integer $\alpha \geq 1$. We suppose the letters of Λ to be naturally ordered as follows:

$$\lambda_1 < \lambda_2 < \cdots < \lambda_\alpha < \cdots,$$

and so we may without loss of generality think of the standard alphabet as consisting of the natural numbers. Then, given any string $\boldsymbol{x}$ on any alphabet A, we define the ***p-canonical*** string $\boldsymbol{x}^*$ corresponding to $\boldsymbol{x}$ to be the lexicographically least string on Λ that is p-equivalent to $\boldsymbol{x}$. Clearly $\boldsymbol{x}^*$ is a unique representative of its p-equivalence class. Further, it is not difficult to show that every string $\boldsymbol{x}^*$ satisfies the following property:

(P) For every positive integer j, if λ_j occurs in $\boldsymbol{x}^*$, then there is no occurrence of λ_{j+1} that precedes the least (leftmost) occurrence of λ_j.

Let $p'[\alpha, n]$ be the number of p-canonical strings of length n formed using exactly the letters of Λ_α (every letter must occur at least once in each string). We imagine these values to be laid out in an infinite two-dimensional array called the p'-array.

Theorem 4.2.2 *For all positive integers α and n:*

(a) $p'[1,n] = 1$;
(b) *if* $\alpha > n$, $p'[\alpha,n] = 0$;
(c) $p'[\alpha,\alpha] = 1$;
(d) *if* $\alpha \geq 2$ *and* $n \geq 2$, $p'[\alpha,n] = p'[\alpha-1,n-1] + \alpha p'[\alpha,n-1]$.

Proof

(a) For $\alpha = 1$, the only p-canonical strings are $\boldsymbol{x}^* = \lambda_1^n$.

(b) By property (P), no p-canonical string can contain a letter λ_α, $\alpha > n$.

(c) Again by property (P), there exists exactly one p-canonical string of length α formed using exactly α distinct letters: $\boldsymbol{x}^* = \lambda_1\lambda_2\cdots\lambda_\alpha$.

(d) Let $\pi_1 = p'[\alpha-1,n-1]$ be the number of distinct p-canonical strings of length $n-1$ that include exactly the $\alpha-1$ letters of $\Lambda_{\alpha-1}$. Denote these strings by

$$S_1 = \{\boldsymbol{x}_1^*, \boldsymbol{x}_2^*, \ldots, \boldsymbol{x}_{\pi_1}^*\}.$$

Then for every integer i satisfying $1 \leq i \leq \pi_1$, each string

$$\boldsymbol{x}_i^*\lambda_\alpha \tag{4.2}$$

is distinct and p-canonical.

Similarly, let $\pi_2 = p'[\alpha,n-1]$ be the number of distinct p-canonical strings of length $n-1$ on exactly α distinct letters Λ_α. Denote these strings by

$$S_2 = \{\boldsymbol{y}_1^*, \boldsymbol{y}_2^*, \ldots, \boldsymbol{y}_{\pi_2}^*\}.$$

Then for every integer i satisfying $1 \leq i \leq \pi_2$, the α strings

$$\{\boldsymbol{y}_i^*\lambda_1, \boldsymbol{y}_i^*\lambda_2, \ldots, \boldsymbol{y}_i^*\lambda_\alpha\} \tag{4.3}$$

must all be distinct and p-canonical. Further, since the distinct final letter occurs at least twice in each string, each of these strings is distinct from any of the strings (4.2). Thus $p'[\alpha,n] \geq p'[\alpha-1,n-1] + \alpha p'[\alpha,n-1]$.

Suppose now that $\boldsymbol{x}^*$ is a p-canonical string of length n formed using exactly the letters Λ_α. Let $\boldsymbol{x}^* = \boldsymbol{y}^*\lambda_i$. If λ_i occurs in $\boldsymbol{y}^*$, then $\boldsymbol{y}^* \in S_2$ and therefore $\boldsymbol{x}^*$ is one of the strings (4.3). Otherwise, by property (P), λ_α cannot occur in $\boldsymbol{y}^*$ either, and so $i = \alpha$, $\boldsymbol{y}^* \in S_1$, and $\boldsymbol{x}^*$ is one of the strings (4.2). We conclude that $p'[\alpha,n] \leq p'[\alpha-1,n-1] + \alpha p'[\alpha,n-1]$, and so the result is proved. □

We now make the pleasing observation that the recurrence relation of Theorem 4.2.2(d) is well-known: with the initial values specified by Theorem 4.2.2(a)–(c), it defines the Stirling numbers $\left\{ {n \atop \alpha} \right\}$ of the second kind [K73a]. Hence, for all positive integers n and α,

$$p'[\alpha, n] = \left\{ {n \atop \alpha} \right\}. \tag{4.4}$$

Table 4.1 displays an upper lefthand corner of the p'-array, showing also the column sums $p[n]$ that count the total number of p-canonical strings of length n. It turns out that the values $p[n]$ are also well known [SP95]: they are called Bell numbers.

We can use an example drawn from Table 4.1 to illustrate in another way the correspondence between Stirling numbers and our p' values that perhaps gives more insight into the relationship. A Stirling number $\left\{ {n \atop \alpha} \right\}$ is usually defined to be the number of ways that a set S of n elements can be decomposed into α nonempty nonintersecting subsets whose union is S. To see how this definition corresponds to $p'[\alpha, n]$, consider as an example the case $n = 4$, $\alpha = 2$. If we write down the seven strings counted by $p'[2, 4] = \left\{ {4 \atop 2} \right\}$ and collect into $\alpha = 2$ subsets the *positions* of identical letters in these strings, we shall find that we have decomposed $\{1, 2, 3, 4\}$ into subsets that are

- unique (because each string is distinct);
- nonempty (because each of the α letters occurs); and
- nonintersecting (because each position contains exactly one letter).

$$\begin{array}{rl}
1234 & \\
aaab & \{1,2,3\}\ \{4\} \\
aaba & \{1,2,4\}\ \{3\} \\
aabb & \{1,2\}\ \{3,4\} \\
abaa & \{1,3,4\}\ \{2\} \\
abab & \{1,3\}\ \{2,4\} \\
abba & \{1,4\}\ \{2,3\} \\
abbb & \{1\}\ \{2,3,4\}
\end{array}$$

n	1	2	3	4	5	6	7	8	9
$\left\{ {n \atop 1} \right\}$	1	1	1	1	1	1	1	1	1
$\left\{ {n \atop 2} \right\}$	0	1	3	7	15	31	63	127	255
$\left\{ {n \atop 3} \right\}$	0	0	1	6	25	90	301	966	2025
$\left\{ {n \atop 4} \right\}$	0	0	0	1	10	65	350	1701	7770
$\left\{ {n \atop 5} \right\}$	0	0	0	0	1	15	140	1050	6951
$\left\{ {n \atop 6} \right\}$	0	0	0	0	0	1	21	266	2646
$\left\{ {n \atop 7} \right\}$	0	0	0	0	0	0	1	28	462
$\left\{ {n \atop 8} \right\}$	0	0	0	0	0	0	0	1	36
$\left\{ {n \atop 9} \right\}$	0	0	0	0	0	0	0	0	1
$p[n]$	1	2	5	15	52	203	877	4140	20147

Table 4.1: The p'-array: strings of length $n \leq 9$ on $\alpha \leq 9$ letters

In this example it is easy to see that the unions of the pairs of sets in the righthand column above exhaust all possible ways of forming $S = \{1, 2, 3, 4\}$ from $\alpha = 2$ nonempty nonintersecting subsets. We leave the proof of the general case to Exercise 4.2.4.

Theorem 4.2.2 provides an iterative method of computing $p'[\alpha, n]$, and various formulæ are known that relate the $\left\{ {n \atop \alpha} \right\}$ to binomial coefficients or to Stirling numbers of the first kind [K73a]. Observe that, for any fixed alphabet size α, the partial column sum $\sum_{i=1}^{\alpha} p'[i, n]$ is the number of p-distinct strings of length n formed from *at most* α letters. Since for n large with respect to α almost all of these strings contain exactly α letters, it follows that

$$\lim_{n \to \infty} \left(\sum_{i=1}^{\alpha} p'[i, n] \bigg/ \frac{\alpha^n}{\alpha!} \right) = 1. \tag{4.5}$$

As we have seen, in the usual meaning of distinctness among strings, the number of distinct strings of length n formed from at most α letters is α^n. Thus (4.5) tells us that using p-distinct strings on an alphabet of fixed cardinality α reduces the number of strings that need to be generated by an asymptotic factor of $1/\alpha!$. For example, for $\alpha = 5$ and $n = 9$, there are $5^9 = 1{,}953{,}125$ distinct strings in the ordinary sense, whereas we find, summing the first five values in column 9 of Table 4.1, that there are only 17,002 p-distinct strings, a reduction by a factor of $115 \approx 5!$.

Of course, as we have seen, the complete column sum

$$p[n] = \sum_{i=1}^{n} p'[i, n]$$

is of particular interest and turns out to be computable in various ways, most notably in terms of previous column sums,

$$p[n] = \sum_{j=0}^{n-1} \binom{n-1}{j} p[j], \ n \geq 2, \tag{4.6}$$

and as a surprising infinite sum,

$$p[n] = \mathrm{e}^{-1} \sum_{j \geq 1} \frac{j^{n-1}}{(j-1)!}, \tag{4.7}$$

in which only $\Theta(n)$ terms actually need to be evaluated, since it can easily be shown, for example, that

$$p[n] - 1 > \mathrm{e}^{-1} \sum_{j=1}^{2n} \frac{j^{n-1}}{(j-1)!}. \tag{4.8}$$

What we have done so far enables us to count p-distinct strings by computing the values $p'[\alpha, n]$ and $p[n]$. We now show how to go further and actually generate p-canonical strings that can be used as test data. Referring to the proof of Theorem 4.2.2(d), we see that, in order to generate all the strings counted by $p'[\alpha, n]$, it suffices to

- append λ_α to the strings counted by $p'[\alpha-1, n-1]$;
- append $\lambda_1, \lambda_2, \ldots, \lambda_\alpha$ to the strings counted by $p'[\alpha, n-1]$.

This simple observation gives rise to a straightforward iterative algorithm to compute the $p[n]$ p-canonical strings of length n. As explained below, we can generate these strings in constant time per string by constructing a rooted tree T_n. In fact, as we discover by referring to the discussion of Problem 2.4 in Section 2.1, T_n is actually a variant of a trie with nodes labelled rather than edges.

The nodes of T_n are labelled with pairs (λ, α), where λ is a letter of Λ and α is the number of distinct letters found in the labels of the nodes that lie on the path to the current node from the root of T_n. The trie T_1 consists of a single root node with label $(\lambda_1, 1)$, and for every integer $n \geq 2$, T_n is formed by adding the following children to every leaf node (λ, α) of T_{n-1}:

$$(\lambda_1, \alpha), (\lambda_2, \alpha), \ldots, (\lambda_\alpha, \alpha); (\lambda_{\alpha+1}, \alpha+1).$$

Figure 4.2 shows the complete trie T_3 with letters λ_i replaced by i in the node labels.

Using Theorem 4.2.2, it is easy to see that T_n has exactly $p[n]$ leaf nodes and that the letters found on the paths to these nodes from the root give exactly the $p[n]$ p-canonical strings $\boldsymbol{x}^*$ of length n. Thus the generation of these strings is accomplished by generating T_n. Observe that, for every integer $n \geq 2$, T_n is formed from T_{n-1} by appending $p[n]$ leaf nodes, a task requiring constant time per node, thus $\Theta(p[n])$ time altogether. Since by (4.6) $p[n] \geq 2p[n-1]$, it follows that T_n can be constructed in $\Theta(p[n])$ time.

Theorem 4.2.3 *For every positive integer n, all p-canonical strings of length n can be computed in $\Theta(p[n])$ time and represented in $\Theta(p[n])$ space.* □

It should be remarked that the algorithm described here is "on-line" in the weak sense discussed in Section 4.1: T_n is formed from T_{n-1}, so that the problem is solved for $n-1$ on the way to solving it for n. This on-line solution is equivalent to a breadth-first construction of T_n, and it means that the solution is easily extendible: if it turns out that T_{n+1} is after all required, it can easily be formed by applying a single step of the same algorithm to T_n.

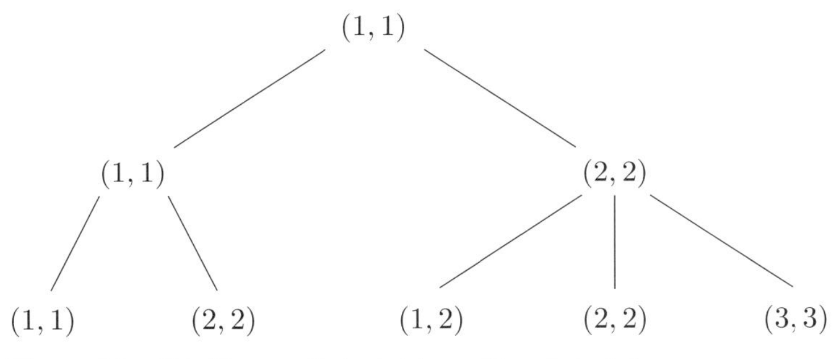

Figure 4.1. Trie T_3 — p-distinct strings of length $k \leq 3$

We shall see in the next section that an efficient breadth-first solution may sometimes be a more difficult matter.

A result similar to Theorem 4.2.3 can be established [MSM99] for the generation of all p-canonical strings counted by $p'[\alpha, n]$. In this case we generate only the subtree of T_n whose paths of length n terminate at a leaf node whose label takes the form $(*, \alpha)$; these paths represent the p-canonical strings of length n that contain exactly α letters.

We conclude this section with the remark that T_n may be traversed in various ways corresponding to various orderings of the p-canonical strings. For example, preorder traversal of T_n (or any subtree of it generated by $p'[\alpha, n]$) yields the strings in lexicographic order; so also does postorder traversal if the empty letter is assumed to sort largest. In fact, if each string of T_n can be discarded after generation, then the strings determined by T_n can actually be generated using only $\Theta(n)$ storage, corresponding to either preorder or postorder traversal of T_n. Since by (4.6) $p[n] \leq 2^n$, this reduces the storage requirement to $\Theta(\log p[n])$.

Exercises 4.2

1. Reformulate Definition 4.2.1 so as to avoid the explicit requirement that $\boldsymbol{x}$ and $\boldsymbol{y}$ have the same length.
2. Recall that an equivalence relation is reflexive, symmetric and transitive. Show that p-equivalence is an equivalence relation.
3. Prove property (P).
4. Use the "usual" definition of the Stirling numbers rather than Theorem 4.2.2 to show that $\left\{ {n \atop \alpha} \right\}$ is the number of p-canonical strings of length n formed from exactly α letters.
5. Prove formulæ (4.6)–(4.8).
6. Construct T_4 and verify that it has $p[4] = 15$ leaf nodes.
7. Specify a recursive algorithm that constructs T_n for a given positive integer n.
8. Show that the subtree of T_n corresponding to $p'[\alpha, n]$ can be constructed in $O(\alpha p'[\alpha, n])$ time using $O(\alpha p'[\alpha, n])$ space.
9. In the text it was claimed without proof that preorder traversal of T_n yields the p-canonical strings of length at most n in lexicographic order. Prove or disprove this claim.

4.3 Distinct Borders

There's not a string attuned to mirth
But has its chord in melancholy.

— Thomas HOOD (1799–1845), *Ode to Melancholy*

We have seen in Sections 1.2 and 1.3 how fundamental the border array is in describing the periodic structure of strings. It is therefore not surprising that the border array plays an important role in many algorithms whose performance or whose output depends upon this structure. For example, the KMP Algorithm (Section 7.1) that computes specific patterns in a string (Problem 2.8) depends on the computation of the border array for the pattern. Algorithms that compute all the covers of a given string (Problem 2.18) depend even more heavily on the border array: effectively, the string is replaced at the start of the algorithm by its border array and the subsequent processing is of the border array only, not the string.

Thus, for these and other algorithms, the test strings of primary interest may well be those that give rise to distinct border arrays. Formally,

Definition 4.3.1 *Two strings $\boldsymbol{x}$ and $\boldsymbol{y}$ are said to be* ***border-equivalent*** *(**b-equivalent**) if and only if their border arrays are identical. Otherwise, $\boldsymbol{x}$ and $\boldsymbol{y}$ are said to be **b-distinct**.*

To make this definition clear, consider the strings

$$\boldsymbol{x} = ababb,\ \boldsymbol{x}' = ababc,\ \boldsymbol{x}'' = abacb.$$

$\boldsymbol{x}$ and $\boldsymbol{x}'$ are p-distinct but b-equivalent, since each has border array $\boldsymbol{\beta} = 00120$. We see then that b-equivalence does not imply p-equivalence. On the other hand, $\boldsymbol{x}$ and $\boldsymbol{x}''$ are b-distinct, since $\boldsymbol{x}''$ has border array 00100, as well as being p-distinct. In fact, a little reflection tells us that b-distinct strings are necessarily also p-distinct; in other words,

Lemma 4.3.2 *If two strings are p-equivalent, they are also b-equivalent.*

Proof See Exercise 4.3.1. □

It follows from this lemma that every class of b-equivalent strings consists of one or more classes of p-equivalent strings, hence that the number of classes of b-equivalent strings is, in a well-defined sense, less than the number of p-equivalent classes. Further, since p-equivalence is clearly also an equivalence relation, we may here again identify a ***b-canonical*** representative $\boldsymbol{x}^*$ of each equivalence class: the lexicographically least among those strings on the standard alphabet that are in the class. According to our definitions so far, every class of b-equivalent strings on Λ is of infinite cardinality, but in view of Lemma 4.3.2, we can without loss of generality simplify matters by restricting such classes only to strings that are also p-canonical. Then, for example, the class of p-canonical b-equivalent strings corresponding to $\boldsymbol{\beta} = 00100$ is

$$S = \{\lambda_1\lambda_2\lambda_1\lambda_3\lambda_2,\ \lambda_1\lambda_2\lambda_1\lambda_3\lambda_3,\ \lambda_1\lambda_2\lambda_1\lambda_3\lambda_4\},$$

with b-canonical element $\boldsymbol{x}^* = \lambda_1\lambda_2\lambda_1\lambda_3\lambda_2$.

In order to be able to compute b-canonical strings efficiently, we would like to be able to establish the same kind of tree structure that was used in Section 4.2, computing canonical strings of length n as children of canonical strings of length $n-1$. To do this, we need to understand how $\boldsymbol{\beta}[n]$ is computed from the preceding elements $\boldsymbol{\beta}[1..n-1]$ of the border array $\boldsymbol{\beta}$. Recall from Lemma 1.3.1 that for any string $\boldsymbol{x} = \boldsymbol{x}[1..n]$, the possible values of $\boldsymbol{\beta}[n]$ are the elements of the descending sequence

$$\langle \boldsymbol{\beta}^{1}[n-1]+1, \boldsymbol{\beta}^{2}[n-1]+1, \ldots, \boldsymbol{\beta}^{k}[n-1]+1, 0 \rangle, \tag{4.9}$$

where $\boldsymbol{\beta}^{1}[n-1] \equiv \boldsymbol{\beta}[n-1]$, $\boldsymbol{\beta}^{i}[n-1] = \boldsymbol{\beta}\big[\boldsymbol{\beta}^{i-1}[n-1]\big]$ for every $i \in 2..k$, and k is the least integer such that $\boldsymbol{\beta}^{k}[n-1] = 0$. We now prove a much stronger result, that the set of values *actually* assumed by $\boldsymbol{\beta}[n]$ is independent of the prefix $\boldsymbol{x}[1..n-1]$:

Lemma 4.3.3 *For every integer $n \geq 2$, the values assumed by $\boldsymbol{\beta}[n]$ depend only on $\boldsymbol{\beta}[1..n-1]$ and the size of the alphabet.*

Proof Suppose that there exist two strings $\boldsymbol{x}$ and $\boldsymbol{y}$ of length $n-1$, both defined on alphabets of size α, both with border array $\boldsymbol{\beta}[1..n-1]$. Suppose further that for some letter λ and some integer m, $\boldsymbol{x'} = \boldsymbol{x}\lambda$ has border array $\boldsymbol{\beta} = \boldsymbol{\beta}[1..n] = \boldsymbol{\beta}[1..n-1]m$, but that there exists no letter μ such that $\boldsymbol{y'} = \boldsymbol{y}\mu$ has border array $\boldsymbol{\beta}$. Then m is one of the values specified in (4.9).

First consider the case $m = \boldsymbol{\beta}^{i}[n-1]+1$ for some integer $i \in 1..k$. Since $\boldsymbol{\beta}^{i}[n-1] = m-1$, it follows that

$$\boldsymbol{y}[1..m-1] = \boldsymbol{y}[n+1-m..n-1].$$

Since $\boldsymbol{\beta}^{i}[n] \neq m$, we observe that setting $\boldsymbol{y}[n] = \boldsymbol{y}[m]$ implies

$$\boldsymbol{y}[1..m'] = \boldsymbol{y}[n+1-m'..n]$$

for some $m' > m$. But this means that

$$\boldsymbol{y}[1..m'-1] = \boldsymbol{y}[n+1-m'..n-1],$$

so that $\boldsymbol{\beta}^{i}[n-1] = m'-1 > m-1$, a contradiction. Thus the lemma holds for every $m = \boldsymbol{\beta}^{i}[n-1]+1$.

Now suppose that $m = 0$. Then every one of the α possible choices $\boldsymbol{y'}[n] = \mu$ yields a unique value $\boldsymbol{\beta}[n] > 0$, while at least one choice $\boldsymbol{x'}[n] = \lambda$ gives rise to $\boldsymbol{\beta}[n] = 0$. Hence there exists $m' > 0$ such that $\boldsymbol{y'}[n]$ yields $\boldsymbol{\beta}[n] = m'$, while $\boldsymbol{x'}[n]$ does not yield $\boldsymbol{\beta}[n] = m'$, in contradiction to the case already considered.

We conclude that $\boldsymbol{\beta}$ is a border array of some $\boldsymbol{x'}$ if and only if it is a border array of some $\boldsymbol{y'}$. $\square$

This fundamental result raises the possibility (see [FGLR02, DLL02]) that $\boldsymbol{\beta}[1..n]$ can be computed from $\boldsymbol{\beta}[1..n-1]$ without reference to any specific string. We can use the result immediately, however, to show that every prefix of a b-canonical string $\boldsymbol{x}^{*}_{\boldsymbol{n}} = \boldsymbol{x}^{*}[1..n]$ must also be b-canonical:

Lemma 4.3.4 *For $n \geq 1$, every b-canonical string $\boldsymbol{x}_n^* = \boldsymbol{x}_{n-1}^*\lambda$, where $\boldsymbol{x}_{n-1}^*$ is also b-canonical and λ is some letter of the standard alphabet.*

Proof Suppose that $\boldsymbol{x}_n^* = \boldsymbol{x}_{n-1}\lambda$ with associated border array $\boldsymbol{\beta}_n = \beta[1..n]$, where $\boldsymbol{x}_{n-1}$ is string of length n that is not b-canonical. Suppose that $\boldsymbol{x}_{n-1}$ has border array $\boldsymbol{\beta}_{n-1} = \beta[1..n-1]$. Then there exists a string $\boldsymbol{y}_{n-1} < \boldsymbol{x}_{n-1}$ that also has border array $\boldsymbol{\beta}_{n-1}$. Hence by Lemma 4.3.3 there also exists $\boldsymbol{y}_n = \boldsymbol{y}_{n-1}\lambda'$ with border array $\boldsymbol{\beta}_n$, where $\boldsymbol{y}_n < \boldsymbol{x}_n^*$. But then $\boldsymbol{x}_n^*$ is not b-canonical, a contradiction. □

It is thus clear that *all* of the b-canonical strings $\boldsymbol{x}_n^*$ can be formed from b-canonical strings $\boldsymbol{x}_{n-1}^*$ — no other strings need be considered. This gives us hope that, as for p-canonical strings, it will be possible to generate b-canonical strings using a tree structure in which the strings $\boldsymbol{x}_n^*$ appear as children of the strings $\boldsymbol{x}_{n-1}^*$. Before we can describe exactly how to construct this tree, however, we need more information about how to generate distinct border arrays $\boldsymbol{\beta}_n$ from a given $\boldsymbol{\beta}_{n-1}$, and also about the form of the associated b-canonical strings $\boldsymbol{x}_n^*$.

Lemma 4.3.5 *Suppose a border array $\boldsymbol{\beta}_{n-1}$ corresponds to a b-canonical string $\boldsymbol{x}_{n-1}^*$ on the standard alphabet Λ. Then $\boldsymbol{\beta}_{n-1}$ is a prefix of exactly κ distinct border arrays $\boldsymbol{\beta}_n$ if and only if $\boldsymbol{x}_{n-1}\lambda_\kappa$ is a b-canonical string that corresponds to $\boldsymbol{\beta}_{n-1}0$.*

Proof Suppose first that $\boldsymbol{x}_{n-1}^*\lambda_\kappa$ is a b-canonical string that has only the empty border. Then, since every b-canonical string corresponding to a given border array must be lexicographically least, it follows that there exists no λ_i, $i < \kappa$, such that $\boldsymbol{x}_{n-1}^*\lambda_i$ has only the empty border; that is, for every $i \in 1..\kappa-1$, every $\boldsymbol{x}_{n-1}^*\lambda_i$ has a distinct nonempty border.

Now suppose that for some integer $i > \kappa$, the b-canonical string $\boldsymbol{x}_{n-1}^*\lambda_i$ has a longest border of length $m > 0$, so that $\boldsymbol{\beta}_n = \boldsymbol{\beta}_{n-1}m$. (In fact, since $m \geq i > \kappa \geq 2$, we see that $m \geq 3$.) It follows from Lemma 4.3.4 that $\boldsymbol{x}_{n-1}^*$ has a b-canonical prefix $\boldsymbol{x}_m^* = \boldsymbol{x}_{m-1}^*\lambda_i$ for some b-canonical string $\boldsymbol{x}_{m-1}^*$. Moreover, since $\boldsymbol{x}_{n-1}^*\lambda_\kappa$ has only the empty border, it follows that the prefix $\boldsymbol{x}_{m-1}^*\lambda_\kappa$ also has only the empty border. Then for some positive integer $\kappa' \leq \kappa$, $\boldsymbol{x}_{m-1}^*\lambda_{\kappa'}$ is a b-canonical string with a nonempty border. In other words, we have reduced an instance of a problem for finite positive integers $n-1$ and κ to an instance of exactly the same problem for finite positive integers $m-1$ and κ'. This reduction can therefore be continued indefinitely, an impossibility which persuades us that there exists no $i > \kappa$ such that $\boldsymbol{x}_{n-1}^*\lambda_i$ has a nonempty border. Thus there are exactly κ distinct border arrays $\boldsymbol{\beta}_n$, and so sufficiency is proved.

To prove necessity, suppose that there exist exactly κ distinct border arrays $\boldsymbol{\beta}_n$. But then one of them must be $\boldsymbol{\beta}_{n-1}0$ and, as we have just seen, must correspond to $\boldsymbol{x}_{n-1}^*\lambda_\kappa$. □

Note that Lemma 4.3.5 does not necessarily hold on a finite alphabet Λ_α; in other words, it holds only if the alphabet is sufficiently large. For example, on the alphabet $\Lambda_3 = \{\lambda_1, \lambda_2, \lambda_3\}$, the b-canonical string $\boldsymbol{x}_7^* = \lambda_1\lambda_2\lambda_1\lambda_3\lambda_1\lambda_2\lambda_1$ has border array $\boldsymbol{\beta}_7 = 0010123$, but there exists no $\boldsymbol{x}_8^* = \boldsymbol{x}_7^*\lambda$ on Λ_3 with border array $\boldsymbol{\beta}_8 = 00101230$.

Lemmas 4.3.3–4.3.5 suggest an algorithm for generating b-canonical strings of length n: for every integer $j = 1, 2, \ldots, n-1$, append to each b-canonical string $\boldsymbol{x}_j^*$ single standard letters $\lambda \in \{\lambda_1, \lambda_2, \ldots\}$, until for some integer $\kappa \geq 2$, $\boldsymbol{x}_j^*\lambda_\kappa$ has only the empty border. Then the strings $\boldsymbol{x}_j^*\lambda_1, \boldsymbol{x}_j^*\lambda_2, \ldots, \boldsymbol{x}_j^*\lambda_\kappa$ will be exactly the b-canonical strings derived from $\boldsymbol{x}_j^*$.

To implement this algorithm, we generate a trie T_n', similar to the trie T_n employed in Section 4.2. As in Section 4.2, the nodes of the trie are labelled with pairs, but in T_n' the labels are (λ, β), where $\lambda \in \Lambda$ and β denotes the border array entry for λ in the string defined by the labels in the nodes on the path from the root of T_n' to the current node. Thus T_1' consists of the root node $(\lambda_1, 0)$ and, for every integer $n \geq 2$, T_n' is the trie that could be formed by adding the children

$$(\lambda_1, \beta_1), (\lambda_2, \beta_2), \ldots, (\lambda_\kappa, 0)$$

to every leaf node of T_{n-1}'. Hence each node of T_n' determines a b-canonical string together with its border array. Denoting by $b[n]$ the number of b-canonical strings of length exactly n, we see that T_n' has exactly $b[n]$ leaf nodes. Thus all $b[n]$ b-canonical strings (and their corresponding border arrays) can be represented simply by appending $b[n]$ children to the leaf nodes of T_{n-1}'.

So far we have discussed T_n' as if it were constructed by a breadth-first or on-line process, similar to the earlier construction of T_n: as if T_{n-1}' should be completed before work begins on the $b[n]$ leaf nodes of T_n'. There is a problem, however, with this model of computation: in order to compute the children of a given node N_j, it is necessary to execute one step of the border array calculation, therefore to have access to $\boldsymbol{x}_j^*$ and $\boldsymbol{\beta_j}$ — respectively, the b-canonical string and corresponding border array spelled out by the path from the root of T_n to N_j. In a breadth-first algorithm, we would not have access to $\boldsymbol{x}_j^*$ and $\boldsymbol{\beta_j}$ at the time that the border array calculation needed to be performed: it would be necessary first to trace the path from the root to N_j, at a cost of $\Theta(j)$ in time, in order to recapture these arrays. This extra work is unacceptable if we are to compute the children of N_j in constant time per child: it would lead instead to an average computation time of $\Theta(j/\kappa)$ for each of the κ children of N_j.

In order to avoid tracing out the path to each node N_j, we use a depth-first, rather than a breadth-first, approach to the construction of T_n': a right sibling of the current node N_j is added to T_n' only if all the children of N_j have already been added. In order to implement this approach, it is necessary to hold strings $\boldsymbol{x}^*$ and $\boldsymbol{\beta}$ in working storage whose current value identifies the current node N_j. Thus, for example, in the trie T_3' shown in Figure 4.3, the nodes would be added in the order $ABDECFG$, while $\boldsymbol{x}^*$ and $\boldsymbol{\beta}$ would successively take the corresponding values

$$\begin{aligned}\boldsymbol{x}^* &= \lambda_1,\ \lambda_1\lambda_1,\ \lambda_1\lambda_1\lambda_1,\ \lambda_1\lambda_1\lambda_2,\ \lambda_1\lambda_2,\ \lambda_1\lambda_2\lambda_1,\ \lambda_1\lambda_2\lambda_2;\\ \boldsymbol{\beta} &= 0, 01, 012, 010, 00, 001, 000.\end{aligned}$$

This depth-first algorithm implements a version of inorder where each node is visited at most twice, once as it is created and perhaps once more in order to access its right sibling. The details of this construction are given in Exercise 4.3.5.

Recall that in order to prove Theorem 1.3.3 we demonstrated that, for any string of length n, the interior of the **while** loop in Algorithm 1.3.1, the border array algorithm, was

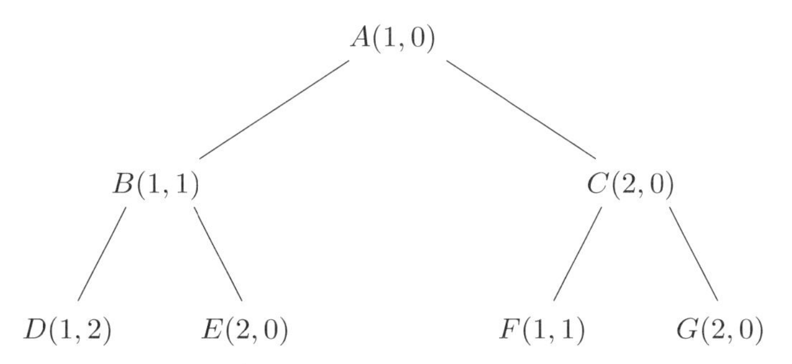

Figure 4.2. Trie T'_3 — b-distinct strings of length $k \le 3$

executed at most $n-2$ times. In the context of T'_n, this means that for every path of length n terminating at a leaf node, a total of at most $\Theta(n)$ time is required to compute the border array for the n strings (prefixes) represented by the path. Thus the average, or ***amortized***, time required for each node on the path is $O(n)/n$ — in other words, constant. We now make an observation about T'_n analogous to that made in Section 4.2 about T_n: since by (4.9) there must be at least two leaf nodes of T'_n corresponding to every leaf node of T'_{n-1}, it follows that $b[n]$ must exceed the total number of nodes in T'_{n-1}. Since the average time required for each node is constant, we have the following analogue to Theorem 4.2.3:

Theorem 4.3.6 *For every positive integer n, all b-canonical strings of length n can be computed in $\Theta(b[n])$ time and represented in $\Theta(b[n])$ space.* □

We may remark that trivial modification to the algorithm outlined above yields an algorithm to compute all the b-canonical strings of length n defined on a bounded alphabet Λ_α: instead of adding children until the border array element becomes zero, children are added until *either* a zero border array element *or* the αth letter occurs. Further, as noted in [FGLR02], the algorithm can easily be modified to generate tries that specify border arrays only or b-canonical strings only. The algorithm's one big drawback is that it is not on-line: T'_{n+1} cannot be easily generated from T'_n. This shortcoming has recently been eliminated by an improved and simplified algorithm [DLL02].

We turn now to a consideration of a b'-array analogous to the p'-array of Section 4.2. For positive integers α and n, $b'[\alpha, n]$ is the number of b-canonical strings of length n formed using exactly the α standard letters of Λ_α. Then the already-defined quantities $b[n]$ are the column sums in the b'-array:

$$b[n] = \sum_{\alpha \ge 1} b'[\alpha, n].$$

In general, it appears to be much more difficult to find well-known expressions for the elements of the b'-array than for those of the p'-array. However, the following theorem provides enough information to allow useful upper bounds to be stated on $b'[\alpha, n]$ and $b[n]$:

Theorem 4.3.7 *For all positive integers α and n:*

(a) $b'[\alpha, n] = 0$ *provided* $\alpha > \lceil \log_2(n+1) \rceil$;
(b) $b'[1, n] = b'[\alpha, 2^{\alpha-1}] = 1$;
(c) $b'[2, n] = p'[2, n] = 2^{n-1} - 1$;
(d) *if j is a positive integer and* $\alpha > 2$,

$$b'[\alpha, 2^{\alpha-1} + j] < p'[\alpha, \alpha + j].$$

Proof (a) The proof is by induction. Observe that the result holds for $n = 1$; that is, $p'[1, 1] = 1$ since the single letter λ_1 is b-canonical, while $p'[\alpha, 1] = 0$ for every $\alpha > 1$ since no string of length 1 can be formed from $\alpha > 1$ letters. Suppose then that the result holds for every n satisfying $2^{\alpha-1} \le n \le 2^{\alpha} - 1$ for some positive integer α. We show that therefore it must hold for values n' satisfying $2^{\alpha} \le n' \le 2^{\alpha+1} - 1$.

By the definition of the b'-array, the inductive assumption is equivalent to supposing that over the range of values n, at most α letters $\lambda_1, \lambda_2, \ldots, \lambda_\alpha$ (in ascending order) are required in order to form the b-canonical string $\boldsymbol{x_n}$ corresponding to each border array $\boldsymbol{\beta_n}$. Thus the letter $\lambda_{\alpha+1}$ does not occur in any position less than 2^{α} of any b-canonical string $\boldsymbol{x}^*_{\boldsymbol{n'}}$, $n' \ge 2^{\alpha}$.

We need to show that for every n' satisfying $2^{\alpha} \le n' \le 2^{\alpha+1} - 1$, no b-canonical string $\boldsymbol{x}^*_{\boldsymbol{n'}}$ contains $\lambda_{\alpha+2}$. Suppose on the contrary that some such $\boldsymbol{x}^*_{\boldsymbol{n'}}$ contains $\lambda_{\alpha+2}$ as its final letter: $\boldsymbol{x}^*_{\boldsymbol{n'}} = \boldsymbol{x}^*_{\boldsymbol{n'-1}}\lambda_{\alpha+2}$. This can only occur if each of the strings

$$\{\boldsymbol{x}^*_{\boldsymbol{n'-1}}\lambda_1, \boldsymbol{x}^*_{\boldsymbol{n'-1}}\lambda_2, \ldots, \boldsymbol{x}^*_{\boldsymbol{n'-1}}\lambda_{\alpha+1}\}$$

is b-canonical with a nonempty border. In particular, let $\boldsymbol{y} = \boldsymbol{x}^*_{\boldsymbol{n'-1}}\lambda_{\alpha+1}$, and let j be the position of the first occurrence of $\lambda_{\alpha+1}$ in $\boldsymbol{y}$. By the inductive hypothesis, $j \ge 2^{\alpha}$, and so the length of the longest border of $\boldsymbol{y}$ must exceed $n'/2$. But this implies that $\boldsymbol{y}\big[j - (n' - j)\big] = \lambda_{\alpha+1}$, contradicting the assumption that j is the first occurrence of $\lambda_{\alpha+1}$. We conclude that $\boldsymbol{x}^*_{\boldsymbol{n'-1}}\lambda_{\alpha+1}$ cannot have a nonempty border, hence by Lemma 4.3.5 that no b-canonical string $\boldsymbol{x}^*_{\boldsymbol{n'}}$ contains $\lambda_{\alpha+2}$, as required.

(b) $b'[1, n] = 1$ corresponding to the strings λ_1^n, while $b'[\alpha, 2^{\alpha-1}] = 1$ corresponding to the strings

$$\widehat{X} = \{\lambda_1, \lambda_1\lambda_2, \lambda_1\lambda_2\lambda_1\lambda_3, \lambda_1\lambda_2\lambda_1\lambda_3\lambda_1\lambda_2\lambda_1\lambda_4, \ldots.\}.$$

Letting $\boldsymbol{\hat{x}_\alpha}$ be the ith string of $\widehat{X}$, $\alpha = 1, 2, \ldots.$, and writing $\boldsymbol{\hat{x}_\alpha} = \boldsymbol{\hat{x}'_\alpha}\lambda_\alpha$, we see that the strings of $\widehat{X}$ can be generated iteratively using

$$\boldsymbol{\hat{x}_\alpha} = \boldsymbol{\hat{x}_{\alpha-1}}\boldsymbol{\hat{x}'_{\alpha-1}}\lambda_\alpha,$$

where by definition $\hat{x}_0 = \hat{x}'_0 = \varepsilon$. The strings $\hat{x}_\alpha$ then all have only an empty border; in addition, for $\alpha \geq 2$, they all have the property that

$$\hat{x}'_\alpha \lambda_i, \ 1 \leq i < \alpha,$$

has a nonempty border. Thus by Lemma 4.3.5 the use of the letter λ_α is required to complete the set of b-canonical strings of length $n = 2^{\alpha-1}$; since no other b-canonical string of this length has this property, it follows that $b'[\alpha, 2^{\alpha-1}] = 1$.

(c) Observe that, in order to form b-canonical strings x^*_n from a given b-canonical string x^*_{n-1}, at least two distinct letters need to be appended. This is because appending $\lambda_1 = x^*_n[1]$ always yields a maximum border length $\beta^* \geq 1$, so that by Lemma 4.3.5 at least λ_2 also needs to be appended in order to construct a string x^*_n with only an empty border. But for $\alpha = 2$ there are *at most* two letters to append, and so we conclude that for $\alpha = 2$, every p-canonical string is also b-canonical. Since by Lemma 4.3.2 every b-canonical string is necessarily p-canonical, the result follows.

(d) For any $\alpha \geq 3$, consider any one of the b-canonical strings $\hat{x}_\alpha$ of length $n = 2^{\alpha-1}$ introduced in (b). This string is also p-canonical and, in the tree T_n, gives rise to α further p-canonical strings of length $n+1$ as a result of appending the letters $\lambda_1, \lambda_2, \ldots, \lambda_\alpha$. However, only two b-canonical strings can be formed from $\hat{x}_\alpha$: $\hat{x}_\alpha \lambda_1$ and $\hat{x}_\alpha \lambda_2$. Thus the entire subtree rooted at the node representing the string $\hat{x}_\alpha \lambda_3$, for instance, occurs in T_n but not in T'_n. Again making use of the fact that every b-canonical string must be p-canonical, we conclude that for every $j > 1$, $b'[\alpha, n+j] < p'[\alpha, \alpha+j]$, as required.

□

Table 4.2 shows an upper lefthand corner of the b'-array. To gain an appreciation of the effect of using p-canonical or b-canonical strings for test data rather than strings that are distinct in the usual sense, note for example that while there are $4^9 = 262{,}144$ strings of length 9 on a 4-letter alphabet, on the other hand from Table 4.1,

$$p'[4, 9] = 7770$$

and from Table 4.2,

$$b'[4, 9] = 4.$$

n	**1**	**2**	**3**	**4**	**5**	**6**	**7**	**8**	**9**	**10**
$b'[1,n]$	1	1	1	1	1	1	1	1	1	1
$b'[2,n]$	0	1	3	7	15	31	63	127	255	511
$b'[3,n]$	0	0	0	1	4	15	46	134	370	997
$b'[4,n]$	0	0	0	0	0	0	0	1	4	16
$b[n]$	1	2	4	9	20	47	110	263	630	1525

Table 4.2: The b'-array: strings of length $n \leq 10$ on $\alpha \leq 4$ letters

More generally, we have already seen from (4.5) that $\sum_{i=1}^{\alpha} p'[\alpha, n]$ is asymptotically less than α^n by a factor $\alpha!$ or more. Theorem 4.3.7(d) implies that there is a further exponential reduction from $p'[\alpha, n]$ to $b'[\alpha, n]$. In fact, this latter result easily yields an upper bound on $b[n]$ expressed in terms of entries in the p'-array: for every positive integer n,

$$b[n] \leq \sum_{k=1}^{k^*} \left\{ \begin{matrix} n - 2^{k-1} + k \\ k \end{matrix} \right\}, \tag{4.10}$$

where $k^* = \lceil \log_2(n+1) \rceil$ and strict inequality holds for $n \geq 3$. In fact, by reducing the value of k^*, we can also use (4.10) to bound the partial column sums of the b'-array.

No closed form expression for $b[n]$ has so far been found, but in [G01] the fascinating result is proved that

$$b[n] \leq f_{2n+1} < \phi^{2n},$$

where f_{2n+1} is the length of the Fibostring $\boldsymbol{f_{2n+1}}$ and $\phi = (1+\sqrt{5})/2 \approx 1.62$ is the ***golden mean*** [K73a].

Exercises 4.3

1. Prove Lemma 4.3.2.
2. Show that b-equivalence is an equivalence relation.
3. Construct T'_5 and verify that it contains 20 nodes.
4. Prove by induction that $b[n] \geq \sum_{1 \leq i < n} b[i]$.
5. Specify a recursive algorithm that constructs T'_n for a given positive integer n in constant time per node.
6. Derive Equation (4.10). Then go on to derive upper bounds on partial column sums in the b'-array.

Part II

Computing Intrinsic Patterns

Part II

To listen to the words of the learned,
and to instill into others the lessons of science,
is better than religious exercises.

— Prophet MOHAMMED (570/1–632)

Chapter 5

Trees Derived from Strings

And the Word became flesh, and dwelt among us, and we beheld His glory . . .

— *John 1:4*

We begin our study of intrinsic patterns by looking at two kinds of trees — border trees and suffix trees — that arise naturally in strings and that have numerous applications. We defer till Section 13.1 discussion of another intrinsic pattern — the cover tree — that, like the border tree, provides information about the periodicity properties of a string. Here we look instead at two other tree-like suffix structures: DAWGs and suffix arrays.

5.1 Border Trees

I don't give a damn for a man that can only spell a word one way.

— Samuel CLEMENS (Mark TWAIN) (1835–1910)

We have already seen in Section 1.2 that the longest border of a string gives rise to a taxonomy of strings based on the normal form. In Section 1.3 we found that all the borders of every prefix of a string — that is, the border array — could be computed in linear time and then used to determine the periods and normal forms of those prefixes. Also in Chapter 1, the idea of a border was used to prove the Periodicity Lemma and the distinctness of the rotations of a nonrepetitive string; it then arose in various contexts in our treatment (Chapter 3) of infamous strings; and we saw in Section 4.3 that it could be used to reduce greatly the test data requirements of various string algorithms. We shall see that borders arise frequently in algorithms that compute specific patterns (Chapters 7 and 8), repetitions (Subsection 12.1.2), and covers (Section 13.1).

Having established then that the border array is an intrinsic pattern of interest, and having remarked (Section 2.1) that it has a natural tree structure, we confine ourselves here to giving an example of this structure, making use of a Fiboexample used previously to explain repeats (Definition 2.3.1):

	1	2	3	4	5	6	7	8	9	10	11	12	13	14	15	16	17	18	19	20	21
$\boldsymbol{f_7} =$	a	b	a	a	b	a	b	a	a	b	a	a	b	a	b	a	a	b	a	b	a
$\boldsymbol{\beta} =$	0	0	1	1	2	3	2	3	4	5	6	4	5	6	7	8	9	10	11	7	8

Figure 5.1 shows how the border array $\boldsymbol{\beta}$ is interpreted as a tree. Observe that the borders of $\boldsymbol{f_7}[1..i]$ are specified by the ancestors of node i in the tree. For example, the borders of $\boldsymbol{f_7}[1..18]$ are $\boldsymbol{f_7}[1..10]$, $\boldsymbol{f_7}[1..5]$, $\boldsymbol{f_7}[1..2]$ and $\boldsymbol{f_7}[1..0] = \varepsilon$.

Exercises 5.1

1. The border array effectively provides a means of traversing the border tree bottom-up — that is, from child to parent — in constant time. Devise a space-efficient data structure that in addition permits access from every parent to all of its children in amortized constant time per child. ("Space-efficient" means use of no more than $3(n + 1)$ storage locations altogether!) Then modify Algorithm 1.3.1 so as to produce this new data structure in time linear in the string length.

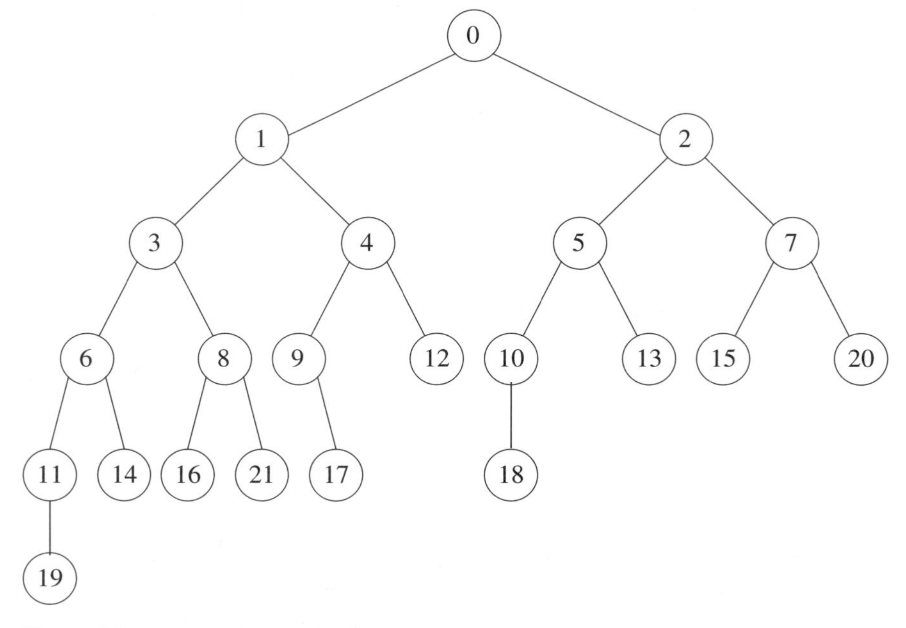

Figure 5.1. The border tree for $\boldsymbol{f_7}$

2. Show that the border tree of a Fibostring is necessarily binary; that is, that every node has at most two children.
3. Show that the subtree rooted at node $i \geq 1$ of the border tree specifies all the repeats of $\boldsymbol{x}[1..i]$ in $\boldsymbol{x}$.

5.2 Suffix Trees

One word sums up probably the responsibility of any vice-president, and that one word is "to be prepared".

— Dan QUAYLE (1947–), *12 June 1989*

As discussed to some degree in Section 2.1, suffix trees have numerous applications to string problems [A85]. They are most commonly used in a "preprocessing phase" that effectively replaces the input string $\boldsymbol{x}$ by its suffix tree $T_{\boldsymbol{x}}$. We have seen that $T_{\boldsymbol{x}}$, once formed, can be searched for any given string $\boldsymbol{u}$ on an alphabet A in $O(|u| \log \alpha)$ time, an attractive feature especially when $|\boldsymbol{u}| \ll |\boldsymbol{x}|$, $\alpha = |A|$ is small, and $\boldsymbol{x}$ is not subject to frequent change — all properties satisfied, for example, by searches of DNA sequences.

We saw also in Section 2.1, especially in Exercise 2.1.7, that $T_{\boldsymbol{x}}$ could be implemented so as to occupy $\Theta(nf(\alpha))$ space, where $f(\alpha)$ might variously be α or some small constant, depending on the nature of the alphabet and the corresponding decision mechanism implemented at each node. In this section we shall see that $T_{\boldsymbol{x}}$ can similarly be constructed in $O(n \log \alpha)$ time when the alphabet is ordered, but in $\Theta(n)$ time when it is indexed (Section 4.1).

There are five subsections, the first four devoted essentially to presenting three suffix tree construction algorithms, the fifth to practical matters. The three algorithms presented, after a preliminary section that enhances our understanding of suffix trees in general, are those of McCreight [M76] and Ukkonen [U92a] for ordered alphabets, followed by that of Farach [F97] for indexed alphabets. In Section 12.1.1 we shall make the rather surprising discovery that an algorithm for the computation of non-right-extendible repeats essentially constructs the suffix tree as a byproduct, thus constituting a fourth algorithm. An earlier algorithm due to Weiner [W73] was the model used for McCreight's algorithm, and also achieves $O(n \log \alpha)$ construction time, but is slightly more complicated and occupies slightly more space, hence is not discussed here. A modified version of Weiner's algorithm [CS85] also permits searches for the reversed substrings of $\boldsymbol{x}$. A recent paper [ALS99] describes an algorithm for another variant of a suffix tree with application to text searching and DNA sequence analysis: $\boldsymbol{x}$ is divided into "words" separated by delimiters, and only those suffixes that begin at word boundaries are included in $T_{\boldsymbol{x}}$.

The implicit assumption is made throughout this section and the next that α is bounded, but this is primarily for convenience of discussion: if alphabet size is not bounded, α can be replaced by the number of distinct letters actually used in the given string $\boldsymbol{x}$. Nevertheless, this replacement will not take place without cost: see the discussion of the Count Occurrences Problem given in Section 4.1.

5.2.1 Preliminaries

Against the World the unstilled world still whirled
About the centre of the silent Word.

— Thomas Stearns ELIOT (1888–1965), *Ash Wednesday*

The algorithms presented later in this section make use of special techniques to facilitate efficient traversal of the suffix tree as it is being built. The purpose of this subsection is to explain some of these techniques and to give a preliminary idea of their roles in the algorithms.

We begin by defining the ***longest common prefix*** $\boldsymbol{u}$ of two given strings $\boldsymbol{x_1}$ and $\boldsymbol{x_2}$ to be the longest string $\boldsymbol{u}$ that is a prefix of both. We write

$$\boldsymbol{u} = \text{LCP}(\boldsymbol{x_1}, \boldsymbol{x_2}).$$

In the context of suffix trees, the situation commonly arises that both $\boldsymbol{x_1}$ and $\boldsymbol{x_2}$ are suffixes of a string $\boldsymbol{x}$, so that $\boldsymbol{x_1} = \boldsymbol{x}[i..n]$ and $\boldsymbol{x_2} = \boldsymbol{x}[j..n]$ for integers i and j in the range $1..n+1$. In this case, we write $\boldsymbol{u}$ in the abbreviated form LCP(i, j). For example, for the Fibostring

$$\begin{array}{rccccccccccccc} & 1 & 2 & 3 & 4 & 5 & 6 & 7 & 8 & 9 & 10 & 11 & 12 & 13 \\ \boldsymbol{f_6} = & a & b & a & a & b & a & b & a & a & b & a & a & b \end{array}$$

we have

$$\text{LCP}(baababaabaab, baabaab) = \text{LCP}(2, 7) = baaba,$$

while LCP$(3, 7) = \varepsilon$ and LCP$(5, 7) = ba$.

These examples are special cases of an elementary, but very important, observation. Let $\boldsymbol{u}$ denote the node of $T_{\boldsymbol{x}}$ that is the root of the minimum size subtree that contains two suffixes i and j of $\boldsymbol{x}$. We call $\boldsymbol{u}$ the ***lowest common ancestor*** of i, j and write $\boldsymbol{u} = \text{LCA}(i, j)$. Then since $\boldsymbol{u}$ is spelled out by the path from the root of $T_{\boldsymbol{x}}$, and since that path is a longest one, we can immediately write down a defining property of a suffix tree, that for any pair of suffixes i and j,

$$\text{LCA}(i, j) = \text{LCP}(i, j). \tag{5.1}$$

We make particular use of this relationship in Farach's algorithm (Subsection 5.2.4).

Of course, as the suffix tree of $\boldsymbol{x}$ is being built, it becomes important to identify, for each suffix $\boldsymbol{u} = \boldsymbol{x}[i..n]\$$, the nodes LCP(i, j) at which the path in $T_{\boldsymbol{x}}$ to $\boldsymbol{u}$ diverges from other paths to other suffixes $\boldsymbol{x}[j..n]\$$. For McCreight's algorithm, which processes suffixes in their natural order $i = 1, 2, \ldots, n$, it is the LCP of *maximum* length over all *longer* suffixes $\boldsymbol{x}[j..n]\$$, $j < i$, that we will be especially interested in: that maximum length will determine the lowest point in $T_{\boldsymbol{x}}$ at which the path to the ith suffix $\boldsymbol{x}[i..n]\$$ could branch from the path to any previously calculated one. We therefore partition the suffix defined by terminal node i, $1 \leq i \leq n+1$, as follows:

$$\boldsymbol{x}[i..n]\$ = \text{head}(i)\text{tail}(i),$$

where $\text{head}(1) = \varepsilon$ and for $i \geq 2$ $\text{head}(i)$ is the $\text{LCP}(i,j)$ of maximum length over all $1 \leq j \leq i-1$. As remarked in Section 2.1 and as illustrated in Figure 5.2, the end-of-string sentinel \$ is introduced merely to ensure that $\text{tail}(i)$ is nonempty, hence that every terminal node is in fact a leaf node of $T_{\boldsymbol{x}}$. This applies also in the case $i = n+1$, where $\boldsymbol{x}[n+1..n]\$$ gives rise to $\text{head}(n+1) = \varepsilon$ and $\text{tail}(n+1) = \$$.

Observe that $\text{head}(i)$ is necessarily always a nonterminal node. Thus, intuitively, $\text{head}(i)$ provides a starting point (prefix) from which a search in $T_{\boldsymbol{x}}$ for the ith suffix $\boldsymbol{x}[i..n]\$$ could be performed. Table 5.1 gives the values of $\text{head}(i)$ for the suffixes of $\boldsymbol{f_6}$ (see Figure 5.2).

A word about the tail: precisely because $\text{head}(i)$ is nonterminal, $\text{tail}(i)$ is represented in $T_{\boldsymbol{x}}$ by a path from the internal node $\text{head}(i)$ to the terminal node i. Exercise 5.2.1 shows that $\text{tail}(i)$ must *also* be a terminal node in $T_{\boldsymbol{x}}$. For example, in Figure 5.2, $\text{tail}(5) = baabaab\$$ is the terminal node 7.

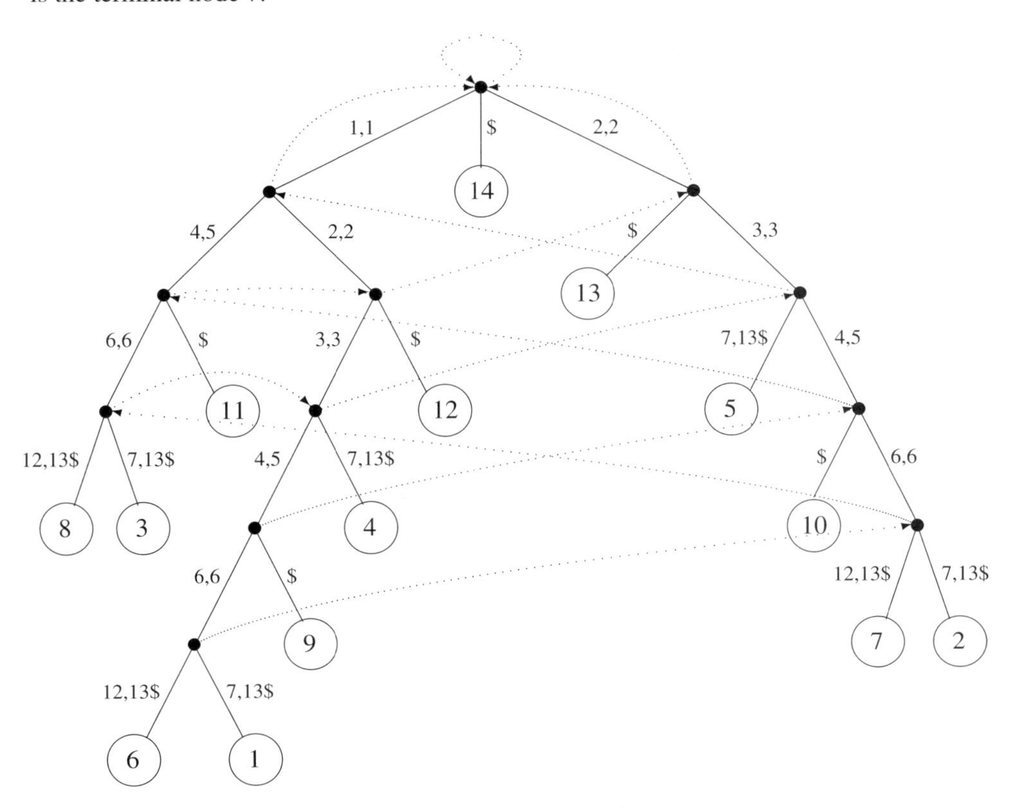

Figure 5.2. The linked suffix tree for $\boldsymbol{f_6} = abaababaabaab\$$

i	**head**(i)	i	**head**(i)
1	ε	8	$aaba$
2	ε	9	$abaab$
3	a	10	$baab$
4	aba	11	aab
5	ba	12	ab
6	$abaaba$	13	b
7	$baaba$	14	ε

Table 5.1: Values of head(i) for $\boldsymbol{f_6}$

The following lemma, used in McCreight's algorithm, shows how knowledge of head(i) can be helpful in locating head$(i+1)$ in $T_{\boldsymbol{x}}$. Informally, it tells us that the longest proper suffix of head(i) is a prefix of head$(i+1)$, hence an ancestor of head$(i+1)$ in $T_{\boldsymbol{x}}$.

Lemma 5.2.1 *Let* head$(i) = \boldsymbol{x}[i..i+h]$ *for integers* $1 \le i \le n$ *and* $-1 \le h \le n-i$. *Then* $\boldsymbol{x}[i+1..i+h]$ *is a prefix of* head$(i+1)$.

Proof The result is trivially true whenever $i+h \le i$, since then the string $\boldsymbol{x}[i+1..i+h] = \boldsymbol{\varepsilon}$ is a prefix of every string, in particular of head$(i+1)$. For $i+h > i$, let $\lambda = \boldsymbol{x}[i]$ and $\boldsymbol{u} = \boldsymbol{x}[i+1..i+h]$, so that head$(i) = \lambda\boldsymbol{u}$. By the definition of head, we know that there exists $j \in 1..i-1$ such that $\lambda\boldsymbol{u}$ is a prefix of both $\boldsymbol{x}[i..n]$ and $\boldsymbol{x}[j..n]$. But then $\boldsymbol{u}$ is a prefix of both $\boldsymbol{x}[i+1..n]$ and $\boldsymbol{x}[j+1..n]$, and so must be a prefix of head$(i+1)$. □

Observe that the heads of $\boldsymbol{f_6}$ in Table 5.1 all satisfy Lemma 5.2.1.

For the algorithms of both McCreight and Ukkonen, it is useful, based on Lemma 5.2.1, to introduce an alternate means of traversing $T_{\boldsymbol{x}}$. We define the ***suffix function*** s on any node $\boldsymbol{u}$ in $T_{\boldsymbol{x}}$ as follows:

- for $\boldsymbol{u} = \varepsilon$, $s(\boldsymbol{u}) = \varepsilon$;
- for $\boldsymbol{u} \ne \varepsilon$ (so that $\boldsymbol{u} = \lambda\boldsymbol{v}$ for a nonempty letter $\lambda \in A$), $s(\boldsymbol{u}) = \boldsymbol{v}$.

The suffix function thus defines a pointer from any nonempty (terminal or nonterminal) node $\boldsymbol{u}$ in $T_{\boldsymbol{x}}$ to the node $\boldsymbol{v} = s(\boldsymbol{u})$ that is its longest proper suffix: Exercise 5.2.3 shows that if $\boldsymbol{u}$ is a node in $T_{\boldsymbol{x}}$, then so is $s(\boldsymbol{u})$. Observe that when $\boldsymbol{u}$ is a terminal node $i \le n$, $s(\boldsymbol{u})$ is necessarily the terminal node $i+1$, while if $\boldsymbol{u}$ is the terminal node $n+1$, $s(\boldsymbol{u}) = \varepsilon$. Observe also that when $\boldsymbol{u} =$ head(i), $s(\boldsymbol{u})$ is by Lemma 5.2.1 a prefix of head$(i+1)$.

A suffix tree that includes links corresponding to the suffix function s is called a ***linked suffix tree***. Figure 5.2 shows the linked suffix tree for $\boldsymbol{f_6}$ with only the links for nonterminal nodes shown explicitly as dotted lines.

We now discuss the mechanisms, briefly mentioned in Section 2.1, for searching for a nonempty string $\boldsymbol{v}$ starting at a node $\boldsymbol{u}$ of $T_{\boldsymbol{x}}$ that we call the ***prefix node***; in other words, given that a prefix node $\boldsymbol{u}$ has already been located, searching for a prefix $\boldsymbol{uv}$ of $\boldsymbol{x}$. We may imagine that the situation is as shown in Figure 5.3, where node $\boldsymbol{u}$ has k distinct children $\boldsymbol{uu_i}$, $i = 1, 2, \ldots, k$, with $\boldsymbol{u_i} = \lambda_i\boldsymbol{u}_i'$ for pairwise distinct letters λ_i. We seek a match with $\boldsymbol{v} = \lambda\boldsymbol{v}'$.

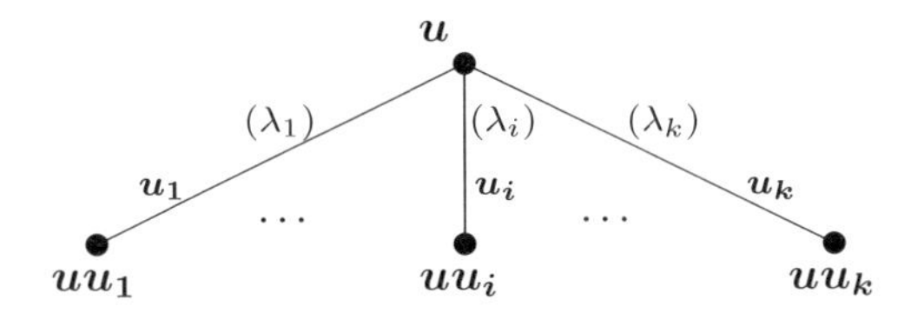

Figure 5.3. Slowscan and fastscan

It may be that we do not know whether or not uv is in fact a prefix of x; in this case, having located the correct downward edge by finding $\lambda = \lambda_i$, say, we have no option but to continue a letter-by-letter comparison of v' with the edge label u_i'. This process we call ***slowscan***.

It can happen however that we already know that uv is a prefix of x: we are being asked only to locate the exact position in x (in other words, the location in T_x) at which the match occurs. In this case, we do not need to check letter by letter, since as before having located the appropriate downward edge u_i, we are faced with only two possibilities:

(1) $|v| > |u_i|$.
$v = u_i v''$ for some nonempty string v'', and the search continues recursively for v'' from a new prefix node uu_i.

(2) $|v| \leq |u_i|$.
$v = u_i[1..|v|]$ and the search is over.

Since the value $|u_i|$ is available from the edge label, it can be determined in constant time which of these two possibilities holds: the first leads to a recursive search for the suffix v'' of v, the other completes the search. Thus, in this case, the time required to search for v is not proportional to $|v|$, but only to the number of nodes that need to be visited in order to spell out v. We therefore call this process ***fastscan***. As we shall see in the next two subsections, the use of fastscan is critical in reducing the time requirement of both McCreight's and Ukkonen's algorithms.

5.2.2 McCreight's Algorithm

Words are like leaves, and where they most abound,
Much fruit of sense beneath is rarely found.

— Alexander POPE (1688–1744), *An Essay on Criticism*

Algorithm McC begins with a tree T_1' that contains only two nodes: the root ε and the terminal node 1 corresponding to the suffix $x\$ = x[1..n]\$$. These nodes are joined by the edge labelled $x\$$. T_1' also includes the suffix link at the root: $s(\varepsilon) = \varepsilon$.

Then, for every $i = 1, 2, \ldots, n$, McC computes a new tree T_{i+1}' from the preceding tree T_i' by adding the suffix $x[i+1..n]\$$. Thus the suffixes are added in descending order of length and, since the terminal nodes $i+2, i+3, \ldots, n+1$ are *not* found in T_{i+1}', none of the trees formed except the final one $T'_{n+1} = T_{\boldsymbol{x}}$ is a suffix tree. Specifically, at the ith iteration, McC adds to T'_i the following:

(1) the terminal node $i+1$;

(2) the nonterminal node head$(i+1)$, provided that this node does not already exist;

(3) the downward edge labelled tail$(i+1)$ from head$(i+1)$ to terminal node $i+1$;

(4) the suffix link $s\big(\text{head}(i)\big)$.

By virtue of addition (4) we see that each tree T'_{i+1} is a linked tree, but one in which only suffix links from nonterminal nodes are computed. It is clear moreover that once the additions for T'_{i+1} are complete, head$(i+1)$ is the only nonterminal node for which a suffix link can possibly not yet be specified.

The additions (1) and (3) computed in step i of McC are trivial once the node head$(i+1)$ has been located, and as we shall see, addition (4) is also a byproduct of the computation of head$(i+1)$. We therefore focus our attention on (2).

We begin by recalling three facts:

- $s(\text{head}(i))$ is a prefix (ancestor in $T_{\boldsymbol{x}}$) of head$(i+1)$ (Lemma 5.2.1);
- if parent$(\boldsymbol{u})$ is the parent in $T_{\boldsymbol{x}}$ of a nonempty node $\boldsymbol{u}$, then either

$$\text{parent}(\boldsymbol{u}) = s(\boldsymbol{u}) = \varepsilon \qquad (5.2)$$

  or $s(\text{parent}(\boldsymbol{u}))$ is a proper prefix of $s(\boldsymbol{u})$ (proved in Exercise 5.2.3);
- in T'_{i+1}, for nonempty head(i), the suffix link $s(\text{parent}(\text{head}(i)))$ has already been computed (as remarked above).

We conclude that in step i of the algorithm, even though $s(\text{head}(i))$ is not available for nonempty head(i), we may nevertheless generally use the node

$$s(\text{parent}(\text{head}(i)))$$

as the starting point for the search for head$(i+1)$. There are two exceptions to this statement:

- Whenever $\boldsymbol{x}[i]$ is the leftmost occurrence in $\boldsymbol{x}$ of some letter λ, it must be true that head$(i) = \varepsilon$; in this case, we use the root node as the starting point for the search, effectively defining parent$(\text{head}(i)) \equiv \varepsilon$;
- Whenever head$(i) \neq \varepsilon$ but nevertheless (5.2) holds, head$(i) = \lambda^k$ for some positive integer k; in this case, we use λ^{k-1} as the starting point — note that for $k=1$, this node is again the root.

Excluding these two cases, suppose that $\boldsymbol{u} = \text{parent}(\text{head}(i))$, and let

$$\boldsymbol{v} = \text{label}(\boldsymbol{u}, \text{head}(i))$$

be the edge label from $\boldsymbol{u}$ to head(i). We can then rewrite head$(i) = \boldsymbol{uv}$, so that the string $\boldsymbol{w} = s(\boldsymbol{u})\boldsymbol{v}$ must be a prefix of head$(i+1)$; that is, even though head$(i+1)$ may not yet be identified as a node in T_i', the substring $\boldsymbol{v}$ must nevertheless be spelled out by some downward path from $s(\boldsymbol{u})$. Thus the conditions given in Section 5.2.1 for fastscan are satisfied, and so we may express the search for $\boldsymbol{v}$ starting at prefix node $s(\boldsymbol{u})$ as

$$\boldsymbol{w} \leftarrow \textit{fastscan}(s(\boldsymbol{u}), \boldsymbol{v}),$$

regarding fastscan as a function that determines the location in T_{i+1}' of $\boldsymbol{w} = s(\boldsymbol{u})\boldsymbol{v}$.

Two cases arise: either $\boldsymbol{w}$ was already a node in T_i', or it was not. If $\boldsymbol{w}$ existed prior to step i of the algorithm, then it may be that some nonempty prefix of $\boldsymbol{x}\big[|\boldsymbol{w}|+1..n\big]$ exists on some downward path from $\boldsymbol{w}$: the longest such prefix (LCP) is precisely head$(i+1)$, which we may compute by executing the function

$$\text{head}(i+1) \leftarrow \textit{slowscan}(\boldsymbol{w}, \text{tail}(i)).$$

Of course, if head$(i+1)$ is not yet identified as a node, then it must be created. Note that of course it may happen that head$(i+1) = \boldsymbol{w}$.

We consider now the case in which $\boldsymbol{w}$ was not already a node in T_i'. Suppose further that head$(i+1) \neq \boldsymbol{w}$, so that the edge downward from $\boldsymbol{w}$ spells out a nonempty prefix of tail(i). But by the definition of head, head(i) has at least two downward edges with no common prefix, tail(i) and one other. Since $\boldsymbol{w} = s(\text{head}(i))$, it follows that $\boldsymbol{w}$ must also have two downward edges with no common prefix, a contradiction. We conclude that in this case head$(i+1) = \boldsymbol{w}$, and so no further searching is required.

Based on this discussion, we summarize McC as shown in Algorithm 5.2.1.

Algorithm 5.2.1 (McC)

– *Compute the linked suffix tree* $T_{\boldsymbol{x}}$

construct T_1'
for $i \leftarrow 1$ **to** n **do** – *construct* T_{i+1}' *from* T_i'
  $\boldsymbol{u} \leftarrow$ parent(head(i)) – $\boldsymbol{u} \leftarrow \varepsilon$ *if* head$(i) = \varepsilon$
  $\boldsymbol{v} \leftarrow$ label$(\boldsymbol{u}$, head$(i))$ – $\boldsymbol{v} \leftarrow \varepsilon$ *if* head$(i) = \varepsilon$
  if $\boldsymbol{u} \neq \varepsilon$ **then**
    $\boldsymbol{w} \leftarrow \textit{fastscan}(s(\boldsymbol{u}), \boldsymbol{v})$
  else
    $\boldsymbol{w} \leftarrow \boldsymbol{v}[1..|\boldsymbol{v}|-1]$
  if $\boldsymbol{w}$ is a new node **then**
    head$(i+1) \leftarrow \boldsymbol{w}$
    add head$(i+1)$ to T_i'
  else
    head$(i+1) \leftarrow \textit{slowscan}\big(\boldsymbol{w}, \text{tail}(i)\big)$
    if head$(i+1)$ is a new node **then**
      add head$(i+1)$ to T_i'
  $s(\text{head}(i)) \leftarrow \boldsymbol{w}$
  add terminal node $i+1$ to T_i'
  label(head$(i+1), i+1) \leftarrow$ tail$(i+1)$

Theorem 5.2.2 *Algorithm 5.2.1 correctly computes the linked suffix tree $T_{\boldsymbol{x}}$.*

Proof It suffices to observe that, by construction,

- for nonempty $\boldsymbol{x}$ there exists no nonterminal node in T'_{n+1} with less than two children;
- the downward edges from each nonterminal node have no common prefix;
- the suffix link of each nonterminal node in T'_{n+1} is correctly set;
- there exists a terminal node i in T'_{n+1} corresponding to every suffix $\boldsymbol{x}[i..n]\$$, $i = 1, 2, \ldots, n+1$.

□

To establish the time complexity of Algorithm McC requires rather more effort and subtlety of argument:

Theorem 5.2.3 *Algorithm 5.2.1 can be implemented so as to compute $T_{\boldsymbol{x}}$ for a string $\boldsymbol{x}[1..n]$ on an ordered alphabet A of size α in $O(n \log \alpha)$ time using $O(n\alpha)$ space.*

Proof Except possibly for fastscan and slowscan, each step in Algorithm 5.2.1 can be executed in constant time using standard techniques for tree traversal and representation. Thus, apart from these two routines, the time requirement is $\Theta(n)$.

First consider fastscan, and let B be the maximum "branch" time over all nodes of the tree — that is, the maximum time required to select the correct matching downward path for the current search string. Then the total time requirement for fastscan, over all n steps of the algorithm, is $O(nB)$, except possibly for recursive calls of fastscan due to the existence of intermediate nodes on the downward path from $s(\text{parent}(\text{head}(i+1))) = s(\boldsymbol{u})$ to $\boldsymbol{uv} = s(\text{head}(i+1))$ that have no counterpart on the path from $\boldsymbol{u}$ to $\boldsymbol{uv}$. Each such intermediate node adds $O(B)$ time to the processing required for fastscan, and so over n steps the existence of m of these nodes adds $O(mB)$ time to the total. But since no path length in $T_{\boldsymbol{x}}$ can exceed n, it follows that $m < n$, and so the total contribution of fastscan to Algorithm McC is $O(nB)$.

Next consider slowscan. Observe that since $s(\text{head}(i))$ is necessarily a prefix of head $(i+1)$, it follows that

$$|\text{head}(i+1)| - |\text{head}(i)| + 1 \geq 0. \tag{5.3}$$

In fact, (5.3) gives the exact number of letters that need to be processed by slowscan in order to locate head$(i+1)$, $i = 1, 2, \ldots, n$. Thus, summing (5.3) over n steps, we get the total number of letters processed by slowscan:

$$|\text{head}(n+1)| - |\text{head}(1)| + n = n,$$

and so slowscan also contributes $O(nB)$ time.

Finally, we consider the magnitude of the branch time B. Since A is ordered, we can implement an ordered array or search tree of letters of A at each node; this data structure may contain as many as $\Theta(\alpha)$ entries and can be updated and searched in $O(\log \alpha)$ time. This completes the proof. □

See Exercise 5.2.5 for further analysis of the time and space complexity of McC.

5.2.3 Ukkonen's Algorithm

Words are but empty thanks.

— Colley CIBBER (1671–1757), *Woman's Wit*

Like McC, Algorithm Ukn performs an iterative computation that begins with an initial tree and ends with $T_{\boldsymbol{x}}$. But Ukn adopts an opposite strategy: whereas McC inserts the suffixes $\boldsymbol{x}[i..n]\$$ in the order $i = 1, 2, \ldots, n+1$, Ukn inserts the *prefixes* $\boldsymbol{x}[1..i]$, $i = 1, 2, \ldots, n$, followed by the prefix $\boldsymbol{x}[1..n]\$$. Thus Ukn is actually an on-line algorithm (Section 4.1): each tree T_i constructed by it is in fact a linked suffix tree for the prefix $\boldsymbol{x}[1..i]$, and the final tree $T_{n+1} = T_{\boldsymbol{x}}$.

It should be noted however that since the end-of-string sentinel \$ is only added in the last iteration, the intermediate trees T_i, $1 \le i \le n$, have the property that terminal nodes are not necessarily leaf nodes: if $\boldsymbol{x}[1..i]$ has two suffixes $\boldsymbol{u}$ and $\boldsymbol{uv}$, then $\boldsymbol{u}$ will appear on the path from the root to $\boldsymbol{uv}$. Furthermore, since it may happen that $\boldsymbol{u}$ occurs in $\boldsymbol{x}$ *only* as a prefix of $\boldsymbol{uv}$, it may therefore happen that $\boldsymbol{u}$ is not a node in T_i at all. To gain an understanding of how Ukn grows trees, study the example for $\boldsymbol{f_4} = abaab$ in Figure 5.4 (where, *for clarity only*, edge labels are given explicitly as substrings rather than implicitly as position pairs "i_1, i_2"). Notice that, similar to Algorithm McC, the only suffix links that are maintained are for internal (branch) nodes. We see in T_4 that Ukn may give rise to nodes that are *both* terminal *and* branch nodes, while T_3 and T_5 both show cases where certain suffixes do not even appear as nodes. Only in T_6 do we find the by now familiar suffix tree form, but as we shall see, the transition from T_5 to T_6 (in general, from T_n to T_{n+1}) follows the same algorithmic rules as T_i to T_{i+1}, $1 \le i \le n-1$.

In general, because Ukn executes on-line, processing one letter of $\boldsymbol{x}$ at a time, our discussion of the algorithm focusses on the transition from T_i to T_{i+1}, $i = 1, 2, \ldots, n$, where it is always assumed that T_1 takes exactly the form shown in Figure 5.4. To simplify this discussion, we incorporate the sentinel symbol \$ into the string, so that from now on $\boldsymbol{x}[n+1] = \$$. If we let $j \in 1..i$ be the jth suffix $\boldsymbol{x}[j..i]$ in T_i, our task in transition i is simply to identify the $i+1$ suffixes $\boldsymbol{x}[j..i+1]$ in T_{i+1}, where now $j \in 1..i+1$. Then the work to be done in the ith transition may be described as follows:

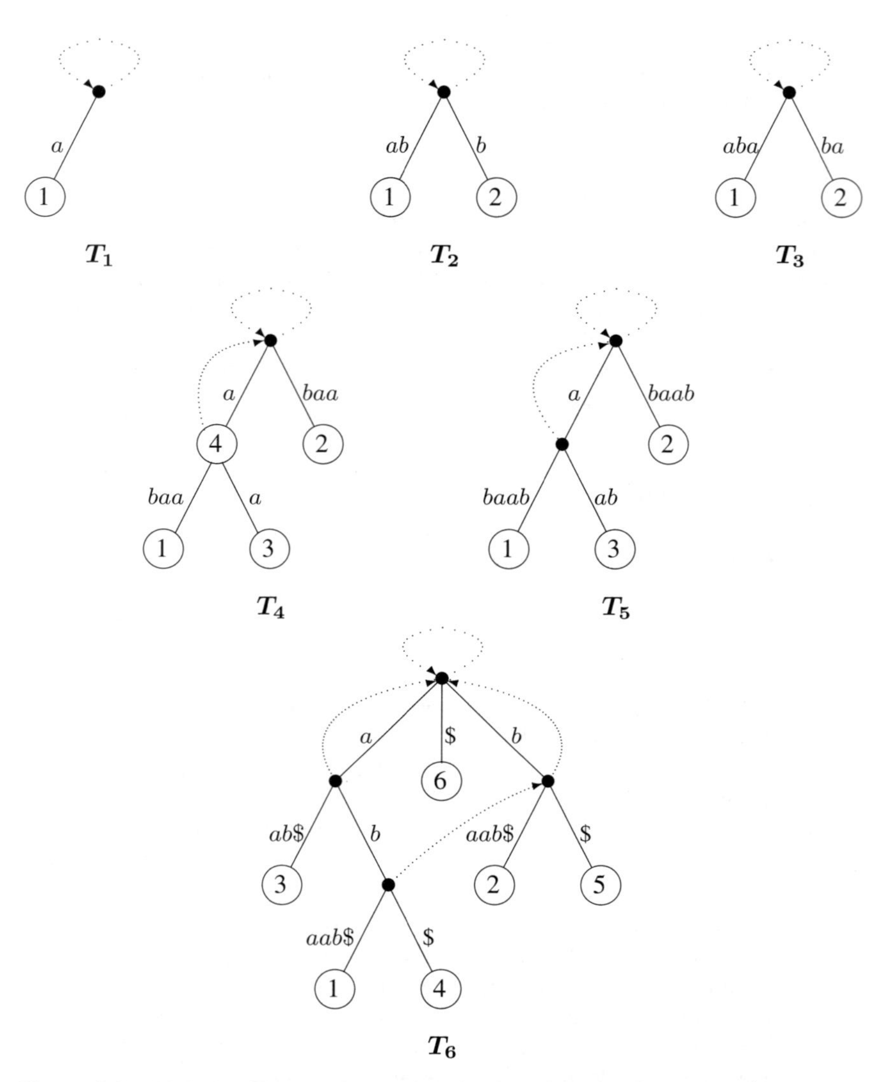

Figure 5.4. Linked suffix trees formed by Algorithm Ukn for $\boldsymbol{f_4} = abaab\$$

(1) Suffix j needs to be extended by the single letter $\boldsymbol{x}[i+1]$. If j is currently a leaf, then all that is required is to append position $i+1$ at the end of the path to j; if not, then we try to extend the existing suffix path $\boldsymbol{x}[j..i]$ in T_i to $\boldsymbol{x}[j..i+1]$. This extension may yield a new path (hence create a new leaf node) or it may simply follow an already existing path; in this latter case, the new suffix is a prefix of an already existing suffix.

(2) A new suffix $i+1$ is identified as a child of the root; if the letter $\boldsymbol{x}[i+1]$ has not occurred previously in $\boldsymbol{x}$, then suffix $i+1$ becomes a new leaf node; otherwise, suffix $i+1$ lies on an existing path from the root. Thus the processing at the root is just a special case of (1).

The following lemma provides a basis for a substantial reduction in the processing required for transition i:

Lemma 5.2.4 *Let j denote the suffix $\boldsymbol{x}[j..n]$ of $\boldsymbol{x}$.*

(a) If $j > 1$ is a leaf node of $T_{\boldsymbol{x}}$, then so is $j-1$.
(b) If from $j \leq n$ there is a downward path in $T_{\boldsymbol{x}}$ that begins with any letter λ, then there is also a downward path from $j+1$ beginning with λ.

Proof See Exercise 5.2.6. □

Although stated as a general result on an arbitrary suffix tree $T_{\boldsymbol{x}}$, this lemma applies in particular to the linked suffix trees T_i produced by Ukn. Since in each T_i, $i = 1, 2, \ldots, n+1$, suffix 1 is always a leaf node, Lemma 5.2.4(a) tells us that there always exists an integer $j_L \in 1..i$ such that every suffix $j \in 1..j_L$ is a leaf node while every suffix $j \in j_L+1..i$ is not a leaf node. Similarly, Lemma 5.2.4(b) implies that for every letter λ there exists an integer $j_\lambda \in 2..i+1$ such that every suffix $j \in j_\lambda..i$ is followed in T_i by λ, while every suffix $j \in 1..j_\lambda - 1$ is not followed by λ. Since of course no leaf node is followed in T_i by any letter, we see also that $j_L < j_\lambda$ for every λ. In Figure 5.4, for the tree T_4, $j_L = 3$ while $j_a = j_b = 4$; in tree T_5, it remains true that $j_L = 3$ and $j_a = 4$, but now $j_b = 6$ reflecting the fact that no suffix is followed by b.

In the context of the transition from T_i to T_{i+1}, we can use the above results to separate the suffixes j into three mutually exclusive classes. These classes are as follows, where λ has been chosen to be $\boldsymbol{x}[i+1]$, the letter added in transition i:

- $1 \leq j \leq j_L$ (the leaf nodes in T_i);
- $j_L + 1 \leq j \leq j_{\boldsymbol{x}[i+1]} - 1$;
- $j_{\boldsymbol{x}[i+1]} \leq j \leq i$ (the suffixes j in T_i that also occur in T_i as prefixes of the substring $\boldsymbol{x}[j..i+1]$).

It turns out that the first and the third of these classes require no processing.

For the first class, we return to the remark made above that for a suffix j that is a leaf node, it is necessary only to append $\boldsymbol{x}[i+1]$ to the path to the suffix. But since a leaf node in T_i remains a leaf node in T_{i+1}, the addition of all subsequent elements

$$\boldsymbol{x}[i+1], \boldsymbol{x}[i+2], \ldots, \boldsymbol{x}[n+1]$$

can be handled simply by storing the index ∞ on the downward edge to j. Specifically, at the time that a leaf node j is created as a child of some node $\boldsymbol{u} = \boldsymbol{x}[1..h]$, the downward edge from $\boldsymbol{u}$ to the leaf is labelled $(h+1, \infty)$, indicating that the entire suffix $\boldsymbol{x}[h+1..n+1]$ is eventually to be appended to the edge.

For the third class, we observe that no update to the tree is required: since there is a downward path from suffix j that begins with $\boldsymbol{x}[i+1]$, the only effect of the transition from T_i to T_{i+1} is that j slips one position down in the tree to denote the string $\boldsymbol{x}[j..i+1]$ that must already have existed in T_i. In later transitions, we may still need to locate suffix j

efficiently, an endeavour that our old ally fastscan will assist in, but in the current transition no formal update to the tree is required.

Thus the processing of transition i of Algorithm Ukn reduces essentially to dealing with "second class" suffixes $j \in j_L+1..j_{\boldsymbol{x}[i+1]}-1$, those that are not leaf nodes in T_i but that do not previously occur in $\boldsymbol{x}[1..i]$ as prefixes of the substring $\boldsymbol{x}[j..i+1]$. Here is an example in which all three of the identified classes occur:

$$\begin{array}{c} \quad\ 1\ 2\ 3\ 4\ 5\ 6\ 7\ 8\ 9\ 10 \\ \boldsymbol{x} = a\ a\ b\ b\ a\ c\ b\ b\ a\ a\ \cdots . \end{array}$$

For $i=9$, $j_L=6$ because

(a) position 6 is the first occurrence of c in $\boldsymbol{x}$;

(b) $\boldsymbol{x}[7..9]$ occurs previously in $\boldsymbol{x}[1..i]$ as $\boldsymbol{x}[3..5]$.

Also $j_{\boldsymbol{x}[i+1]} = j_a = 9$ because $\boldsymbol{x}[9..10] = aa$ occurs earlier in the string while $\boldsymbol{x}[8..10] = baa$ does not. Thus the three classes of suffix j for transition $i=9$ are $\{1..6, 7..8, 9..9\}$.

In transition i from T_i to T_{i+1}, Algorithm Ukn begins by inspecting the first nonleaf suffix j_L+1 in T_i; if j_L+1 does not already occur followed by $\boldsymbol{x}[i+1]$ in $\boldsymbol{x}[1..i]$, then the leaf node j_L+1 in T_{i+1} is created, j_L is incremented by one, and the inspection is repeated. When $j_L > i$, or when j_L turns out to be followed by $\boldsymbol{x}[i+1]$, transition i terminates.

The search for suffix j_L+1 is initiated from a prefix node $\boldsymbol{u}$: initially, in transition 1, $\boldsymbol{u}$ is the root, and thereafter $\boldsymbol{u}$ is maintained as the nearest ancestor of j_L+1 for which a suffix link exists.

To maintain $\boldsymbol{u}$, a slightly modified variant of fastscan, called ***smartscan***, is employed for searches: like fastscan, *smartscan*$(\boldsymbol{u}, \boldsymbol{v})$ identifies the string $\boldsymbol{w} = \boldsymbol{uv}$ in T_i, but it also replaces the prefix node $\boldsymbol{u}$ by the nearest ancestor of $\boldsymbol{w}$ (possibly, in the case that $\boldsymbol{w}$ is already a node, by $\boldsymbol{w}$ itself) so as to speed up possible future searches for the same string $\boldsymbol{w}$. Thus smartscan computes a pair as shown:

$$(\boldsymbol{u}, \boldsymbol{w}) \leftarrow \textit{smartscan}(\boldsymbol{u}, \boldsymbol{v}).$$

As in Algorithm McC, use is made of the fact that if $\boldsymbol{u}$ is a prefix node of j_L, then its suffix link $s(\boldsymbol{u})$ is a prefix node of j_L+1. Each tree T_i is maintained as a linked suffix tree, with suffix links computed for each internal node: these suffix links facilitate the initial search for suffix j_L+1, and any subsequent search for the same suffix uses the prefix node as updated by smartscan.

The details of Ukn are shown in Algorithm 5.2.2.

One tricky point remains to be explained. If in T_i a sequence of leaf nodes is created, the suffix link $s(\boldsymbol{w})$ of the parent of the final leaf node j_L+h in the sequence can be set only when it is discovered that the next suffix j_L+h+1 is *not* a leaf. In this case $s(\boldsymbol{w})$ is set to point to $\boldsymbol{w}' = \boldsymbol{x}[j_L+h+1..i]$, which is assumed to be a node. The basis for this assumption is as follows: $\boldsymbol{w} = \boldsymbol{x}[j_L+h..i]$ gives rise to a downward path labelled $\boldsymbol{\pi}$, say, that does not begin with $\boldsymbol{x}[i+1]$; however, $\boldsymbol{w}'$ has at least two downward paths, $\boldsymbol{\pi}$ *and* a path beginning with $\boldsymbol{x}[i+1]$, and so must already be a node of T_i. This argument establishes Theorem 5.2.5.

Algorithm 5.2.2 (Ukn)

```
– Compute the linked suffix tree T_x on-line

construct T_1  – with suffix link (root) = root
j_L ← 1  – by Lemma 5.2.4(a) leaf nodes added in order 1, 2, ...
u ← ε  – the initial prefix node is the root
for i ← 1 to n do  – transition i from T_i to T_{i+1}
  – the last branch node formed in the repeat loop must have
    its suffix link updated upon exit from the loop
  w_prev ← ε
  exit ← FALSE
  repeat  – create new leaf nodes if necessary
      – locate the suffix w = uv = x[j_L + 1..i] in T_i
        and update the prefix node u if necessary
    (u, w) ← smartscan(u, v)
    if there exists no downward path labelled x[i + 1] from w then
        – by Lemma 5.2.4(b) a new leaf node must be formed
      j_L ← j_L + 1; if j_L > i then exit ← TRUE
        – insert node w on its edge if necessary
      if node w does not exist in T_i then
        create node w
        label downward edge to w
        label downward edge from w
      add leaf node j_L and the "infinity edge" from w to j_L
        – set suffix link from previous node processed
      if w_prev ≠ ε then s(w_prev) ← w
        – set prefix node for next execution of smartscan
      u ← s(u)
    else
        – the suffix j_L + 1 in T_i extends to suffix j_L + 1 in T_{i+1},
          and so by Lemma 5.2.4(b) do all subsequent suffixes

        – set suffix link for last node processed
      if w_prev ≠ ε then
        s(w_prev) ← w  – w must already exist as a node
      exit ← TRUE
    – exit if j_L > i or if no more leaf nodes can be added
  until exit
```

Theorem 5.2.5 *Algorithm 5.2.2 correctly computes the linked suffix trees T_i, $i = 1, 2, \ldots, n + 1$.* □

Proof of the time complexity of Algorithm 5.2.2 follows the pattern established by Algorithm 5.2.1:

Theorem 5.2.6 *Algorithm 5.2.2 can be implemented so as to compute $T_{\boldsymbol{x}}$ for a string $\boldsymbol{x}[1..n]$ on an ordered alphabet A of size α in $O(n \log \alpha)$ time using $O(n\alpha)$ space.*

Proof Observe first that no step in Algorithm 5.2.2 can be called more than $2n - 1$ times: once for each value of i plus one extra time for each of at most $n - 1$ leaf nodes created in the **repeat** loop.

Except possibly for smartscan and the identification of the downward path from $\boldsymbol{w}$, each step in Algorithm 5.2.2 can be executed in constant time using standard techniques for tree traversal and represenation. The downward path from $\boldsymbol{w}$ can be identified in $O(B)$ time where, as in the proof of Theorem 5.2.3, B is the maximum branch time over all nodes of the tree. Thus, except for smartscan, Algorithm 5.2.2 requires $O(nB)$ time.

Now consider smartscan. For each value of j_L, smartscan may be invoked several times, but because of the recomputation of the prefix node $\boldsymbol{u}$, all invocations after the first one require only constant time. Further, after initialization as the empty string, $\boldsymbol{u}$ is updated only by smartscan, which cannot decrease its length, and by the suffix function s, which decreases its length by at most one. It follows that the total decrease in $|\boldsymbol{u}|$, over all invocations of smartscan, cannot exceed n, hence that the total number of intermediate nodes processed by smartscan cannot exceed n. Thus altogether smartscan processes at most $2n$ nodes, each requiring $O(B)$ time. Interpreting B as in the proof of Theorem 5.2.3, we have the desired result. □

In Exercise 5.2.9 at the end of this section, comparison of the McC and Ukn algorithms is encouraged. In practice, McC appears to execute somewhat faster than Ukn [GK95]. On the other hand, the on-line nature of Ukn permits a suffix tree $T_{\boldsymbol{x}_1\boldsymbol{x}_2\cdots\boldsymbol{x}_k}$ to be easily formed from a concatenation of a dynamically growing collection of strings $\boldsymbol{x}_1, \boldsymbol{x}_2, \ldots, \boldsymbol{x}_k$, not all of which were known at the time that the initial suffix tree was formed.

5.2.4 Farach's Algorithm

Words, words, words —
I'm so sick of words!
I get words all day through,
First from him, now from you —
Is that all you blighters can do?

— Alan Jay LERNER (1918–1986) and Frederick LOEWE (1904–1988)
My Fair Lady

Algorithm F constitutes a significant breakthrough: it is the first algorithm for suffix tree construction that truly executes in $\Theta(n)$ time, even if alphabet size is not regarded as a constant. The new algorithm achieves its efficiency by confining itself to strings on an indexed alphabet (Section 4.1) — that is, essentially, an integer alphabet $A = \{1, 2, \ldots, \alpha\}$ —

where it is supposed in addition that $\alpha \in O(n)$. Such alphabets are frequent in practice — as we discover in Section 7.2, the efficiency of the Boyer-Moore pattern-matching algorithm depends critically on the alphabet being indexed and not too large.

The methodology of Algorithm F derives from an algorithm proposed in [FM96], but otherwise, perhaps not surprisingly, departs completely from that of other suffix tree algorithms. It executes in five distinct stages, that we now give an overview of, slightly adapted from the original formulation in [F97], then describe in more detail below. To make the exposition easier to follow, we make use of an example for $n = 12$ on the alphabet $A = \{1, 2\}$, also taken from [F97]:

$$g = 121112212221.$$

(I) Construct $T_{\boldsymbol{x}}^{\text{odd}}$.
$T_{\boldsymbol{x}}^{\text{odd}}$ is a suffix tree of $\boldsymbol{x}$ whose leaf nodes however are restricted to be odd positions $1, 3, 5, \ldots$ of $\boldsymbol{x}\$$. Figure 5.5 displays $T_{\boldsymbol{g}}^{\text{odd}}$.

The primary mechanism for the efficient construction of $T_{\boldsymbol{x}}^{\text{odd}}$ is a $\Theta(n)$-time radix sort of all ordered pairs

$$\pi_i = \big(\boldsymbol{x}[2i-1], \boldsymbol{x}[2i]\big),$$

$i = 1, 2, \ldots, \lceil n/2 \rceil$, where for odd n the final pair is $\big(\boldsymbol{x}[n], \$\big)$, with \$ by convention equal to $\alpha + 1$. In fact, a new string $\boldsymbol{x'}$ is formed from the ***ranks*** $\rho(\pi_i)$ of the pairs in the radix sort. In our example, the pairs

$$(1,2),\ (1,1),\ (1,2),\ (2,1),\ (2,2),\ (2,1)$$

have ranks $2, 1, 2, 3, 4, 3$, respectively, so that $\boldsymbol{g'} = 212343$. The observation is then made that in the suffix tree $T_{\boldsymbol{x'}}$,

- the leaf nodes i correspond to leaf nodes $2i-1$ in $T_{\boldsymbol{x}}^{\text{odd}}$;
- the internal nodes representing substrings of length k correspond to internal nodes in $T_{\boldsymbol{x}}^{\text{odd}}$ that represent substrings of length $2k$.

$T_{\boldsymbol{g'}}$ is shown in Figure 5.6.

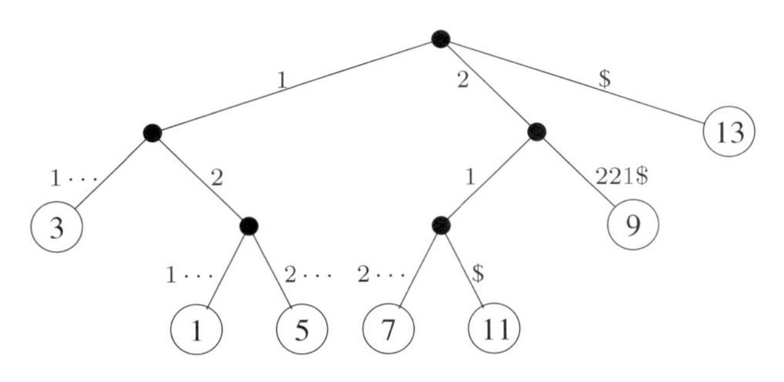

Figure 5.5. Odd tree of $\boldsymbol{g} = 121112212221\$$

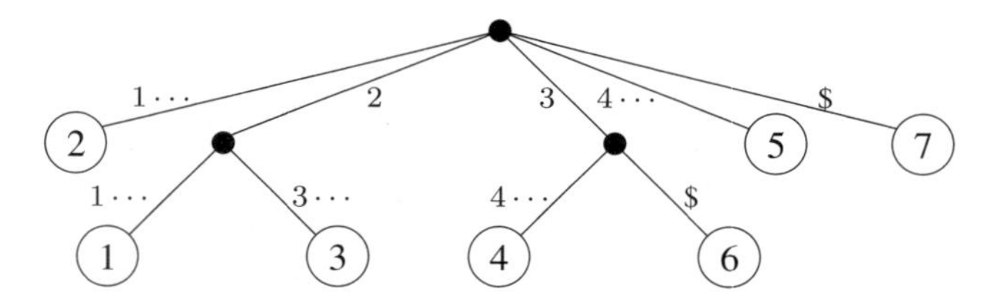

Figure 5.6. $T_{\boldsymbol{g}'}$ where $\boldsymbol{g}' = 212343\$$

It is then shown that $T_{\boldsymbol{x}'}$ can easily be massaged into $T_{\boldsymbol{x}}^{\text{odd}}$ in $\Theta(n)$ time. Thus if we denote by $t(n)$ the maximum time, over all strings $\boldsymbol{x}$ of length n, required to compute $T_{\boldsymbol{x}}$ using this method (all five stages of it), then $T_{\boldsymbol{x}}^{\text{odd}}$ can be computed in time at most

$$t(n/2) + \Theta(n) \tag{5.4}$$

by computing $T_{\boldsymbol{x}'}$ together with the pre- and post-processing just described. If moreover we are able to show that the time requirement for the remaining four stages of Algorithm F is $\Theta(n)$, it will then follow that

$$t(n) - t(n/2) \in \Theta(n), \tag{5.5}$$

hence that $t(n) \in \Theta(n)$.

(II) Construct $T_{\boldsymbol{x}}^{\text{even}}$ from $T_{\boldsymbol{x}}^{\text{odd}}$.
The key observation here is that if the lexicographic order of suffixes beginning at odd positions $2i+1$ of $\boldsymbol{x}$ is known, $i = 1, 2, \ldots, \lceil n/2 \rceil - 1$, then we can determine the lexicographic order of suffixes beginning at even positions $2i$ merely by another radix sort based on the first element of the pairs

$$\big(\boldsymbol{x}[2i], 2i+1\big),$$

provided the sort is ***stable*** — that is, that the initial order of the odd positions is maintained. Stability is of course achieved naturally by the usual radix sort, and the time requirement is $\Theta(n)$.

In our example, we can obtain the ordering $3, 1, 5, 7, 11, 9, 13$ of the odd positions in $\boldsymbol{g}$ by a simple preorder traversal of $T_{\boldsymbol{x}}^{\text{odd}}$. Eliminating position 1 and picking up $\boldsymbol{g}[2i]$, $i = 1, 2, \ldots, 6$, we therefore need to sort

$$(2,3),\ (1,5),\ (2,7),\ (2,11),\ (1,9),\ (1,13)$$

on the first entry of the pair, obtaining

$$(1,5),\ (1,9),\ (1,13),\ (2,3),\ (2,7),\ (2,11).$$

Subtracting one from the second entry of the pair then yields

$$4, 8, 12, 2, 6, 10,$$

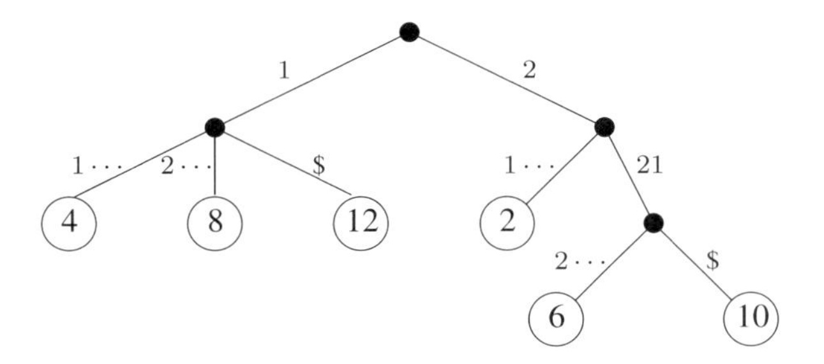

Figure 5.7. Even tree of $\boldsymbol{g} = 1\underline{2}1\underline{1}1\underline{2}2\underline{1}2\underline{2}2\underline{1}\$$

the lexicographic order of the even suffixes $\boldsymbol{g}[2i..n+1]$ of $\boldsymbol{g}$, hence the correct order of the leaf nodes of $T_{\boldsymbol{x}}^{\text{even}}$.

As explained later, we may now determine in $\Theta(n)$ time the length

$$\text{lcp}(2i, 2j) = |\text{LCP}(2i, 2j)|$$

of the least common prefix of adjacent leaf nodes $2i, 2j$ in $T_{\boldsymbol{x}}^{\text{even}}$, and hence construct $T_{\boldsymbol{x}}^{\text{even}}$. Figure 5.7 shows the result for $\boldsymbol{g}$.

(III) Overmerge $T_{\boldsymbol{x}}^{\text{odd}}$ and $T_{\boldsymbol{x}}^{\text{even}}$.

Of course what is now required is to find a way to merge the odd and even trees into $T_{\boldsymbol{x}}$. It turns out to be efficient to "overmerge" the two trees, then adjust the result.

A search is carried out simultaneously of $T_{\boldsymbol{x}}^{\text{odd}}$ and $T_{\boldsymbol{x}}^{\text{even}}$, beginning with the root. We suppose that at each node of $T_{\boldsymbol{x}}^{\text{odd}}$ and $T_{\boldsymbol{x}}^{\text{even}}$ the downward edges are identified in a list that gives, in ascending lexicographical order, the substrings to which the edges correspond. The overmerge algorithm looks only at the first letter of each substring in both $T_{\boldsymbol{x}}^{\text{odd}}$ and $T_{\boldsymbol{x}}^{\text{even}}$, say λ^{odd} and λ^{even}:

- If $\lambda^{\text{odd}} \neq \lambda^{\text{even}}$, then the subtree corresponding to the lesser of the two letters is distinct and should therefore be attached unchanged to the parent node.
- If $\lambda^{\text{odd}} = \lambda^{\text{even}}$ and the *lengths* of the substrings represented by the two downward edges are equal (though not necessarily the substrings themselves), then the two child nodes, one from the odd tree, one from the even, are merged into the same node in the overmerge tree.
- If $\lambda^{\text{odd}} = \lambda^{\text{even}}$ and the lengths of the downward edges are unequal, then in the overmerge tree both child nodes are included on a single downward path, the node corresponding to the shorter length preceding the other.

Having performed this procedure for the roots of the odd and even trees, we recursively perform it for the roots of all subtrees that may possibly contain both odd and even nodes — these roots are nodes that have already been placed in the overmerge tree because λ^{odd} and λ^{even} were found to be equal. Since each edge of each of the two trees requires only constant-time processing, the overmerge can be performed in $\Theta(n)$ time.

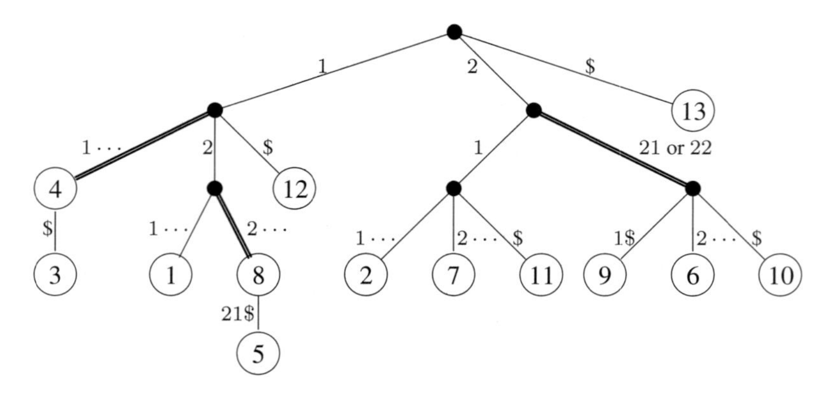

Figure 5.8. Overmerged tree $M_{\boldsymbol{g}}$ of $\boldsymbol{g} = 121112212221\$$

Clearly the ultimate effect of this procedure is to build a tree $M_{\boldsymbol{x}}$ in which subtrees are all merged that should be merged, but in which there may be subtrees that should not have been merged. The task that remains then is to "unmerge" $M_{\boldsymbol{x}}$ so as to yield $T_{\boldsymbol{x}}$. Figure 5.8 shows the overmerged tree for $\boldsymbol{g}$, where heavy lines indicate edges that have perhaps been overmerged. Note that it may now occur, as a result of overmerging, that suffixes (such as 4 and 8 in Figure 5.8) may be internal nodes of $M_{\boldsymbol{x}}$.

(IV) Construct the LCP tree.
As a prelude to the unmerging of $M_{\boldsymbol{x}}$, it is helpful to make some simple preliminary observations:

(O1) During the construction of $M_{\boldsymbol{x}}$, it is straightforward to record, for every node $\boldsymbol{u}$ in $M_{\boldsymbol{x}}$, whether or not $\boldsymbol{u}$ is in a portion of $M_{\boldsymbol{x}}$ that was merged — that is, formed from both $T_{\boldsymbol{x}}^{\text{even}}$ and $T_{\boldsymbol{x}}^{\text{odd}}$.

(O2) A node $\boldsymbol{u}$ is in a merged portion of $M_{\boldsymbol{x}}$ if and only if it has descendants $2i$ and $2j-1$ (possibly including $\boldsymbol{u}$ itself) that are leaf nodes of $T_{\boldsymbol{x}}^{\text{even}}$ and $T_{\boldsymbol{x}}^{\text{odd}}$, respectively.

(O3) Let $\boldsymbol{u} = \text{LCA}(2i, 2j-1)$ in $M_{\boldsymbol{x}}$ and $\boldsymbol{u}' = \text{LCA}(2i, 2j-1)$ in $T_{\boldsymbol{x}}$ for some choice of $2i, 2j-1$ in the subtree of $M_{\boldsymbol{x}}$ rooted at $\boldsymbol{u}$. Then, by virtue of the overmerge property, $\boldsymbol{u}'$ is a prefix of $\boldsymbol{u}$. Hence *every* pair $(2i, 2j-1)$ with LCA $\boldsymbol{u}$ in $M_{\boldsymbol{x}}$ has the same LCA ($\boldsymbol{u}'$) in $T_{\boldsymbol{x}}$.

(O4) Note in particular that if $\boldsymbol{u} = \text{LCA}(2i, 2j-1)$ in $M_{\boldsymbol{x}}$ is not the root of $M_{\boldsymbol{x}}$, then $\text{LCP}(2i, 2j-1) \neq \varepsilon$, and so $\boldsymbol{u}' = \text{LCA}(2i, 2j-1)$ in $T_{\boldsymbol{x}}$ is not the root of $T_{\boldsymbol{x}}$.

(O5) Let (i, i') be a pair of leaf nodes of $M_{\boldsymbol{x}}$ such that both are even or both are odd. Then by virtue of the corresponding property in $T_{\boldsymbol{x}}^{\text{even}}$ and $T_{\boldsymbol{x}}^{\text{odd}}$, $\text{LCA}(i, i') = \text{LCP}(i, i')$ — that is, (5.1) holds.

(O6) In view of (O2), (O3) and (O5), $M_{\boldsymbol{x}} = T_{\boldsymbol{x}}$ if and only if for every internal node $\boldsymbol{u} = \text{LCA}(2i, 2j-1)$ of $M_{\boldsymbol{x}}$,

$$\boldsymbol{u} = \text{LCP}(2i, 2j-1).$$

This is just a restatement of (5.1) for the special case of even nodes $2i$ and odd nodes $2j-1$.

Observation (O6) tells us that in order to unmerge $M_{\boldsymbol{x}}$ correctly so that it is transformed into $T_{\boldsymbol{x}}$, we need to ensure that (5.1) holds for every pair $(2i, 2j-1)$ of even/odd leaf nodes. This means that we must be able to compute LCP$(2i, 2j-1)$; that is, by (O3), the prefix of

$$\boldsymbol{u} = \text{LCA}(2i, 2j-1) \text{ in } M_{\boldsymbol{x}}$$

of length $\text{lcp}(2i, 2j-1)$ that identifies

$$\boldsymbol{u'} = \text{LCA}(2i, 2j-1) \text{ in } T_{\boldsymbol{x}}.$$

We remark that

$$\begin{aligned} \text{lcp}(2i, 2j-1) &= \text{lcp}(2i+1, 2j) + 1, \text{ if } \boldsymbol{x}[2i] = \boldsymbol{x}[2j-1]; \\ &= 0, \text{ otherwise.} \end{aligned} \tag{5.6}$$

But, by (O4), $\text{lcp}(2i, 2j+1) = 0$ if and only if $\boldsymbol{u} = \text{LCA}(2i, 2j-1)$ is the root of $M_{\boldsymbol{x}}$. Thus, for all other internal nodes $\boldsymbol{u}$, (5.6) together with (O3) implies that $\text{lcp}(2i, 2j-1)$ can be computed from $\text{lcp}(2i+1, 2j)$ for *all* even/odd pairs $2i, 2j-1$ in the subtree rooted at $\boldsymbol{u}$. Accordingly, if we let $\boldsymbol{v} = \text{LCA}(2i+1, 2j)$ in $M_{\boldsymbol{x}}$, we can define a tree on the internal nodes of $M_{\boldsymbol{x}}$ such that $\boldsymbol{v}$ is the parent of $\boldsymbol{u}$ in the tree. Provided $\boldsymbol{v}$ is not the root of $M_{\boldsymbol{x}}$, it also will have a parent in the tree, and so on: ultimately the root of the tree will be the root of $M_{\boldsymbol{x}}$. We call this tree the ***LCP tree***, and we observe that $\text{lcp}(2i, 2j-1)$ can be computed for all even/odd pairs $2i, 2j-1$ simply by determining the depth of the vertex $\boldsymbol{u} = \text{LCA}(2i, 2j-1)$ in the LCP tree.

It is shown below that the LCP tree can be computed in $\Theta(n)$ time. The LCP tree for the example string $\boldsymbol{g}$ is shown in Figure 5.9: observe, for example, that since node 8 is at depth 3 in the LCP tree, therefore $\text{lcp}(5, 8) = 3$; similarly,

$$\text{lcp}(3,4) = \text{lcp}(1,8) = \text{lcp}(2,7) = \text{lcp}(2,11) = \text{lcp}(6,9) = \text{lcp}(9,10) = 2.$$

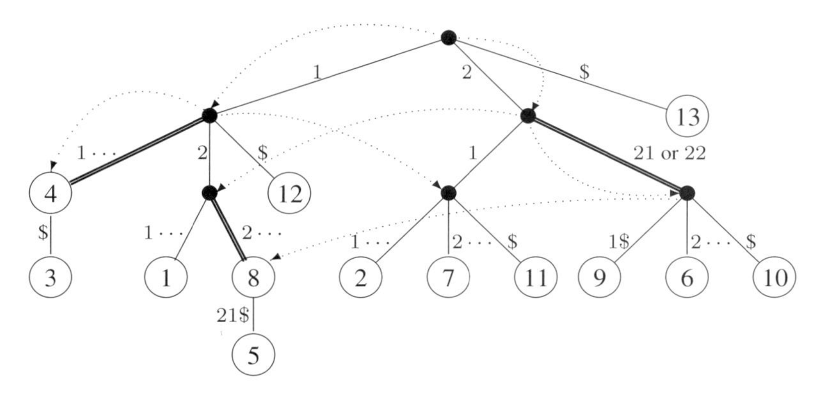

Figure 5.9. Overmerged tree $M_{\boldsymbol{g}}$ of $\boldsymbol{g} = 121112212221\$$ with LCP tree

(V) Compute $T_{\boldsymbol{x}}$ from $M_{\boldsymbol{x}}$ and the LCP tree.
In order to transform $M_{\boldsymbol{x}}$ into $T_{\boldsymbol{x}}$, we need first to determine for each node $\boldsymbol{u}$ in the LCP tree whether or not its depth $d(\boldsymbol{u})$ in the tree satisfies

$$d(\boldsymbol{u}) = |\boldsymbol{u}|, \tag{5.7}$$

where $|\boldsymbol{u}|$ is the length of the substring spelled out by the path from the root to $\boldsymbol{u}$ in $M_{\boldsymbol{x}}$. The depth can be stored for each node by a simple traversal of the LCP tree, and $|\boldsymbol{u}|$ can be stored for each node as $M_{\boldsymbol{x}}$ is formed from $T_{\boldsymbol{x}}^{\text{even}}$ and $T_{\boldsymbol{x}}^{\text{odd}}$. Further, because $M_{\boldsymbol{x}}$ is formed by overmerging, we know that $d(\boldsymbol{u}) \leq |\boldsymbol{u}|$.

Suppose that $\boldsymbol{v}$ is a vertex such that, for every ancestor $\boldsymbol{u}$ of $\boldsymbol{v}$ in the LCP tree, (5.7) holds. If moreover $d(\boldsymbol{v}) = |\boldsymbol{v}|$, then there is no action to be taken; but if $d(\boldsymbol{v}) < |\boldsymbol{v}|$, two processes must be performed:

(P1) If we denote by $p(\boldsymbol{v})$ the parent of $\boldsymbol{v}$ in $M_{\boldsymbol{x}}$, then a new node $\boldsymbol{v}'$ needs to be inserted as the child of $p(\boldsymbol{v})$ such that

$$|\boldsymbol{v}'| = d(\boldsymbol{v}).$$

$\boldsymbol{v}'$ is in fact LCP$(2i, 2j-1)$ for every even/odd pair $2i, 2j-1$ in the subtree of $M_{\boldsymbol{x}}$ rooted at $\boldsymbol{v}$, and so $\boldsymbol{v}'$ becomes the parent in $T_{\boldsymbol{x}}$ of the even and odd subtrees of $T_{\boldsymbol{x}}^{\text{even}}$ and $T_{\boldsymbol{x}}^{\text{odd}}$, respectively, that in Stage (III) were merged and rooted at $\boldsymbol{v}$. $\boldsymbol{v}$ now disappears.

(P2) Since it has now been discovered that the node $\boldsymbol{v}$ was indeed a proper overmerge of the odd and even subtrees, it follows that all the odd and even subtrees that were merged in the subtree of $M_{\boldsymbol{x}}$ rooted at $\boldsymbol{v}$ should now be unmerged. These odd and even trees should actually be disjoint beginning at $\boldsymbol{v}'$. Rather than go through a complicated recursive unmerging process, it is much simpler to store at each node of $M_{\boldsymbol{x}}$ pointers to the odd and even subtrees that were merged at that node. Then the correct subtree at $p(\boldsymbol{v})$ can be formed as shown in Figure 5.10.

Here ℓ^{even} (respectively, ℓ^{odd}) is the length of the downward edge to the even (respectively, odd) subtree of $T_{\boldsymbol{x}}^{\text{even}}$ (respectively, $T_{\boldsymbol{x}}^{\text{odd}}$) that was originally merged into $M_{\boldsymbol{x}}$ at $\boldsymbol{v}$.

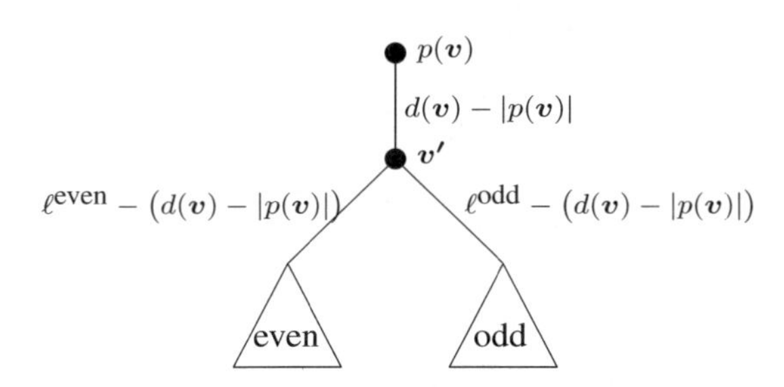

Figure 5.10. Forming the correct subtree at $p(\boldsymbol{v})$

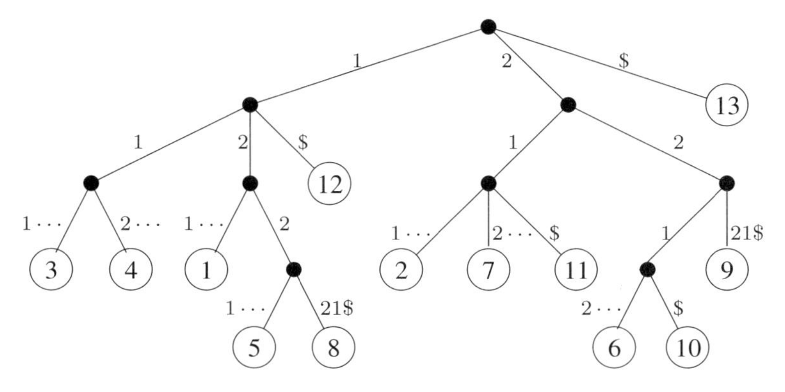

Figure 5.11. Suffix tree $T_{\boldsymbol{g}}$ of $\boldsymbol{g} = 121112212221\$$

We can understand more clearly how (P1) and (P2) operate by studying the transformation of the overmerged tree $M_{\boldsymbol{g}}$ of Figure 5.9 into $T_{\boldsymbol{g}}$, shown in Figure 5.11.

Let $\boldsymbol{v}$ be the parent of $9, 6, 10$ in $M_{\boldsymbol{g}}$, and as before let $p(\boldsymbol{v})$ be the parent of $\boldsymbol{v}$. Because

$$2 = d(\boldsymbol{v}) < |\boldsymbol{v}| = 3,$$

a new node $\boldsymbol{v}'$ is introduced as a child of $p(\boldsymbol{v})$ with a downward edge of length

$$d(\boldsymbol{v}) - |p(\boldsymbol{v})| = 1. \tag{5.8}$$

This quantity represents the length of the prefix that needs to be removed from the downward edges to the merged subtrees of $T_{\boldsymbol{x}}^{\text{even}}$ and $T_{\boldsymbol{x}}^{\text{odd}}$. Referring back to Figures 5.7 and 5.5, we see that the original even and odd subtrees were as shown in Figure 5.12.

Thus $\ell^{\text{even}} = 2$ and $\ell^{\text{odd}} = 4$. We see in Figure 5.11 that the downward edges have been appended with the first letter (2) removed from each path in accordance with (5.8): both ℓ^{even} and ℓ^{odd} have been reduced by 1. Of course we need to ensure that access to the two subtrees is governed by the lexicographic ordering of the first letter on the downward path from $\boldsymbol{v}'$; we therefore show the even tree on the left and the odd tree on the right in Figure 5.10.

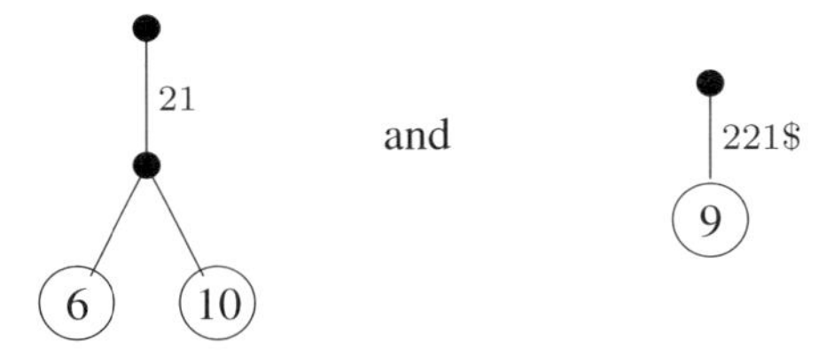

Figure 5.12. Fragments of even and odd subtrees

Note that it may happen that $\ell^{\text{even}} = 1$ or $\ell^{\text{odd}} = 1$ (but not both). In this case the corresponding downward edge disappears and the root of the corresponding subtree is merged into $\boldsymbol{v}'$.

This completes our "overview" of Algorithm F — the reader may feel rather that every nook and cranny of it has been subjected to intense scrutiny! But there remain some details to consider, most of them related to implementation.

Implementation

Here we deal with some details omitted in the preceding overview of Algorithm F.

(I) Construct $T_{\boldsymbol{x}}^{\text{odd}}$.
In the overview some details of the sorting process were left vague, as were also details of the transformation of $T_{\boldsymbol{x}'} \to T_{\boldsymbol{x}}^{\text{odd}}$.

In order to complete a sort of the ordered integer pairs

$$\pi_i = \big(\boldsymbol{x}[2i-1], \boldsymbol{x}[2i]\big),$$

two radix sorts are required, a first one by the second letter $\boldsymbol{x}[2i]$, a second one by the first letter $\boldsymbol{x}[2i-1]$. Each of these sorts requires $\Theta(\alpha + n/2)$ time and, as shown in Exercise 5.2.12, the stability property of the radix sort, employed also in Stage (II), ensures that the correct order results.

As shown in Exercise 5.2.13, the ordering of the ranks $\rho_i = \rho(\pi_i)$ of these sorted pairs determined by the suffix tree of the string

$$\boldsymbol{x}' = \rho_1 \rho_2 \cdots \rho_{\lceil n/2 \rceil}$$

corresponds exactly to the ordering of the leaf nodes $2i-1$ of $T_{\boldsymbol{x}}^{\text{odd}}$. Similarly, every internal node $\boldsymbol{u}$ of $T_{\boldsymbol{x}'}$ that defines a substring of $\boldsymbol{x}'$ of length $|\boldsymbol{u}|$ corresponds to a node of $T_{\boldsymbol{x}}^{\text{odd}}$ of length $2|\boldsymbol{u}|$ — the single letters in the alphabet of $\boldsymbol{x}'$ are just replaced by the corresponding substring $\boldsymbol{x}[2i-1..2i]$ of $\boldsymbol{x}$. Implementing these changes to the example $T_{\boldsymbol{g}'}$ of Figure 5.6 yields the tree $T_{\boldsymbol{g}'}^{\text{mod}}$ shown in Figure 5.13.

It is evident from the example that $T_{\boldsymbol{x}'}^{\text{mod}}$ is not necessarily even a trie, since during the substitution of one alphabet for another it can easily happen that

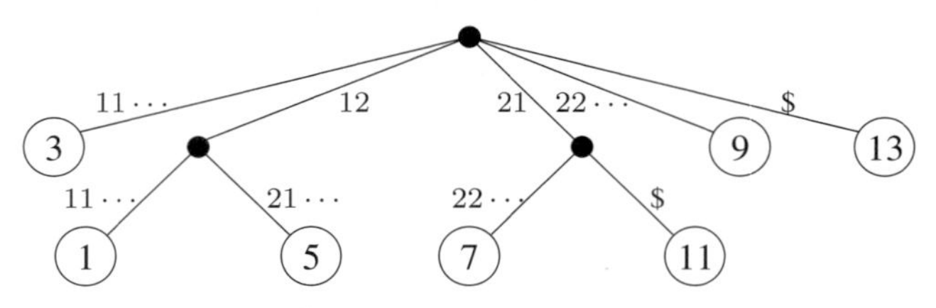

Figure 5.13. $T_{\boldsymbol{g}'}^{\text{mod}}$ where $\boldsymbol{g}' = 212343\$$, $\boldsymbol{g} = 121112212221\$$

downward edges from some node end up with the same initial letter. Thus by the substitutions done so far we seem to have taken a backward step.

But on the other hand we know that it is *only* the initial letters on two downward paths that can be identical: if the second letters also were the same, there would have been two paths in $T_{\boldsymbol{x}'}$ with the same initial letter ρ in the rank alphabet, contrary to the requirement that $T_{\boldsymbol{x}'}$ be a suffix tree. Thus transforming $T_{\boldsymbol{x}'}^{\text{mod}}$ into $T_{\boldsymbol{x}}^{\text{odd}}$ is easy: at each internal node, with downward edges given in lexicographical order, simply compare the initial letter of each edge with the initial letter of the preceding one — in case they are equal, introduce a new downward edge of length one with a new node at the end of it. It is easy to check that following this procedure in Figure 5.13 yields $T_{\boldsymbol{g}}^{\text{odd}}$ as shown in Figure 5.5.

We remark that the construction of $T_{\boldsymbol{x}'}^{\text{mod}}$ requires a single traversal of $T_{\boldsymbol{x}'}$ with constant-time substitutions required for terminal nodes and all edges — thus $\Theta(n)$ time. The subsequent changes to $T_{\boldsymbol{x}'}^{\text{mod}}$ again require a single traversal of the tree with constant-time processing required for each edge — also $\Theta(n)$ time. We have thus justified (5.4).

(II) Construct $T_{\boldsymbol{x}}^{\text{even}}$ from $T_{\boldsymbol{x}}^{\text{odd}}$.
Recall that by performing a second radix sort based on our knowledge of the lexicographic ordering of the odd suffixes of $\boldsymbol{x}$, we were able to obtain the lexicographic ordering of the even suffixes. We then make the observation, analogous to (5.6), that for arbitrary even vertices $2i, 2j$,

$$\begin{aligned} \text{lcp}(2i, 2j) &= \text{lcp}(2i+1, 2j+1) + 1, \text{ if } \boldsymbol{x}[2i] = \boldsymbol{x}[2j]; \\ &= 0, \text{ otherwise.} \end{aligned} \tag{5.9}$$

Thus, given $\text{lcp}(2i+1, 2j+1)$, we can always compute $\text{lcp}(2i, 2j)$ in constant time. But we have $T_{\boldsymbol{x}}^{\text{odd}}$ available to us, a suffix tree in which by (5.1) $\text{LCA}(2i+1, 2j+1) = \text{LCP}(2i+1, 2j+1)$ for any pair of leaf nodes. A celebrated algorithm [HT84] enables us to compute the LCA of any pair of nodes in any rooted tree in constant time, based on preprocessing linear in the number of nodes in the tree. Thus, using this algorithm we can compute $\text{lcp}(2i, 2j)$ in constant time for each pair of adjacent suffixes in the lexicographical ordering of $T_{\boldsymbol{x}}^{\text{even}}$. But in $T_{\boldsymbol{x}}^{\text{even}}$ we already know, using (5.1) again, that $\text{lcp}(2i, 2j)$ determines $\text{LCA}(2i, 2j)$, and thus provides us with the information we need to construct $T_{\boldsymbol{x}}^{\text{even}}$.

To summarize, by making use of $\Theta(n)$-time preprocessing of $T_{\boldsymbol{x}}^{\text{odd}}$, we can compute $T_{\boldsymbol{x}}^{\text{even}}$ in $\Theta(n)$ time.

(III) Overmerge $T_{\boldsymbol{x}}^{\text{odd}}$ and $T_{\boldsymbol{x}}^{\text{even}}$.
This should be a straightforward $\Theta(n)$-time algorithm!

(IV) Construct the LCP tree.
In order to perform this calculation, we again make use of the algorithm [HT84] that computes the LCA of any pair of nodes in constant time. As a byproduct of the construction of the overmerge tree $M_{\boldsymbol{x}}$, we have available the neighbouring even/odd (or odd/even) pairs $2i, 2j-1$ that were leaf nodes in $T_{\boldsymbol{x}}^{\text{even}}$ and $T_{\boldsymbol{x}}^{\text{odd}}$,

respectively. For each such pair we compute $\boldsymbol{u} = \text{LCA}(2i, 2j-1)$ in constant time. As discussed earlier, the parent of $\boldsymbol{u}$ in the LCP tree is just

$$\boldsymbol{v} = \text{LCA}(2i+1, 2j),$$

also computable in constant time. Thus, in time proportional to the number of neighbouring heterogeneous pairs, shown in Exercise 5.2.15 to be n, we can construct the LCP tree.

In our example, we consider the $n+1 = 13$ nodes

$$4, 3, 1, 8, 5, 12, 2, 7, 11, 9, 6, 10, 13$$

in their order of occurrence in $M_{\boldsymbol{g}}$, and we consider each of the 12 neighbouring heterogeneous pairs

$$(4,3),\ (4,1),\ (1,8),\ (8,5),\ (5,12),\ (5,2),$$
$$(2,7),\ (2,11),\ (2,9),\ (9,6),\ (9,10),\ (10,13),$$

computing the LCA of each pair. Of course an individual node $\boldsymbol{u} = \text{LCA}(2i, 2j-1)$ may possibly be visited more than once during this calculation, but the total number of steps is still n.

(V) Compute $T_{\boldsymbol{x}}$ from $M_{\boldsymbol{x}}$ and the LCP tree.
As we have seen, top-down processing of the LCP tree enables us to identify the first node $\boldsymbol{v}$ on each downward path such that (5.7) is not satisfied. At such nodes $\boldsymbol{v}$, processes (P1) and (P2) are performed, each requiring only a bounded number of pointer resettings in order to replace $\boldsymbol{v}$ by $\boldsymbol{v}'$ and correctly append the even and odd subtrees that had previously been overmerged. Thus the processing required for Stage (V) is proportional to the number of nodes in the LCP tree — in other words, $O(n)$.

Correctness and Efficiency

From the preceding discussion of implementation, we have seen that the time $t(n)$ required for the first stage of Algorithm F has an upper bound (5.4), while the time required for the other four phases is $\Theta(n)$. Thus (5.5) holds, and we may formally state our main result:

Theorem 5.2.7 *Algorithm F computes the suffix tree $T_{\boldsymbol{x}}$ of a string $\boldsymbol{x}[1..n]$ on an indexed alphabet in time $\Theta(n)$.* □

Despite its asymptotically optimal time bound, it remains unclear whether Algorithm F will be useful in practice. Even though, as we have seen, the underlying ideas are simple, the algorithm as a whole is complicated, and its implementation is likely to suffer severely from that *bête noire* of suffix tree algorithms, excessive use of space. However, as we shall see particularly with exact pattern-matching algorithms (Chapters 7 and 8), good algorithms have a way of giving birth to even better algorithms.

5.2.5 Application and Implementation

The word we had not sense to say —
Who knows how grandly it had rung?

— Edward Rowland SILL (1841–1887), *The Fool's Prayer*

Having looked in detail at suffix tree construction, we are now better placed to understand how suffix trees could be used. To determine whether or not a given string $\boldsymbol{u}$ is a substring of $\boldsymbol{x}$ is clearly an application of slowscan; the location of the first occurrence of $\boldsymbol{u}$ in $\boldsymbol{x}$ can be determined if a little care is taken in the selection of the integers used for edge labelling; and, if all the suffixes of $\boldsymbol{x}$ are leaf nodes in $T_{\boldsymbol{x}}$, the locations of *all* occurrences of $\boldsymbol{u}$ can be efficiently computed by applying standard tree traversal methods to the subtree of $T_{\boldsymbol{x}}$ rooted at $\boldsymbol{u}$ (if $\boldsymbol{u}$ is a node) or at the child node on the edge that spells out the prefix $\boldsymbol{u}$ (if not). Indeed, if $\boldsymbol{u}$ is already known to be a substring of $\boldsymbol{x}$, its first occurrence can be determined with particular efficiency using fastscan. Consult Exercises 5.2.16 and 5.2.17 for further details about these applications.

We have not so far mentioned a rather startling application of suffix trees, one that depends however on a fundamental property: every distinct substring of $\boldsymbol{x}$ is spelled out exactly once along a path from the root of $T_{\boldsymbol{x}}$. Thus an inventory of all the distinct substrings in $\boldsymbol{x}$ can be produced by listing all the strings spelled out along each such path. Even with edge labels "i_1, i_2" representing the substring $\boldsymbol{x}[i_1..i_2]$, this inventory must in some cases require $\Theta(n^2)$ time, since as we saw in Exercise 1.2.10 there may be as many as $\binom{n}{2}$ distinct substrings in $\boldsymbol{x}$. But suppose we are required only to count the number of these substrings. By the property quoted above, the number of distinct substrings on any path is just the number of letters on the path; that is, the sum of $i_2 - i_1 + 1$ over all edge labels "i_1, i_2" on the path. For each edge let us call $i_2 - i_1 + 1$ the ***weight*** of the edge — that is, the number of letters spanned by the edge's label. Then the number of distinct substrings in $\boldsymbol{x}$ is just the sum of all the edge weights in $T_{\boldsymbol{x}}$; using standard tree traversal techniques, this sum can be computed in time proportional to the number of edges, in turn proportional to n. Thus we have the satisfying result that, *once $T_{\boldsymbol{x}}$ has been formed*, the number of distinct substrings in $\boldsymbol{x}$ is computable in $\Theta(n)$ time.

As a final application, suppose that we are given two strings $\boldsymbol{x_1}$ and $\boldsymbol{x_2}$ on an alphabet A of size α, and asked to find the longest common factor LCF$(\boldsymbol{x_1}, \boldsymbol{x_2})$, a problem first raised in Exercise 2.2.16. Making use of a hint and our more sophisticated understanding of suffix trees, we may now be able to show that this problem can be solved in time $\Theta\big(|\boldsymbol{x_1}| + |\boldsymbol{x_2}|\big)$, again based on the supposition that a suitable suffix tree has already been constructed: see Exercise 5.2.20 below.

Such marvellous facilities as these do not come without cost, initially expressed in terms of space. We have seen in Theorems 5.2.3 and 5.2.6 that $O(\log \alpha)$ search time can be achieved at each node, but only by storing a data structure at each node that must include $O(\alpha)$ pointers and perhaps other auxiliary pointers and data. For an indexed alphabet, constant search time can be achieved at each node, but again at a cost of $O(\alpha)$ space. In addition, each node of the tree may require a parent pointer, a suffix link, and a thread or other pointer to facilitate tree traversal. There will also be $2 \log n$ bits required per edge for label storage. Thus even for small α, implementation of a suffix tree is difficult to achieve without substantial, perhaps unacceptable, space penalties. For very small alphabets (say, $\alpha \le 4$), $T_{\boldsymbol{x}}$ can efficiently be implemented as a corresponding binary tree, thus eliminating much of the above overhead. As noted in Digression 2.1, suffix trees are especially valuable in cases where $\boldsymbol{x}$ is not subject to change, searches are frequent, and α is small: conditions often satisfied by DNA sequences.

A final remark: even for very small alphabets, the search efficiency promised by the theory may, as a practical matter, turn out to be illusory. With each letter occupying only two bits, a DNA sequence of length n could be stored as an array of $n/4$ bytes, whereas the corresponding suffix tree, storing at least two pointers plus ancillary information at each node together with edge labels, could not hope to consume less than $10n$ bytes — in fact, a recent sophisticated implementation of McCreight's algorithm [K99] does well to use only $20n$ bytes of storage for its execution. Since it is primarily for large values of n that suffix trees can provide significant *theoretical* benefit, this difference by a factor of 40 or more can be devastating — the expected speed-up can be lost many times over as the operating system pages sections of the suffix tree in and out of memory during the course of the search. See [C01] for further discussion of suffix tree space requirements.

Considerations such as these have led researchers to investigate alternative suffix-based search structures, two of which are discussed in the next section.

Exercises 5.2

1. Show that for every integer $i \in 1..n+1$, tail(i) occurs as a terminal node j in $T_{\boldsymbol{x}}$, where $j = i + |\text{head}(i)|$.
2. The entries in Table 5.1 satisfy the curious inequality

$$|\text{head}(i+1)| \ge |\text{head}(i)| - 1,$$

for every $i \in 1..13$. Show that this inequality holds in general. Derive a corresponding inequality for the tails.

3. Suppose $\boldsymbol{u}$ is a node in a suffix tree $T_{\boldsymbol{x}}$ but not the root, and let s denote the suffix function. Show that:

   (a) $s(\boldsymbol{u})$ is also a node in $T_{\boldsymbol{x}}$;

   (b) $s(\boldsymbol{u})$ is not a node in the subtree rooted at $\boldsymbol{u}$;

   (c) if $\boldsymbol{v}$ is the parent of $\boldsymbol{u}$ in $T_{\boldsymbol{x}}$, then either $s(\boldsymbol{v}) = s(\boldsymbol{u}) = \boldsymbol{\varepsilon}$ or $s(\boldsymbol{v})$ is a proper ancestor of $s(\boldsymbol{u})$.

4. A ***source*** is a node $\boldsymbol{v}$ in $T_{\boldsymbol{x}}$ such that, for every node $\boldsymbol{u}$ in $T_{\boldsymbol{x}}$, $\boldsymbol{v} \ne s(\boldsymbol{u})$. Show that

   (a) the terminal node 1 is always a source;

   (b) if $|\boldsymbol{x}| > 1$, there exists a nonterminal source node;

   (c) if α' is the number of distinct letters other than \$ that occur in $\boldsymbol{x}$, there are exactly α' nonterminal source nodes.

5. Give an example of a string $\boldsymbol{x}$ for which, if slowscan is used in Algorithm McC instead of fastscan, $\Theta(n^2)$ time is required to construct $T_{\boldsymbol{x}}$.
6. Prove Lemma 5.2.4.
7. For the string $\boldsymbol{x} = aabbacbbaa$, show the suffix tree T_9 produced by transition 8 of Ukn, and describe the processing required in transition 9 to yield the suffix tree T_{10}.
8. In Theorems 5.2.3 and 5.2.6 it was suggested that an ordered array or search tree of letters together with an appropriate search technique be implemented at each node of the tree. Discuss these and alternate strategies for the data structure implemented at each node and indicate circumstances in which they might be used. Can you reduce the storage (respectively, processing time) required to $\Theta(n)$, even on a unordered alphabet, and, if so, what is the effect on processing time (repectively, storage)?
9. Algorithm Ukn is clearly more "complex" in an intellectual sense than McC is, but what of the computational complexity? Given that McC and Ukn both produce the same suffix tree in the end (decorated let us say with its terminal $ symbols), and given that both algorithms make use of the same branching data structure at each node, what can be said about the *actual* time requirements of the two algorithms? Rather more generally, can you give a specific example of a circumstance in which the on-line property of Ukn would confer significant benefit?
10. Explain why it is important for the efficiency of Algorithm F that $\alpha \in O(n)$.

    Hint: How does a radix sort work?
11. Use (5.5) to show that for Algorithm F, $t(n) \in \Theta(n)$.
12. Show how two radix sorts can be implemented to sort the ordered integer pairs $(\boldsymbol{x}[2i-1], \boldsymbol{x}[2i])$ correctly into ascending order.
13. Show that the order of the suffixes $2i-1$ determined by $T_{\boldsymbol{x}}^{\text{odd}}$ is identical to the order of the ranks ρ_i determined by the suffix tree of the string $\rho_1\rho_2\cdots\rho_{\lceil n/2\rceil}$.
14. Show that, given the lexicographical ordering of the suffixes of $\boldsymbol{x}$, the suffix tree $T_{\boldsymbol{x}}$ can be constructed from knowledge of the LCA of adjacent pairs in the ordering.
15. Show that the number of adjacent odd/even (heterogeneous) pairs in a sequence of $n+1$ integers is exactly n if and only if the integers are not homogeneous (not all even, not all odd).
16. Explain how the edge labels "i_1, i_2" can be chosen during suffix tree construction so as to permit the first occurrence of any substring $\boldsymbol{u}$ to be identified. Then explain how fastscan could be used to determine the first occurrence of $\boldsymbol{u}$ in the case that $\boldsymbol{u}$ is already known to be a substring — what is the time requirement of this search?
17. Explain how all occurrences of a given substring $\boldsymbol{u}$ in $\boldsymbol{x}$ can be specified using appropriate edge labels and traversal techniques. Specify the time and space requirements of this facility.
18. The author states that, even for an alphabet size of only four, the use of suffix trees may introduce a 20-fold, or even a 40-fold, increase in the storage requirement for a long string. Since he provides no justification, his view could be pessimistic, perhaps even alarmist. For strings $\boldsymbol{x}$ of length $n = 100K, 1M, 5M$ and alphabet size $\alpha = 4$, estimate the minimum storage required for a suitably implemented $T_{\boldsymbol{x}}$.

19. A ***position tree*** $P_{\boldsymbol{x}}$ [W73] is a modified $T_{\boldsymbol{x}}$ that displays, as a leaf node i, the shortest prefix of $\boldsymbol{x}[i..n]$ that occurs nowhere else in $\boldsymbol{x}$, whenever such a prefix exists.

(a) Show that i is a leaf node of $P_{\boldsymbol{x}}$ if and only if $i \leq j_L$; that is, if and only if i is also a leaf node of $T_{\boldsymbol{x}}$.

(b) Hence show that $P_{\boldsymbol{x}}$ has n leaf nodes if and only if the letter $\boldsymbol{x}[n]$ occurs exactly once in $\boldsymbol{x}$.

(c) Describe a simple algorithm to construct $P_{\boldsymbol{x}}$ from $T_{\boldsymbol{x}}$.

20. Outline an algorithm that uses a suffix tree to compute the LCF of two given strings $\boldsymbol{x_1}$ and $\boldsymbol{x_2}$ in time proportional to $|\boldsymbol{x_1}| + |\boldsymbol{x_2}|$. Show that your algorithm can be extended to handle $k \geq 2$ strings.

Hint: Form the string $\boldsymbol{x} = \boldsymbol{x_1}\$_1\boldsymbol{x_2}\$_2$ using distinct end-of-string sentinels $\$_1$ and $\$_2$, then compute $T_{\boldsymbol{x}}$. Observe that LCF($\boldsymbol{x_1}, \boldsymbol{x_2}$) is the node $\boldsymbol{u}$ in $T_{\boldsymbol{x}}$ of greatest length such that the subtree rooted at $\boldsymbol{u}$ contains the symbol $\$_1$.

5.3 Alternative Suffix-Based Structures

You see it's like a portmanteau —
there are two meanings packed up into one word.

— Charles DODGSON (Lewis CARROLL) (1832–1898)
Through the Looking Glass

In this section we consider two alternatives to suffix trees, both of them suffix-based data structures designed to yield theoretical benefits that may be more likely to be realized in practice. Although these structures are quite different from each other, they have in common the property of being derivable, in an algorithmic sense, from suffix trees.

5.3.1 Directed Acyclic Word Graphs

Word is a shadow of deed.

— DEMOCRITUS (c. 460–400 BC), *Fragment 145*

The invention of the DAWG derives from a simple observation that the reader has no doubt already made: suffix trees have a tendency to contain isomorphic subtrees that are repeated, it seems redundantly — corresponding nodes in the repeated subtrees give rise to corresponding edges with identical labels, and the only differences between the isomorphic copies are the labels of the leaf nodes. For example, in Figure 5.2, the subtree $T'_{\boldsymbol{v}}$ corresponding to $\boldsymbol{v} = aababaabaab\$$ occurs twice, appended both to nodes b and ab; while the subtree $T'_{\boldsymbol{w}}$ corresponding to $\boldsymbol{w} = abaabaab\$$ occurs three times, appended to the nodes $aaba$, $baaba$ and $abaaba$. Similarly, in suffix tree T_6 of Figure 5.4, the subtree $T'_{aab\$}$ occurs twice, appended again to nodes b and ab. The DAWG was introduced [BBEH85, C86] essentially as a mechanism for eliminating this redundancy.

Before discussing the DAWG itself, therefore, we pave the way by specifying the circumstances in which isomorphic subtrees of $T_{\boldsymbol{x}}$ are repeated. As in the description of Ukkonen's algorithm, we assume, in order to simplify the discussion, that $\boldsymbol{x} = \boldsymbol{x}[1..n+1]$ where $\boldsymbol{x}[n+1] = \$$. Further, since DAWGs are most easily explained as transformations of "decompacted" suffix trees, we also assume for the time being that every substring $\boldsymbol{u}$ of $\boldsymbol{x}$ is represented in $T_{\boldsymbol{x}}$ by a node, even if the node has but one child. See Figure 5.14(a).

For each nonempty substring $\boldsymbol{u}$ of $\boldsymbol{x}$, denote by $\Pi_{\boldsymbol{u}}$ the set of all positions in $\boldsymbol{x}$ at which occurrences of $\boldsymbol{u}$ terminate; for reasons that will soon appear, we let $\Pi_{\boldsymbol{\varepsilon}} = \{1, 2, \ldots, n+1\}$. As we now explain, these ***terminator sets*** reflect the structure of the suffix tree $T_{\boldsymbol{x}}$. Let $\boldsymbol{u}$ be a nonempty substring of $\boldsymbol{x}$ occurring at exactly k positions $j_1, j_2, \ldots, j_k$ in $\boldsymbol{x}$. These positions are therefore the labels of the leaf nodes in $T_{\boldsymbol{x}}$ (of course representing suffixes of $\boldsymbol{x}$) that have prefix $\boldsymbol{u}$. The terminator set of $\boldsymbol{u}$ can then be specified as follows:

$$\Pi_{\boldsymbol{u}} = \{j_1 + |\boldsymbol{u}| - 1, j_2 + |\boldsymbol{u}| - 1, \ldots, j_k + |\boldsymbol{u}| - 1\}.$$

In the case that $\boldsymbol{u} = \boldsymbol{\varepsilon}$, the set of positions $\Pi_{\boldsymbol{\varepsilon}}$ is just the set of labels of the leaf nodes that have prefix $\boldsymbol{\varepsilon}$; that is, all the leaf labels. Note that every $\Pi_{\boldsymbol{u}}$, $\boldsymbol{u} \neq \boldsymbol{\varepsilon}$, is a proper subset of $\Pi_{\boldsymbol{\varepsilon}}$.

As a simple example, if λ is any single letter, Π_λ is just the set of all positions in $\boldsymbol{x}$ at which λ occurs (as well of course as the set of all leaf nodes with prefix λ). As another example, if $\boldsymbol{u} = \boldsymbol{x}[j..n+1]$ for any $1 \le j \le n+1$, then $\Pi_{\boldsymbol{u}} = \{n+1\}$, indicating that these substrings of $\boldsymbol{x}$ are prefixes only of a single suffix $\boldsymbol{x}[j..n+1]$. More generally, if $\boldsymbol{x}[j]$ is a letter that occurs only once in $\boldsymbol{x}$ and $\boldsymbol{u}$ is any substring of $\boldsymbol{x}$ containing $\boldsymbol{x}[j]$, then $|\Pi_{\boldsymbol{u}}| = 1$. In these latter examples, $\boldsymbol{u}$ corresponds either to a leaf node of $T_{\boldsymbol{x}}$ or to a node with exactly one child.

For the construction of DAWGs, we will be particularly interested in substrings $\boldsymbol{u}_1$ and $\boldsymbol{u}_2$ of $\boldsymbol{x}$ that have identical terminator sets since, as we shall discover, these substrings are exactly the nodes in $T_{\boldsymbol{x}}$ that are the roots of isomorphic subtrees. When $\Pi_{\boldsymbol{u}_1} = \Pi_{\boldsymbol{u}_2}$, we say that $\boldsymbol{u}_1$ and $\boldsymbol{u}_2$ are **Π-*equivalent***. Exercise 5.3.1 shows that Π-equivalence is in fact an equivalence relation.

Lemma 5.3.1 *Let $\boldsymbol{u}_1$ and $\boldsymbol{u}_2$ be two nonempty substrings of $\boldsymbol{x}$ such that $|\boldsymbol{u}_1| \le |\boldsymbol{u}_2|$. Then $\boldsymbol{u}_1$ and $\boldsymbol{u}_2$ are Π-equivalent if and only if $\boldsymbol{u}_1$ occurs in $\boldsymbol{x}$ only as a suffix of $\boldsymbol{u}_2$.*

Proof If $\Pi_{\boldsymbol{u}_1} = \Pi_{\boldsymbol{u}_2}$, then by the definition of terminator set, $\boldsymbol{u}_1$ and $\boldsymbol{u}_2$ terminate at the same set of positions; since $|\boldsymbol{u}_1| \le |\boldsymbol{u}_2|$, $\boldsymbol{u}_1$ is a suffix of $\boldsymbol{u}_2$. Conversely, if $\boldsymbol{u}_1$ occurs only as a suffix of $\boldsymbol{u}_2$, its terminator set must be identical to that of $\boldsymbol{u}_2$. □

We can interpret this result in terms of our old ally the suffix function, introduced in Section 5.2.1. Let s^k denote $k \ge 1$ compounded applications of the suffix function s: thus if

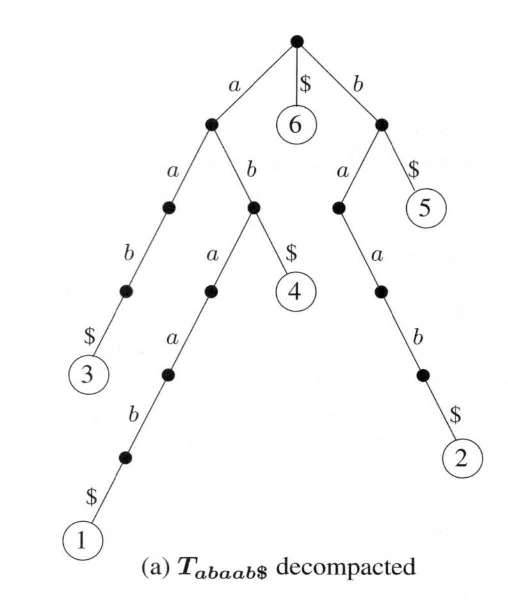

(a) $\boldsymbol{T_{abaab\$}}$ decompacted

(b) $\boldsymbol{D_{abaab\$}}$ uncompacted (7 nodes, 10 edges)

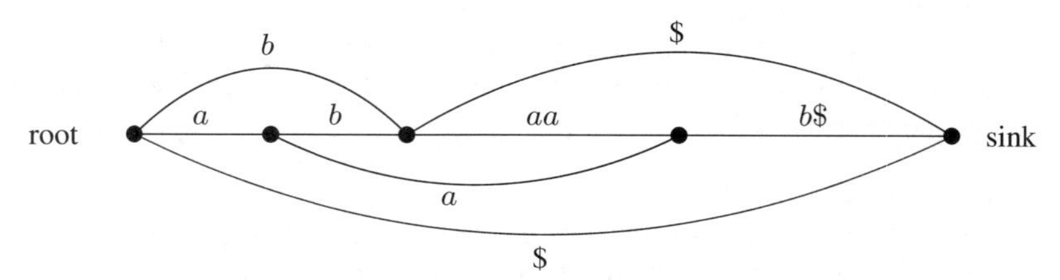

(c) $\boldsymbol{D_{abaab\$}}$ recompacted (5 nodes, 8 edges)

Figure 5.14. Transforming $T_{\boldsymbol{f_4}}$ into $D_{\boldsymbol{f_4}}$ for $\boldsymbol{f_4} = abaab\$$ (see Figure 5.4)

$s(\boldsymbol{u_3}) = \boldsymbol{u_2}$ and $s(\boldsymbol{u_2}) = \boldsymbol{u_1}$, we write $s^2(\boldsymbol{u_3}) = \boldsymbol{u_1}$. Lemma 5.3.1 therefore tells us that if $\boldsymbol{u_1}$ and $\boldsymbol{u_2}$ are Π-equivalent and $|\boldsymbol{u_1}| < |\boldsymbol{u_2}|$, it will follow that $s^k(\boldsymbol{u_2}) = \boldsymbol{u_1}$ for some k. Thus Π-equivalent substrings of $\boldsymbol{x}$ occur in "runs" determined by the suffix function:

Lemma 5.3.2 *Let $\boldsymbol{u_1}$ and $\boldsymbol{u_2}$ be two nonempty Π-equivalent substrings of $\boldsymbol{x}$ such that $|\boldsymbol{u_1}| < |\boldsymbol{u_2}|$. Then for some positive integer k, $s^k(\boldsymbol{u_2}) = \boldsymbol{u_1}$, and $\boldsymbol{u_1}$ is Π-equivalent to every $s^i(\boldsymbol{u_2})$, $i = 1, 2, \ldots, k$.*

Proof Let $\boldsymbol{u} = s^i(\boldsymbol{u_2})$ for some $i \in 1..k$. Since by Lemma 5.3.1, $\boldsymbol{u_1}$ occurs only as a suffix of $\boldsymbol{u_2}$, it follows that $\boldsymbol{u}$ also occurs only as a suffix of $\boldsymbol{u_2}$. Thus by Lemma 5.3.1, $\boldsymbol{u}$ is Π-equivalent to $\boldsymbol{u_2}$, hence to $\boldsymbol{u_1}$. $\square$

We show next that terminator sets cannot overlap; in other words, they correspond to a tree structure.

Lemma 5.3.3 *Let $\boldsymbol{u_1}$ and $\boldsymbol{u_2}$ be two substrings of $\boldsymbol{x}$ such that $|\boldsymbol{u_1}| < |\boldsymbol{u_2}|$. Then exactly one of the following statements is true:*

(a) $\Pi_{\boldsymbol{u_1}} \cap \Pi_{\boldsymbol{u_2}} = \emptyset$;

(b) $\Pi_{\boldsymbol{u_2}} \subset \Pi_{\boldsymbol{u_1}}$.

Proof The result is trivially true for $\boldsymbol{u_1} = \varepsilon$, and so we may suppose $\boldsymbol{u_1}$ to be nonempty.

Suppose $\Pi_{\boldsymbol{u_1}}$ and $\Pi_{\boldsymbol{u_2}}$ have an element in common. This element is therefore the end position of both $\boldsymbol{u_1}$ and $\boldsymbol{u_2}$, so that $\boldsymbol{u_1}$ is a suffix of $\boldsymbol{u_2}$. But then every element of $\Pi_{\boldsymbol{u_2}}$ must also be an element of $\Pi_{\boldsymbol{u_1}}$. $\square$

We are now in a position, as promised, to identify Π-equivalence with isomorphic subtrees of $T_{\boldsymbol{x}}$.

Lemma 5.3.4 *Let $\boldsymbol{u_1}$ and $\boldsymbol{u_2}$ be arbitrary nodes of $T_{\boldsymbol{x}}$. Then the subtrees $T'_{\boldsymbol{u_1}}$ and $T'_{\boldsymbol{u_2}}$ of $T_{\boldsymbol{x}}$ are isomorphic if and only if $\boldsymbol{u_1}$ and $\boldsymbol{u_2}$ are Π-equivalent.*

Proof Since the result is trivially true for $\boldsymbol{u_1} = \boldsymbol{u_2}$, we may assume that $\boldsymbol{u_1} \neq \boldsymbol{u_2}$. Since ε is Π-equivalent only to itself, we may also assume that neither node is empty.

Suppose first that $\Pi_{\boldsymbol{u_1}} = \Pi_{\boldsymbol{u_2}}$ and further, without loss of generality, that $|\boldsymbol{u_1}| \leq |\boldsymbol{u_2}|$. By Lemma 5.3.1, $\boldsymbol{u_1}$ therefore occurs only as a suffix of $\boldsymbol{u_2}$. Hence the subtree $T'_{\boldsymbol{u_2}}$ rooted at node $\boldsymbol{u_2}$ is also the subtree for the suffix $\boldsymbol{u_1}$ of $\boldsymbol{u_2}$, and so must be isomorphic to $T'_{\boldsymbol{u_1}}$.

Suppose next that $T'_{\boldsymbol{u_1}}$ and $T'_{\boldsymbol{u_2}}$ are isomorphic. Then there exists a suffix $\boldsymbol{u_1x}[j..n+1]$ of $\boldsymbol{x}$ if and only if there also exists a suffix $\boldsymbol{u_2x}[j..n+1]$. For each such integer j, $\boldsymbol{u_1}$ and $\boldsymbol{u_2}$ both terminate at the same position $j-1$ of $\boldsymbol{x}$; since neither one can terminate at any other position, it follows that $\Pi_{\boldsymbol{u_1}} = \Pi_{\boldsymbol{u_2}}$, as required. $\square$

We can now provide an intelligible definition of the ***DAWG*** $D_{\boldsymbol{x}}$ of a given string $\boldsymbol{x}$, showing how $D_{\boldsymbol{x}}$ can be constructed from a "decompacted" $T_{\boldsymbol{x}}$ as illustrated in Figure 5.14:

(1) For each run $u_1, u_2, \ldots, u_k$ of Π-equivalent nodes, merge the subtrees T'_{u_i}, $i = 1, 2, \ldots, k$, into a single subgraph. This can be done by merging corresponding nodes of each subtree.

Since as we have seen $\Pi_{x[j..n+1]} = \{n+1\}$ for every leaf node $j \in 1..n+1$, it follows that all the leaf nodes are Π-equivalent. Thus by Lemma 5.3.4 this step includes merging all the leaf nodes into a single node called the ***sink***.

In Figure 5.14 there are four sets of Π-equivalent nodes: all the leaf nodes, $\{aa, baa, abaa\}$, $\{ba, aba\}$ and $\{b, ab\}$.

(2) Orient the arcs of the graph from the root in accordance with the implied parent-child orientation in T_x. In Figure 5.14 the orientation in D_x is not shown explicitly: it is assumed to be from left to right, just as in T_x it is assumed to be from top to bottom.

An optional step (3), illustrated in Figure 5.14(c), is the "recompaction" of D_x, easily accomplished by the elimination of nodes that have both indegree 1 and outdegree 1. But it turns out that even the uncompacted D_x can be stored in linear space, and so this is the form of the DAWG that we will initially discuss.

Observe that by construction a DAWG spells out exactly the same $n+1$ suffixes spelled out by the corresponding suffix tree, and that moreover, like the suffix tree, the labels on the outarcs from each node are pairwise distinct. Thus a DAWG contains exactly $n+1$ paths from root to sink of lengths $1, 2, \ldots, n+1$, and it can be searched in exactly the same way that a suffix tree is searched. It is clear that D_x is acyclic, since all the paths from root to sink are oriented according to the parent-child orientation of T_x.

This definition of DAWG, though given in a semi-algorithmic form, is not yet an algorithm; in fact, since it depends on processing a decompacted T_x, which may require as much as $\Theta(n^2)$ storage, it would not be a very desirable one. We defer till later the question of efficient construction of D_x, first dealing instead with its space requirement.

Theorem 5.3.5 *For a given string $x = x[1..n]\$$ of length $n+1$, D_x has $N \leq 2n+1$ nodes and at most $N + n - 1 \leq 3n$ arcs.*

Proof Each node of D_x corresponds to one or more terminator sets of nodes of T_x. We have seen from Lemma 5.3.3 that these sets cannot overlap, and we have also seen that for the root node, $\Pi_\varepsilon = \{1, 2, \ldots, n+1\}$. Thus the nodes of D_x must correspond exactly to the nodes of a tree rooted at $\{1, 2, \ldots, n+1\}$ in which siblings are nonempty disjoint sets whose union is included in the parent. As proved in Exercise 5.3.4, such a tree can contain at most $2n+1$ nodes.

Now consider a spanning tree ST_x of D_x that includes the longest path labelled x from root to sink as illustrated in Figure 5.15. ST_x has N nodes and $N-1$ edges. Then suppose that edges of D_x are added to ST_x one by one. Since the addition of each such edge adds at least one new path, and since altogether there are exactly n paths to be added, it follows that the number of edges in D_x is at most $(N-1)+n$. □

It was a relatively easy task to show that a DAWG D_x occupies space proportional to the length of the string x. Although it is more difficult to devise an algorithm that constructs D_x

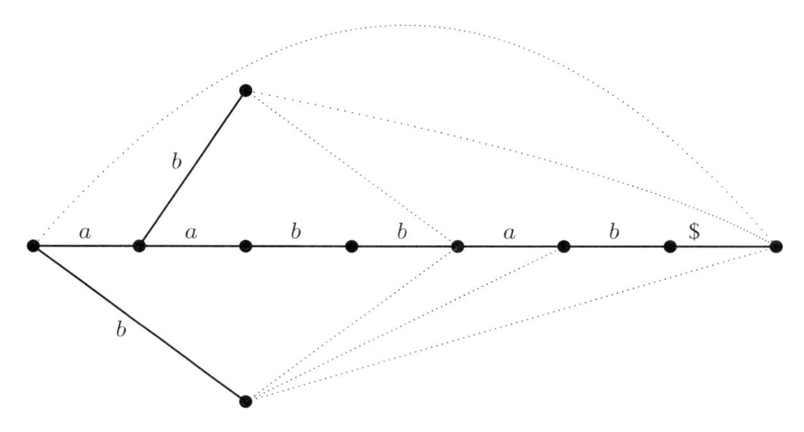

Figure 5.15. The spanning tree of $D_{aabbab\$}$ ($N = 10$, $n + 1 = 7$)

in time proportional to $|\boldsymbol{x}|$, this new task is nevertheless an interesting one: we shall discover that the suffix function s has a nice relationship, hinted at in Lemma 5.3.2, to the isomorphism of subtrees of the suffix tree $T_{\boldsymbol{x}}$, hence to the notion of Π-equivalence introduced in this section. Our algorithm will transform $T_{\boldsymbol{x}}$, of course in its original compacted form, into $D_{\boldsymbol{x}}$ by a two-stage process that can be simply stated:

(1) The first stage uses the suffix links of $T_{\boldsymbol{x}}$ to create a **D**irected **A**cyclic **G**raph or ***DAG***, not yet quite a DAWG. This process is illustrated in Figure 5.16(a)–(b) for $\boldsymbol{x} = abaab\$$.

(2) The second stage minimizes the arc weights (introduced in Section 5.2.5) on the inarcs at each node, thus yielding a DAWG. This process is illustrated in Figure 5.16(b)–(c).

The following lemma is the basis of Stage (1):

Lemma 5.3.6 *Let $\boldsymbol{u_1}$ and $\boldsymbol{u_2}$ be nodes of $T_{\boldsymbol{x}}$ such that $|\boldsymbol{u_1}| < |\boldsymbol{u_2}|$. Then the subtrees $T'_{\boldsymbol{u_1}}$ and $T'_{\boldsymbol{u_2}}$ are isomorphic if and only if*

(a) they contain the same number of leaf nodes; and

(b) $s^k(\boldsymbol{u_2}) = \boldsymbol{u_1}$ for some integer $k \geq 1$.

Proof The result is trivially true for $\boldsymbol{u_1} = \boldsymbol{\varepsilon}$, since $T'_{\boldsymbol{\varepsilon}} = T_{\boldsymbol{x}}$ is isomorphic only to itself. We assume therefore that $\boldsymbol{u_1} \neq \boldsymbol{\varepsilon}$.

Suppose that $T'_{\boldsymbol{u_1}}$ and $T'_{\boldsymbol{u_2}}$ are isomorphic. By Lemma 5.3.4 $\boldsymbol{u_1}$ and $\boldsymbol{u_2}$ are therefore Π-equivalent, and so by Lemma 5.3.2 there exists an integer k such that $s^k(\boldsymbol{u_2}) = \boldsymbol{u_1}$. It follows from Exercise 5.3.3 that $T'_{\boldsymbol{u_1}}$ and $T'_{\boldsymbol{u_2}}$ have the same number of leaf nodes. This proves sufficiency.

Suppose that conditions (a) and (b) both hold, and let $\nu_{\boldsymbol{u}}$ be the number of leaf nodes in $T'_{\boldsymbol{u}}$. Observe that $\nu_{\boldsymbol{u}}$ is just the number of occurrences of $\boldsymbol{u}$ in $\boldsymbol{x}$. Since by (b) $\boldsymbol{u_1}$ is a suffix

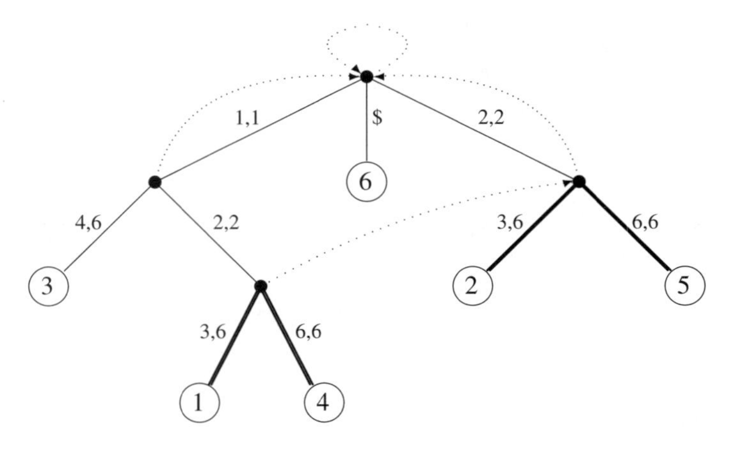

(a) $\boldsymbol{T_{abaab\$}}$ linked and compacted
(**heavy lines** for isomorphic subtrees)

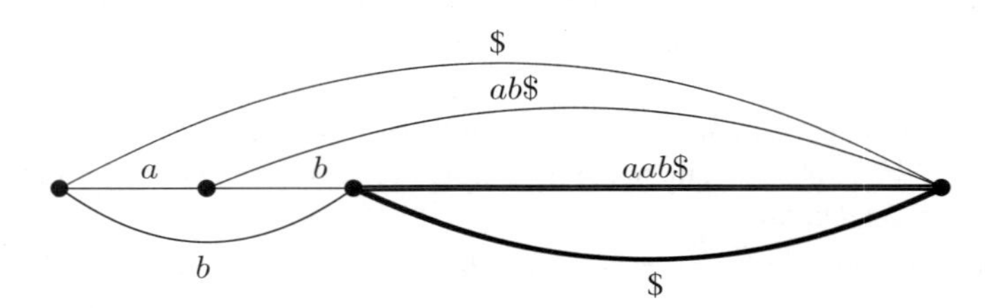

(b) Compacted DAG for $\boldsymbol{x} = \boldsymbol{abaab\$}$
(**heavy lines** for merged subtrees)

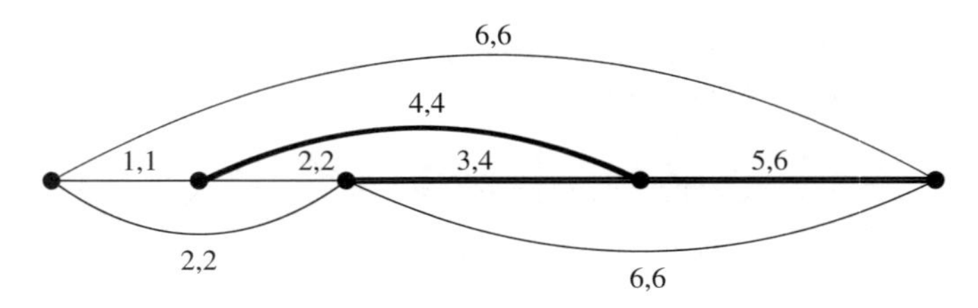

(c) Compacted $\boldsymbol{D_{abaab\$}}$
(**heavy lines** for changed arcs)

Figure 5.16. Efficiently transforming $T_{\boldsymbol{f}_4}$ into $D_{\boldsymbol{f}_4}$ (see Figure 5.14)

of $\boldsymbol{u_2}$, every leaf node in $T'_{\boldsymbol{u_2}}$ is also a leaf node in $T'_{\boldsymbol{u_1}}$. Since by (a) $\nu_{\boldsymbol{u_2}} = \nu_{\boldsymbol{u_1}}$, there can be no occurrences of $\boldsymbol{u_1}$ in $\boldsymbol{x}$ other than as a suffix of $\boldsymbol{u_2}$. Therefore by Lemma 5.3.1 $\Pi_{\boldsymbol{u_1}} = \Pi_{\boldsymbol{u_2}}$, and so by Lemma 5.3.4 $T'_{\boldsymbol{u_1}}$ and $T'_{\boldsymbol{u_2}}$ are isomorphic. □

Observe that this result holds also in the special case that $\boldsymbol{u_1}$ and $\boldsymbol{u_2}$ are leaf nodes. Thus merging isomorphic subtrees includes merging all the leaf nodes of $T_{\boldsymbol{x}}$ into the sink of $D_{\boldsymbol{x}}$. All the other isomorphic subtrees can be located simply by following suffix links in $T_{\boldsymbol{x}}$ and merging subtrees if and only if they contain the same number of leaf nodes. One way to implement this computation in time proportional to n is to first preprocess $T_{\boldsymbol{x}}$ as outlined in Algorithm 5.3.1. (This preprocessing could have been performed as a byproduct of the original construction of $T_{\boldsymbol{x}}$.)

As is evident, for example, by comparing Figure 5.14(c) with Figure 5.16(b), the compacted graph derived from Stage (1) may not be a DAWG: it may contain arcs, such as $ab\$$ in Figure 5.16(b), whose labels have not been identified as suffixes of the labels of other arcs, such as $aab\$$ in the same figure. This problem is a consequence of the fact that $T_{\boldsymbol{x}}$ has been used in its compacted form, so that certain isomorphic subtrees, such as the edge $b\$$ in Figure 5.14(a), have not been found.

As illustrated in Figure 5.17, Stage (2) locates these missing suffixes by considering the inarcs to every node $\boldsymbol{v}$ of the DAG. First an inarc $(\boldsymbol{u}, \boldsymbol{v})$ of longest label $\boldsymbol{w}$ (that is, largest weight $|\boldsymbol{w}|$) is identified, and then each of the other inarcs of label $\boldsymbol{w}'$, $2 \leq |\boldsymbol{w}'| \leq |\boldsymbol{w}|$, is considered one by one. Since each $\boldsymbol{w}'$ must be a suffix of $\boldsymbol{w}$, each such inarc can be deleted and replaced by an arc $(\boldsymbol{u}', \boldsymbol{v}')$ of unit weight, as shown in the figure.

Algorithm 5.3.1 (TD)

– *Transform a suffix tree $T_{\boldsymbol{x}}$ into a DAWG $D_{\boldsymbol{x}}$*

– *preprocessing*
traverse $T_{\boldsymbol{x}}$:
  (a) store at each node $\boldsymbol{u}$ the number of leaf nodes in $T'_{\boldsymbol{u}}$
  (b) store the source node of each chain of suffix links

– *Stage (1): Construct the DAG from $T_{\boldsymbol{x}}$*
for every stored source node $\boldsymbol{u}$ **do**
  – *explore each chain of suffix links to the root*
  while $s(\boldsymbol{u}) \neq \varepsilon$ **do**
    if $T'_{\boldsymbol{u}}$ and $T'_{s(\boldsymbol{u})}$ have the same number of leaf nodes **then**
      merge $T'_{\boldsymbol{u}}$ and $T'_{s(\boldsymbol{u})}$
    $\boldsymbol{u} \leftarrow s(\boldsymbol{u})$

– *Stage (2): Construct $D_{\boldsymbol{x}}$ from the DAG*
for every vertex $\boldsymbol{v}$ **do**
  determine an inarc $(\boldsymbol{u}, \boldsymbol{v})$ of greatest weight $|\boldsymbol{w}|$
  for every other inarc $(\boldsymbol{u}', \boldsymbol{v})$ of weight $|\boldsymbol{w}'| > 1$ **do**
    delete arc $(\boldsymbol{u}', \boldsymbol{v})$
    create arc $(\boldsymbol{u}', \boldsymbol{v}')$ – *as shown in Figure 5.17*

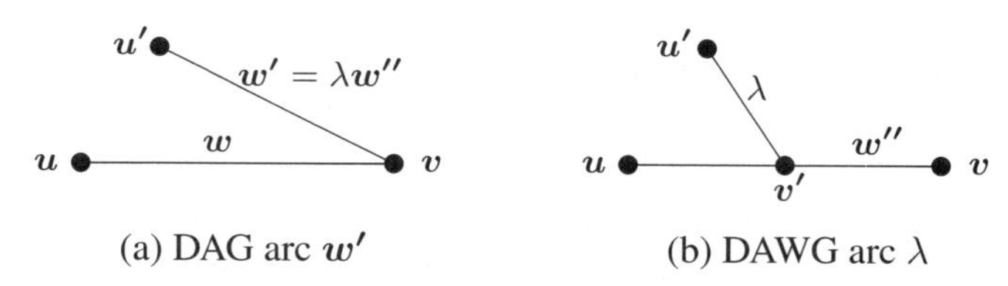

(a) DAG arc w' (b) DAWG arc λ

Figure 5.17. Transforming DAG inarcs into DAWG arcs

Based on this discussion, we state the second main result of this subsection:

Theorem 5.3.7 *Given the suffix tree $T_{\boldsymbol{x}}$ of a given string $\boldsymbol{x}$, Algorithm 5.3.1 correctly computes the DAWG $D_{\boldsymbol{x}}$ of $\boldsymbol{x}$.* □

It could be anticipated that the time requirement for Algorithm 5.3.1 would display the same dependence on alphabet size that suffix tree construction and use generally do. This turns out not to be so, however: if we suppose that the suffix tree $T_{\boldsymbol{x}}$ has already been formed, and that it incorporates provision for efficient tree traversal, the transformation of $T_{\boldsymbol{x}}$ into $D_{\boldsymbol{x}}$ can actually be performed in time linear in $|\boldsymbol{x}|$.

Theorem 5.3.8 *Algorithm 5.3.1 transforms the suffix tree $T_{\boldsymbol{x}}$ of a given string $\boldsymbol{x} = \boldsymbol{x}[1..n]$ into the corresponding DAWG $D_{\boldsymbol{x}}$ in $\Theta(n)$ time.*

Proof In Algorithm 5.3.1 the preprocessing requires that each node of $T_{\boldsymbol{x}}$ be visited: to compute the number of leaf nodes of each $T'_{\boldsymbol{u}}$, a standard postorder traversal suffices. During this traversal, a list can be formed of nodes that are either leaf nodes or pointed to by a suffix link from some other node; the nodes of $T_{\boldsymbol{x}}$ *not* in this list are the source nodes required. This processing can be performed using standard tree traversal techniques that require no searches for edge labels at each node; thus, since $T_{\boldsymbol{x}}$ has $\Theta(n)$ nodes, the time requirement of this phase is $\Theta(n)$.

In Stage (1), each node of each suffix link chain from source to root is inspected: if the number of leaf nodes stored in the preprocessing phase equals the number for the previous node in the chain, a "merger" of subtrees takes place that requires only constant time for resetting of pointers. Since the number of nodes with suffix links is $\Theta(n)$, Stage (1) also requires only $\Theta(n)$ time.

In Stage (2) constant time is spent processing every arc in the DAG; the number of these arcs is equal to the number of arcs in $D_{\boldsymbol{x}}$, shown in Theorem 5.3.5 to be $\Theta(n)$. Thus the total time spent in Stage (2) is $\Theta(n)$. □

We remark that two recent papers [CV97, IHST01] have described algorithms that compute a compact DAWG directly, without the requirement to compute an intervening suffix tree.

As recent studies indicate [K99, IHST01, HC02], the careful implementation of DAWGs can substantially reduce storage requirement below that needed by suffix trees. But perhaps just as important a reason for discussing DAWGs is the insight that they provide into suffix trees and other similar search structures, particularly in Lemmas 5.3.1–5.3.4 and 5.3.6. In the next subsection we investigate another data structure that also provides major space savings over suffix trees and that at the same time avoids any dependence on alphabet size.

5.3.2 Suffix Arrays — Saving the Best till Last?

Yet each man kills the thing he loves,
By each let this be heard,
Some do it with a bitter look,
Some with a flattering word.

— Oscar WILDE (1854–1900), *The Ballad of Reading Gaol*

As usual, we suppose a string $\boldsymbol{x} = \boldsymbol{x}[1..n]$ is given. As in all our discussion of suffix trees, we let every suffix $\boldsymbol{x}[j..n]$ of $\boldsymbol{x}$, $j = 1, 2, \ldots, n$, be denoted simply by the label j of its leaf node in $T_{\boldsymbol{x}}$. Then a ***suffix array*** (or suffix string!) $\boldsymbol{\sigma} = \boldsymbol{\sigma_x}$ is just an array of n integers $j = \boldsymbol{\sigma}[i]$ each specifying the suffix of $\boldsymbol{x}$ that occurs ith in lexicographical order among all the suffixes of $\boldsymbol{x}$ [MM90, MM93]. For example, the string

$$\begin{matrix} & {\scriptstyle 1} & {\scriptstyle 2} & {\scriptstyle 3} & {\scriptstyle 4} & {\scriptstyle 5} & {\scriptstyle 6} & {\scriptstyle 7} & {\scriptstyle 8} \\ \boldsymbol{g} = & a & b & a & a & b & a & a & b \end{matrix}$$

introduced in Section 2.1 has suffixes

6	aab	1	$abaabaab$
3	$aabaab$	8	b
7	ab	5	$baab$
4	$abaab$	2	$baabaab$

listed here in lexicographical order, so that $\boldsymbol{\sigma_g} = 63741852$. It is instructive to compare $\boldsymbol{\sigma_g}$ with the suffix tree $T_{\boldsymbol{g}}$ given in Figure 2.2: notice in particular the grouping of the leaf nodes into subtrees. This grouping reflects the fact that all suffixes of $\boldsymbol{x}$ with the same prefix occur consecutively in $\boldsymbol{\sigma}$, a very strong property that of course is important for searching: all k occurrences of a given substring $\boldsymbol{u}$ in $\boldsymbol{x}$ correspond to consecutive positions in $\boldsymbol{\sigma}$. In our example, $\boldsymbol{u} = ab$ corresponds to $\boldsymbol{\sigma}[3..5]$, while $\boldsymbol{u} = baa$ corresponds to positions $\boldsymbol{\sigma}[7..8]$.

Given $T_{\boldsymbol{x}}$, the computation of $\boldsymbol{\sigma_x}$ is straightforward: a depth-first-search (DFS) of $T_{\boldsymbol{x}}$ is performed with the downward edges from each node being chosen in lexicographical order, based on the first letter of each edge label. Provided the sentinel $ is defined to be lexicographically least, this search will visit the leaf nodes in lexicographical order. Thus

Theorem 5.3.9 *The suffix array $\boldsymbol{\sigma_x}$ of a given string $\boldsymbol{x} = \boldsymbol{x}[1..n]$ on an alphabet A of size α can be computed from the suffix tree $T_{\boldsymbol{x}}$ in $O(n \log \alpha)$ time.* □

In fact, depending on the nature of the data structure implemented at each node of $T_{\boldsymbol{x}}$, the construction of $\boldsymbol{\sigma_x}$ can in certain cases be accomplished in worst-case time $\Theta(n)$. This subject is explored further in Exercise 5.3.7. Moreover, as shown in [MM93], $\boldsymbol{\sigma_x}$ can actually be computed in $\Theta(n)$ expected time over all possible strings $\boldsymbol{x}$ without any reference to suffix trees at all; this direct approach has a relatively low space requirement with respect to suffix tree construction, with perhaps some sacrifice of processing time [MM93, La96, BS97, Sa98].

Now that we know what a suffix array is and have seen how to compute it, we devote ourselves in the rest of this section to learning how to make effective and efficient use of it for searching. As with other suffix-based structures, the fundamental problem is to locate in $\boldsymbol{x} = \boldsymbol{x}[1..n]$ the $k \geq 0$ occurrences of a given substring $\boldsymbol{u} = \boldsymbol{u}[1..m]$. We shall present three algorithms that compute all k occurrences of $\boldsymbol{u}$ in $\boldsymbol{x}$, evocatively named **SANaïve**, **SASimple** and **SAComplex**.

SANaïve and SASimple make use only of the array $\boldsymbol{\sigma}$, n integer values at $\log n$ bits per value, and have identical guaranteed upper time bounds; however, SASimple is in practice faster than SANaïve and, according to Manber and Myers [MM93], also faster in practice than SAComplex. In addition to $\boldsymbol{\sigma}$, SAComplex uses an auxiliary integer array containing up to $3n$ positions (computable as we shall see as a byproduct of the calculation of $\boldsymbol{\sigma}$) in order to be able to guarantee a *theoretical* worst-case performance very close to that of suffix trees and DAWGs — of course its performance *in practice* will generally be better. Table 5.2 summarizes the main characteristics of these algorithms, under the assumptions that the suffix array has already been computed in a preprocessing phase, and that the suffix tree used for this purpose has been deleted from computer memory.

The basic idea of all three algorithms is elementary, one that every student learns in Computer Science 101: binary search! SANaïve is in fact nothing but a binary search that uses $\boldsymbol{\sigma}$ as an index to select the Mth largest suffix in lexicographical order. As presented in Algorithm 5.3.2, it returns a nonzero value if and only if $\boldsymbol{u}$ occurs in the nonempty string $\boldsymbol{x}$: if indeed $\boldsymbol{u}$ does occur, the value returned is one of a set of values j such that $\boldsymbol{u} = \boldsymbol{x}[j..j+m-1]$. The algorithm performs $O(\log n)$ comparisons each requiring $O(m)$ time, and so can be executed in time $O(m \log n)$. By virtue of the fact, mentioned above, that all k occurrences of $\boldsymbol{u}$ in $\boldsymbol{x}$ will be indexed by neighbouring positions in $\boldsymbol{\sigma}$, it is only necessary to inspect adjacent positions in the $[L, R]$ interval surrounding M in order to locate the remaining $k-1$ occurrences. This requires an additional $O(mk)$ time, and so the time bound given in Table 5.2 is correct.

The author offers his apology to the reader whose sensibilities are rightly offended by the annoyingly opaque and inefficient formulation of binary search given here. His excuse is a good intention: to ease the transition to the presentation of SASimple and SAComplex.

Algorithm	**Additional storage (bytes)**	**Theoretical time bound**
SANaïve	$n \log n/8$	$O\big(m(\log n + k)\big)$
SASimple	$n \log n/8$	$O\big(m(\log n + k)\big)$
SAComplex	$4n \log n/8$	$O(m + \log n + k)$

Table 5.2: Suffix array algorithms for k occurrences of $\boldsymbol{u}$ in $\boldsymbol{x}$

Algorithm 5.3.2 (SANaïve)

– *Naïvely use a suffix array to locate* $\boldsymbol{u}$ *in* $\boldsymbol{x}$

$j \leftarrow 0;\ L \leftarrow 1;\ R \leftarrow n$
repeat
  $M \leftarrow \lceil (R+L)/2 \rceil$
  if $\boldsymbol{u} = \boldsymbol{x}\big[\boldsymbol{\sigma}[M]..\boldsymbol{\sigma}[M]+m-1\big]$ **then**
    $j \leftarrow \boldsymbol{\sigma}[M]$
  elsif $\boldsymbol{u} > \boldsymbol{x}\big[\boldsymbol{\sigma}[M]..\boldsymbol{\sigma}[M]+m-1\big]$ **then**
    $L \leftarrow M$
  else
    $R \leftarrow M$
until $L = M$ or $j \neq 0$

Improvement of Algorithm 5.3.2 depends essentially on devising techniques to improve the efficiency of the two **if** statements. Exercise 5.3.8 suggests an immediately obvious improvement. The basis of more significant improvement is the observation, already made above in a slightly different context, that if suffixes $\boldsymbol{\sigma}[L]$ and $\boldsymbol{\sigma}[R]$ both have a certain prefix, then so does every $\boldsymbol{\sigma}[K]$, $K = L, L+1, \ldots, R$. Let us call this the ***cluster property***. To make use of this property, we return to another familiar idea, the LCP introduced in Section 5.2.1. Adopting the notation lcp $= |\text{LCP}|$, we modify Algorithm 5.3.2 so as to have $P_L = \text{lcp}\big(\boldsymbol{u}, \boldsymbol{\sigma}[L]\big)$ and $P_R = \text{lcp}\big(\boldsymbol{u}, \boldsymbol{\sigma}[R]\big)$ available at the beginning of each iteration. Let $P = \min\{P_L, P_R\}$. By the cluster property, comparisons of $\boldsymbol{u}$ with prefixes $\boldsymbol{\sigma}[K]$ need only deal with $\boldsymbol{u}[P+1..m]$, since the prefix $\boldsymbol{u}[1..P]$ must be common across the entire interval. We are thus led to formulate Algorithm 5.3.3, shown below.

Algorithm 5.3.3 (SASimple)

– *Simply use a suffix array to locate* $\boldsymbol{u}$ *in* $\boldsymbol{x}$

$j \leftarrow 0;\ L \leftarrow 1;\ R \leftarrow n$
$P_L \leftarrow 0;\ P_R \leftarrow 0$
repeat
  $P \leftarrow \min\{P_L, P_R\}$
  $M \leftarrow \lceil (R+L)/2 \rceil$
  – *Compute* $\text{lcp}\big(\boldsymbol{u}, \boldsymbol{\sigma}[M]\big)$
  $P_M \leftarrow P + \text{lcp}\Big(\boldsymbol{u}[P+1..m], \boldsymbol{x}\big[\boldsymbol{\sigma}[M]+P..n\big]\Big)$
  if $P_M = m$ **then**
    $j \leftarrow \boldsymbol{\sigma}[M]$
  elsif $\boldsymbol{u}[P_M+1] > \boldsymbol{x}\big[\boldsymbol{\sigma}[M]+P_M\big]$ **then**
    $L \leftarrow M;\ P_L \leftarrow P_M$
  else
    $R \leftarrow M;\ P_R \leftarrow P_M$
until $L = M$ or $j \neq 0$

Since every instruction in this algorithm except the lcp calculation requires only constant time, the upper bound on time complexity given in Table 5.2 is again correct. It is not difficult to find examples of strings $\boldsymbol{x}$ and $\boldsymbol{u}$ for which this upper bound is also a lower bound, so that SASimple is in the worst case not a theoretical improvement on SANaïve. In practice, however, since P is monotone nondecreasing, the letter comparisons required are greatly reduced in the average-case from those required by SANaïve: SASimple is perhaps the fastest sequential algorithm known for computing occurrences of a given substring $\boldsymbol{u}$ in a given string $\boldsymbol{x}$.

We turn finally to Algorithm SAComplex. The key idea of this algorithm is to identify in advance all the possible nontrivial $[L, R]$ search intervals, $L < R$, that can arise during the binary search, and to precompute $\text{lcp}(\boldsymbol{\sigma}[L], \boldsymbol{\sigma}[R])$ for each one of them. As we shall soon discover, there are altogether only $2n - 3$ possible search intervals, for each of which the lcp value can be precomputed into a new array (or string) $\boldsymbol{\pi} = \boldsymbol{\pi}_{\boldsymbol{x}}$ of maximum dimension $3n - 8$, $n \geq 5$. We shall also see that $\boldsymbol{\pi}$ can be computed as a byproduct of the original computation of the suffix array from $T_{\boldsymbol{x}}$. Thus at an additional cost in time and space that is essentially only linear in $|\boldsymbol{x}|$, we shall be able to determine $\text{lcp}(\boldsymbol{\sigma}[L], \boldsymbol{\sigma}[R])$ in constant time simply by accessing the appropriate position in $\boldsymbol{\pi}_{\boldsymbol{x}}$. It turns out that this knowledge can be used to reduce the total time for the binary search to $O(m + \log n)$.

The easiest of the above statements to justify is that the number of search intervals is exactly $2n - 3$: this is illustrated for $n = 7$ in Figure 5.18, where each of 11 nodes corresponds to a search interval, and left for the reader to prove in general in Exercise 5.3.12. In the figure the labels within the nodes represent positions in the lcp array $\boldsymbol{\pi}$; these are selected according to the "heap rule", so that the interval whose lcp is stored in $\boldsymbol{\pi}[i]$ breaks down into two subintervals whose lcp is stored in $\boldsymbol{\pi}[2i]$ and $\boldsymbol{\pi}[2i + 1]$. The reader will not find it difficult to verify that, based on this storage scheme, the maximum dimension required for $\boldsymbol{\pi}$ is $3n - 8$ for every $n \geq 5$.

Next, assuming that $\boldsymbol{\pi}$ has been computed, we describe the execution of SAComplex, focussing on a single step that corresponds to an interval $[L, R]$. We suppose as in SASimple

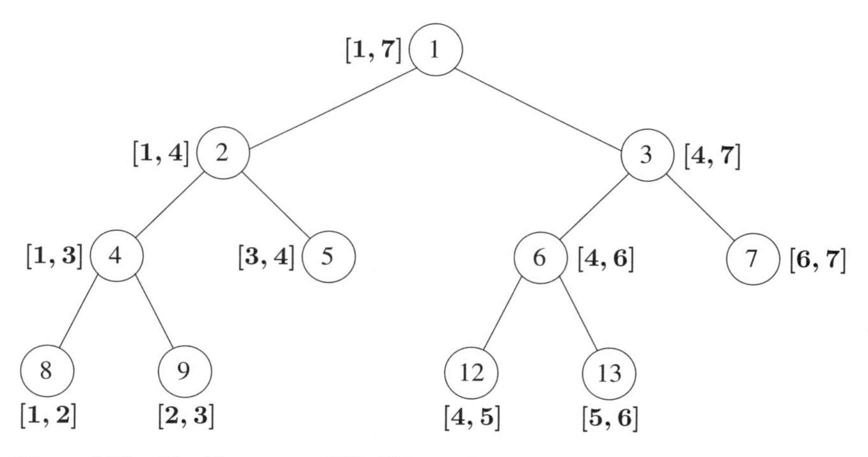

Figure 5.18. The binary tree of $[L, R]$ intervals, $n = 7$

that P_L and P_R are available, and that M is computed in the same way. We suppose also that a value i is determined such that

$$P_{LR} \equiv \text{lcp}\big(\boldsymbol{\sigma}[L], \boldsymbol{\sigma}[R]\big) = \boldsymbol{\pi}[i].$$

All of these values (L, R, P_L, P_R, i) are known for the first step; we show that they can be computed for the next step of the algorithm in time proportional to the increase in $\max\{P_L, P_R\}$ plus a constant.

First observe that if $P_L = P_R$, we can compute P_M exactly as in SASimple using $P = P_L$; the lcp calculation requires $P_M - P_L + 1$ letter comparisons and the increase in $\max\{P_L, P_R\}$ in the next iteration will be exactly $P_M - P_L$. As noted earlier, all instructions other than the lcp require only constant time.

Now we suppose that $P_L \neq P_R$. For the purpose of analysis, we suppose further that $P_L > P_R$; the argument in the case $P_R > P_L$ is entirely symmetrical. Three cases arise depending on the value of $P_{LM} = \boldsymbol{\pi}[2i]$:

(1) $P_{LM} > P_L$
For every $K \in L..M$, $\boldsymbol{\sigma}[K]$ has a prefix of length $P_{LM} > P_L = \text{lcp}\big(\boldsymbol{u}, \boldsymbol{\sigma}[L]\big)$. Therefore, by the cluster property, $\boldsymbol{u}$ cannot occur in the interval $[L, M]$, but may possibly occur in $[M, R]$. Thus no further tests are necessary: since already $\boldsymbol{\sigma}[M]$ has a prefix of length P_L, the substitutions

$$L \leftarrow M, \;\; i \leftarrow 2i+1$$

suffice to define the parameters required for the next iteration with the value of $\max\{P_L, P_R\}$ unchanged.

(2) $P_{LM} < P_L$
Here it must be true that $\text{lcp}(\boldsymbol{u}, \boldsymbol{\sigma}[L]) > \text{lcp}(\boldsymbol{u}, \boldsymbol{\sigma}[M])$. Suppose that $\boldsymbol{u}$ occurs at some position K in the interval $[M, R]$. Then $\boldsymbol{\sigma}[K]$ has a prefix of length P_L, as does $\boldsymbol{\sigma}[L]$, while at an intermediate position $M \in L..K$, $\boldsymbol{\sigma}[M]$ does not have a prefix of length P_L, a violation of the cluster property. We conclude that $\boldsymbol{u}$ cannot lie in the interval $[M, R]$ and so, if it occurs at all, it must lie in the interval $[L, M]$. Thus the substitutions

$$R \leftarrow M, \;\; P_R \leftarrow P_{LM}, \;\; i \leftarrow 2i$$

define the parameters for the next iteration with the value of $\max\{P_L, P_R\}$ again unchanged.

(3) $P_{LM} = P_L$
In this case $\text{lcp}(\boldsymbol{u}, \boldsymbol{\sigma}[L]) = \text{lcp}(\boldsymbol{u}, \boldsymbol{\sigma}[M])$, and so we may once more use $P = P_L$ to compute P_M exactly as in SASimple, and then recompute either (L, P_L, i) or (R, P_R, i) in constant time.

We have thus shown that each iteration of SAComplex computes the five parameters (L, R, P_L, P_R, i) of the next iteration in time proportional to the increase in $\max\{P_L, P_R\}$ plus a constant. Since the quantity $\max\{P_L, P_R\}$ is monotone nondecreasing and since the interval size decreases monotonely to a final interval $[L, L+1]$, the algorithm locates an occurrence of $\boldsymbol{u}$, if it exists, either at some position M or in the final interval. Thus, using a somewhat modified termination condition that deals appropriately with the final interval, Algorithm SAComplex does indeed compute an occurrence of $\boldsymbol{u}$ in $\boldsymbol{x}$ in $O(m + \log n)$ time, as predicted in Table 5.2. We leave the specification of SAComplex in algorithmic form to Exercise 5.3.16. Exercise 5.3.17 demonstrates that all k occurrences of $\boldsymbol{u}$ in $\boldsymbol{x}$ can be determined in an additional $\Theta(k)$ time using the $\boldsymbol{\pi}$ array.

Our one remaining task in this subsection is to show how to compute the $\boldsymbol{\pi}$ array as a byproduct of the computation of $\boldsymbol{\sigma}$. This is a simple two-stage process:

(1) During the lexicographical DFS of $T_{\boldsymbol{x}}$, the path from suffix $\boldsymbol{\sigma}[i]$ to $\boldsymbol{\sigma}[i+1]$ visits a node $\boldsymbol{v}$ that is the root of a minimum size subtree containing both $\boldsymbol{\sigma}[i]$ and $\boldsymbol{\sigma}[i+1]$. As remarked in Section 5.2.1,

$$\text{lcp}\big(\boldsymbol{\sigma}[i], \boldsymbol{\sigma}[i+1]\big) = |\boldsymbol{v}|.$$

Thus for every $i = 1, 2, \ldots, n-1$, the lcp for interval $[i, i+1]$ can be computed as a byproduct of the DFS; since as shown in Exercise 5.3.14 the position j in $\boldsymbol{\pi}$ corresponding to the interval $[i, i+1]$ can also be computed, the lcps for these intervals are stored using assignments of the form

$$\boldsymbol{\pi}[j] \leftarrow |\boldsymbol{v}|.$$

(2) It is shown in Exercise 5.3.10 that for any interval $[L, R]$, $R \geq L$,

$$lcp\big(\boldsymbol{\sigma}[L], \boldsymbol{\sigma}[R]\big) = \min\Big\{lcp\big(\boldsymbol{\sigma}[L], \boldsymbol{\sigma}[M]\big),\ lcp\big(\boldsymbol{\sigma}[M], \boldsymbol{\sigma}[R]\big)\Big\}, \tag{5.10}$$

for any $M \in L..R$. Thus by processing the interval tree from leaf to root (that is, processing the corresponding locations in $\boldsymbol{\pi}$), and applying this relationship, all the entries in $\boldsymbol{\pi}$ can be computed in $O(n)$ time.

This concludes our introduction to suffix arrays, perhaps the most attractive of the suffix-based structures: efficient processing is provided at a relatively low cost in terms of space, and without direct dependence on alphabet size. There may remain, however, an indirect dependence, since the algorithms we have presented for computation of the arrays $\boldsymbol{\sigma}$ and $\boldsymbol{\pi}$ depend on the prior computation of the suffix tree. But as noted earlier, the use of suffix trees can be avoided altogether, at some cost in processing time [MM93]. In fact, several more recent algorithms applicable to direct suffix array construction [La96, BS97, Sa98] appear to yield results quite competitive in execution time with suffix tree construction using McCreight's algorithm [Sa98].

In Section 13.2 we make use of suffix arrays again to provide an efficient solution of quite a different problem: computation of all the nonextendible repeating substrings in $\boldsymbol{x}$.

Exercises 5.3

1. Prove that Π-equivalence is an equivalence relation on the substrings of x.
2. If v is a substring of x, is the subtree T'_v of T_x rooted at node v necessarily a suffix tree?
3. If u_1 and u_2 are Π-equivalent with $|u_1| < |u_2|$, show that j is a label of a leaf node in T'_{u_1} if and only if $j + |u_2| - |u_1|$ is a label of a leaf node in T'_{u_2}.
4. Let T be a tree whose nodes are nonempty sets of positive integers. Suppose that the root node of T is the set $\{1, 2, \ldots, n\}$, and suppose that the children of every node are disjoint sets whose union is included in the parent. Show that T can contain at most $2n - 1$ nodes.
5. Prove based on the original definition of DAWG that at most one inarc to any given node of a DAWG has weight greater than one.
6. The preprocessing stage of Algorithm TD "traverses" T_x in order to perform certain computations. Explain carefully how this traversal could be implemented so as to avoid dependence on alphabet size and therefore how the required computations can be performed in time proportional only to $|x|$.
7. In the text the statement is made that the DFS search of T_x that creates σ_x can "in certain cases" be performed in time proportional only to n. Explain what these cases are, and describe the corresponding processing.
8. Algorithm SANaïve as shown could require duplicate testing of positions of u in the case that it shares a common prefix with leaf node $\sigma[M]$ of x. Revise the algorithm so as to avoid this duplication.
9. Explain the author's disparaging remarks about Algorithm 5.3.2, and explain what changes should be made to yield a better general-purpose binary search.
10. Prove that for any two positions L and R in x and any string u,

$$\min\Big\{|u|, \mathrm{lcp}\big(\sigma[L], \sigma[R]\big)\Big\} = \min\{P_L, P_R\}.$$

11. Give examples of classes of strings x and u for which Algorithm SASimple executes in $\Omega(m \log n)$ time.
12. Suppose that a binary search of an array of length n is carried out using Algorithm 5.3.2. Show that

    (a) for every $n \geq 2$, the number of possible $[L, R]$ intervals formed, $L < R$, is exactly $2n - 3$;

    (b) for every $n \geq 5$, the maximum dimension of the π array, if constructed as described in the text, is $3n - 8$.

    Characterize those values of n for which the maximum dimension given in (b) is attained.

13. Compute π for $\boldsymbol{g} = abaabaab$.

14. Writing $n = 2^k + h$ for integers $k \geq 1$ and $h \in 0..2^k - 1$, express the exact dimension of the array π as a function of k and h. Then show how to locate in π the position corresponding to the interval $[L, L+1]$, $L \in 1..n-1$.

15. Instead of storing the array π with up to $3n - 8$ elements, a binary search tree with only $2n - 3$ nodes could be used instead. Discuss these options and explain which one is preferable.

16. Based on the discussion in the text, write out SAComplex in algorithmic form.

17. Given that one occurrence of $\boldsymbol{u}$ in $\boldsymbol{x}$ has already been found using SAComplex, show how any other occurrences of $\boldsymbol{u}$, if any, can be located in time proportional to the number of occurrences.

Hint: Make use of the cluster property and the fact that π contains positions corresponding to all intervals $[L, L+1]$, $L = 1, 2, \ldots, n-1$.

18. Prove Equation (5.10).

19. Write out Stage (2) of the algorithm to compute π, then show that it executes in time $\Theta(n)$.

20. Exercise 5.2.18 asked for estimates of space required by suffix trees for strings of length $n = 100K, 1M, 5M$ in the case $\alpha = 4$. Using Table 5.2 provide the same estimates for suffix arrays and work out the ratios between these estimates and the storage required only for $\boldsymbol{x}$. What conclusions can be reached? What is the effect of halving or doubling the alphabet size on the relative desirability of suffix trees and suffix arrays?

Chapter 6

Decomposing a String

The difference between the almost right word and the right word is really a large matter — it's the difference between the lightning bug and the lightning.

— Samuel CLEMENS (Mark TWAIN) (1835–1910)
letter to George Bainton, 15 October 1888

In Section 1.4 we learned about Lyndon words and some of their properties, and we proved that Lyndon words form the basis of an intrinsic pattern called Lyndon decomposition that is unique for every string. What we did not learn was how to efficiently compute the Lyndon decomposition

$$\boldsymbol{w}_1^{q_1}\boldsymbol{w}_2^{q_2}\cdots\boldsymbol{w}_k^{q_k}$$

of a given string $\boldsymbol{x}$. In Section 6.1 we present two elegant linear-time decomposition algorithms due to Duval [D83], followed in Section 6.2 by two interesting applications. One further application has already been discussed in Section 1.4: a Lyndon decomposition of $\boldsymbol{x}$ immediately gives rise to the canonical form of the corresponding necklace $C(\boldsymbol{x})$, thus permitting efficient solution of a problem that arises frequently in computational geometry. We shall briefly discuss in Section 6.2 an even more efficient solution of this problem.

We go on in Section 6.3 to describe another, completely different, decomposition of $\boldsymbol{x}$ called the s-factorization [LZ76], also computable in linear time. Originally employed as a technique for data compression [ZL77], the s-factorization turns out to have important application to computing the repetitions in $\boldsymbol{x}$ (Section 12.2).

Throughout this chapter, as throughout the previous chapter, the alphabet A will be assumed to be ordered. For Lyndon decomposition, we will find in addition that all algorithms are independent of alphabet size: Lyndon decomposition of a string on an unbounded alphabet raises no more difficulties than on an alphabet of size two. This property does not however hold for s-factorization, which as we shall see depends on the use of suffix trees for its efficient calculation.

6.1 Lyndon Decomposition: Duval's Algorithm

And naked to the hangman's noose
The morning clocks will ring
A neck God made for other use
Than strangling in a string.

— A. E. HOUSMAN (1859–1936), *A Shropshire Lad IX*

We first describe Dvl, a linear-time algorithm that processes $\boldsymbol{x} = \boldsymbol{x}[1..n]$ letter-by-letter from left to right with a bounded amount of backtracking. We then go on to describe Dvl*, a simple modification of Dvl that reduces backtracking at the cost of $\Theta(n)$ additional storage. These algorithms are based on an interesting connection between two apparently quite dissimilar ideas: normal form and Lyndon words. In order to give an overview of Dvl, some notation will be useful:

- Let L be the set of all Lyndon words.
- In honour of Chen, Fox and Lyndon [CFL58], let $\text{CFL}(\boldsymbol{x})$ be the Lyndon decomposition of $\boldsymbol{x}$.
- Let $\text{NF}(\boldsymbol{x})$ denote the normal form $\boldsymbol{u}^r\boldsymbol{u'}$ of $\boldsymbol{x}$ introduced in Section 1.2, where $\boldsymbol{u}$ is the generator of $\boldsymbol{x}$ of minimum-length, $\boldsymbol{u'}$ a proper prefix of $\boldsymbol{u}$, and $r \geq 1$. We may then say that $\boldsymbol{u}$ ***generates*** $\boldsymbol{x}$, and we let L^* denote the set of all strings generated by Lyndon words.

Recalling from Lemma 1.4.5 that every Lyndon word is primitive, hence generated by itself with $r = 1$ and $\boldsymbol{u'} = \varepsilon$, we see that $L \subset L^*$. Of course not every string with a primitive generator is an element of L^*; for example, $\text{NF}(baaba) = (baa)(ba)$ with primitive generator baa, but since $baa \notin L$, therefore $baaba \notin L^*$. On the other hand, by Theorem 1.4.2, we know that every string that is not a repetition has a unique rotation that is a Lyndon word; thus $aab = R_1(baa) \in L$, so that for example $(aab)^r aa \in L^*$ for every $r \geq 1$.

At any point in the execution of Dvl, the given string $\boldsymbol{x}$ is a concatenation $\boldsymbol{x_1x_2x_3}$ of three factors, where

- $\boldsymbol{x_1}$ is a prefix of $\boldsymbol{x}$ for which $\text{CFL}(\boldsymbol{x_1})$ has already been computed;
- $\boldsymbol{x_2}$ is a substring of $\boldsymbol{x}$ whose positions have already been visited by Dvl, but whose CFL has not yet been computed;
- $\boldsymbol{x_3}$ is a suffix of $\boldsymbol{x}$ that has not yet been processed.

Thus at each step of Dvl, $\boldsymbol{x_1x_2}$ is increased on the right by a single letter, say μ, that is at the same time removed from the left of $\boldsymbol{x_3}$. From step to step, Dvl maintains the invariant that $\boldsymbol{x_2} \in L^*$: this is easy if it happens that $\boldsymbol{x_2}\mu \in L^*$; if not, however, then $\boldsymbol{x_2}\mu$ is separated into factors $\boldsymbol{x'_1x'_2}$ such that $\text{CFL}(\boldsymbol{x'_1})$ can be computed and $\boldsymbol{x'_2} \in L^*$. In this latter case, the new prefix of $\boldsymbol{x}$ for which CFL has been computed becomes $\boldsymbol{x_1x'_1}$.

We now provide more details about Algorithm Dvl together with a justification of it. The basis of this justification resides in the following three lemmas, themselves based on the three Lyndon word lemmas of Section 1.4. The first of these lemmas tells us that simple tests of the next letter μ permit a classification of the substring $\boldsymbol{x_2}\mu$ to be made.

Lemma 6.1.1 *Suppose that $\boldsymbol{x_2} \in L^*$. Specifically, let* $\text{NF}(\boldsymbol{x_2}) = \boldsymbol{w}^r\boldsymbol{w}'$ *for some $\boldsymbol{w} \in L$, and write $\boldsymbol{w} = \boldsymbol{w}'\lambda\boldsymbol{w}''$ where $\lambda \in A$. Then for $\mu \in A$:*

(a) $\mu > \lambda \ \Rightarrow\ \boldsymbol{x_2}\mu \in L;$

(b) $\mu = \lambda \ \Rightarrow\ \boldsymbol{x_2}\mu \in L^* - L;$

(c) $\mu < \lambda \ \Rightarrow\ \boldsymbol{x_2}\mu \notin L^*.$

Proof To prove (a), we show first that $\boldsymbol{w}'\mu \in L$. This is true by definition when $\boldsymbol{w}' = \varepsilon$. Then for nonempty $\boldsymbol{w}'$, since $\boldsymbol{w} \in L$, Lemma 1.4.6 implies that

$$\boldsymbol{w} = \boldsymbol{w}'\lambda\boldsymbol{w}'' < \text{suff}(\boldsymbol{w}')\lambda\boldsymbol{w}''$$

for every nonempty proper suffix $\text{suff}(\boldsymbol{w}')\lambda$ of $\boldsymbol{w}'\lambda$. Since $\lambda\boldsymbol{w}'' < \mu$,

$$\boldsymbol{w}'\lambda\boldsymbol{w}'' < \text{suff}(\boldsymbol{w}')\lambda\boldsymbol{w}'' < \text{suff}(\boldsymbol{w}')\mu,$$

so that $\boldsymbol{w}' < \text{suff}(\boldsymbol{w}')\mu$. Therefore, for every nonempty proper suffix $\text{suff}(\boldsymbol{w}')\mu$ of $\boldsymbol{w}'\mu$,

$$\boldsymbol{w}'\mu < \text{suff}(\boldsymbol{w}')\mu$$

and so by Lemma 1.4.6, $\boldsymbol{w}'\mu \in L$.

Now induction based on Lemma 1.4.7 establishes (a): since $\boldsymbol{w} \in L$ and $\boldsymbol{w}'\mu \in L$ with $\boldsymbol{w} < \boldsymbol{w}'\mu$, it follows that $\boldsymbol{w}\boldsymbol{w}'\mu \in L$; in general, if $\boldsymbol{w} \in L$ and $\boldsymbol{w}^s\boldsymbol{w}'\mu \in L$ for any nonnegative integer s, the fact that $\boldsymbol{w} < \boldsymbol{w}^s\boldsymbol{w}'\mu$ implies $\boldsymbol{w}^{s+1}\boldsymbol{w}'\mu \in L$.

To prove (b), we make the observation that $\boldsymbol{x_2}\mu = \boldsymbol{x_2}\lambda$ has nonempty border $\boldsymbol{w}'\lambda$; thus by Lemma 1.4.5 $\boldsymbol{x_2}\mu \notin L$. However, $\boldsymbol{x_2}\mu = \boldsymbol{w}^r\boldsymbol{w}'\lambda$ is certainly generated by $\boldsymbol{w}$, and so $\boldsymbol{x_2}\mu \in L^*$.

To prove (c), first observe that

$$\lambda > \mu \ \Rightarrow\ \boldsymbol{w}'\lambda > \boldsymbol{w}'\mu \ \Rightarrow\ \boldsymbol{x_2}\mu > \boldsymbol{w}'\mu,$$

so that $\boldsymbol{x_2}\mu$ is lexicographically greater than a proper prefix of itself; thus by Lemma 1.4.6, $\boldsymbol{x_2}\mu \notin L$. Suppose therefore that $\boldsymbol{x_2}\mu$ is generated by $\overline{\boldsymbol{w}} \in L$. Since $\boldsymbol{x_2}\mu \notin L$, $\overline{\boldsymbol{w}}$ must be a proper prefix of $\boldsymbol{x_2}\mu$; that is, a prefix of $\boldsymbol{x_2}$. But every prefix of $\boldsymbol{x_2}$ of length greater than $\boldsymbol{w}$ is not primitive; thus by Lemma 1.4.5, $|\overline{\boldsymbol{w}}| \le |\boldsymbol{w}|$. Further, it cannot be true that $|\overline{\boldsymbol{w}}| < |\boldsymbol{w}|$, since in that case $\overline{\boldsymbol{w}}$ would also generate $\boldsymbol{x_2}$, in contradiction to the NF requirement that $\boldsymbol{w}$ be the minimum-length generator of $\boldsymbol{x_2}$. The only remaining possibility is that $\overline{\boldsymbol{w}} = \boldsymbol{w}$, also false by virtue of the fact that, since $\mu \neq \lambda$, $\boldsymbol{w}'\mu$ is not a prefix of $\boldsymbol{w}$. We conclude that the generator of $\boldsymbol{x_2}\mu$ is not in L, so that $\boldsymbol{x_2}\mu \notin L^*$. □

To enhance our understanding of this result, consider the example $\boldsymbol{x_2} = (aabb)^2aab$ with $\boldsymbol{w} = aabb \in L$, $r = 2$, $\boldsymbol{w}' = aab$, and $\lambda = b$. If the next letter $\mu = c > \lambda$, then the string $\boldsymbol{x_2}c$ will surely be lexicographically less than any rotation of itself, so that $\boldsymbol{x_2}c \in L$, as predicted in case (a) of the lemma. For $\mu = b = \lambda$, $\boldsymbol{x_2}b = (aabb)^3$, clearly in $L^* - L$ as predicted by case (b). And for $\mu = a < \lambda$, $\text{NF}(\boldsymbol{x_2}a) = (aabbaabbaab)a$; since the generator $(aabb)^2aab$ is not primitive, $\boldsymbol{x_2}a \notin L^*$, in accordance with case (c).

The next two lemmas show how $\text{CFL}(\boldsymbol{x})$ can be built up a Lyndon word at a time: the first lemma is a general result that is then applied in the second to show how the CFL of a prefix $\boldsymbol{w}^* \in L^*$ can be computed.

Lemma 6.1.2 *Let $\boldsymbol{x} = \boldsymbol{w}\boldsymbol{x}'$ where $\boldsymbol{w}$ is the longest prefix of $\boldsymbol{x}$ that is a Lyndon word. Then*

$$\text{CFL}(\boldsymbol{x}) = \boldsymbol{w}\text{CFL}(\boldsymbol{x}').$$

Proof Let $\text{CFL}(\boldsymbol{x}') = \boldsymbol{w}'_1\boldsymbol{w}'_2\cdots\boldsymbol{w}'_k$. Since $\boldsymbol{w}$ is longest, $\boldsymbol{w}\boldsymbol{w}'_1 \notin L$. Therefore by Lemma 1.4.7, $\boldsymbol{w} \geq \boldsymbol{w}'_1$, and so $\boldsymbol{w}\boldsymbol{w}'_1\boldsymbol{w}'_2\cdots\boldsymbol{w}'_k$ is by definition a Lyndon decomposition of $\boldsymbol{x}$, unique by Theorem 1.4.9. Then $\text{CFL}(\boldsymbol{x}) = \boldsymbol{w}\text{CFL}(\boldsymbol{x}')$. □

The next result shows how components of $\text{CFL}(\boldsymbol{x})$ are computed in the case that the next letter μ is "small" — case (c) of Lemma 6.1.1.

Lemma 6.1.3 *Suppose $\boldsymbol{x} = \boldsymbol{x_2}\mu\boldsymbol{x}'$ with $\boldsymbol{x_2} \in L^*$ and $\mu \in A$. Let $\text{NF}(\boldsymbol{x_2}) = \boldsymbol{w}^r\boldsymbol{w}'$ with $\boldsymbol{w} = \boldsymbol{w}'\lambda\boldsymbol{w}''$ for $\lambda \in A$. If $\mu < \lambda$, then*

$$\text{CFL}(\boldsymbol{x}) = \boldsymbol{w}^r\text{CFL}(\boldsymbol{w}'\mu\boldsymbol{x}').$$

Proof Observe that since $\boldsymbol{x_2} \in L^*$, $\boldsymbol{w} \in L$. We show first that $\boldsymbol{w}$ is the longest prefix of $\boldsymbol{x}$ that is a Lyndon word. Suppose that there exists a prefix $\overline{\boldsymbol{w}} \in L$ of $\boldsymbol{x}$ such that $|\overline{\boldsymbol{w}}| > |\boldsymbol{w}|$. But if $|\overline{\boldsymbol{w}}| \leq |\boldsymbol{x_2}|$, $\overline{\boldsymbol{w}}$ has a nonempty border, hence is not primitive, hence by Lemma 1.4.5 is not a Lyndon word. Thus $|\overline{\boldsymbol{w}}| > |\boldsymbol{x_2}|$, so that $\overline{\boldsymbol{w}} = \boldsymbol{x_2}\mu\overline{\boldsymbol{x}}$ for some possibly empty prefix $\overline{\boldsymbol{x}}$ of $\boldsymbol{x}'$. Then $\overline{\boldsymbol{w}}$ has the suffix $\boldsymbol{w}'\mu\overline{\boldsymbol{x}} < \boldsymbol{w}'\lambda$, where $\boldsymbol{w}'\lambda$ is a prefix of $\overline{\boldsymbol{w}}$. Therefore

$$\overline{\boldsymbol{w}} \geq \boldsymbol{w}'\lambda > \boldsymbol{w}'\mu\overline{\boldsymbol{x}} = \text{suff}(\overline{\boldsymbol{w}}),$$

in contradiction to Lemma 1.4.6. We conclude that there exists no prefix $\overline{\boldsymbol{w}} \in L$ such that $|\overline{\boldsymbol{w}}| > |\boldsymbol{w}|$.

Now we can apply Lemma 6.1.2 to $\boldsymbol{x} = \boldsymbol{w}^r\boldsymbol{w}'\mu\boldsymbol{x}'$, yielding

$$\text{CFL}(\boldsymbol{x}) = \boldsymbol{w}\text{CFL}\big(\boldsymbol{w}^{r-1}\boldsymbol{w}'\mu\boldsymbol{x}'\big),$$

where either $r - 1 = 0$ or else $\boldsymbol{w}$ is the longest prefix of $\boldsymbol{w}^{r-1}\boldsymbol{w}'\mu\boldsymbol{x}'$ that is a Lyndon word. Successive applications of Lemma 6.1.2 thus produce the desired result. □

Referring back to our previous example $\boldsymbol{x}\mu = (aabb)^2(aab)a$, we see that by this result we can form

$$\text{CFL}(\boldsymbol{x}\mu) = (aabb)^2\text{CFL}(aaba)$$

in the case that $\mu = a$. Observe more generally that if we suppose μ to be a special end-of-string letter $ that is lexicographically less than all other letters, Lemma 6.1.3 applies also at the end of the string $\boldsymbol{x}$. This fact is actually used in Algorithm Dvl.

We are now in a position to understand how Algorithm Dvl executes: as $\boldsymbol{x} = \boldsymbol{x_1 x_2 x_3}$ is scanned letter-by-letter from left to right, $\boldsymbol{x_1}$ at any point corresponds to a CFL breakdown of a prefix of $\boldsymbol{x}$; $\boldsymbol{x_2} \in L^*$ is maintained in normal form; and the initial character μ of $\boldsymbol{x_3}$ is used to control NF$(\boldsymbol{x_2})$ and to determine when a prefix $\boldsymbol{w}^r$ of $\boldsymbol{x_2}$ should be appended to CFL$(\boldsymbol{x_1})$. The details are shown as Algorithm 6.1.1, where now the variables h, i and j are the positions of the right end of $\boldsymbol{x_1}$, λ, and μ, respectively. Specifically, the assignment $i \leftarrow h+1$ corresponds to case (a) of Lemma 6.1.1 and $i \leftarrow i+1$ to case (b).

Algorithm 6.1.1 (Dvl)

– *Given the input string* $\boldsymbol{x}\$$, *output in order the end positions of every Lyndon word in* CFL$(\boldsymbol{x})$

$h \leftarrow 0$ – $h = |\boldsymbol{x_1}|$, *the current length of CFL already output*
while $h < n$ **do** – *continue as long as* $|\text{CFL}(\boldsymbol{x_1})| < n$
  $i \leftarrow h+1$ – $\boldsymbol{x}[i] = \lambda$
  $j \leftarrow h+2$ – $\boldsymbol{x}[j] = \mu$

  – *this loop continues as long as* $\boldsymbol{x_2} \in L^*$
  while $\boldsymbol{x}[j] \geq \boldsymbol{x}[i]$ **do** – *while* $\mu \geq \lambda$
    if $\boldsymbol{x}[j] > \boldsymbol{x}[i]$ **then** $i \leftarrow h+1$ **else** $i \leftarrow i+1$
    $j \leftarrow j+1$

  – *for* $\mu < \lambda$, *based on Lemma 6.1.3, this loop outputs all* r *occurrences of the current generator* $\boldsymbol{w}$ *of length* $j-i+1$

  repeat
    $h \leftarrow h+(j-i)$; **output** h
  until $h \geq i$

Note that some backtracking occurs in Algorithm Dvl when $\mu < \lambda$: in this case, after r copies of the Lyndon word $\boldsymbol{w}$ are output, Dvl restarts at the beginning of $\boldsymbol{w'}\mu$ by setting $i \leftarrow h+1$, $j \leftarrow h+2$. Since $\boldsymbol{w'}$ is a prefix of $\boldsymbol{w}$, this will duplicate processing already performed whenever $|\boldsymbol{w'}| > 1$. In terms of the example $\boldsymbol{x}\mu = (aabb)^2(aab)a$ given above, the string $\boldsymbol{w'} = aab$ would have to be processed again, even though it would already have been processed as a prefix of $\boldsymbol{w} = aabb$ at the beginning of the preceding execution of the outer **while** loop. As we shall see later, Dvl* eliminates this reprocessing by maintaining a record of the periodicity of $\boldsymbol{w}$.

We state the first main result of this section:

Theorem 6.1.4 *Algorithm 6.1.1 correctly computes the Lyndon decomposition of a given string* $\boldsymbol{x}$ *in* $\Theta(n)$ *time using constant additional space.*

Proof The correctness follows from Lemmas 6.1.1 and 6.1.3 in the light of the preceding discussion. The space bound follows from the fact that no auxiliary storage is required by the algorithm, apart from the position counters h, i, j.

To analyze the time requirement, observe that each output consists of $r \geq 1$ copies of $\boldsymbol{w} \in L$ and that backtracking occurs over a proper prefix $\boldsymbol{w'}$ of $\boldsymbol{w}$. Since $\boldsymbol{w}$ is the generator of the current $\boldsymbol{x_2} = \boldsymbol{w}^r\boldsymbol{w'}$, its length $p = |\boldsymbol{w}|$ is the period of $\boldsymbol{x_2}$. Thus in order to append r strings of total length rp to the current CFL, exactly $rp + |\boldsymbol{w'}| - 1$ iterations of the inner **while** loop are required. In the best case ($|\boldsymbol{w'}| = 0$), there will be $rp - 1$ iterations of the inner **while** loop for each set of rp letters output; in the worst case ($r = 1$ and $|\boldsymbol{w'}| = p-1$), there will be $2p - 2$ iterations of the inner **while** loop for each output of p letters. Since the processing for each iteration requires only constant time, and since the total time required for output is $O(n)$, it follows that Algorithm 6.1.1 executes in $\Theta(n)$ time. □

Before we move on to Algorithm Dvl*, it will be helpful to discuss a little further the connections between normal form and Lyndon words that appear in the current algorithm. Dvl may be thought of as an algorithm that computes the normal form of maximal elements of L^* that occur in a given string $\boldsymbol{x}$. Thus it is a way of calculating NF without explicit reference to the border array. But the connection is there implicitly, as we now explain.

For the Dvl variables h, i, j, let $i' = i - h$ and $j' = j - h$. Then since at any time $h = |\boldsymbol{x_1}|$, it follows that $\lambda = \boldsymbol{x}[i] = \boldsymbol{x_2}[i']$ and $\mu = \boldsymbol{x}[j] = \boldsymbol{x_2}[j']$. Observe that every time an iteration of the outer **while** loop begins, the following statement is trivially true: i' is the largest integer less than j' such that

$$\boldsymbol{x_2}[1..i'-1] = \boldsymbol{x_2}[j'-i'+1..j'-1]. \tag{6.1}$$

In other words, $\boldsymbol{x_2}[1..i'-1]$ is the longest border of $\boldsymbol{x_2}[1..j'-1]$ — or, in the notation of Section 1.3, $\boldsymbol{\beta_2}[j'-1] = i'-1$, where $\boldsymbol{\beta_2}$ is the border array of $\boldsymbol{x_2}$. It is only in the inner **while** loop that the values of i and j are changed, while h is untouched — thus, if (6.1) is maintained by this loop, it will always be maintained:

- If $\boldsymbol{x_2}[j'] = \boldsymbol{x_2}[i']$, then both i' and j' increase by one, so that the invariant (6.1) is maintained.
- If $\boldsymbol{x_2}[j'] > \boldsymbol{x_2}[i']$, then $i' \leftarrow 1$ while j' is incremented by one. We know from Lemma 6.1.1(a) that in this case the new string $\boldsymbol{x_2}$ is a Lyndon word, therefore primitive, therefore with longest border ε. Thus in this case also the invariant (6.1) is correctly updated.

Thus (6.1) must be maintained throughout the execution of Dvl, and so we have proved

---

Lemma 6.1.5 *At every entry to, and exit from, the inner* **while** *loop in Algorithm 6.1.1,*

$$\boldsymbol{\beta_2}[j'-1] = i'-1,$$

where $j' = j - h$, $i' = i - h$, and $\boldsymbol{\beta_2}$ is the border array of $\boldsymbol{x_2}[1..j']$. □

Reflecting on this result, we realize that if we maintain the array β_2 for every new value j' that is processed, we will have a record of the corresponding value of i' for which the comparison $\boldsymbol{x_2}[j'] : \boldsymbol{x_2}[i']$ took place. This observation is the basis of Dvl*, shown here as Algorithm 6.1.2. In order to reduce the redundant processing, Dvl* stores $n/2$ locations of β_2 in a slightly altered form, as explained below.

Algorithm 6.1.2 (Dvl*)

- *Given the input string* $\boldsymbol{x}\$$, *output in order the end positions of every Lyndon word in* $\text{CFL}(\boldsymbol{x})$

$h \leftarrow 0$ – $h = |\boldsymbol{x_1}|$, *the current length of CFL already output*
$j \leftarrow 2$; $\boldsymbol{\beta'}[2] \leftarrow 1$ – *initialize* j *and* $\boldsymbol{\beta'}[j]$
while $h < n$ **do** – *continue as long as* $|\text{CFL}(\boldsymbol{x_1})| < n$
  $i \leftarrow h + \boldsymbol{\beta'}[j-h]$ – $\boldsymbol{x}[i] = \lambda$

  – *this loop continues as long as* $\boldsymbol{x_2} \in L^*$
  while $\boldsymbol{x}[j] \geq \boldsymbol{x}[i]$ **do** – *while* $\mu \geq \lambda$
    if $\boldsymbol{x}[j] > \boldsymbol{x}[i]$ **then** $i \leftarrow h+1$ **else** $i \leftarrow i+1$
    $j \leftarrow j+1$
      – *since* i *and* j *have both been updated,*
      $\boldsymbol{\beta'}[j-h] = i-h \iff \boldsymbol{\beta_2}[j-h-1] = i-h-1$
    if $j-h \leq n/2$ **then** $\boldsymbol{\beta'}[j-h] \leftarrow i-h$

    – *for* $\mu < \lambda$, *based on Lemma 6.1.3, this loop outputs all* r *occurrences of the current generator* $\boldsymbol{w}$ *of length* $j-i+1$

  repeat
    $h \leftarrow h + (j-i)$; **output** h
  until $h \geq i$
  – μ *moves right one position when* $\boldsymbol{w'} = \varepsilon$
  if $h = j-1$ **then** $j \leftarrow j+1$

Comparing Dvl* with Dvl, we observe that the initialization of h, i, j yields the same values and that the loop structures are identical. The differences between the two algorithms all relate to the recomputation of i after a set of r Lyndon words $\boldsymbol{w}$ has been output; in Dvl* this calculation makes use of an array $\boldsymbol{\beta'}$ that is updated in the inner **while** loop. As stated in a comment, the update of $\boldsymbol{\beta'}$ is equivalent to maintaining the border array $\boldsymbol{\beta_2}$ of $\boldsymbol{x_2}$; thus $\boldsymbol{\beta'}$ records for every $j'-1$ the corresponding border $i'-1 = \boldsymbol{\beta_2}[j'-1] = \boldsymbol{\beta'}[j']$, so that these values can be recaptured as required when the suffix $\boldsymbol{w'}$ of $\boldsymbol{x_2}$ is processed after the output of $\boldsymbol{w}^r$. The value of i is correct during this recapture because it is the same as it was when $\boldsymbol{w'}$ was first processed as a prefix of the first occurrence of $\boldsymbol{w}$; otherwise, the execution of Dvl* is just a simulation of Dvl that preserves the invariant of Lemma 6.1.5. There are two small points that require further explanation:

(1) The update of $\boldsymbol{\beta'}$ is performed only if $j-h \leq n/2$. Recall that $j-h = |\boldsymbol{x_2}|$, where $\boldsymbol{x_2} = \boldsymbol{w}^r\boldsymbol{w'}$. The elements of $\boldsymbol{\beta'}$ are used only for the purpose of specifying positions in $\boldsymbol{w'}$ after $\boldsymbol{w}^r$ has been output. But

$$|\boldsymbol{w}'| < |\boldsymbol{x_2}|/2 = (j-h)/2 \le (n+1)/2.$$

Thus only elements of $\boldsymbol{\beta}'$ up to position $n/2$ will ever possibly be required.

(2) If $j = h+1$ when output of $\boldsymbol{w}^r$ is complete, j is incremented by one. This is just the case $\boldsymbol{w}' = \varepsilon$, and so the initial conditions are reestablished with $j = h+2$, $i = h + \boldsymbol{\beta}[2] = h+1$.

Observing that Dvl* must have the same asymptotic time requirement that Dvl has, we have thus established

Theorem 6.1.6 *Algorithm 6.1.2 correctly computes the Lyndon decomposition of a given string $\boldsymbol{x}$ in $\Theta(n)$ time using $\Theta(n)$ additional space.* □

We conclude our study of these algorithms by attempting a more precise estimate of their worst-case time requirements than has been provided so far. We have seen that Dvl* avoids redundant processing of nonempty suffixes $\boldsymbol{w}'$, but we have not so far made any effort to quantify the benefits that might be expected. A criterion often employed to measure the efficiency of string algorithms is the number of comparisons made between letters of the string. For example, there is a huge literature [CH92, GG91, GG92, CHPZ95] devoted to determining upper and lower bounds on the number of letter comparisons required in the worst case to locate a specific pattern $\boldsymbol{u}$ in a given string $\boldsymbol{x}$. In keeping with this tradition, then, we will count letter comparisons in Algorithms 6.1.1 and 6.1.2.

To make such an analysis meaningful, it is first necessary to specify what is meant by a "letter comparison". In this book we shall use this term in the sense of ***binary letter comparison*** — that is, a comparison whose possible result is constrained to be either TRUE or FALSE. Thus, for example, a single comparison between μ and λ could determine whether or not $\mu = \lambda$, or whether or not $\mu \ge \lambda$, but two comparisons would be required in the worst case to distinguish among the three cases

$$\mu > \lambda,\ \ \mu = \lambda,\ \ \mu < \lambda.$$

This convention is of course in accordance with the usual binary logic of digital computers.

Recall that for Dvl it was shown in the proof of Theorem 6.1.4 that at most $2p-2$ iterations of the inner **while** loop could be required for every p letters covered by the output. To translate this bound into letter comparisons, observe that each iteration involves two of them — one in the **while** statement, the other in the **if**. Thus $4p-4$ letter comparisons could be required in the inner **while** loop, plus one additional letter comparison to recognize that $\mu < \lambda$ and so terminate the loop. Thus altogether Dvl could require in the worst case as many as $4n-3$ letter comparisons to output all the Lyndon words in $\boldsymbol{x}$.

In Dvl*, on the other hand, there can be at most one iteration of the inner **while** loop, hence at most two letter comparisons, corresponding to each value of $j = 2, 3, \ldots, n$ — thus at most $2n-2$ letter comparisons in total. In addition, one letter comparison (the one in the inner **while** statement) will be required every time a substring of $\boldsymbol{w}'$ is output: this is the case in which the prefix $\boldsymbol{w}'$ of the Lyndon word just output itself contains as substrings one or more Lyndon words that form part of the Lyndon decomposition of $\boldsymbol{x}$. Since in the case that $\boldsymbol{w}' = \varepsilon$ an increment to j occurs, it follows that this extra comparison could be required for no more than half the values of j. Thus altogether Dvl* could require in the worst case no more than $2.5n - 2$ letter comparisons, substantially fewer than the total for Dvl.

Dvl* is therefore a big improvement on Dvl — or is it?

Digression 6.1.1 It is of course of interest, from both practical and theoretical points of view, to attempt to establish more precise bounds on the time requirements of string algorithms. But it should be pointed out nevertheless that algorithms with a small or minimum number of letter comparisons may well not be the most efficient in practice. It may well be that the "housekeeping" required to avoid letter comparisons consumes more time than the comparisons themselves. For example, it was remarked in Section 4.1 that Algorithm Easy (2.2.1) for pattern-matching might on average execute faster than algorithms such as KMP (Algorithm 7.1.1) that greatly reduce the letter comparisons required in the worst case: like other primitive creatures, Easy has the overriding merit of being simple.

These issues arise also in the case of Dvl and Dvl*. Even though, as we have shown, the former requires in the worst case more letter comparisons than the latter, it nevertheless remains quite unclear whether Dvl* should *in practice* be preferred to Dvl, as the following remarks suggest:

- Algorithm Dvl* requires $n/2$ additional storage locations and associates additional processing with each new letter μ as well as with each set of identical Lyndon words. Are these costs justified on average by the benefits conferred?
- As shown in Exercise 6.1.3, output time can be reduced to a constant for each distinct Lyndon word found in $\boldsymbol{x}$. Is this simple change not on average more effective than trying to reduce letter comparisons?
- As suggested by Exercise 6.1.5, the upper bounds given on letter comparisons by both Dvl and Dvl* are not sharp. Thus, even without considering the average case, it may well be that the worst-case difference between the two algorithms is entirely misrepresented by the upper bounds; therefore, in the absence of a more precise analysis, no valid conclusions can be based on these bounds.
- Dvl* has an advantage over Dvl only in cases where $|\boldsymbol{w}'| > 1$ — for random strings, even on a small alphabet, this condition should be a rare one. Perhaps Dvl* has been invented to fix a problem that hardly ever exists!
- In general, for problems that require $\Theta(n)$ time for their solution, is it not more important to be concerned about the average-case behaviour of algorithms rather than worst-case? The differences between any two $\Theta(n)$ algorithms will in the worst case be relatively small, and so the average-case behaviour becomes correspondingly more important.
- The influence of alphabet size needs to be considered. In cases where alphabet size is very small, say $\alpha = 2$ or 4, $\boldsymbol{x}$ requires only $n/8$ or $n/4$ bytes of storage, respectively; on the other hand, the $n/2$ elements in the new $\boldsymbol{\beta}'$ array required for Dvl* are integers in the range $0..n/2$ and so altogether will require at least $\lceil n' \log n' \rceil/8$ bytes of

additional storage, where $n' = n/2 + 1$. For large n, this extra storage may be an unacceptable cost.

- The characteristics of the machine on which the algorithms are implemented need to be considered. For example, the relative efficiency of comparisons between bit strings on the one hand and comparisons between bytes or full words on the other may in certain cases have a major influence on whether Dvl or Dvl* executes faster: for an alphabet of size 2, letter comparisons might involve only single bits, while for long strings operations on β' may involve several bytes.

The purpose in putting forward these *caveats* is not to downgrade the essential role played by theoretical computer scientists in devising new and ever more elegant/efficient algorithms; rather, it is to make it clear that many factors can influence the behaviour of an algorithm *as it is implemented*, and that the net effect of these factors can often be determined only by experiment on a case-by-case basis. □

Exercises 6.1

1. Show that a string $\boldsymbol{x}$ is a Lyndon word if and only if in the suffix array $\boldsymbol{\sigma_x}$, $\boldsymbol{\sigma_x}[1] = 1$.
2. In the terminology of Section 4.2, show that every Lyndon word on the standard alphabet (4.1) is p-canonical. Does the converse hold?
3. In Lemma 1.4.7 the integer variable r was used to count the number of occurrences of the output Lyndon word $\boldsymbol{w}$. Express r in terms of the variables h, i, j used in Algorithm 6.1.1, and observe that for $r > 1$ what is being output is simply a repetition. Hence apply an encoding of the output already encountered several times in this book to enable all r occurrences of $\boldsymbol{w}$ to be specified by a single output tuple. Specify corresponding modifications to the algorithm. Is the asymptotic time complexity affected by your changes?
4. For alphabet sizes $\alpha = 2, 4, 8$ and string lengths $n = 100K, 1M, 5M$, estimate the storage requirements for Dvl and Dvl*. Then, as well as you can without doing extensive experiments, prepare an analysis of the costs/benefits of using Dvl/Dvl* in each of these nine cases.
5. Can you find examples of worst-case (or at least "bad-case") strings for Dvl/Dvl*? From your examples, can you make inferences or draw conclusions about the bounds given in the text on the number of letter comparisons required by the two algorithms?
6. As Algorithm Dvl* is presented in the text, the update of β' is guarded by a test to ensure that $j - h \leq n/2$. Thus β' need contain only $n/2$ positions. Another option would have been to dispense with the guarding test altogether, while at the same time expanding β' to n positions. Discuss the pros and cons of these two options.
7. Suppose the three possible outcomes $\{>, =, <\}$ of comparing μ with λ are equally likely. On this basis determine upper and lower bounds on the *expected* number of letter comparisons required by Dvl and Dvl*.
8. Is Algorithm Dvl* an on-line algorithm?

9. In [D83] P was defined to be the set of all prefixes of the Lyndon words on a given alphabet A. If A contains a maximum letter ω, let P' be the set of all repetitions of ω; otherwise, let $P' = \emptyset$. Show that $L^* = P \cup P'$.

Hint: Make use of Lemma 6.1.1, especially case (a).

6.2 Lyndon Applications

I have tried my best sweet words to combine
To tell you all I feel.

— Cole PORTER (1892–1964), *Out of This World*

In this section we show how Algorithm Dvl can be modified in a simple way so as to compute in $\Theta(n)$ time the following intrinsic patterns of a given string $\boldsymbol{x}$:

- the minimum nonempty suffix $s_{\min}(\boldsymbol{u})$,
- the maximum suffix $s_{\max}(\boldsymbol{u})$,

of every nonempty prefix $\boldsymbol{u}$ of $\boldsymbol{x}$. Of course corresponding modifications may also be made to Dvl*.

The basis of both of these applications is found in two lemmas. The first lemma considers the CFL breakdown of $\boldsymbol{x}$ such as occurs in the execution of Algorithm Dvl:

$$\boldsymbol{x}[1..j] = \boldsymbol{x_1}\boldsymbol{x_2}[1..j-h],$$

where $|\boldsymbol{x_1}| = h$ and CFL$(\boldsymbol{x_1})$ has already been computed. Essentially, the lemma assures us that

$$s_{\min}\big(\boldsymbol{x}[1..j]\big) = s_{\min}\big(\boldsymbol{x_2}[1..j-h]\big);$$

that is, the computation of $s_{\min}$ for positions in $\boldsymbol{x_2}$ is independent of the prefix $\boldsymbol{x_1}$.

Lemma 6.2.1 *Let $\boldsymbol{x} = \boldsymbol{x_1}\boldsymbol{w}$ for nonempty strings $\boldsymbol{x_1}$ and $\boldsymbol{w}$. Then* CFL$(\boldsymbol{x})$ = CFL$(\boldsymbol{x_1})\boldsymbol{w}$ *if and only if $s_{\min}(\boldsymbol{x}) = \boldsymbol{w}$.*

Proof If $CFL(\boldsymbol{x})$ = CFL$(\boldsymbol{x_1})\boldsymbol{w}$, then $\boldsymbol{w}$ is a Lyndon word and we may write

$$\text{CFL}(\boldsymbol{x}) = \boldsymbol{w_1}\boldsymbol{w_2}\cdots\boldsymbol{w_k},$$

where $\boldsymbol{w_k} = \boldsymbol{w}$ and $\boldsymbol{w_1} \geq \boldsymbol{w_2} \geq \cdots \geq \boldsymbol{w_k}$. Without loss of generality, let $s_{\min}(\boldsymbol{x}) = \boldsymbol{w}_i'\boldsymbol{w_{i+1}}\cdots\boldsymbol{w_k}$ for some integer $i \in 1..k$ and $\boldsymbol{w}_i'$ a nonempty suffix of

w_i. Since by Lemma 1.4.6, $w'_i \geq w_i$ and since $w_i \geq w_k$, it follows that $i = k$ and $s_{\min}(x) = w_k = w$ as required.

Conversely, suppose that $s_{\min}(x) = w$ but that $\text{CFL}(x) = \text{CFL}(x_1)\overline{w}$ for some $\overline{w} \neq w$. Then by the above argument, $s_{\min}(x) = \overline{w} \neq w$, a contradiction. □

The second lemma has a structure that mimics that of Lemma 6.1.1 and that therefore leads naturally to an algorithm whose structure exactly parallels that of Dvl.

Lemma 6.2.2 *Suppose that $x_2 \in L^*$. Specifically, let $\text{NF}(x_2) = w^r w'$ for some $w \in L$, and write $w = w'\lambda w''$ where $\lambda \in A$. Also let $s'\lambda = s_{\min}(w'\lambda)$. Then for $\mu \in A$:*

(a) $\mu > \lambda \Rightarrow s_{\min}(x_2\mu) = x_2\mu$;

(b) $\mu = \lambda \Rightarrow s_{\min}(x_2\mu) = s'\mu$;

(c) $\mu < \lambda \Rightarrow s_{\min}(x_2\mu x') = s_{\min}(s'\mu x')$ *for any string* x'.

Proof To prove (a), observe by Lemma 6.1.1 that $x_2\mu \in L$, so that by Lemma 1.4.6 $x_2\mu$ is less than any proper suffix of itself. To prove (b), recall from Lemma 6.1.2 that $\text{CFL}(x_2\lambda) = w^r\text{CFL}(w'\lambda)$ with $w \geq w'\lambda$ and $w < R_j(w)$ for every $j \in 1..n-1$.

To prove (c), observe by Lemma 1.4.7 that

$$\text{CFL}(x_2\mu x') = w^r\text{CFL}(w'\mu x'),$$

from which it follows as in (b) that

$$s_{\min}(x_2\mu x') = s_{\min}(w'\mu x').$$

Let S' be a suffix of w' such that $|S'| > |s'|$. Then by the definition of s', $s'\lambda < S'\lambda$. But since $|s'\lambda| \leq |S'|$, we see that in fact $s'\lambda \leq S'$ and so $s'\lambda < S'\mu x'$. Thus

$$s'\mu x' < s'\lambda < S'\mu x',$$

so that $S'\mu x'$ cannot be a minimum suffix of $w'\mu x'$. Therefore any such minimum suffix S' must satisfy $|S'| \leq |s'|$ and so be a suffix of s'. □

It is not difficult to see that Lemmas 6.2.1 and 6.2.2 provide us with the information we need to compute the function $s_{\min}$ for every prefix of x. As remarked above, Lemma 6.2.1 allows us to ignore x_1, the prefix for which the CFL is already computable (and for which $s_{\min}$ has already been computed). In view of Lemma 6.1.1(a), Lemma 6.2.2(a) is just a special case of Lemma 6.2.1: a Lyndon word is its own minimum suffix. In cases (b) and (c) of Lemma 6.2.2, we find that $s_{\min}(x_2\mu)$ is determined completely by the prior computation of $s_{\min}(w'\lambda)$, where $w'\lambda$ is a prefix of w and a proper prefix of $x_2\mu$.

The details of the MinSuff calculation are shown as Algorithm 6.2.1. In order to economize on output, the mimimum nonempty suffixes are compressed into an integer array $\sigma' = \sigma'[1..n]$, where for every integer $j \in 1..n$, $\sigma'[j] = i$ if and only if $s_{\min}(x[1..j]) = x[i..j]$. Computing this array is equivalent to computing the initial element $\boldsymbol{\sigma}_{\boldsymbol{x}[1..j]}[1]$ in n suffix arrays defined by the n nonempty prefixes of $\boldsymbol{x}$. Here is an example of the minimum suffix calculation, with gaps indicating the breakdown into Lyndon words:

	1	2	3	4	5	6	7	8	9	10	11	12	13
$\boldsymbol{f_6} =$	a	b	a	a	b	a	b	a	a	b	a	a	b
$\boldsymbol{\sigma}' =$	1	1	3	4	3	6	3	8	9	8	11	12	11

Algorithm 6.2.1 (MinSuff)

– *Given the input string $\boldsymbol{x}\$$, output an array $\boldsymbol{\sigma}'[1..n]$ such that for every $j \in 1..n$, $\boldsymbol{\sigma}'[j] = i$ if and only if $\boldsymbol{x}[1..j]$ has minimum nonempty suffix $\boldsymbol{x}[i..j]$*

```
h ← 0; σ'[1] ← 1
while h < n do
  i ← h+1  - x[i] = λ
  j ← h+2  - x[j] = μ

  while x[j] ≥ x[i] do  - while μ ≥ λ
    if x[j] > x[i] then
        - for μ > λ, s_min(x₂μ) = x₂μ
      σ'[j] ← h+1; i ← h+1
    else
        - for μ = λ, s_min(x₂μ) = s_min(w'μ)
      σ'[j] ← σ'[i] + (j−i); i ← i+1
    j ← j+1

  - μ < λ : recompute h
  r ← ⌈(i−h)/(j−i)⌉; h ← h + r * (j−i)
  if h = j−1 then
    σ'[j] ← h+1
```

Theorem 6.2.3 *Algorithm 6.2.1 correctly computes the array $\boldsymbol{\sigma}'[1..n]$ (the minimum nonempty suffix of every nonempty prefix) for a given string $\boldsymbol{x}$ in $\Theta(n)$ time.*

Proof Observe that the calculation and use of the variables h, i, j in MinSuff are identical to Dvl except for the calculation of h when $\mu < \lambda$. But in this case, the update of h using r yields the same value produced by the **repeat** loop in Dvl (see Exercise 6.2.1); and the increment to h when $h = j-1$ simply allows for the special case in which the single remaining letter $\boldsymbol{x}[j]$ is the least letter in $\boldsymbol{x}[1..j]$, so that $s_{\min}(\boldsymbol{x}[1..j]) = \boldsymbol{x}[j]$. (This is the same case that arises for $h = 0$ and $j = 1$.) Thus MinSuff executes in exactly the same way that Dvl does, and so requires $\Theta(n)$ time.

Now consider the four updates to $\boldsymbol{\sigma}'$. We have already discussed the case $h = j - 1$, and the update $\boldsymbol{\sigma}'[1] \leftarrow 1$ is trivially correct. The two other updates, when $\mu > \lambda$ and when $\mu = \lambda$, follow from Lemma 6.2.2, cases (a) and (b), respectively. Thus MinSuff correctly computes $\boldsymbol{\sigma}'$. $\square$

We now turn to the second problem mentioned at the beginning of this section: computing the maximum suffix $s_{\max}(\boldsymbol{u})$ of every nonempty prefix of $\boldsymbol{x}$. An obvious approach to this problem is to replace lexicographic order $<$ with a reverse lexicographic order $<_{\mathcal{R}}$, in that hope that Algorithm MinSuff could be run using $<_{\mathcal{R}}$ to yield the maximum rather than the minimum. The first step would be to define

$$\mu <_{\mathcal{R}} \lambda \iff \lambda < \mu \tag{6.2}$$

over all pairs of letters $\lambda, \mu \in A$; the second would be to extend the new ordering to all strings on A. But here a difficulty arises: for $\boldsymbol{u} = a$ and $\boldsymbol{v} = ab$, for example, $\boldsymbol{u} < \boldsymbol{v}$ but $\boldsymbol{v} \not<_{\mathcal{R}} \boldsymbol{u}$. In fact,

---

Lemma 6.2.4 *For all strings $\boldsymbol{u}$ and $\boldsymbol{v}$ on an ordered alphabet A,*

$$\boldsymbol{u} < \boldsymbol{v} \text{ and } \boldsymbol{u} <_{\mathcal{R}} \boldsymbol{v}$$

if and only if $\boldsymbol{u}$ is a prefix of $\boldsymbol{v}$.

Proof See Exercise 6.2.6. $\square$

Thus the order $<_{\mathcal{R}}$ extends in a nice way to arbitrary strings $\boldsymbol{u}$ and $\boldsymbol{v}$ only when one is not a prefix of the other. The consequences of Lemma 6.2.4 may be seen by considering the examples $\boldsymbol{x} = cab$ and $\boldsymbol{y} = bab$, both strings whose maximum suffix is the string itself. For $\boldsymbol{x}$ there is no problem: with respect to $<_{\mathcal{R}}$, it is a Lyndon word, primitive, and lexicographically less than any of its suffixes; obligingly, with respect to $<$, it is also lexicographically greater than any of its suffixes. For $\boldsymbol{y}$, however, the situation is different: even though, with respect to $<_{\mathcal{R}}$, it is not primitive, not a Lyndon word, and not lexicographically less than any of its suffixes (since $bab \not<_{\mathcal{R}} b$), nevertheless, with respect to $<$, it is greater than all of its suffixes (since $bab > b > ab$). Thus Algorithm MinSuff, applied using $<_{\mathcal{R}}$ rather than $<$, would yield the desired result for $\boldsymbol{x} = cab$ but not for $\boldsymbol{y} = bab$. In general, Lemma 6.2.4 tells us that every substring with a nonempty border would be processed incorrectly.

So a naïve approach to the maximum suffix problem will not work. In fact, it turns out that, in order to deal with the problem raised by nonempty borders, it is necessary to establish a rather remarkable identity. We recall of course the set L^* of all strings generated by Lyndon words. Let $L^{\mathcal{R}}$ denote the set of all strings $\boldsymbol{x}$ such that, using order $<_{\mathcal{R}}$, $s_{\max}(\boldsymbol{x}) = \boldsymbol{x}$. The next scattering of lemmas will show that $L^{\mathcal{R}} = L^*$.

Lemma 6.2.5 *If $\boldsymbol{x} \in L^{\mathcal{R}}$ and $\boldsymbol{x}'$ is a nonempty prefix of $\boldsymbol{x}$, then $\boldsymbol{x}' \in L^{\mathcal{R}}$.*

Proof Let $\boldsymbol{x} = \boldsymbol{x}'\boldsymbol{y}$ and let $\boldsymbol{u}'$ be a proper prefix of $\boldsymbol{x}'$. Therefore, since $\boldsymbol{x} \in L^{\mathcal{R}}$, $\boldsymbol{u}' <_{\mathcal{R}} \boldsymbol{x} = \boldsymbol{x}'\boldsymbol{y}$. But since $|\boldsymbol{u}'| < |\boldsymbol{x}'|$, it follows that $\boldsymbol{u}' <_{\mathcal{R}} \boldsymbol{x}'$ and so, using $<_{\mathcal{R}}$, $s_{\max}(\boldsymbol{x}') = \boldsymbol{x}'$. Thus $\boldsymbol{x}' \in L^{\mathcal{R}}$, as required. □

Lemma 6.2.6 *If $\boldsymbol{x} \in L^{\mathcal{R}}$ and $\boldsymbol{x}'$ is a prefix of $\boldsymbol{x}$, then $\boldsymbol{x}\boldsymbol{x}' \in L^{\mathcal{R}}$.*

Proof The result is trivially true for $\boldsymbol{x}' = \varepsilon$, so suppose that $\boldsymbol{x}'$ is nonempty and let $\boldsymbol{u}$ be a proper suffix of $\boldsymbol{x}\boldsymbol{x}'$. There are exactly two cases:

(a) $\boldsymbol{u} = \boldsymbol{x}''\boldsymbol{x}'$, where $\boldsymbol{x}''$ is a nonempty proper suffix of $\boldsymbol{x}$.
Since $\boldsymbol{x} \in L^{\mathcal{R}}$, $\boldsymbol{x}'' <_{\mathcal{R}} \boldsymbol{x}$; and since $|\boldsymbol{x}''| < |\boldsymbol{x}|$, $\boldsymbol{u} = \boldsymbol{x}''\boldsymbol{x}' <_{\mathcal{R}} \boldsymbol{x}\boldsymbol{x}'$.

(b) $\boldsymbol{u}$ is a suffix of $\boldsymbol{x}'$.
If $\boldsymbol{u} = \boldsymbol{x}'$, then $\boldsymbol{u}$ is a proper prefix of $\boldsymbol{x}\boldsymbol{x}'$, so that $\boldsymbol{u} <_{\mathcal{R}} \boldsymbol{x}\boldsymbol{x}'$. Suppose therefore that $\boldsymbol{u}$ is a proper suffix of $\boldsymbol{x}'$, and observe that by Lemma 6.2.5, $\boldsymbol{x}' \in L^{\mathcal{R}}$. Then $\boldsymbol{u} <_{\mathcal{R}} \boldsymbol{x}' <_{\mathcal{R}} \boldsymbol{x}\boldsymbol{x}'$.

In both cases, $\boldsymbol{x}\boldsymbol{x}' \in L^{\mathcal{R}}$. □

Lemma 6.2.7 *If $\boldsymbol{x} \in L$, then $\boldsymbol{x} \in L^{\mathcal{R}}$.*

Proof By Lemma 1.4.5, $\boldsymbol{x}$ is primitive, so that every nonempty proper suffix $\boldsymbol{u}$ is guaranteed not to be a prefix of $\boldsymbol{x}$. Further, by Lemma 1.4.6, $\boldsymbol{x} < \boldsymbol{u}$, so that by Lemma 6.2.4, $\boldsymbol{u} <_{\mathcal{R}} \boldsymbol{x}$. But then, in $<_{\mathcal{R}}$ order, $s_{\max}(\boldsymbol{x}) = \boldsymbol{x}$, as required. □

Lemma 6.2.8 *If $\boldsymbol{x} \in L^*$, then $\boldsymbol{x} \in L^{\mathcal{R}}$.*

Proof An immediate consequence of Lemmas 6.2.6 and 6.2.7. □

Lemma 6.2.9 $L^{\mathcal{R}} = L^*$.

Proof Lemma 6.2.8 tells us that $L^* \subset L^{\mathcal{R}}$. Suppose therefore that $\boldsymbol{x} \in L^{\mathcal{R}}$ and let $\boldsymbol{u}$ be the longest prefix of $\boldsymbol{x}$ that is in L^*. Hence write $\text{NF}(\boldsymbol{u}) = \boldsymbol{w}^r\boldsymbol{w}'$ for some $\boldsymbol{w} = \boldsymbol{w}'\lambda\boldsymbol{w}'' \in L$. Suppose $\boldsymbol{u} \neq \boldsymbol{x}$ so that $\boldsymbol{x} = \boldsymbol{u}\mu\boldsymbol{x}'$ for some possibly empty suffix $\boldsymbol{x}'$. Since $\boldsymbol{u}\mu \notin L^*$, Lemma 6.1.1 implies that $\mu < \lambda$. Hence $\lambda <_{\mathcal{R}} \mu$, so that $\boldsymbol{w}'\lambda <_{\mathcal{R}} \boldsymbol{w}'\mu$, and

$$\boldsymbol{x} = \boldsymbol{u}\mu\boldsymbol{x}' <_{\mathcal{R}} \boldsymbol{u}\mu\boldsymbol{x}',$$

contradicting the assumption that $\boldsymbol{x} \in L^{\mathcal{R}}$. We conclude that $\boldsymbol{u} = \boldsymbol{x} \in L^*$. □

In order to be able to apply this result to an algorithm, we now need to disentangle what we have learned. Lemma 6.2.9 tells us that a string $\boldsymbol{u} \in L^*$ if and only if $\boldsymbol{u}$ is its own maximum suffix with respect to $<_{\mathcal{R}}$. Since $\left(<_{\mathcal{R}}\right)_{\mathcal{R}} \equiv <$, the "reverse" of this statement is also true: $\boldsymbol{u} \in L^*$ with respect to $<_{\mathcal{R}}$ if and only if $\boldsymbol{u}$ is its own maximum suffix with respect to $<$. Thus by running a variant of Dvl with respect to $<_{\mathcal{R}}$, we might hope somehow to be able to identify strings $\boldsymbol{u}$ such that $\boldsymbol{u} = s_{\max}(\boldsymbol{u})$.

This brings us to a critical observation: a string $\boldsymbol{u}$ is a maximum suffix of a given string $\boldsymbol{x}$ if and only if it is the *longest* suffix of $\boldsymbol{x}$ that is its own maximum suffix — that is, the longest suffix of $\boldsymbol{x}$ such that $\boldsymbol{u} = s_{\max}(\boldsymbol{u})$. But by Lemma 6.2.9, every $\boldsymbol{u}$ that is its own maximum suffix computed using one order is an element of L^* computed using the reverse order. Thus we are led to design an algorithm that, using $<_{\mathcal{R}}$, computes elements of L^* that are suffixes of $\boldsymbol{x}$, then simply chooses the longest such suffix $\boldsymbol{u}$ as the maximum suffix of $\boldsymbol{x}$ with respect to $<$. But this is precisely what Algorithm Dvl does: for every prefix $\boldsymbol{x}[1..j]$, it identifies the elements of L^* that terminate at position j — that are, in other words, suffixes of j. To select the longest such element of L^* for each j is easy since Dvl processes $\boldsymbol{x}$ from left to right: simply select the leftmost one. This is the basis of MaxSuff, presented here as Algorithm 6.2.2.

Algorithm 6.2.2 (MaxSuff)

– *Given the input string $\boldsymbol{x}\$$, output an array $\boldsymbol{\sigma}''[1..n]$ such that for every $j \in 1..n$, $\boldsymbol{\sigma}''[j] = i$ if and only if $\boldsymbol{x}[1..j]$ has maximum suffix $\boldsymbol{x}[i..j]$*

for $j \in 1..n$ **do** $\boldsymbol{\sigma}''[j] \leftarrow 0$
$h \leftarrow 0;\ \boldsymbol{\sigma}''[1] \leftarrow 1$
while $h < n$ **do**
  $i \leftarrow h+1$ – $\boldsymbol{x}[i] = \lambda$
  $j \leftarrow h+2$ – $\boldsymbol{x}[j] = \mu$

  while $\boldsymbol{x}[j] \geq_{\mathcal{R}} \boldsymbol{x}[i]$ **do** – *while* $\mu \leq \lambda$
    if $\boldsymbol{x}[j] >_{\mathcal{R}} \boldsymbol{x}[i]$ **then** $i \leftarrow h+1$ **else** $i \leftarrow i+1$
    if $\boldsymbol{\sigma}''[j] = 0$ **then** $\boldsymbol{\sigma}''[j] \leftarrow h+1$

$j \leftarrow j+1$

– $\mu > \lambda$: *recompute* h

$r \leftarrow \lceil (i-h)/(j-i) \rceil$; $h \leftarrow h + r * (j-i)$

if $h = j-1$ **then**

  if $\boldsymbol{\sigma}''[j] = 0$ **then** $\boldsymbol{\sigma}''[j] \leftarrow h+1$

As with MinSuff, we introduce an array to provide a compact representation of the information computed: in the integer array $\boldsymbol{\sigma}''[1..n]$, $\boldsymbol{\sigma}''[j] = i$ for every integer $j \in 1..n$ if and only if $s_{\max}(\boldsymbol{x}[1..j]) = \boldsymbol{x}[i..j]$ — that is, as we have just seen, if and only if $\boldsymbol{x}[i..j]$ is the longest suffix of $\boldsymbol{x}[1..j]$ such that $s_{\max}(\boldsymbol{x}[i..j]) = \boldsymbol{x}[i..j]$. Thus for each position j — that is, for each prefix $\boldsymbol{x}[1..j]$ of $\boldsymbol{x}$ — we seek the least value i such that $\boldsymbol{x}[1..j] \in L^*$ with respect to $<_{\mathcal{R}}$. Since position $h+1$ always marks the beginning of the element of L^* that is currently being processed, we will need to execute

$$\boldsymbol{\sigma}''[j] \leftarrow h+1$$

in order to achieve this. Of course position j may occur within several overlapping words of L^*: to ensure that the smallest appropriate value of h is used, we suppress this assignment statement after its first execution by making it conditional on $\boldsymbol{\sigma}''[j] = 0$, as shown in the algorithm.

We present here as an example of the maximum suffix calculation the same string $\boldsymbol{f_6}$ used above for minimum suffix. Note that now the gaps between substrings reflect the breakdown into Lyndon words with respect to the reverse order $<_{\mathcal{R}}$. Another way to think of this is that the gaps now reflect the Lyndon decomposition using the original order $<$ of the conjugate string $\overline{\boldsymbol{f_6}}$ formed by interchanging a and b in $\boldsymbol{f_6}$.

	1	2	3	4	5	6	7	8	9	10	11	12	13
$\boldsymbol{f_6} =$	a	b	a	a	b	a	b	a	a	b	a	a	b
$\overline{\boldsymbol{f_6}} =$	b	a	b	b	a	b	a	b	b	a	b	b	a
$\boldsymbol{\sigma}'' =$	1	2	2	2	2	2	5	5	5	5	5	5	5

In this example, note that for positions $j = 5$ and 6, $\boldsymbol{\sigma}''[j] = 2$, the starting point of the "reverse Lyndon word" baa, even though these positions are also included in a later reverse Lyndon word, $babaabaa$ — thus the longest suffix of $\boldsymbol{f_6}[1..j]$ that is in L^* with respect to $<_{\mathcal{R}}$ is identified.

Theorem 6.2.10 *Algorithm 6.2.2 correctly computes the array $\boldsymbol{\sigma}''[1..n]$ (the maximum suffix of every nonempty prefix) for a given string $\boldsymbol{x}$ in $\Theta(n)$ time.*

Proof The only differences between Algorithms 6.2.1 and 6.2.2 relate to the calculation of $\boldsymbol{\sigma}'$ and $\boldsymbol{\sigma}''$, respectively. The correctness of the calculation of $\boldsymbol{\sigma}''$ follows from Lemma 6.2.9 and the above discussion. □

In this section we have focussed on variations of Algorithm Dvl, but, as often occurs with interesting string problems, there are actually several quite different algorithms that compute either the Lyndon decomposition of $\boldsymbol{x}$ or the closely related canonical form for the necklace $C(\boldsymbol{x})$ that is determined by MSP($\boldsymbol{x}$) (see Section 1.4). Algorithms due to Booth [B80] and Shiloach [S81] compute MSP($\boldsymbol{x}$), the former using a generalized border array calculation, the latter a sieve technique that efficiently eliminates possible starting positions of the canonical form. Another CFL algorithm [IS95] is competitive with Dvl in the average-case, but may require as much as $\Theta(n \log n)$ time in the worst case. Yet another algorithm due to Shiloach [S79] deals specifically with the comparison of two necklaces $C(\boldsymbol{x_1})$ and $C(\boldsymbol{x_2})$, and so applies directly to the congruent polygons problem mentioned in Section 1.4.

Perhaps the most remarkable of the canonical form algorithms is that of Apostolico and Crochemore [AC91]. They first show that Algorithm Dvl can be modified to compute MSP($\boldsymbol{x}$) with no increase in the number of letter comparisons required. They then show that in fact, with the help of a modified version of the $\boldsymbol{\beta}''$ array used for the KMP algorithm (Section 7.1), Algorithm Dvl* can be transformed into an on-line algorithm that computes the MSPs of every prefix of $\boldsymbol{x}$ in $\Theta(n)$ time — and that furthermore requires only 50% more letter comparisons than are required by Algorithm Dvl. Applying this algorithm to every suffix of $\boldsymbol{x}$ thus enables the MSPs of every *substring* of $\boldsymbol{x}$ to be computed in $\Theta(n^2)$ time using $\Theta(n)$ additional space.

Exercises 6.2

1. Show that in Algorithm 6.2.1, the line

$$r \leftarrow \lceil (i-h)/(j-i) \rceil; \ h \leftarrow h + r * (j-i)$$

can be replaced by

$$h \leftarrow \boldsymbol{\sigma}'[i] + (j-i-1).$$

2. In Algorithm 6.2.1, is the end-of-string sentinel $ still required in the input? With reference to the example $\boldsymbol{x} = aba$, show that there may be unnecessary iterations of the outer **while** loop performed before the algorithm terminates. Hence replace the condition $h < n$ for the outer **while** loop with a condition on j that solves this problem and also allows the $ to be removed from the input.
3. Show how to modify Algorithm 6.1.2 so as to compute the array $\boldsymbol{\sigma}'$.
4. Characterize the Lyndon decomposition of $\boldsymbol{f_n}$.
5. Establish necessary and sufficient conditions that

   (a) $\boldsymbol{\sigma}'[j] = 1$;
   (b) $\boldsymbol{\sigma}'[j] = j$;
   (c) $\boldsymbol{\sigma}'[j+1] = \boldsymbol{\sigma}'[j] + 1$.

6. Prove Lemma 6.2.4.
7. Prove Lemma 6.2.8.

8. Show how to modify Algorithm 6.1.2 so as to compute the array $\boldsymbol{\sigma}''$. Your modified algorithm should eliminate the requirements to initialize $\boldsymbol{\sigma}''$ and to test $\boldsymbol{\sigma}''[j]$.

9. Prove that $\boldsymbol{\sigma}''[j+1] \geq \boldsymbol{\sigma}''[j]$ for every $j \in 1..n-1$.

10. In Exercise 5.2.19 the idea of a position tree was introduced. Show how a position tree can be used to compute the canonical form of a necklace $C(\boldsymbol{x})$.

6.3 s-Factorization: Lempel-Ziv

He can compress the most words into the smallest ideas of any man I ever met.

— Abraham LINCOLN (1809–1865)

We consider now an alternative decomposition of a given string $\boldsymbol{x} = \boldsymbol{x}[1..n]$ that depends on the pattern of repeating occurrences of substrings within $\boldsymbol{x}$. Unlike Lyndon decomposition, it does not depend in any way on lexicographical order, and so can be defined on an unordered alphabet.

Definition 6.3.1 *A decomposition $\boldsymbol{x} = \boldsymbol{w_1 w_2} \cdots \boldsymbol{w_k}$ is an* ***s-factorization*** *if and only if each $\boldsymbol{w_j}$, $j = 1, 2, \ldots, k$, is*

(a) *a letter that does not occur in $\boldsymbol{w_1 w_2} \cdots \boldsymbol{w_{j-1}}$;* or otherwise

(b) *the substring of greatest length that occurs at least once in $\boldsymbol{w_1 w_2} \cdots \boldsymbol{w_{j-1}}$.*

The s-factorization of $\boldsymbol{x}$ thus depends on a left-to-right scan of $\boldsymbol{x}$ that recursively computes the current factor $\boldsymbol{w_j}$ from the already-computed factors $\boldsymbol{w_1 w_2} \cdots \boldsymbol{w_{j-1}}$. Of course by part (a) of the definition, $\boldsymbol{w_1} = \boldsymbol{x}[1]$. Thereafter, at step j, if the letter in position

$$p_j = |\boldsymbol{w_1}| + |\boldsymbol{w_2}| + \cdots + |\boldsymbol{w_{j-1}}| + 1$$

of $\boldsymbol{x}$ does not occur in $\boldsymbol{x}[1..p_j - 1]$, then we again invoke part (a); if however it does occur, then according to part (b) we must locate the longest substring occurring at position p_j that also occurs in $\boldsymbol{x}[1..p_j - 1]$.

As an example of an s-factorization, consider the Fibostring

$$\boldsymbol{f_6} = abaababaabaab,$$

with s-factorization

$$\boldsymbol{w_1} = a,\ \boldsymbol{w_2} = b,\ \boldsymbol{w_3} = a,\ \boldsymbol{w_4} = aba,$$
$$\boldsymbol{w_5} = baaba,\ \boldsymbol{w_6} = ab,$$

that can compactly be represented as $a/b/a/aba/baaba/ab$.

Note that case (b) of Definition 6.3.1 permits a previous occurrence of $\boldsymbol{w_j}$ to overlap with the suffix $\boldsymbol{w_j}$. Thus, for example, the s-factorization of a^n is a/a^{n-1}.

Curiously, even though the definition of s-factorization is not dependent on the nature of the alphabet, the only known efficient algorithms to compute an s-factorization depend on the use of suffix trees, and therefore depend on an alphabet that is ordered, known in advance, and preferably not too large (see Section 2.1).

In order to compute the s-factorization of $\boldsymbol{x}$, we suppose then that the suffix tree $T_{\boldsymbol{x}}$ of $\boldsymbol{x}$ has already been computed (Section 5.2). We suppose further that each internal node i of $T_{\boldsymbol{x}}$ has been labelled with the least label over all terminal nodes that are descendants of i in the tree — a calculation that can be carried out by a standard postorder traversal of $T_{\boldsymbol{x}}$ that visits each of the $\Theta(n)$ nodes exactly once [K73a, AHU74]. Thus the label at each internal node is the leftmost starting position in $\boldsymbol{x}$ of the substring of $\boldsymbol{x}$ represented by the path to the node from the root. For convenience, the root node is given label 0. The labelled suffix tree for the Fibostring $\boldsymbol{f_7}$ is shown in Figure 6.1.

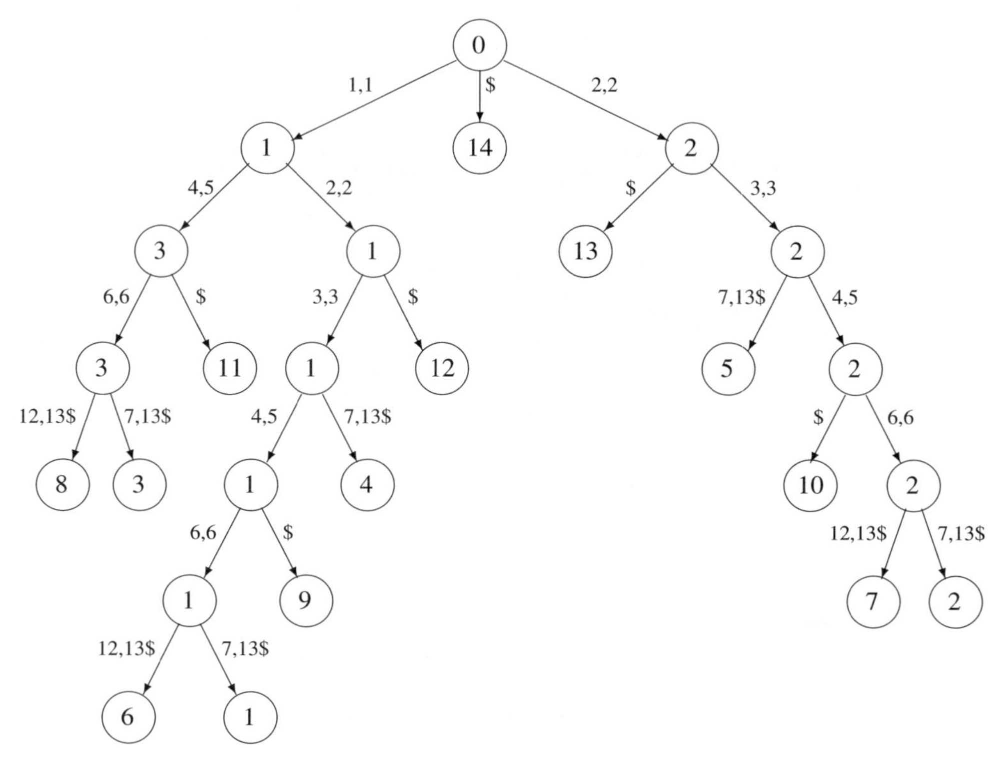

Figure 6.1. The labelled suffix tree for $\boldsymbol{f_6} = abaababaabaab\$$

In order to facilitate left-to-right processing of the s-factorization, we specify each factor $\boldsymbol{w_j}$ in the s-factorization of $\boldsymbol{x}$ by a pair (i_L, ℓ), where i_L is the starting position of the leftmost occurrence of the repeating substring $\boldsymbol{w_j}$ in $\boldsymbol{x}$ and $\ell = |\boldsymbol{w_j}|$ is the substring's length. Let i_0 be the current position in $\boldsymbol{x}$ — that is, the first position following the prefix $\boldsymbol{w_1 w_2} \cdots \boldsymbol{w_{j-1}}$. For each i_0 we imagine computing the correct value of ℓ by searching $T_{\boldsymbol{x}}$ from the root for the suffix $\boldsymbol{x}[i_0..n]$ that will of course lead to the terminal node i_0. In order to search, we initialize $i \leftarrow i_0$, then match on $\boldsymbol{x}[i]$, incrementing i until matching starts along a downward edge of $T_{\boldsymbol{x}}$ whose lower end is a node labelled i_0. If at that point the last node traversed has a nonzero label, say ν, then we set $i_L \leftarrow \nu$ and set ℓ equal to the length of the substring represented by ν. If the last node traversed was the root node of label 0, we set $(i_L, \ell) \leftarrow (i_0, 0)$, indicating that a new letter has been identified. Algorithm LZ, presented as Algorithm 6.3.1, displays the simple logic of the search, based on a *match* function whose details are left to Exercise 6.3.2. The output of the algorithm corresponding to input $\boldsymbol{f_6}$ would be

$$(1,0), (2,0), (1,1), (1,3), (2,5), (1,2). \tag{6.3}$$

Algorithm 6.3.1 (LZ)

- *Using the labelled suffix tree $T_{\boldsymbol{x}}$, compute the s-factorization of $\boldsymbol{x}$ as a sequence of pairs (i_L, ℓ)*

$i_0 \leftarrow 1$
while $i_0 \le n$ **do**
  $(i_L, \ell) \leftarrow match(i_0, T_{\boldsymbol{x}})$
  output (i_L, ℓ)
  $i_0 \leftarrow i_0 + \ell$

Using Algorithm Ukn (Subsection 5.2.3) for suffix tree construction, Algorithm LZ can actually be implemented on-line, so that the s-factorization of $\boldsymbol{x}$ is computed at the same time that $T_{\boldsymbol{x}}$ is created [G97, ZL77].

In order to use the output of Algorithm LZ for compression, slight modifications are convenient. For example, to facilitate decompression, the first occurrence of each letter could be given explicitly, and the use of separators could be reduced. Thus, for the Fibostring $\boldsymbol{f_6} = abaababaabaab$, the output corresponding to (6.3) could be rewritten as

$$a/b/1,1/1,3/2,5/1,2 \tag{6.4}$$

In this case there is actually an increase in the number of symbols used, from 13 to 19, a typical result for short strings. Generally, however, for longer strings, the s-factorization tends to be more compressed due to the effect of longer and more frequent repeating substrings. For example, the next Fibostring,

$$\boldsymbol{f_7} = abaababaabaababaababa$$

of length 21 has an s-factorization

$$a/b/1, 1/1, 3/2, 5/4, 8/2, 2 \tag{6.5}$$

of length 23, scarcely longer than (6.4).

s-factorization forms the basis of Ziv-Lempel data compression and its variants, one of which is implemented in the Unix routines *compress* and *uncompress*. Typically, for long English-language text files, *compress* will reduce the length of a file to one-half or one-third of its original size. We leave as Exercise 6.3.5 the details of the complementary algorithm *uncompress* that reconverts an s-factorization such as (6.5) back to its original form.

In Section 12.2 we study the application of s-factorization to the computation of the repetitions in a given string $\boldsymbol{x}$.

Exercises 6.3

1. Prove that every string $\boldsymbol{x}$ has one and only one s-factorization.
2. Write the function *match* used by Algorithm 6.3.1. Then prove the algorithm's correctness.
3. Draw the suffix tree $T_{\boldsymbol{x}}$ for $\boldsymbol{x} = abaababcabab$ and then use it to compute the s-factorization of $\boldsymbol{x}$.
4. Write down the s-factorization of $\boldsymbol{x} = (ab)^n$, $n \geq 2$, in the form illustrated by (6.4). What is the compression ratio in this case — that is, the number of symbols in the s-factorization divided by the string length $2n$?

   Hint: Assume that each symbol, in both $\boldsymbol{x}$ and its s-factorization, can be encoded into a half-byte (four bits). This includes the letters of the alphabet, separators, and each decimal digit of any integer.
5. Write a linear-time algorithm *uncompress* that converts an s-factorization of $\boldsymbol{x}$ into $\boldsymbol{x}$ itself.
6. An alternative form of s-factorization is sometimes used in which case (b) of Definition 6.3.1 is replaced by

   (b) the substring of maximum-length that occurs in $\boldsymbol{w}_1\boldsymbol{w}_2\cdots\boldsymbol{w}_{j-1}$.

   Rewrite the function *match* and the algorithm *uncompress* in accordance with this revised definition. What will the compression ratio now be for $\boldsymbol{x} = (ab)^n$?

Part III

Computing Specific Patterns

Part III

For I am a Bear of Very Little Brain,
and long words Bother me.

— A. A. MILNE (1882–1956), *Winnie-the-Pooh*

Computer science as a *recognized* intellectual discipline should no doubt date back to Alan Turing (1935), but recognition generally comes slowly: as Winnie-the-Pooh observes, things take time. Perhaps then we could assert more cautiously that our discipline has been established for 40 years or so. Thus the term "classical" — often used also in other fields such as physics or music or mathematics to refer to accomplishments, perhaps hundreds or even thousands of years ago, of illustrious pioneers — can in the context of computer science seem quite inappropriate: phrases such as "classical approach" or "classical algorithms" could describe work done less than 25 years ago by people who are still very much alive! This is certainly true in the area of computer science studied in this book: the earliest references to computing patterns in strings date to the 1960s, and interest in computing specific patterns — the "classical approach" to pattern-matching — really took off in the 1970s, particularly with the publication in 1977 of the "classical" Boyer-Moore (BM) and Knuth-Morris-Pratt (KMP) algorithms, all of whose illustrious authors are fortunately still with us.

With this *caveat*, then, we describe the subject matter of Part III as "classical": the area of pattern-matching that was the first to be intensively studied, and that surely remains today the most heavily researched overall. Of course this is not an accident: the location of specific patterns in strings is certainly an obvious focus of research, giving rise as it does to by far the most widespread applications. In Chapters 7 and 8 we consider algorithms to locate exact single patterns, extended in Chapters 9 and 10 to deal with single patterns that are in some sense "approximate", as explained in Section 2.2. Finally, in Chapter 11, we discuss the use of finite automata in pattern-matching, specifically for the matching of regular expressions and for the simultaneous location of multiple distinct patterns in a given string, both exact and approximate.

Chapter 7

Basic Algorithms

Words ought to be a little wild,
for they are the assault of thoughts
on the unthinking.

— John Maynard KEYNES (1883–1946)
New Statesman & Nation (15 July 1933)

In this chapter we describe four quite different "classical" approaches to computing all the occurrences of a given nonempty pattern $\boldsymbol{p} = \boldsymbol{p}[1..m]$ in a given nonempty text $\boldsymbol{x} = \boldsymbol{x}[1..n]$. The first two of these approaches (though the second only with some modification) give rise to algorithms guaranteed to execute in $O(n + m)$ time; the second two do not provide quite such an iron-clad guarantee, but nevertheless have desirable properties that could in certain cases make them algorithms of choice. In the next chapter we present a partial taxonomy of exact pattern-matching algorithms that includes several of the many dozens of variants of the classical ones that have been proposed over the last 20 years or so. Excellent coverage of pattern-matching algorithms is provided at web site [ChL97], including brief descriptions, C code, and Java applets.

7.1 Knuth-Morris-Pratt

Use soft words and hard arguments.

— *Anonymous*

The main idea of Algorithm KMP (and, as we shall see, also a main idea of Algorithm BM) is to shift the pattern $\boldsymbol{p}$ along the string at each mismatch so as to avoid the kind of retesting that occurred in Algorithm Easy (Section 2.2) with $\boldsymbol{x} = a^n$, $\boldsymbol{p} = a^{m-1}b$, $2 \leq m \leq n$.

In that example, observe that after comparing $a^{m-1}b$ with $\boldsymbol{x}[1..m]$, we have available the information that

$$\boldsymbol{x}[1..m-1] = \boldsymbol{p}[1..m-1] = a^{m-1}.$$

In order to check whether $\boldsymbol{p}$ matches with $\boldsymbol{x}[2..m+1]$, it is unnecessary to make any further comparison with positions $\boldsymbol{x}[2..m-1] = a^{m-2}$: we need only compare

$$\boldsymbol{x}[m] : \boldsymbol{p}[m-1]$$

and, upon finding equality, discover then the mismatch

$$\boldsymbol{x}[m+1] \neq b = \boldsymbol{p}[m].$$

Applying this strategy at each position $i \leq n-m+1$ in $\boldsymbol{x}$ ensures that exactly $2n-m$ letter comparisons are performed, rather than $m(n-m+1)$, a result left to Exercise 7.1.4.

The key insight of KMP [MP70, KMP77] extends this example. In general, a (partial or full) match

$$\boldsymbol{x}[i..i+h-1] = \boldsymbol{p}[1..h],\ 1 \leq h \leq m,$$

can be used to avoid further testing of the substring $\boldsymbol{u} = \boldsymbol{x}[i..i+h-1]$. If $\boldsymbol{p}$ is actually to occur at some position $i' \in i..i+h-1$ of $\boldsymbol{x}$, then there must exist a nonempty proper prefix $\boldsymbol{p}[1..h']$ of $\boldsymbol{u} = \boldsymbol{p}[1..h]$, $h' = i+h-i'$, that is also a suffix $\boldsymbol{x}[i'..i+h-1]$ of $\boldsymbol{u}$. In other words (be sure you are sitting down), $\boldsymbol{p}[1..h']$ must be a *border* of $\boldsymbol{u} = \boldsymbol{p}[1..h]$! See Figure 7.1.

So the only possible positions i' at which $\boldsymbol{p}$ can occur are those determined by the borders of $\boldsymbol{u}$. But since each $\boldsymbol{u} = \boldsymbol{p}[1..h]$ for some $h \in 1..m$, we do not need to compute the borders of $\boldsymbol{u}$ during the execution of our algorithm — they can be *pre*computed for every position h in $\boldsymbol{p}$ and stored in a border array $\boldsymbol{\beta} = \boldsymbol{\beta}[1..m]$ for use as required. Thus the KMP algorithm has a preprocessing phase in which $\boldsymbol{\beta}$ is computed in $\Theta(m)$ time using Algorithm Border Array (1.3.1), followed by the matching algorithm itself.

In fact, the precomputed array is a very slight variant of $\boldsymbol{\beta}$. Let j be the position in $\boldsymbol{p}$ of the first mismatch. Then the prefix of $\boldsymbol{p}$ that does not need to be checked in the next attempted match is just the longest border of $\boldsymbol{p}[1..j-1]$, and so the next letter comparison should take place with

$$\boldsymbol{p}\big[\boldsymbol{\beta}[j-1]+1\big],$$

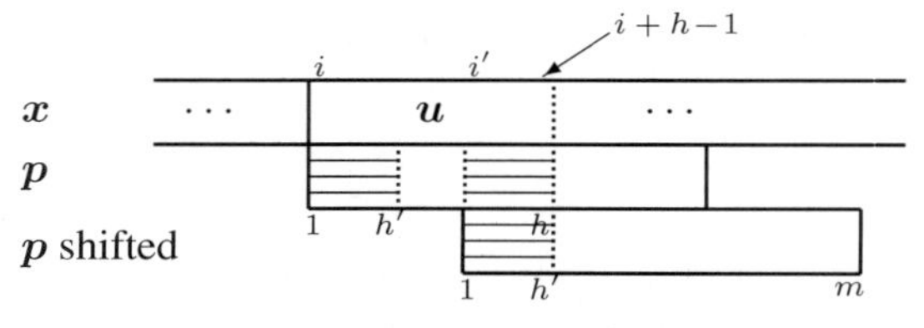

Figure 7.1. $\boldsymbol{p}[1..h']$ must be a border of $\boldsymbol{p}[1..h]$

where of course $\beta[j-1] < j-1$ is the length of the longest border of $\boldsymbol{p}[1..j-1]$. Thus the array that is actually precomputed is

$$\beta'[j] = \beta[j-1] + 1$$

for every $j \in 2..m+1$, with a special value $\beta'[1] = 0$. Note that for $j \geq 2$, $\beta[j-1]+1 < (j-1)+1 = j$, so that $\beta'[j] < j$. When $j = 1$, there has been no partial match at all, and so in this case it is the value of i that needs to be changed, rather than the value of j. With this understanding, KMP can be simply described as shown in Algorithm 7.1.1. Observe that the assignment $j \leftarrow \beta'[j]$ resulting from a mismatch $\boldsymbol{p}[j] \neq \boldsymbol{x}[i]$ corresponds to a right shift of $\boldsymbol{p}$ along $\boldsymbol{x}$ by $j - \beta'[j]$ positions.

Algorithm 7.1.1 (KMP)

– *Find all occurrences of* $\boldsymbol{p}$ *in* $\boldsymbol{x}$

$i \leftarrow 1; j \leftarrow 1$
while $i \leq n-m+j$ **do**
  $(i, j) \leftarrow$ *match*(i, j, m)
  if $j = m+1$ **then output** $i-m$
  if $j = 1$ **then** $i \leftarrow i+1$
  else $j \leftarrow \beta'[j]$

Note that the *match* routine in KMP executes just like its counterpart in Algorithm Easy: a character-by-character comparison from left to right. But the KMP *match* routine differs slightly in that it starts by comparing $\boldsymbol{x}[i]$ with $\boldsymbol{p}[j]$, for some $1 \leq j \leq m$, rather than always starting with $\boldsymbol{p}[1]$; further, *match* needs to return a value of i as well as of j to indicate where the mismatch has occurred in both $\boldsymbol{x}$ and $\boldsymbol{p}$.

Based on this discussion, we conclude:

Theorem 7.1.1 *Algorithm 7.1.1 correctly computes all occurrences of a given nonempty pattern* $\boldsymbol{p}$ *in a given string* $\boldsymbol{x}$. □

In order to establish the complexity of Algorithm KMP, consider the effect of the letter comparisons carried out in routine *match*. If equality is found in such a comparison, so that

$$\boldsymbol{x}[i] = \boldsymbol{p}[j],$$

then both i and j are incremented by one: in other words, i is incremented while $\boldsymbol{p}$ remains stationary with respect to $\boldsymbol{x}$. If on the other hand

$$\boldsymbol{x}[i] \neq \boldsymbol{p}[j],$$

then for $j = 1$ *both* an increment of i *and* a right shift of $\boldsymbol{p}$ take place, while for $j > 1$ only a right shift of $\boldsymbol{p}$, by $j - \beta'[j] \geq 1$ positions, occurs. Observe that i can be incremented at most n times, while $\boldsymbol{p}$ can be shifted right at most $n - m + 1$ times; however, these two upper bounds cannot be attained simultaneously, since attaining either one forces exit from the **while** loop. It follows that the maximum number of letter comparisons is at most

$$n + (n - m + 1) - 1 = 2n - m.$$

Since all other steps in KMP require only constant time for each execution of the **while** loop, we have

Theorem 7.1.2 *Algorithm 7.1.1 executes in $O(n)$ time and uses $\Theta(m)$ additional space, while performing at most $2n - m$ comparisons on elements of $\boldsymbol{x}$.* □

The example $\boldsymbol{x} = a^n$, $\boldsymbol{p} = a^{m-1}b$ shows that the upper bound on letter comparisons is least possible. If the $\Theta(m)$ preprocessing time is added to the time for the execution of Algorithm 7.1.1, the total time required to compute all occurrences of $\boldsymbol{p}$ in $\boldsymbol{x}$ becomes $O(n + m)$. In the context of Digression 1.2.1, observe that this time bound depends on being able to report each occurrence of $\boldsymbol{p}$ in constant time, using only an integer to specify each location of $\boldsymbol{p}$ in $\boldsymbol{x}$.

It is instructive to consider also the example $\boldsymbol{x} = a^n$, $\boldsymbol{p} = a^m$, where the pattern $\boldsymbol{p}$ will be found $n - m + 1$ times in $\boldsymbol{x}$. Because $\boldsymbol{p}$ has border a^{m-1}, $\boldsymbol{\beta}'[m+1] = m$, and so in order to report each occurrence of $\boldsymbol{p}$ after the first one, only one letter comparison is required: $\boldsymbol{p}[m]$ against $\boldsymbol{x}[i]$. Thus for this example KMP has the remarkable property that it needs only n letter comparisons to report $n - m + 1$ occurrences of $\boldsymbol{p}$ — constant time per occurrence as n grows large with respect to m. We shall see in the next section that this case (a highly periodic pattern occurring frequently in the text) provides much more difficulty for KMP's main competitor, Algorithm BM.

Having introduced Algorithm KMP, we now show how to improve upon it — rather, how to improve the calculation of the auxiliary array $\boldsymbol{\beta}'$. Recall that in Algorithm 7.1.1 the discovery that $\boldsymbol{x}[i] \neq \boldsymbol{p}[j]$, $j > 1$, leads to the replacement $j \leftarrow \boldsymbol{\beta}'[j]$. But this replacement is useless if in fact $\boldsymbol{p}[j] = \boldsymbol{p}\big[\boldsymbol{\beta}'[j]\big]$ — it will still on the next iteration be true that $\boldsymbol{x}[i] \neq \boldsymbol{p}[j]$! Thus what we are really interested in (if it exists) is the longest border of $\boldsymbol{p}[1..j-1]$ that is *not* followed by the letter $\boldsymbol{p}[j]$. In other words, we seek the largest integer j', if it exists, such that

- $\boldsymbol{p}[1..j'-1]$ is a border of $\boldsymbol{p}[1..j-1]$;
- $\boldsymbol{p}[1..j']$ is *not* a border of $\boldsymbol{p}[1..j]$.

This observation leads to the construction of a new auxiliary array $\boldsymbol{\beta}''$ whose jth element is defined as follows:

If $j = m+1$, $\boldsymbol{\beta}''[j] \leftarrow \boldsymbol{\beta}'[j]$. Otherwise, for $j \in 1..m$, $\boldsymbol{\beta}''[j] \leftarrow j'$ where $j' - 1$ is the length of the longest border of $\boldsymbol{p}[1..j-1]$ such that $\boldsymbol{p}[j'] \neq \boldsymbol{p}[j]$; if no such border exists, $\boldsymbol{\beta}''[j] \leftarrow 0$.

Recall from Section 1.3 that the borders of $\boldsymbol{p}[j-1]$ may be specified in descending order of length by the set

$$S_{j-1} = \{\boldsymbol{\beta}[j-1], \boldsymbol{\beta}^{\boldsymbol{2}}[j-1], \ldots, \boldsymbol{\beta}^{\boldsymbol{k}}[j-1]\}, \tag{7.1}$$

where k is the unique integer for which $\boldsymbol{\beta}^{\boldsymbol{k}}[j-1] = 0$. Thus, for $j \in 2..m$, a nonzero value j' of $\boldsymbol{\beta}''[j]$ occurs if and only if there exists a length $j'-1 \in S_{j-1}$ that satisfies the condition given above.

Note that the above definition applies even when the length $j'-1$ of the longest suitable border is zero. When for some $j < m+1$ and *every* border length $j'-1$, $\boldsymbol{p}[j'] = \boldsymbol{p}[j]$, the special value $\boldsymbol{\beta}''[j] = 0$ is used to indicate that the next letter comparison should be

$$\boldsymbol{x}[i+1] : \boldsymbol{p}[1],$$

just as in the case when *match* returns $j = 1$. The details of the relatively minor changes required to transform $\boldsymbol{\beta}$ (or $\boldsymbol{\beta}'$) into $\boldsymbol{\beta}''$ and to update Algorithm 7.1.1 are left to Exercises 7.1.9 and 7.1.10. Also left as an exercise (7.1.11) is the demonstration that the new algorithm, which we designate by KMP*, has exactly the same worst-case behaviour that KMP does. Thus

Theorem 7.1.3 *Algorithm KMP* correctly computes all occurrences of a given nonempty pattern* $\boldsymbol{p}[1..m]$ *in a given string* $\boldsymbol{x}[1..n]$ *using* $O(n)$ *time,* $\Theta(m)$ *additional space, and performing at most* $2n-m$ *letter comparisons.* □

It is of interest to examine the three arrays $\boldsymbol{\beta}$, $\boldsymbol{\beta}'$ and $\boldsymbol{\beta}''$ as they apply to a Fiboexample last seen in Section 5.1. We find that the differences between $\boldsymbol{\beta}'$ and $\boldsymbol{\beta}''$ can be substantial; in particular, notice that for $\boldsymbol{f_7}$ all occurrences at position j of a preceded by b must yield $\boldsymbol{\beta}''[j] = 0$, since every Fibostring begins with a and contains no occurrence of bb.

	1	2	3	4	5	6	7	8	9	10	11	12	13	14	15	16	17	18	19	20	21	22
$\boldsymbol{f_7} =$	a	b	a	a	b	a	b	a	a	b	a	a	b	a	b	a	a	b	a	b	a	
$\boldsymbol{\beta} =$	0	0	1	1	2	3	2	3	4	5	6	4	5	6	7	8	9	10	11	7	8	
$\boldsymbol{\beta}' =$	0	1	1	2	2	3	4	3	4	5	6	7	5	6	7	8	9	10	11	12	8	9
$\boldsymbol{\beta}'' =$	0	1	0	2	1	0	4	0	2	1	0	7	1	0	4	0	2	1	0	12	0	9

Algorithms KMP and KMP* are elegant and interesting, and they have stimulated much research into methods for computing specific patterns in strings. However, these algorithms are unfortunately not efficient: both theoretical [BY89b] and empirical [S82, DB86] studies show that KMP* is not significantly faster on the average than Algorithm Easy, even though its worst-case performance is of course much better. In the next section we introduce new ideas that do turn out to lead to a significant improvement in average search efficiency.

Exercises 7.1

1. Write the KMP routine *match* and prove its correctness.
2. Suppose that appended to $\boldsymbol{x}$ and $\boldsymbol{p}$ there are special distinct letters (called ***sentinels***) $\$_1$ and $\$_2$, respectively, not equal to any other letters in the alphabet. Rewrite *match* so as to make use of these appendages, and comment on the new routine's efficiency compared to the one in the previous exercise.
3. Suppose that Algorithm KMP is applied to $\boldsymbol{x} = a^n, \boldsymbol{p} = a^m, 1 \leq m \leq n$, using:

   (a) the *match* routine of the previous exercise;

   (b) the *match* routine of the previous exercise but one.

   In the text it was shown that in case (b) only n letter comparisons were required. Compute the number of letter comparisons required in case (a), and hence draw a conclusion about the use of the minimum number of letter comparisons as a criterion for designing efficient pattern-matching algorithms.
4. Suppose that Algorithm KMP (with its original *match* routine) is executed to find all occurrences of $\boldsymbol{p} = a^{m-1}b$ in $\boldsymbol{x} = a^n$, $2 \leq m \leq n$. Show that exactly $2n - m$ letter comparisons are performed.
5. Prove the statement made in the text that the upper bound on increments and the upper bound on right shifts of $\boldsymbol{p}$ cannot be simultaneously attained in KMP.
6. Nothing was said in the text about the *lower* bound on the number of letter comparisons required by Algorithm KMP. Show that this lower bound is exactly $n - m + 1$. What then can be said then about the *average* number of letter comparisons required by KMP, taking into account the preprocessing of $\boldsymbol{p}$?
7. For distinct letters $\lambda_1, \lambda_2, \ldots, \lambda_k$, how many letter comparisons does KMP require to compute all occurrences of a pattern $\boldsymbol{p} = (\lambda_1\lambda_2\cdots\lambda_k)^{m/k}$ in a string $\boldsymbol{x} = (\lambda_1\lambda_2\cdots\lambda_k)^{n/k}$?
8. Prove that the offset $\boldsymbol{\beta}''[j]$ corresponding to a mismatch at position j equals the offset $\boldsymbol{\beta}''[\boldsymbol{\beta}'[j]]$ corresponding to a mismatch at position $\boldsymbol{\beta}'[j]$ if and only if $\boldsymbol{p}[j] = \boldsymbol{p}[\boldsymbol{\beta}'[j]]$. Hence show that

$$\boldsymbol{\beta}''[j] = \boldsymbol{\beta}'[j] \text{ or } \boldsymbol{\beta}''\left[\boldsymbol{\beta}'[j]\right].$$

9. Assuming that $\boldsymbol{p}$ is padded on the right with the unique letter \$, revise Algorithm 1.3.1 to compute

   (a) $\boldsymbol{\beta}'$

   (b) $\boldsymbol{\beta}''$

   in $\Theta(m)$ time. Prove the correctness of your revised algorithms based on the assumption that Algorithm 1.3.1 is correct.

Hints: If done correctly, Revision (a) should actually be *simpler* than Algorithm 1.3.1. Then, using the preceding exercise, Revision (b) should be a straightforward modification of Revision (a). But can you find a way to compute β'' without having to store the array β' as well?

10. Revise Algorithm KMP to make use of β'' rather than β'.

11. Compute β'' for $p = a^{m-1}b$, hence show that Algorithm KMP* requires at most $2n - m$ letter comparisons in the worst case.

12. Compute β' and β'' for

(a) $\boldsymbol{p} = abcdabce$;
(b) $\boldsymbol{p} = abcabcacab$;
(c) $\boldsymbol{p} = abacabadabacabae$.

Is there a string for which $\beta'' = \beta'$?

7.2 Boyer-Moore

Kind words can be short and easy to speak,
but their echoes are truly endless.

— Mother TERESA (1910–1997)

Like KMP, Algorithm BM [BM77] shifts the pattern $\boldsymbol{p}$ along the text string $\boldsymbol{x}$ from left to right; unlike KMP, however, BM tests the letters of the pattern itself *from right to left* — that is, in the order $\boldsymbol{p}[m], \boldsymbol{p}[m-1], \ldots, \boldsymbol{p}[1]$. As we shall see, this apparently innocuous change revolutionizes the strategy for determining, in case of mismatch, the next placement of $\boldsymbol{p}$ with respect to $\boldsymbol{x}$. This is true even though, again like KMP, BM relies upon preprocessed arrays in order to compute each shift of $\boldsymbol{p}$ in constant time: in fact, we shall find that one of the preprocessed arrays of BM turns out to be closely connected to the β'' array described in the previous section.

Before dealing with the details of Algorithm BM, we should make clear some of the limitations on its use. Even though many pattern-matching problems that arise in practice are consistent with these restrictions, nevertheless they are relatively severe, and there are also many problems that are not consistent with them:

- One of BM's preprocessed arrays, usually called δ_1, depends upon knowledge of occurrences in $\boldsymbol{p}$ of each letter of the alphabet of $\boldsymbol{x}$. Thus the alphabet of $\boldsymbol{x}$ must be finite and it must be known in advance.
- Furthermore, the access to elements in δ_1 is based on letters of the alphabet; that is, it is supposed that the array reference $\delta_1[\lambda]$ will access in constant time a quantity that

has been precomputed for the specified letter λ. This means that, in the terminology of Section 4.1, the alphabet should be *indexed*. Recall also from Section 4.1 that if the alphabet is ordered but not indexed, as much as $\Theta(\log m)$ time may be required to identify a particular letter λ in $\boldsymbol{p}[1..m]$ and to access a quantity associated with it. In such cases, therefore, Algorithm BM would probably not be the algorithm of choice.

- Finally, since the pattern $\boldsymbol{p}$ is searched from right to left, the entries in $\boldsymbol{p}$ need to be accessible in reverse order. In some cases, as suggested in Digression 1.2.1, this may require some linear-time preprocessing of $\boldsymbol{p}$ to copy it into an array.

Algorithm 7.2.1 (BM)

– *Find all occurrences of* $\boldsymbol{p}$ *in* $\boldsymbol{x}$

$i \leftarrow m$
while $i \leq n$ **do**
  $(i, j) \leftarrow hctam(i, m)$
  if $j = 0$ **then output** $i + 1$
  $i \leftarrow i + \max\{\boldsymbol{\delta_1}[\boldsymbol{x}[i]], \boldsymbol{\delta_2}[j]\}$

Without further elaboration, we present BM as shown in Algorithm 7.2.1. I hope that the reader will not think me frivolous for giving the name *hctam* to the routine that looks for a match of $\boldsymbol{p}$ with $\boldsymbol{x}$ by processing $\boldsymbol{p}$ character-by-character from the right. Observe that now a match with $\boldsymbol{p}$ is returned when j, initially set to m, is reduced to zero, contrary to the KMP algorithm where j was increased to $m+1$ from some initial value — thus *hctam* is in a sense simpler than *match* because it does not depend on the previous value of j. The algorithm as shown is transparently correct, provided of course that *hctam* does its job, and provided that the shift of the pattern determined by reference to the $\boldsymbol{\delta_1}$ and $\boldsymbol{\delta_2}$ arrays is correct (does not shift past any possible matches) and increments i by at least one. Much of the remainder of this section is devoted to a discussion of the computation and use of these arrays. Meanwhile, we state

Theorem 7.2.1 *Subject to correct implementation of* hctam, $\boldsymbol{\delta_1}$ *and* $\boldsymbol{\delta_2}$, *Algorithm 7.2.1 correctly computes all occurrences of a given nonempty pattern* $\boldsymbol{p}[1..m]$ *in a given string* $\boldsymbol{x}[1..n]$. □

The main idea of the $\boldsymbol{\delta_1}$ array is a simple one. We suppose that the routine *hctam* has found a (possibly empty) partial match

$$\boldsymbol{x}[i+1..i+h] = \boldsymbol{p}[m-h+1..m]$$

for some $i \in 0..n-h$ and some $h \in 0..m$, thus returning the pair $(i, j) = (i, m-h)$. As for the KMP *match* routine, and as suggested in Exercise 7.2.2, it is convenient to prepend

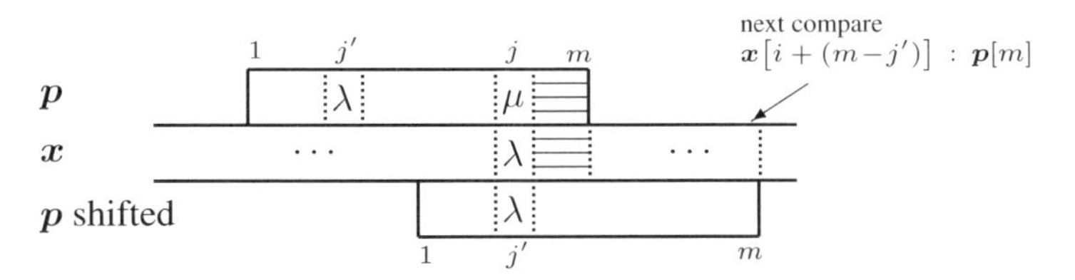

Figure 7.2. $j' < j$: $\boldsymbol{p}$ shifted to match the rightmost λ in $\boldsymbol{p}$

sentinels to both $\boldsymbol{x}$ and $\boldsymbol{p}$, so that $\boldsymbol{x}[0] = \$_1$ and $\boldsymbol{p}[0] = \$_2$, and thus we may assume that in every case, including $i = 0$ or $j = 0$, a mismatch

$$\lambda = \boldsymbol{x}[i] \neq \boldsymbol{p}[j] = \mu \tag{7.2}$$

has been found. We now make the observation that if any substring that includes position i of $\boldsymbol{x}$ is to match with $\boldsymbol{p}$, it must be true that $\boldsymbol{p}$ includes an occurrence of the letter $\lambda = \boldsymbol{x}[i]$. Further, if we know the position, say j', of the *rightmost* occurrence of λ in $\boldsymbol{p}$, then we can make the *minimum* shift of $\boldsymbol{p}$ consistent with the requirement that $\boldsymbol{x}[i] = \lambda$. As shown in Figure 7.2, in the case that $j' < j$, this leads to a right shift of $\boldsymbol{p}$ by $j - j'$ positions, so that the next call of *hctam* will begin by comparing $\boldsymbol{p}[m]$ with $\boldsymbol{x}\big[i + (m - j')\big]$.

Thus, in this case, the required change to i is

$$i \leftarrow i + (m - j'), \tag{7.3}$$

and so for each letter λ (including $\$_1$) in the alphabet A, we must compute the position j' of the rightmost occurrence of λ in $\boldsymbol{p}$, setting $j' \leftarrow 0$ if λ does not occur in $\boldsymbol{p}$ at all. In our preprocessing, we then set

$$\boldsymbol{\delta_1}[\lambda] \leftarrow m - j', \tag{7.4}$$

so that whenever a mismatch (7.2) occurs *and* $j' < j$ (that is, $\boldsymbol{\delta_1}[\lambda] > m - j$), we may recompute the next position to be matched using

$$i \leftarrow i + \boldsymbol{\delta_1}[\lambda],$$

as in Algorithm 7.2.1.

Using δ_1 and δ_2

"All very well," the careful reader is no doubt muttering, "but what happens when $\boldsymbol{\delta_1}[\lambda] < m - j$? What then?" A good point: this circumstance may arise for any $j < m$ and in fact *must* occur for $j \leq 1$. All that we can safely do in this case is to move $\boldsymbol{p}$ one position to the right, so that $\boldsymbol{p}[m]$ is compared with $\boldsymbol{x}\big[i + (m - j) + 1\big]$, equivalent to the assignment

$$i \leftarrow i + (m - j) + 1. \tag{7.5}$$

Comparing (7.3) and (7.5), we find that for $j' > j$ $\big(\boldsymbol{\delta_1}[\lambda] < m - j\big)$,

$$(m - j) + 1 > m - j',$$

so that as expected (7.5) would always be preferred. But we shall discover below that in fact

$$\boldsymbol{\delta_2}[j] \geq (m-j)+1 \tag{7.6}$$

for all j, and so the assignment (7.5) is rendered unnecessary by the assignment

$$i \leftarrow i + \max\left\{\boldsymbol{\delta_1}[\boldsymbol{x}[i]], \boldsymbol{\delta_2}[j]\right\}$$

of Algorithm 7.2.1. Pending a proof of (7.6), so much for $\boldsymbol{\delta_1}$: as we find out in Exercises 7.2.4 and 7.2.5, the computation of this array is straightforward, at least for an indexed alphabet.

Like $\boldsymbol{\delta_1}$, the array $\boldsymbol{\delta_2} = \boldsymbol{\delta_2}[0..m]$ is employed after a partial match

$$\boldsymbol{x}[i+1..i+h] = \boldsymbol{p}[m-h+1..m],$$

for some $i \in 0..n-h$ and some $h \in 0..m$, is terminated by a mismatch (7.2), where as before $j = m-h$. But on the other hand the calculation of $\boldsymbol{\delta_2}$ also has something in common with the array $\boldsymbol{\beta''}$ used in KMP*: it is another border-like calculation with additional constraints applied.

More precisely, after the partial match we look for the largest $j' < j$ such that

$$\boldsymbol{p}[j'+1..j'+h] = \boldsymbol{p}[j+1..m]$$

and

$$\mu' = \boldsymbol{p}[j'] \neq \boldsymbol{p}[j] = \mu.$$

If there exists such a j', we assign

$$i \leftarrow i + (m-j'), \tag{7.7}$$

as shown in Figure 7.3. Of course, the amount $\boldsymbol{\delta_2}[j] = m-j'$ of the shift can be precomputed for each $j \in 1..m$. Note the similarity between Figures 7.2 and 7.3.

If there is no j' that satisfies this condition (including the case in which $j = 0$), then we look instead for the largest $j' < m$ such that $\boldsymbol{p}[1..j']$ is a suffix of $\boldsymbol{p}[j+1..m]$ — thus $\boldsymbol{p}[1..j']$ will be the longest border of $\boldsymbol{p}$ such that $j' \leq h$. In this case, as shown in Figure 7.4, we must assign

$$i \leftarrow i + (m-j) + (m-j'), \tag{7.8}$$

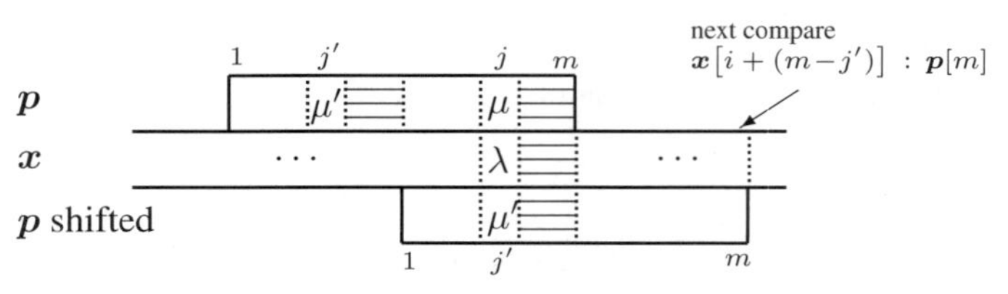

Figure 7.3. Type I shift: $\boldsymbol{\delta_2}[j] \leftarrow m-j'$

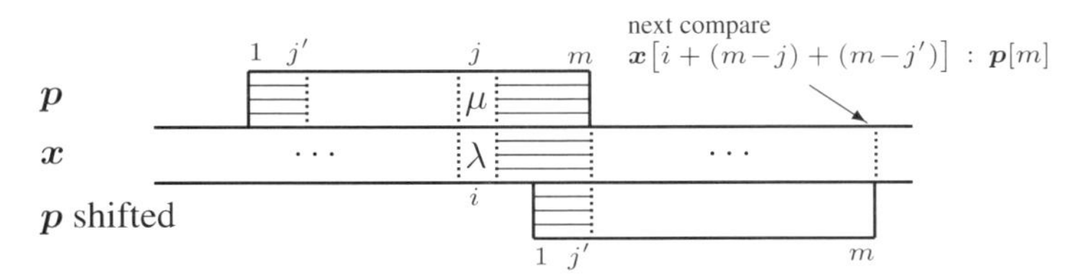

Figure 7.4. Type II shift: $\delta_2[j] \leftarrow (m-j)+(m-j')$

and of course we will again precompute the displacement $\delta_2[j] = (m-j)+(m-j')$ for every j.

We claim that the calculation of $\delta_2[j]$ described in Figures 7.3 and 7.4 satisfies the following two conditions:

- $\delta_2[j]$ is ***safe***; that is, it shifts the pattern to the right along the string by a minimum amount consistent with the requirements that a proper substring of p should match with $x[i..i+h-1]$ (Figure 7.3) or else that a proper prefix of p should match with a suffix of $x[i..i+h-1]$ (Figure 7.4). The minimum shift is ensured by choosing the largest j' in the corresponding assignments (7.7) and (7.8).
- $\delta_2[j] \geq (m-j)+1$ for every j, as claimed in (7.6). This follows in (7.7) from the fact that $j' < j$, and in (7.8) from the fact that $m - j' \geq 1$. Thus it does indeed suffice to consider only the shift $\max\{\delta_1[\lambda], \delta_2[j]\}$. Note further that this shift ensures that p is moved right at least one position at each step, and so Algorithm BM must terminate.

In view of these facts, we conclude that the use of $\delta_1[\lambda]$ and $\delta_2[j]$ ensures the correctness of Algorithm 7.2.1.

Computing δ_2

We turn now to the computation of δ_2, not as easy as that of δ_1, but therefore more interesting: a real test, as we shall see, of our "borderarraymanship". Essentially, we are asked to compute a position j' as a function of a given position $j \in 0..m$. The computation of course breaks down into two parts: a first one (which we call ***Type I***) based on Figure 7.3, valid only for those values of j for which a corresponding j' actually exists (that is, valid only where there exists a border of the required length preceded by $\mu' \neq \mu$); then a second one (***Type II***) based on Figure 7.4, invoked for those values of j whose corresponding j' can*not* be computed in the first part. An important distinction between Type I and Type II shifts is this: the latter shifts p to the right of the current position i of mismatch, while the former does not. For clarity of discussion, suppose that a Type I computation yields an array $g_I = g_I[1..m]$, where $j' = g_I[j]$ is the value corresponding to j, whenever it exists; while a Type II one similarly yields $g_{II} = g_{II}[0..m]$. The calculation of g_{II} is easier, and so we consider it first.

Suppose that the border array $\beta = \beta[1..m]$ of $p[1..m]$ has been computed in linear time using Algorithm Border Array (1.3.1). Then, given a value $j \in 1..m$, it is clear from Figure 7.4 that the value of $j' = g_{II}[j]$ should simply be the length of the longest border of

$\boldsymbol{p}$ whose length is at most $h = m - j$. Thus to fill in the values in $\boldsymbol{g}_{II}$, we need only access elements of the descending sequence

$$S_m = \langle \boldsymbol{\beta}[m], \boldsymbol{\beta}^2[m], \ldots, \boldsymbol{\beta}^k[m] = 0 \rangle,$$

identified in Section 1.3. That is, for every $0 < j \leq m - \boldsymbol{\beta}[m]$, we set

$$\boldsymbol{g}_{II}[j] = \boldsymbol{\beta}[m];$$

while for every $m - \boldsymbol{\beta}[m] < j \leq m - \boldsymbol{\beta}^2[m]$, we set

$$\boldsymbol{g}_{II}[j] = \boldsymbol{\beta}^2[m];$$

and so on. For $j = 0$, the case in which a match with $\boldsymbol{p}[1..m]$ has been found, we set $\boldsymbol{g}_{II}[j] = \boldsymbol{\beta}[m]$. Clearly the array $\boldsymbol{g}_{II}$ can be computed in linear time.

To compute $\boldsymbol{g}_I[j]$, $1 \leq j \leq m$, we need to find the largest integer $j' < j$, if it exists, such that

- $\boldsymbol{p}[j'+1..j'+(m-j)]$ is a border of $\boldsymbol{p}[j'+1..m]$;
- $\boldsymbol{p}[j'..j'+(m-j)]$ is *not* a border of $\boldsymbol{p}[j'..m]$.

If such a j' exists, we set $\boldsymbol{g}_I[j] \leftarrow j'$; if not, $\boldsymbol{g}_I[j] \leftarrow 0$.

In order to understand this computation better, consider the borders of suffixes of $\boldsymbol{p}$, a topic the diligent and retentive reader will recall being introduced in Exercise 1.3.10 as the "right border array". Let us denote this array by $\boldsymbol{\rho} = \boldsymbol{\rho}[1..m]$, where $\boldsymbol{\rho}[j]$ gives the length of the longest border of $\boldsymbol{p}[j..m]$. Recall from the same exercises that if $\boldsymbol{\beta}_T$ denotes the border array of the transposed string

$$\boldsymbol{p}_T = \boldsymbol{p}[m]\boldsymbol{p}[m-1]\cdots\boldsymbol{p}[1],$$

then $\boldsymbol{\beta}_T[m-i+1] = \boldsymbol{\rho}[i]$ for every $i \in 1..m$. Hence the lengths of the borders of $\boldsymbol{p}[i..m]$ are elements of the descending sequence

$$S^T_{m-i+1} = \langle \boldsymbol{\beta}_T[m-i+1], \boldsymbol{\beta}^2_T[m-i+1], \ldots, \boldsymbol{\beta}^k_T[m-i+1] \rangle,$$

where as before k is the integer for which $\boldsymbol{\beta}^k_T[m-i+1] = 0$. Then the calculation of $\boldsymbol{g}_I[j]$ requires us to find the largest integer $j' < j$ such that $m-j$ is an element of $S^T_{m-j'}$ while $m-j+1$ is not an element of $S^T_{m-j'+1}$. Observe that if b^* is the largest element of $\boldsymbol{\beta}_T$, then $\boldsymbol{g}_I[j] \leftarrow 0$ for every j such that $m-j > b^*$.

The calculation of $\boldsymbol{g}_I$ is thus very similar, but not identical, to the calculation of $\boldsymbol{\beta}''$ described in Section 7.1: with some adjustments to the details, similar processing can be used, also requiring only $\Theta(m)$ time.

We leave these details to Exercise 7.2.8: recall that in the details is where the devil hides! Pending completion of that exercise, we state formally the main result of this subsection:

Theorem 7.2.2 *Corresponding to a given pattern* $\boldsymbol{p} = \boldsymbol{p}[1..m]$*, the BM array* $\boldsymbol{\delta_2} = \boldsymbol{\delta_2}[0..m]$ *can be computed in* $\Theta(m)$ *time.* □

Finally, we refer the reader interested in this diabolical calculation to [R80], where the first correct linear-time calculation of $\boldsymbol{\delta_2}$ was presented, and more recently to [CHL02], where a simpler version is proposed.

BM Examples

In order to better understand the calculation and use of $\boldsymbol{\delta_1}$ and $\boldsymbol{\delta_2}$ in Algorithm BM, we consider again the same Fiboexample introduced in the previous section:

	0	1	2	3	4	5	6	7	8	9	10	11	12	13	14	15	16	17	18	19	20	21
$\boldsymbol{f_7} =$	$\$_2$	a	b	a	a	b	a	b	a	a	b	a	a	b	a	b	a	a	b	a	b	a
$\boldsymbol{\beta} =$		0	0	1	1	2	3	2	3	4	5	6	4	5	6	7	8	9	10	11	7	8
$\boldsymbol{\rho} =$		8	7	6	5	4	3	2	1	8	7	6	5	4	3	2	1	3	2	1	0	0
$\boldsymbol{\beta_T} =$		0	0	1	2	3	1	2	3	4	5	6	7	8	1	2	3	4	5	6	7	8
$\boldsymbol{g_I} =$		0	0	0	0	0	0	0	0	0	0	0	0	8	0	0	0	0	16	0	16	20
$\boldsymbol{g_{II}} =$	8	8	8	8	8	8	8	8	8	8	8	8	8	8	3	3	3	3	3	1	1	0
$\boldsymbol{\delta_2} =$	34	33	32	31	30	29	28	27	26	25	24	23	22	13	25	24	23	22	5	22	5	1

The computation of $\boldsymbol{\delta_1}$ is easy:

$$\boldsymbol{\delta_1}[a] = 0; \quad \boldsymbol{\delta_1}[b] = 1; \quad \boldsymbol{\delta_1}[\lambda] = 21, \text{ for } \lambda \neq a, b.$$

Thus in searching any text string $\boldsymbol{x}$ on $\{a, b\}$ for matches with $\boldsymbol{f_7}$, there is no advantage in using $\boldsymbol{\delta_1}$, since as we observe it is always true that, for every $j \in 0..m$,

$$\max\left\{\boldsymbol{\delta_1}[\lambda], \boldsymbol{\delta_2}[j]\right\} = \boldsymbol{\delta_2}[j]. \tag{7.9}$$

On the other hand, if $\boldsymbol{x}$ contains other letters, there can be some advantage, for $j = 13, 18, 20, 21$, in using $\boldsymbol{\delta_1}$.

To illustrate the case in which however the use of $\boldsymbol{\delta_1}$ is more important, it is hard to improve upon the example given in the original BM paper [BM77]: the pattern $\boldsymbol{p} =$ `AT-THAT` is matched against the text

$$\boldsymbol{x} = \cdots \texttt{WHICH-FINALLY-HALTS.--AT-THAT-POINT} \cdots.$$

In this case, $\boldsymbol{\delta_2}[0..7] = 12\ 11\ 10\ 9\ 8\ 7\ 4\ 1$ while

$$\boldsymbol{\delta_1}[\texttt{A}] = 1, \quad \boldsymbol{\delta_1}[\texttt{T}] = 0, \quad \boldsymbol{\delta_1}[\texttt{-}] = 4 \quad \boldsymbol{\delta_1}[\texttt{H}] = 2,$$

with $\boldsymbol{\delta_1}[\lambda] = 7$ for all other letters λ. The effects of $\boldsymbol{\delta_1}$ and $\boldsymbol{\delta_2}$ during the matching process are indicated below with vertical arrows pointing to the starting positions for the letter comparisons performed by *hctam*. Note that altogether only 15 letter comparisons are required to locate $\boldsymbol{p}$, of which eight are consumed matching AT-THAT and finding the final mismatch of $\boldsymbol{p}[0] = \$_2$.

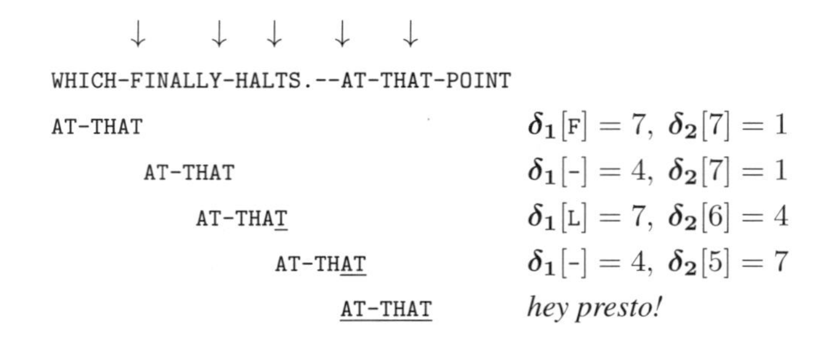

Efficiency of BM

Unlike KMP, Algorithm BM represents more than a theoretical improvement on Algorithm Easy (2.2.1) — as suggested by the preceding example, BM also executes more quickly in practice. In fact, both empirical [S82] and theoretical [BYGR90, BYR92] studies confirm that the average number of letter comparisons executed by BM is sublinear in the length of the text, even taking into account the letter comparisons performed during the computation of $\boldsymbol{\delta_1}$ and $\boldsymbol{\delta_2}$. Typically, for natural language text of length n, the average number of letter comparisons is about $0.3n$, including preprocessing overheads, for patterns of length $m \geq 10$ with $m \ll n$ [S82]. This compares favourably with the KMP algorithm where the corresponding average is in excess of $1.0n$ (see Exercise 7.1.6).

It turns out that, in the average case, the use of $\boldsymbol{\delta_1}$ in BM is important in ensuring that the number of letter comparisons is "sublinear" in n — especially when the average is taken over English-language and other large alphabets. In the worst-case analysis, however, the effect of $\boldsymbol{\delta_1}$ can be made to disappear entirely simply by using a binary alphabet. This is shown above in the example $\boldsymbol{x} = \boldsymbol{f_7}$ and also below in the example $\boldsymbol{x} = (a^k b)^r$ — essentially, we ensure that (7.9) always holds. Thus, for the worst-case analysis, we always make the simplifying assumption that $\boldsymbol{\delta_1}$ can be ignored.

Unfortunately, whether or not $\boldsymbol{\delta_1}$ is included in BM, the worst-case behaviour is no better than that of Algorithm Easy: it turns out that $m(n-m+1)$ letter comparisons may be required. This fact can be inferred from the example $\boldsymbol{x} = a^n$, $\boldsymbol{p} = a^m$, so successfully dealt with by Algorithm KMP: now for each one of the $n-m+1$ occurrences of $\boldsymbol{p}$ in $\boldsymbol{x}$, BM requires m letter comparisons to identify it. This phenomenon reflects a sharp and important distinction between Algorithms KMP and BM: given repeated occurrences of a periodic pattern in $\boldsymbol{x}$, KMP needs to perform comparisons only on the period of $\boldsymbol{p}$; BM, on the other hand, must always perform a comparison on every position of every occurrence of $\boldsymbol{p}$ that it discovers. We shall see in Section 8.4 how to modify BM so as to guarantee linear-time behaviour even for highly periodic patterns. Meanwhile, we investigate the complexity properties of the original BM.

However, even avoiding periodic patterns and ignoring $\boldsymbol{\delta_1}$ still leaves a complicated problem: one clear advantage that KMP has over BM is that its worst-case behaviour is much easier to analyze! A difficult proof in [KMP77] showed that at most $7n$ letter comparisons are required by BM to find all occurrences of an aperiodic pattern; a few years later another difficult result [GO80] reduced the upper bound to $4n$. It was not until 1994 that a sharp upper bound of roughly $3n$ letter comparisons was published [C94]. The proof of this latter result is however also technical and tedious, and so we shall content ourselves here with presenting an easier proof, based on one given in [C94], that subject to a restriction on the periodicity of the pattern, BM requires at most $4n$ letter comparisons.

Before embarking on this proof, we first present a class of examples for which roughly $3n$ comparisons are required. Consider a text string $\boldsymbol{x} = (a^k b)^r$, where $k \geq 2$ and $r \geq 1$ are integers. We imagine using Algorithm BM to match a pattern $\boldsymbol{p} = a^{k-1}ba^{k-1}$ against $\boldsymbol{x}$: exactly $r-1$ occurrences of $\boldsymbol{p}$ will be found with $2k-1$ letter comparisons required for each occurrence. It is left to Exercise 7.2.9 to show that for $r \geq 3$, BM will perform exactly $(3k-1)(r-1)$ letter comparisons against $\boldsymbol{p}$. Observe then that as $k \to \infty$ and $r \to \infty$, the ratio

$$\frac{(3k-1)(r-1)}{n} = \frac{(3k-1)(r-1)}{(k+1)r} \to 3.$$

Hence for every real number $\varepsilon > 0$, there exist matching problems on text strings of length $n = n(\varepsilon)$ for which Algorithm BM requires more than $3n - \varepsilon$ letter comparisons. In view of the result [C94] quoted above, these problems essentially constitute worst-case behaviour for BM.

In order to establish the $4n$ worst-case upper bound on letter comparisons, it is convenient to introduce some new notation. We let $\boldsymbol{t}$ be the substring of $\boldsymbol{x}$ that *hctam* has currently found a match with; thus in fact $\boldsymbol{t} = \boldsymbol{p}[j+1..m]$, the suffix of $\boldsymbol{p}$ of length $t = |\boldsymbol{t}| = m - j$. Observe that the number of letter comparisons carried out by the current invocation of *hctam* is exactly $t+1$ (t matches together with one mismatch). Further, we let s be the length of the shift of $\boldsymbol{p}$ computed using $\boldsymbol{\delta_2}$ only — thus either $s = (m-j') - (m-j) = j - j'$ (Figure 7.3) or $s = m - j'$ (Figure 7.4). The suffix of $\boldsymbol{p}$ of length s is denoted by $\boldsymbol{s}$.

We consider now the string $\boldsymbol{ts}$. Observe from Figure 7.3 that for Type II shifts, $\boldsymbol{ts} = \boldsymbol{p}[j'+1..m]$ has border $\boldsymbol{p}[m-j]$, hence period $s = (m-j') - (m-j)$. On the other hand, we see from Figure 7.4 that for Type II shifts, $\boldsymbol{ts}$ also has border $\boldsymbol{p}[m-j]$; since in this case $|\boldsymbol{ts}| = (m-j) + (m-j')$, $\boldsymbol{ts}$ again has period $s = m - j'$.

For the current invocation of *hctam*, let f be the number of positions in $\boldsymbol{x}$ that are compared for the first time — that have not been compared by any previous invocation of *hctam*. Since at each step of Algorithm BM $\boldsymbol{p}$ is always shifted right at least one position, it follows that $f \geq 1$. We shall show that for *every* attempted match — that is, every invocation of *hctam* — the number $t+1$ of letter comparisons satisfies

$$t + 1 \leq f + 3s, \tag{7.10}$$

provided that in normal form $\boldsymbol{p} = \boldsymbol{u}^k\boldsymbol{u}'$, where $k < 3$. (In this case we say that $\boldsymbol{p}$ is ***not 3-periodic***.) Since over all attempted matches, neither of the sums $\sum f$, $\sum s$ can exceed n, we can conclude that the total number of letter comparisons, over all invocations of *hctam*, is at most $n + 3n = 4n$ for non-3-periodic patterns $\boldsymbol{p}$.

It is natural now to distinguish between a ***long shift*** ($3s \geq t$) and a ***short shift*** ($3s < t$). Observe then that for every attempted match that gives rise to a long shift, (7.10) must hold, and so it is necessary only to consider attempted matches that give rise to short shifts. Further, since the Type II shift of Figure 7.4 implies that $\boldsymbol{p}$ has a period $m - j'$, a short shift would require that

$$m - j' = s < t/3 = (m-j)/3 \leq m/3,$$

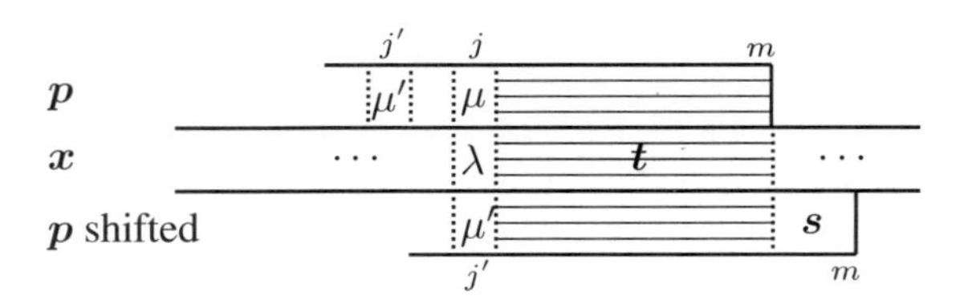

Figure 7.5. A short Type I shift

and so p would necessarily be 3-periodic, the case excluded by (7.10). Thus in order to prove (7.10), we need only consider the Type I shifts $s = j - j'$ of Figure 7.3 with the additional constraint that $3(j - j') < m - j$. In particular, since those cases in which there is a complete match of p with the text always lead to a Type II shift we may therefore in addition assume that no complete match occurs ($j > 0$ or, equivalently, $t < m$). The situation may be represented as shown in Figure 7.5, where we can verify the relationship stated above, that suffix $ts = p[j' + 1..m]$ has a border of length $t = m - j$, hence period $s = j - j'$.

It is possible that the repeated substring s of ts is itself a repetition, and so we may suppose without loss of generality that $s = v^k$ for some $k \geq 1$, where v is *not* a repetition. Letting $v = |v|$, we see then that both the substring t of x and the suffix ts of p have period v. Thus in right normal form (see Exercise 1.2.24), $ts = v'v^r$ where $r = \lfloor (s + t)/v \rfloor$ and v' is a possibly empty proper suffix of v, while $t = v'v^{r-k}$.

Having now thoroughly characterized the current match of t, we go on to reflect on previous matches — previous invocations of *hctam*.

Lemma 7.2.3 *If the current shift is of Type I (hence partial, with $t < m$), then in any previous match:*

(a) $p[m]$ *was not compared with any occurrence of* $v[v]$ *in* t *— that is, not compared with any position* $t[t - iv]$, $i = 0, 1, \ldots, r - k - 1$;

(b) *at most v comparisons took place between positions of* p *and* t;

(c) $p[m]$ *could only have been compared with some position either in the prefix* $t[1..v - 1]$ *or in* $t[t - (v - 1)..t - 1]$.

Proof (a) Suppose that such a comparison did occur. Since the current match compares $p[m]$ with $t[t]$, the rightmost occurrence of $v[v]$ in t, we may suppose that $p[m]$ was compared with $t[t - iv]$ for some integer $i \in 1..r - k - 1$, and that it was therefore found that

$$p[m - (t - iv) + 1..m] = t[1..t - iv]$$

with a mismatch $p[m - (t - iv)] = \mu \neq \lambda$, where as before λ is the letter occurring to the left of the occurrence of t in x.

We now ask ourselves what shift would take place as a result of the mismatch: certainly a short Type I shift (Figure 7.5), since the value j' used for the current match would be feasible. Some suitable larger value j' could perhaps be found, but it would always have to be true that $j' < j$, due to the fact that a complete prefix of $\boldsymbol{t}$ is matched before the mismatch occurs. We observe however that, just as in Figure 7.5, every such choice of j' would lead to a shift of $\boldsymbol{p}$ such that $\boldsymbol{p}[m]$ would lie to the right of $\boldsymbol{t}$. Thus the current match could not have taken place, a contradiction. We conclude that $\boldsymbol{p}[m]$ was not previously compared with any occurrence of $\boldsymbol{v}[v]$ in $\boldsymbol{t}$.

(b) Suppose that $\boldsymbol{p}[m]$ was aligned for comparison with $\boldsymbol{t}[q]$ for some integer $q \in 1..t$. It follows from (a) that the position q does not coincide with any occurrence of $\boldsymbol{v}[v]$. Thus there can be no more than v comparisons of $\boldsymbol{p}$ with $\boldsymbol{t}$, the last one being a mismatch, since otherwise $\boldsymbol{v}$ would equal a rotation of itself and hence, by Theorem 1.4.2, be a repetition, contrary to assumption.

(c) Suppose further that q is *not* in the specified ranges. Then q must be a position in some occurrence of $\boldsymbol{v}$; denote by $\bar{\boldsymbol{v}}$ this occurrence, and observe that after the mismatch, $\boldsymbol{p}[m]$ is at most shifted to be compared with $\bar{\boldsymbol{v}}[v]$. Therefore after a finite number of partial matches, $\boldsymbol{p}[m]$ will finally be compared with $\bar{\boldsymbol{v}}[v]$, in contradiction to (a). This completes the proof.

□

Note that this result does not depend on any assumptions about the periodicity of $\boldsymbol{p}$: it holds for all Type I matches. We use this lemma and the preceding discussion to prove the main result of this subsection:

Theorem 7.2.4 *If the pattern* $\boldsymbol{p}$ *is not 3-periodic, then Algorithm 7.2.1 requires at most* $4n$ *letter comparisons.*

Proof As discussed earlier, we may confine ourselves to a consideration of Type I shifts where only a partial match occurs ($t < m$). Thus Lemma 7.2.3 applies. In particular, by Lemma 7.2.3(b)–(c), at most the positions $[1..v-1]$ and $\left[t-(2v-2)..t-1\right]$ of $\boldsymbol{t}$ can have been previously compared over all previous attempted matches — a total of $3v-3 \leq 3s-3$ positions. Thus the number f of positions in $\boldsymbol{t}$ compared for the first time satisfies

$$f + 3s \geq t - (3s - 3) + 3s = t + 3.$$

Therefore (7.10) holds and so, as already explained, at most $4n$ letter comparisons are required. □

As noted above, an unconditional upper bound of $3n$ letter comparisons can be established using similar, but more detailed, arguments.

This completes for the time being our discussion of the Boyer-Moore algorithm. In the next chapter, however, we consider some of its many variants, including one (Section 8.4) whose linear-time behaviour is easily established, even for periodic patterns.

Exercises 7.2

1. Describe a data structure to retrieve values $\delta_1(\lambda)$ in $O(\log \alpha)$ time for letters λ in an ordered (but not indexed) alphabet A of size α.
2. Based on the assumption that the text $\boldsymbol{x}$ (respectively, the pattern $\boldsymbol{p}$) is padded in position 0 with a special sentinel $\$_1$ (respectively, $\$_2$), write the BM routine *hctam* and prove its correctness.
3. In the discussion of the function $\boldsymbol{\delta_1}$, the cases $j' < j$ and $j' > j$ are considered, but not $j' = j$. Account for this omission.
4. Given an indexed alphabet A and a pattern $\boldsymbol{p}$, describe an algorithm to compute the array $\boldsymbol{\delta_1}$ for $\boldsymbol{p}$. Determine the time complexity of your algorithm.
5. Solve the previous exercise in the case that the alphabet is not indexed, but only finite and ordered, by introducing a "pre-preprocessing" stage that converts the given alphabet A into an indexed alphabet A' and also converts $\boldsymbol{p}$ into a string on A'.
6. Compute $\boldsymbol{\delta_1}$ for the indexable alphabet $A = \{a, b, c, d, e\}$ and the pattern $\boldsymbol{p} = abcabacad$. For arbitrary $\boldsymbol{p}$, what value does $\delta_1[\boldsymbol{p}[m]]$ always take? Give an example of a string $\boldsymbol{x}$ such that use of $\boldsymbol{\delta_1}$ only for shifts would in fact lead to leftward shifts of $\boldsymbol{p}$ along $\boldsymbol{x}$.
7. Show how the idea of the $\boldsymbol{\delta_1}$ array can be adapted to apply to Algorithm KMP. Describe in detail the new computation of $\boldsymbol{\delta_1}$ and the modifications required to KMP. Do you think the extra processing required for these adaptations is justified?
8. Describe a $\Theta(m)$-time algorithm to compute the array $\boldsymbol{\delta_2}$ for $\boldsymbol{p}$. Prove your algorithm's correctness and time complexity. In addition to the $\Theta(m)$ storage for $\boldsymbol{p}$ and $\boldsymbol{\delta_2}$, how much space does your algorithm require? Do you think this additional space can be reduced to constant size?
9. Show that for $\boldsymbol{x} = (a^k b)^r$ and $\boldsymbol{p} = a^{k-1}ba^{k-1}$, exactly $(3k-1)(r-1)$ letter comparisons are required for $k \geq 2$, $r \geq 3$. How many are required for $r = 1, 2$? Assume that sentinels $\$_1$ and $\$_2$ are used for $\boldsymbol{x}$ and $\boldsymbol{p}$, respectively.
10. In the proof of Lemma 7.2.3 it was stated that "$\boldsymbol{p}[m]$ is at most shifted to be compared with $\bar{\boldsymbol{v}}[v]$". Justify this assertion.
11. How many letter comparisons does BM require to compute all occurrences of a pattern $\boldsymbol{p} = (\lambda_1\lambda_2\cdots\lambda_k)^{m/k}$ in a string $\boldsymbol{x} = (\lambda_1\lambda_2\cdots\lambda_k)^{n/k}$?

7.3 Karp-Rabin

Words are, of course, the most powerful drug used by mankind.

— Rudyard KIPLING (1865–1936), *The Times, London, 15 February 1923*

Almost all of the dozens of pattern-matching algorithms that have been proposed since the 1970s are variants either of KMP or of BM, or both. One notable exception to this general rule is the Karp-Rabin algorithm, described briefly in this section, that makes use of quite a different approach and quite different ideas: rather than trying to shift the pattern $\boldsymbol{p}$ more quickly along the text string $\boldsymbol{x}$, Algorithm KR tries to reduce the time required for comparison at each position i of $\boldsymbol{x}$.

The fundamental idea of KR is that of a ***signature***; that is, a representative of a given string that satisfies the following requirements:

(1) it is unlikely that two distinct strings have the same signature;

(2) a signature can be efficiently computed;

(3) it can be efficiently compared with other signatures.

A signature thus plays the same role in pattern-matching that a ***hash function*** plays in data retrieval [AHU74]. In fact this is exactly the application proposed in [KR87]: a signature σ of $\boldsymbol{x}$ is an integer hash function computed from the letters of $\boldsymbol{x}$. Specifically, $\sigma(\boldsymbol{p})$ is computed in an initialization phase, followed by the successive computation, for every $i \in 1..n-m+1$, of $\sigma(\boldsymbol{x}[i..i+m-1])$. If for some value of i,

$$\sigma(\boldsymbol{x}[i..i+m-1]) = \sigma(\boldsymbol{p}),$$

the routine *match*(i, m) of Algorithm Easy (Section 2.2) is used to check in brute force fashion whether or not $\boldsymbol{x}[i..i+m-1] = \boldsymbol{p}$.

With a "good" choice of σ, this approach satisfies the requirements (1) and (3) given above. Then only rarely will a ***false match*** occur; that is, that

$$\sigma\big(\boldsymbol{x}[i..i+m-1]\big) = \sigma(\boldsymbol{p}) \;\text{ but }\; \boldsymbol{x}[i..i+m-1] \neq \boldsymbol{p}.$$

Hence almost always, by a simple comparison of two integers, it can be determined in constant time — in fact, with only a couple of machine instructions — whether or not $\boldsymbol{x}[i..i+m-1] = \boldsymbol{p}$. The contribution of [KR87] is to show that $\sigma(\boldsymbol{x}[i..i+m-1])$ can be efficiently computed, thus also satisfying requirement (2).

In order to discuss this approach, let us simplify things by abbreviating $\sigma(\boldsymbol{x}[i..i+m-1])$ to $\sigma(i)$. Of course, if it were necessary to compute $\sigma(i)$ in its entirety at the ith step, the hash function approach would be impractical: since presumably each position of $\boldsymbol{x}[i..i+m-1]$ would need to be accessed, time $\Theta(m)$ would be required at each step, so that, for example, just to determine that $\boldsymbol{p}$ does not occur in $\boldsymbol{x}$ would always require overall time $\Theta(m(n-m+1))$.

But Algorithm KR finds a way of computing $\sigma(i)$ *incrementally* — so that $\sigma(i+1)$ is computed from $\sigma(i)$. First the simplifying observation is made that both $\boldsymbol{p}$ and $\boldsymbol{x}$ can be treated as strings of bits $\{0,1\}$: for if indeed they were actually defined on another alphabet, each of that alphabet's letters would necessarily be encoded in the computer as a bit string. It then makes sense to form the following sums:

$$\Sigma(i) = \sum_{j=1}^{m} 2^{m-j}\boldsymbol{x}[i+j-1]; \;\; \Sigma(\boldsymbol{p}) = \sum_{j=1}^{m} 2^{m-j}\boldsymbol{p}[j]. \tag{7.11}$$

These sums simply re-express strings of length m as binary integers of the same length (including leading zeros). Since therefore $\Sigma(i) = \Sigma(\boldsymbol{p})$ if and only if $\boldsymbol{x}[i..i+m-1] = \boldsymbol{p}$, an algorithm for locating $\boldsymbol{p}$ in $\boldsymbol{x}$ might simply compute and compare the Σ values. However, for long patterns (perhaps $m > 32$), the Σ values might become too large for easy comparison in the computer; accordingly [KR87] proposes using an odd prime number q to define the signatures as residues modulo q:

$$\sigma(i) = \Sigma(i) \bmod q; \quad \sigma(\boldsymbol{p}) = \Sigma(\boldsymbol{p}) \bmod q. \tag{7.12}$$

This is a hash function that maps the 2^m possible strings of length m on $\{0, 1\}$ into the integers $0..q-1$ with an average of $2^m/q$ strings corresponding to each integer. There is then a probability $1/q$ of ***collision*** between two distinct strings.

To define the incremental computation of $\sigma(i)$, we introduce the operators $+_q$, $-_q$ and $\times_q$, denoting respectively addition, subtraction and multiplication, modulo q. Basic properties of these operators are left as Exercises 7.3.1 and 7.3.2. These properties allow us to show fairly easily that for every $i \geq 1$, $\sigma(i+1)$ can be computed in constant time from $\sigma(i)$:

Lemma 7.3.1 $\sigma(i+1) = 2 \times_q \left(\sigma(i) -_q (2^{m-1} \bmod q) \times_q \boldsymbol{x}[i]\right) +_q \boldsymbol{x}[i+m].$

Proof See Exercise 7.3.3. □

Since $2^{m-1} \bmod q$ can be precomputed, this result tells us that the calculation of $\sigma(i+1)$ from $\sigma(i)$ requires only four operations, all of which can be efficiently implemented in a computer. With this in mind, KR takes the form shown in Algorithm 7.3.1.

Algorithm 7.3.1 (KR)

```
– Find all occurrences of p in x

choose q
compute σ(p)   – using (7.12)
for i ← 1 to n−m+1 do
   compute σ(i)   – by Lemma 7.3.1, constant time for i > 1
   if σ(i) = σ(p) then
      j ← match(i, m)
   if j = m+1 then output i
```

There is no doubt that Algorithm KR is correct, since it reduces essentially to Algorithm Easy whenever $\sigma(i) = \sigma(\boldsymbol{p})$, and since $\sigma(i) \neq \sigma(\boldsymbol{p})$ implies that $\boldsymbol{x}[i..i+m-1] \neq \boldsymbol{p}$. Hence

Theorem 7.3.2 *Algorithm 7.3.1 correctly computes all occurrences of a given nonempty pattern* $\boldsymbol{p}$ *in a given string* $\boldsymbol{x}$. □

But the efficiency of Algorithm KR can vary widely depending on the particular choice of q. A very bad or unlucky choice of q could result in *match*(i, m) being called $n - m + 1$ times (every match a false one) and so result in $\Theta(mn)$ execution time, no better than Algorithm Easy. On the other hand, a lucky choice of q could mean that there were no false matches, so that KR would require time $O(n + km)$, where k is the number of occurrences of $\boldsymbol{p}$ in $\boldsymbol{x}$. In [KR87] number theoretic arguments are used to show that if the prime number q is initially chosen pseudorandomly in the range $1..mn^2$, and chosen pseudorandomly again whenever a false match occurs, then the expected execution time of KR actually attains the minimum, $\Theta(n + km)$. Thus the expected time requirement of Algorithm KR, modified in this way, is almost comparable to that of BM. It should be observed however that the dependence of the execution time on k can be a serious defect of KR in cases where $\boldsymbol{p}$ occurs often in $\boldsymbol{x}$; indeed, for $k \in \Theta(n)$, execution time becomes $\Theta(mn)$.

In practice, it is suggested in [KR87] that KR is competitive primarily for long patterns, say $m \geq 200$. The reason given is that, unlike KMP and BM, Algorithm KR requires only constant additional storage; thus an implementation of KR can make use of registers for fast access to data no matter how large m becomes, whereas the execution of KMP and BM could be slowed somewhat for larger m due to the need to access data in arrays of size $\Theta(m)$. On the other hand, these access delays may not be severe for certain machine architectures (cache, for example), and KR itself can be slow as a result of its intensive use of integer arithmetic [BYG92], even in efficient implementations such as the one described in [GBY91].

As remarked in [KR87], Algorithm KR can be adapted to use with two-dimensional patterns, even those of irregular shape, and it also lends itself to efficient parallel implementation.

Exercises 7.3

1. Let $\bullet$ denote any one of the operators $+$, $-$, $\times$, and let $\bullet_q$ denote the corresponding operator $+_q$, $-_q$ or $\times_q$, respectively. Show that for all positive integers q, r and s,

$$(r \bullet s) \bmod q = (r \bmod q) \ \bullet_q \ (s \bmod q).$$

2. For large m, the computations (7.12) could overflow word boundaries in the computer if the sums (7.11) were formed prior to determining the residues modulo q. Using the operators $+_q$ and $\times_q$, describe an efficient iterative routine to compute σ that keeps the partial sums bounded by $O(q)$.
3. Prove Lemma 7.3.1.
4. Suppose that it is known that $m \leq 32$. Design an efficient KR-type algorithm that uses the sums Σ rather than σ. What is the worst-case upper time bound on your algorithm's performance? The best-case lower bound? The average-case complexity?
5. Suppose that $\boldsymbol{p}$ and $\boldsymbol{x}$ are strings on an integer alphabet $\{0, 1, \ldots, t\}$. Show how the computation of Σ can be reformulated accordingly, then state and prove a generalized version of Lemma 7.3.1.

7.4 Dömölki-(Baeza-Yates)-Gonnet

Man does not live by words alone, despite the fact that sometimes he has to eat them.

— Adlai STEVENSON (1900–1965)

Before discussing this algorithm, we need to discuss its history. Its recent history begins in 1992, when Baeza-Yates and Gonnet, in a celebrated paper [BYG92], proposed the bit-mapping approach described in this section. Their work has since inspired dozens, perhaps even hundreds, of other papers that explore one variant or another of their methodology. But it turns out that there is an unsuspected ancient history as well: a 1968 paper [D68] by Dömölki describes essentially the same algorithm and, furthermore, refers to a 1965 paper [D65] in which the "technique is described in detail". It thus appears that Dömölki's algorithm, even though presented in English in a reputable computer science journal, nevertheless simply lay dormant for 27 years before being discovered all over again by somebody else!

This extraordinary fact was brought to my attention by a participant [N01] in a series of lectures given by the author at the University of Debrecen in October–November 2001. The name originally chosen for this section's algorithm was BYG — to recognize Dömölki's precedence, the name has been changed to DBG.

Algorithm DBG is similar to KR in that, unlike both KMP and BM, it makes no attempt to shift the pattern $\boldsymbol{p}$ more quickly along the text $\boldsymbol{x}$, but rather plods along a step at a time, making a comparison at each position of $\boldsymbol{x}$. Thus, like KR, DBG seeks to minimize the processing required at each step rather than to decrease the number of steps: it seeks to sprint rather than plod! In fact, DBG finds a way to compare the current letter $\boldsymbol{x}[i]$ with *all* of the m positions of $\boldsymbol{p}$ in a single operation that under favourable circumstances requires only constant time.

The key to Algorithm DBG is its use of an array (or string) $\boldsymbol{s} = \boldsymbol{s}[1..m]$ of bits $\{0,1\}$ that describes the current ***state*** of the computation as i successively takes the values $1, 2, \ldots, n$ [BYG92]. This bit vector technique seems to have been applied earlier in pattern-matching algorithms by Allison and Dix [AD86], who also made use of the array $\boldsymbol{t}$ defined below in an approach to the LCS problem (Problem 2.10), discussed in Sections 9.3 and 9.4.

Specifically, for every $j \in 1..m$, the following invariant is maintained:

$$\boldsymbol{s}[j] = 0 \;\Leftrightarrow\; \boldsymbol{x}[i-j+1..i] = \boldsymbol{p}[1..j], \tag{7.13}$$

where i is the position currently being considered in $\boldsymbol{x}$. We may think of $\boldsymbol{s}[j]$ as specifying whether ($\boldsymbol{s}[j] = 0$) or not ($\boldsymbol{s}[j] = 1$) $\boldsymbol{p}[1..j]$ matches the current substring of $\boldsymbol{x}$. Then an occurrence $\boldsymbol{x}[i-m+1..i]$ of $\boldsymbol{p}$ is found in $\boldsymbol{x}$ whenever $\boldsymbol{s}[m] = 0$. Clearly we can initialize $\boldsymbol{s} = 1^m$ to reflect the fact that initially no position of $\boldsymbol{p}$ is matched. We now show how to update $\boldsymbol{s}$ according to the value of $\boldsymbol{x}[i]$ to maintain the invariant (7.13).

First we introduce an array that has something in common with the $\boldsymbol{\delta_1}$ array used in Algorithm BM. In fact, in order to use this array, we are required to make the same assumptions previously made for BM, that the alphabet of $\boldsymbol{x}$ is finite, known in advance, and *indexed* (Section 4.1). Then, without loss of generality (Exercise 4.1.3), we may suppose that the alphabet consists of the natural numbers $N_\alpha = \{1, 2, \ldots, \alpha\}$. We can now define the two-dimensional array $\boldsymbol{t} = \boldsymbol{t}[1..\alpha, 1..m]$, where for every $h \in 1..\alpha$ and every $j \in 1..m$,

$$\begin{aligned} \boldsymbol{t}[h,j] &= 0, \ \text{ if } \boldsymbol{p}[j] = h; \\ &= 1, \ \text{ otherwise.} \end{aligned} \tag{7.14}$$

Thus for every letter $h \in N_\alpha$, $\boldsymbol{t}[h, 1..m]$ may be thought of as a string of m bits $\{0, 1\}$ in which entries 0 identify positions in $\boldsymbol{p}$ at which h occurs. The following lemma shows how $\boldsymbol{t}$ can be used to update the state $\boldsymbol{s}$ from position $i - 1$ in the text $\boldsymbol{x}$ to position i:

Lemma 7.4.1 *Let $\boldsymbol{s}^{(i)}$ denote the string $\boldsymbol{s}$ that defines the state corresponding to position i in the text $\boldsymbol{x}$. Define the initial state $\boldsymbol{s}^{(0)}[0..m] = 0(1^m)$. Then for every $i \in 1..n$ and every $j \in 1..m$,*

$$\boldsymbol{s}^{(i)}[j] = \boldsymbol{s}^{(i-1)}[j-1] \ \vee \ \boldsymbol{t}\big[\boldsymbol{x}[i], j\big], \tag{7.15}$$

where $\vee$ is the logical OR *operator. (Recall that a logical OR yields 1 if either operand is 1; otherwise, 0.)*

Proof Let i be the current position in $\boldsymbol{x}$. Observe that, according to Lemma 7.4.1, no calculation is performed on $\boldsymbol{s}^{(i-1)}[m]$. Suppose first that for any $j \in 1..m$, $\boldsymbol{s}^{(i-1)}[j-1] = 0$, so that by (7.13)

$$\boldsymbol{x}[i-j+1..i-1] = \boldsymbol{p}[1..j-1].$$

If $\boldsymbol{t}\big[\boldsymbol{x}[i], j\big] = 0$, then by (7.14) it is moreover true that

$$\boldsymbol{x}[i] = \boldsymbol{p}[j],$$

so that we should also set $\boldsymbol{s}^{(i)}[j] \leftarrow 0$. Conversely, if $\boldsymbol{t}\big[\boldsymbol{x}[i], j\big] = 1$, then by (7.14)

$$\boldsymbol{x}[i] \neq \boldsymbol{p}[j],$$

and we should set $\boldsymbol{s}^{(i)}[j] \leftarrow 1$. Thus in each case the correct result is ensured by the logical OR in (7.15).

A similar argument applies when $\boldsymbol{s}^{(i-1)}[j-1] = 1$. □

To see how (7.15) can be used to compute states $\boldsymbol{s}^{(i)}$, $i = 1, 2, \ldots, n$, consider the simple example

$$\boldsymbol{p} = 212, \quad \boldsymbol{x} = 12112121$$

on the alphabet $A = \{1, 2\}$, so that $m = 3$, $n = 8$, and

$$\boldsymbol{t} = \begin{pmatrix} 1\,0\,1 \\ 0\,1\,0 \end{pmatrix}.$$

Using $s^{(0)} = 0111$, we compute

$$\begin{aligned} s^{(1)}[1] &= 0 \vee t[1,1] = 1, \\ s^{(1)}[2] &= 1 \vee t[1,2] = 1, \\ s^{(1)}[3] &= 1 \vee t[1,3] = 1, \end{aligned}$$

and similarly for $i = 2, 3, \ldots, 8$, we find

$$\begin{aligned} &s^{(2)} = 011, \quad s^{(3)} = 101, \\ &s^{(4)} = 111, \quad s^{(5)} = 011, \\ &s^{(6)} = 101, \quad s^{(7)} = 010, \\ &s^{(8)} = 101. \end{aligned}$$

The value $s^{(7)}[3] = 0$ identifies the match $p = x[5..7]$.

Algorithm 7.4.1 (DBG)

– *Find all occurrences of* p *in* x

$s[0..m] \leftarrow 0(1^m)$ – *initial state*
for $i \leftarrow 1$ **to** n **do**
  – *Shift each bit of* s *right one position,*
  – *leaving* $s[0] = 0$ *unchanged, then compute*
  – *bitwise logical* OR *of* $s[1..m]$ *with* $t\big[x[i], 1..m\big]$
  $s \leftarrow \mathit{rightshift}(s, 1) \vee t\big[x[i], 1..m\big]$
  if $s[m] = 0$ **then output** $i - m + 1$

DBG is presented as Algorithm 7.4.1; since it is a straightforward application of Lemma 7.4.1, we formally claim its correctness:

Theorem 7.4.2 *Algorithm 7.4.1 correctly computes all occurrences of a given nonempty pattern* p *in a given string* x. □

In order to discuss the time and space complexity of DBG, we need to introduce a parameter w, the length of a computer word. We see from Algorithm 7.4.1 that, if $m \leq w$, $s[1..m]$ and $t[h, 1..m]$ each occupy at most one computer word, so that the calculation of s can be viewed as a pair of constant-time operations: a one-bit shift followed by a logical OR in a one-word register. More generally, for arbitrary m, $s[1..m]$ and $t[h, 1..m]$ each occupy $\lceil m/w \rceil$ computer words and therefore each calculation of s requires $\Theta\big(\lceil m/w \rceil\big)$ time. Hence

Theorem 7.4.3 *Algorithm 7.4.1 computes all occurrences of* $\boldsymbol{p}$ *in* $\boldsymbol{x}$ *in time* $\Theta(\lceil m/w \rceil n)$ *using* $\Theta((\alpha+1)\lceil m/w \rceil)$ *additional space.* □

Thus for long patterns DBG requires $\Theta(mn)$ time and for a large alphabet as much as $\Theta(mn)$ space: it is therefore not the algorithm of choice in these cases. However, for short patterns and small alphabets, Algorithm DBG can provide very fast and efficient pattern-matching. We shall see in Chapter 10 that DBG adapts very easily to the calculation of approximate patterns, and becomes one of the most flexible and useful algorithms in this context [WM92].

Observe that the array $\boldsymbol{t} = \boldsymbol{t}[1..\alpha, 1..m]$ can be computed in a preprocessing phase that initially sets every word to 1^w, then resets position $\boldsymbol{t}[\boldsymbol{p}[j], j] \leftarrow 0$ for every $j \in 1..m$. Hence

Theorem 7.4.4 *The array* $\mathbf{t}$ *can be computed in time* $\Theta(\alpha\lceil m/w \rceil + m)$.

Proof See Exercise 7.4.2. □

The asymptotic time complexity of Algorithm DBG is therefore not affected by the preprocessing phase.

Finally, we recall to the reader's mind Digression 1.3.1, in which we discussed the relationship between problem size n and word-length w, concluding that in practice, for $w \geq 32$, we were safe in supposing that $\log_2 n/w \leq 2$. In other words, we could suppose that $w \in \Omega(\log_2 n)$. It follows then from Theorem 7.4.3 that the time requirement for Algorithm DBG can be expressed as $O(mn/\log n)$, an upper bound surely valid for any text string $\boldsymbol{x}[1..n]$ that we will ever want to search in an electronic computer!

Exercises 7.4

1. Complete the proof of Lemma 7.4.1.
2. Specify an algorithm that computes the array $\boldsymbol{t}$ used in Algorithm DBG. Hence prove Theorem 7.4.4.
3. Given an alphabet $A = \{a, b, c, d, e\}$ and pattern $\boldsymbol{p} = abacabadabacabae$, compute $\boldsymbol{t}$.
4. Can you restate the problem of this section so that the algorithm used to solve it uses AND rather than OR?

7.5 Summary

In three words I can sum up everything
I've learned about life: it goes on.

— Robert FROST (1874–1963)

Of the four algorithms discussed in this chapter, it is clear that overall, at least for exact pattern-matching, the advantage lies with Algorithm BM: it is fast in practice over the full range of pattern-matching parameters (string length, pattern length, number of complete matches found), while at the same time, for aperiodic patterns, it is guaranteed in theory never to require more than $3n$ letter comparisons and $\Theta(n)$ time. Furthermore, as we have seen, the average number of letter comparisons required by BM is only about one-third of the average required by KMP. Algorithm KMP nevertheless remains interesting since, despite its slower average-case performance, it guarantees at most $2n$ letter comparisons in the worst case, and it does handle multiple occurrences of periodic patterns with particular ease. Periodic patterns aside, the only disadvantages of BM, rather theoretical ones, are the restrictions on its use stated at the beginning of Section 7.2. On balance, as a robust general-purpose algorithm for exact pattern-matching, BM is hard to beat, a fact that has of course not stopped computer scientists from trying to beat it! Thus it is that BM has given rise to many redoubtable offspring, some of whom can even claim to outperform their famous parent. In the next chapter we look at a few of these sturdy competitors. In particular, we shall see how a simple modification to BM enables it to deal with multiple periodic patterns with the same facility that KMP does.

Chapter 8

Son of BM Rides Again!

And stringing pretty words that make no sense,
And kissing full sense into empty words.

— Elizabeth Barrett BROWNING (1806–1861), *Aurora Leigh*

In this chapter we consider a few of the numerous variants of the Boyer-Moore algorithm (Section 7.2) that have been proposed over the last quarter-century or so with a view to improving it either in theory (reduced number of letter comparisons) or in practice (reduced execution time). In fact, the extensive literature on this subject divides fairly sharply between theory and practice: those who want to establish tight bounds on the worst-case number of letter comparisons required, and those who want to design the fastest general-purpose algorithm. In this chapter I shall favour the latter approach, not because I think that the theory is uninteresting or unimportant, but simply because to do justice to the history of ideas that has led to the current theoretical understanding of pattern-matching would nearly require a book in itself. The goal then in this chapter is to provide good coverage of the main ideas — many of them very simple heuristics — that have been proposed to improve the practical efficiency of Algorithm BM. As we shall see, several of these ideas also lead to theoretical improvement — reduction in the number of letter comparisons — as well. We also consider here an interesting variant of Algorithm KMP (Section 7.1) that reduces the worst-case number of letter comparisons from $2n - m$ to $1.5n - m/2$. In a final section, a brief survey of theoretical advances is given.

8.1 The BM Skip Loop

A string of excited, fugitive, miscellaneous pleasures is not happiness . . .

— Jorge Augustín Nicolás Ruiz de SANTAYANA (1863–1952)
Realms of Being

We begin with a variant that was in fact proposed in the original Boyer-Moore paper [BM77], but then largely ignored for several years. The basic idea of what is generally called a ***skip loop*** is to shift the pattern $\boldsymbol{p}$ right along the string $\boldsymbol{x}$ in large "skips" whose length is determined by some easily-tested condition that *hopefully* will permit the inference to be made that no occurrences of $\boldsymbol{p}$ exist in the skipped portion of $\boldsymbol{x}$.

In particular, the BM skip loop makes repeated use of a slightly modified $\boldsymbol{\delta_1}$ array (which we call $\boldsymbol{\delta'_1}$) to provide the skip condition. The modification is simple: for every letter λ, $\boldsymbol{\delta'_1}[\lambda] = \boldsymbol{\delta_1}[\lambda]$ *except* for $\lambda = \boldsymbol{p}[m]$ — in that case, instead of setting

$$\boldsymbol{\delta_1}\left[\boldsymbol{p}[m]\right] \leftarrow 0$$

we set

$$\boldsymbol{\delta'_1}\left[\boldsymbol{p}[m]\right] \leftarrow N$$

for any fixed choice of $N > m + n$. Then the skip loop

$$\textbf{repeat } i \leftarrow i + \boldsymbol{\delta'_1}\left[\boldsymbol{x}[i]\right] \textbf{ until } i > n$$

is executed as the first instruction in the **while** loop of Algorithm BM.

Let i_0 be the value of i prior to entering the skip loop. If on exit from the skip loop it is true that $i > N$, we can make the following inferences:

(1) $\boldsymbol{x}[i - N] = \boldsymbol{p}[m]$;
(2) there exists no position $i' \in i_0..(i - N) - 1$ such that $\boldsymbol{x}[i'] = \boldsymbol{p}[m]$.

Thus $i - N$ is the first position in $\boldsymbol{x}$ to the right of position $i_0 - 1$ at which the letter $\boldsymbol{p}[m]$ occurs. In this case then, we may return to normal BM processing (that is, execute the routine $hctam(i, m - 1)$) after performing the following instruction:

$$\textbf{if } i > N \quad \textbf{then } i \leftarrow (i - N) - 1.$$

Suppose on the other hand that on exit from the skip loop, $i \leq N$. In this case, since $i > n$, we can infer that there exists no position i' such that $i_0 \leq i' \leq n$ and $\boldsymbol{x}[i'] = \boldsymbol{p}[m]$ — in other words, $\boldsymbol{p}$ does not occur in $\boldsymbol{x}[i - m + 1..n]$ and the search may be terminated.

Based on the snippets of code given above, it is not difficult to modify Algorithm BM so as to make use of a skip loop: the resulting algorithm is usually called BMFAST, since it turns out that in practice the use of the skip loop, especially when efficiently implemented, improves the average performance of the algorithm by a factor of about three over a wide range of data sets and computer platforms [HS91]. Thus the Boyer-Moore skip loop is a very useful and important enhancement to the original BM algorithm. The specification of BMFAST and the characterization of cases in which it is much more, as well as much less, efficient than BM are left to Exercises 8.1.1–8.1.5.

We can get a qualitative understanding of the benefit of the skip loop by observing that, in effect, BMFAST replaces a single letter comparison

$$\boldsymbol{x}[i_0] \; : \; \boldsymbol{p}[m]$$

by one or more skips:

- If in fact $\boldsymbol{x}[i_0] = \boldsymbol{p}[m]$, then the instructions executed in the skip loop will require somewhat more time than the letter comparison, and so in this case there will be a net loss in efficiency from using BMFAST.
- If $\boldsymbol{x}[i] = \boldsymbol{p}[m]$ for some $i > i_0$, then the identification of such an i will generally take place more quickly than would have been the case using BM.
- If on the other hand there exists no i for which $\boldsymbol{x}[i] = \boldsymbol{p}[m]$, BMFAST will generally be much more efficient than BM, especially for larger alphabets.

Other forms of skip loop have been proposed, most notably a straightforward search through $\boldsymbol{x}$ for a single specified letter of $\boldsymbol{p}$ [H80], if possible a letter that is known to occur infrequently in $\boldsymbol{x}$. Although this approach does speed up BM somewhat, it appears nevertheless to be much slower than BMFAST [HS91] over a wide range of patterns, text strings, and alphabets.

Exercises 8.1

1. Explain clearly why the inferences (1) and (2) can be made.
2. Write out Algorithm BMFAST and prove its correctness given that BM is correct.
3. Characterize in as general a way as you can those cases in which BMFAST is

   (a) much faster

   (b) much slower

   than BM. Don't forget to include the case $\boldsymbol{x} = a^n$, $\boldsymbol{p} = a^m$!
4. Recall the worst-case example $\boldsymbol{x} = (a^k b)^r$, $\boldsymbol{p} = a^{k-1}ba^{k-1}$ introduced in Section 7.2. For large k and r, estimate the number of letter comparisons required by BMFAST to match $\boldsymbol{p}$ against $\boldsymbol{x}$. Do you think BMFAST will actually execute faster than BM in this case?
5. Will BMFAST always require fewer letter comparisons than BM? Justify your answer carefully.
6. Write out the algorithm mentioned at the end of the section, that implements a skip loop to search for occurrences of a specific letter, say $\boldsymbol{p}[k]$, of $\boldsymbol{p}$.

8.2 BM-Horspool

Sing, sing! What shall I sing?
The cat's run away with the pudding-bag string.

— *Anonymous*

In the previous section we saw how to make use of a simple change to $\boldsymbol{\delta_1}\left[\boldsymbol{p}[m]\right]$ to implement long "skips" of the pattern along the string. In this section we investigate another way of altering $\boldsymbol{\delta_1}\left[\boldsymbol{p}[m]\right]$, this time in order to permit an implementation of Algorithm BM that does not require $\boldsymbol{\delta_2}$ — thus we avoid all the complications involved in computing and using $\boldsymbol{\delta_2}$ that infested much of Section 7.2 (and that will occupy us again in Section 8.4). The benefit is a simple, practical algorithm whose average-case efficiency is comparable in practice to the original BM; the cost is that the algorithm may, in unusual cases beyond those already identified in Section 7.2, require time proportional to mn.

Recall that the use of $\boldsymbol{\delta_2}$ achieves two main purposes: it ensures that BM never backtracks — that is, that every shift moves the pattern $\boldsymbol{p}$ to the right — and it also ensures that, except for 3-periodic patterns, BM executes in time $O(n)$ — in establishing theoretical upper bounds on the running time of BM, we do not even take $\boldsymbol{\delta_1}$ into account. The achievement of Horspool [H80] was to show that the first of these goals, at least, could be achieved using nothing but $\boldsymbol{\delta_1}$.

As defined in Section 7.2, $\boldsymbol{\delta_1}[\lambda] = m - j'$ for every letter $\lambda \in A$, where j' is either the position of the rightmost occurrence of λ in $\boldsymbol{p}$ or zero if there is no such λ. The main idea of Algorithm BMH is that when a mismatch occurs, $\boldsymbol{p}$ should be shifted not according to the position of the mismatch, but rather according to the rightmost position in $\boldsymbol{p}$. Thus, suppose that a partial match

$$\boldsymbol{x}[i+1..i+h] = \boldsymbol{p}[m-h+1..m]$$

has been found for some $h \in 0..m$, terminating with a mismatch

$$\lambda = \boldsymbol{x}[i] \neq \boldsymbol{p}[m-h] = \mu.$$

Instead of making the shift $i \leftarrow i + \boldsymbol{\delta_1}\left[\boldsymbol{x}[i]\right]$ prescribed by BM, Algorithm BMH makes instead the shift

$$i \leftarrow (i+h) + \boldsymbol{\delta_1}\left[\boldsymbol{x}[i+h]\right];$$

that is, $\boldsymbol{p}[m]$ is compared next with a position i in $\boldsymbol{x}$ that is $\boldsymbol{\delta_1}\left[\boldsymbol{x}[i+h]\right]$ positions to the right of the position $\boldsymbol{p}[m]$ is currently compared with. This strategy solves one problem that arose with the $\boldsymbol{\delta_1}$ shift of BM: it is no longer possible that $\boldsymbol{p}$ might be shifted *left* along the pattern, because every position in $\boldsymbol{\delta_1}$ is nonnegative. On the other hand, the new approach appears to exacerbate the zero shift problem $\left(\boldsymbol{\delta_1}(\lambda) = 0\right)$ that arises whenever $\lambda = \boldsymbol{p}[m]$: now every time that $h > 0$, there must have been a match $\boldsymbol{p}[m] = \boldsymbol{x}[i+h]$, so that in fact

$$\boldsymbol{\delta_1}\left[\boldsymbol{x}[i+h]\right] = m - m = 0.$$

As suggested at the beginning of this section, BMH resolves this difficulty by redefining $\boldsymbol{\delta_1}[\lambda]$ (as $\boldsymbol{\delta_1''}[\lambda]$) if and only if $\lambda = \boldsymbol{p}[m]$. First consider the situation that arises when $\boldsymbol{p}[m] = \boldsymbol{x}[i+h]$ is the only occurrence of λ in $\boldsymbol{p}$. In this case there can be no match between $\boldsymbol{p}$ and $\boldsymbol{x}$ unless $\boldsymbol{p}$ is shifted entirely past position $i+h$ of $\boldsymbol{x}$, so that the next invocation of *hctam* begins by comparing $\boldsymbol{p}[m]$ with $\boldsymbol{x}\big[(i+h)+m\big]$. We ensure that this case is handled correctly by setting

$$\boldsymbol{\delta_1''}[\lambda] \leftarrow m.$$

More generally, if λ exists elsewhere in $\boldsymbol{p}$, let $j' < m$ be the position of the rightmost occurrence of λ *to the left of* position m. So the correct shift is ensured by setting

$$\boldsymbol{\delta_1''}[\lambda] \leftarrow m - j'.$$

Thus BMH makes use of a new array $\boldsymbol{\delta_1''}$ identical to $\boldsymbol{\delta_1}$ for every letter except $\lambda = \boldsymbol{p}[m]$, while $\boldsymbol{\delta_1''}[\lambda]$ is computed as just described. BMH can easily be specified by making minor straightforward changes to BM, to *hctam*, and to the calculation of $\boldsymbol{\delta_1}$. We leave these modifications to Exercise 8.2.1. Also left as Exercise 8.2.3 is the determination of strings $\boldsymbol{p}$ and $\boldsymbol{x}$ such that BMH requires $\Theta(mn)$ time for its execution.

The relative simplicity of BMH (compared to BM) not only makes it an attractive alternative in practice, but also renders analysis of its average-case behaviour much easier. It turns out [BY89b, BYGR90, BYR92] that, for an alphabet of size α and text length n large with respect to pattern length m, the expected number of letter comparisons is somewhat more than

$$(n-m+1)/\alpha,$$

an approximation that becomes increasingly precise as m and α become larger. Thus for large alphabets, Algorithm BMH, despite its unfortunate worst-case behaviour, appears to execute more quickly on average than the standard BM algorithm described in Section 7.2: as noted there, BM can be expected to require about $0.3n$ letter comparisons on average, even for large alphabets.

Exercises 8.2

1. In accordance with the description given in this section, specify Algorithm BMH, including updates to routine *hctam* and to the computation of $\boldsymbol{\delta_1}$.
2. Prove the correctness of BMH as specified in the preceding exercise.
3. Describe infinite classes of strings $\boldsymbol{p} = \boldsymbol{p}[1..m]$ and $\boldsymbol{x}[1..n]$ such that BMH requires time $\Theta(mn)$ to match $\boldsymbol{p}$ against $\boldsymbol{x}$.
4. The first two sections of this chapter describe extraordinarily different modifications to BM, both of them based on a recomputation of a single value, $\boldsymbol{\delta_1}\big[\boldsymbol{p}[m]\big]$. What if these two modifications were combined into a single algorithm that did not use $\boldsymbol{\delta_2}$ at all while using a revised $\boldsymbol{\delta_1}$ for both skips and shifts? Specify such an algorithm, prove its correctness, and say what you can about its asymptotic complexity and its efficiency in practice.

Remark: As far as the author of this book knows, the hybrid algorithm proposed here has never been studied before. Your research into its properties should break new ground! (As indeed it already has: see [J02].)

8.3 Frequency Considerations and BM-Sunday

A newspaper consists of just the same number of words, whether there be any news in it or not.

— Henry FIELDING (1707–1754)

Even though BM is designed around a right-to-left scan of $\boldsymbol{p}$, there is nothing sacred about this particular ordering of its elements: the real object always is to find mismatches as quickly as possible, so that $\boldsymbol{p}$ can be moved (using skips or shifts) along $\boldsymbol{x}$ in the shortest time possible. In this section we consider a collection of modifications to BM that relate to circumstances in which it is possible to estimate the frequency of occurrence in $\boldsymbol{x}$ of each letter of the alphabet — and where these frequencies vary widely from one letter to another. In such cases we become particularly interested in the letters of $\boldsymbol{p}$ that occur less frequently in $\boldsymbol{x}$: these letters will be less likely to match with $\boldsymbol{x}$, and so will be more likely to give rise quickly to a long skip or shift of the pattern along $\boldsymbol{x}$. In particular, pattern-matching on natural language texts, where letter frequencies are usually well known, can be significantly speeded up using these techniques.

Matching with a Guard

Suppose that a preprocessing step has identified a position j_{rarest} such that $\boldsymbol{p}[j_{\text{rarest}}]$ is the letter of $\boldsymbol{p}$ that is expected to occur least frequently in a given (randomly selected) text string on some alphabet. Then the routine *hctam* is invoked only if a match is found between the ***guard*** $\boldsymbol{p}[j_{\text{rarest}}]$ and the corresponding position in $\boldsymbol{x}$:

if $\boldsymbol{x}[i-(m-j_{\text{rarest}})] = \boldsymbol{p}[j_{\text{rarest}}]$ **then**
  $(i,j) \leftarrow hctam(i,m)$
  if $j=0$ **then output** $i+1$
else
  $j \leftarrow j_{\text{rarest}}$
$i \leftarrow i + \max\{\boldsymbol{\delta_1}[\boldsymbol{x}[i]], \boldsymbol{\delta_2}[j]\}$

We leave to Exercise 8.3.1 the proof of the correctness of this modification to BM (it is not obvious).

The idea of a guard seems to have originated in [KM84], then to have been reinvented in [HS91]. According to the tests reported in [HS91], use of a guard provides considerable benefit in natural-language searches.

More Effective Skipping

It has already been suggested, in Section 8.1, that a straightforward search through $\boldsymbol{x}$ for a specified letter $\boldsymbol{p}[j]$ of $\boldsymbol{p}$ will be expedited by choosing $j = j_{\text{rarest}}$. More interesting perhaps is the suggestion made in Exercise 8.3.5 for this section that a guard algorithm can more effectively be implemented as a skip loop that uses j_{rarest} to avoid testing every position of $\boldsymbol{x}$ during the skip.

Here we outline a third approach to skipping, introduced in [HS91]: modification of the BMFAST skip loop to replace the use of $\boldsymbol{p}[m]$ by some less frequent letter $\boldsymbol{p}[j]$.

Suppose we were asked to locate a pattern $\boldsymbol{p} = marzipan$ in a text $\boldsymbol{x}$ of lower-case English-language text, where for ease of discussion we assume that spaces and punctuation are removed. BMFAST would perform skips $\boldsymbol{\delta}_1'\left[\boldsymbol{x}[i]\right]$ according to the letter $\boldsymbol{x}[i]$ currently aligned with $\boldsymbol{p}[8] = n$. These skips would range in length from 1 ($\boldsymbol{x}[i] = a$) through 8 ($\boldsymbol{x}[i]$ not in $\boldsymbol{p}$) to N ($\boldsymbol{x}[i] = n$). Since n is the sixth most frequent letter in English, this latter skip could occur quite often; since moreover this skip is the wasteful one (considerably less efficient than simply recognizing that $\boldsymbol{x}[i] = \boldsymbol{p}[m] = n$), we would be happy to reduce its frequency of occurrence.

One possible way to reduce the number of skips of length N is to search, not for the letter of $\boldsymbol{x}$ aligned with $\boldsymbol{p}[8] = m$, but rather for the letter of $\boldsymbol{x}$ aligned with the rightmost occurrence of a lower frequency letter such as $\boldsymbol{p}[6] = p$ or (especially) $\boldsymbol{p}[4] = z$. (We ignore $\boldsymbol{p}[7] = a$ and $\boldsymbol{p}[5] = i$ because they are actually higher frequency than n: third and fifth most frequent, respectively.) This search could be implemented by computing $\boldsymbol{\delta}_1'$ as if $\boldsymbol{p}$ were only six characters long (using $\boldsymbol{p}[6] = p$), or four characters long (using $\boldsymbol{p}[4] = z$). Of course this would mean that the average skip length would be reduced, since apart from skips of length N, the longest skip would now be only 6 (respectively, 4) positions. Thus there is a trade-off that needs to be evaluated between the cost of reduced skip length and the benefit of reduced occurrence of skips of length N. Based on an analysis of this trade-off for given pattern $\boldsymbol{p}$ and given probabilities of occurrence of letters in A, [HS91] describes a methodology for selecting the position of $\boldsymbol{p}$ that "probably" minimizes skip time. The cost of this preprocessing is $\Theta(m)$ and yields, according to the tests reported in [HS91], a relatively small average decrease in search time (3–5%).

BM-Sunday(1)

As Exercise 8.3.5 makes clear, a guard is really just a disguised skip. Thus, what we have studied so far in this section could be characterized simply as a collection of skip techniques based on letter frequency. Now however we study a much more revolutionary idea: redefining the BM $\boldsymbol{\delta}_1$ array in such a way as to entirely free the resulting BMS1 algorithm from the need to perform letter comparisons on $\boldsymbol{p}$ in right-to-left order.

Like virtually all the proposed modifications to Algorithm BM, the generating idea of BMS1 is a simple one [S90]. Suppose that in the current attempted match, $\boldsymbol{p}[m]$ is aligned with $\boldsymbol{x}[i_0]$. Then after a mismatch occurs, the next attempted match cannot succeed unless $\boldsymbol{x}[i_0+1] = \lambda$ matches some occurrence of λ in $\boldsymbol{p}$. In particular, if j' is the *rightmost* position at which λ occurs in $\boldsymbol{p}$, it is safe to shift $\boldsymbol{p}$ right by $m-j'+1$ positions. The preprocessing for Algorithm BMS1 then computes an array $\boldsymbol{\Delta_1}$ as follows: for every letter $\lambda \in A$,

$$\boldsymbol{\Delta_1}[\lambda] \leftarrow m - j' + 1,$$

where j' is the rightmost position of occurrence for λ in $\boldsymbol{p}$, zero if $\boldsymbol{p}$ does not contain λ. Observe that for every λ,

$$\boldsymbol{\Delta_1}[\lambda] = \boldsymbol{\delta_1}[\lambda] + 1.$$

Observe also that, for every mismatch, $\boldsymbol{\Delta_1}\left[\boldsymbol{x}[i_0+1]\right]$ is the amount that $\boldsymbol{p}$ should be shifted to the right for the next attempted match, in which $\boldsymbol{p}[m]$ will be aligned with

$$\boldsymbol{x}\Big[i_0 + \boldsymbol{\Delta_1}\left[\boldsymbol{x}[i_0+1]\right]\Big].$$

As pointed out in [S90], the $\boldsymbol{\Delta_1}$ array has several significant advantages over the $\boldsymbol{\delta_1}$ array:

(1) Like the elements of the $\boldsymbol{\delta_1''}$ array of Algorithm BMH, $\boldsymbol{\Delta_1}[\lambda]$ is always greater than zero and so suffices on its own to shift $\boldsymbol{p}$ to the right along $\boldsymbol{x}$.

(2) In fact, we can expect that shifts resulting from the use of $\boldsymbol{\Delta_1}$ will be substantially greater than those based on $\boldsymbol{\delta_1}$. Recall that the rule applied in BM is to increment the position i of mismatch by $\boldsymbol{\delta_1}\left[\boldsymbol{x}[i]\right]$, leading to a shift of $\boldsymbol{p}$ along $\boldsymbol{x}$ of

$$\boldsymbol{\delta_1}\left[\boldsymbol{x}[i]\right] - (m-j) = \boldsymbol{\Delta_1}\left[\boldsymbol{x}[i]\right] - 1 - (m-j) \tag{8.1}$$

positions, where $m-j$ is the length of the partial match prior to the mismatch at i. Since on average it may be expected that

$$\boldsymbol{\Delta_1}\left[\boldsymbol{x}[i_0+1]\right] = \boldsymbol{\Delta_1}\left[\boldsymbol{x}[i]\right],$$

(8.1) tells us that on average the $\boldsymbol{\Delta_1}$ shift of BMS1 can be expected to exceed the $\boldsymbol{\delta_1}$ shift of BM by $m-\hat{j}+1$ positions, where $m-\hat{j}$ is the average length of a partial match.

(3) Even more significant is the realization that, because the shift of $\boldsymbol{p}$ is now independent of the position of mismatch, the shift is therefore independent of the order in which the letters of $\boldsymbol{p}$ are compared. At a stroke, the dependency of BM on right-to-left processing of $\boldsymbol{p}$ is gone!

In view of advantage (1), BMS1 can easily be implemented so as to make use only of $\boldsymbol{\Delta_1}$ for shifts, a strategy similar to that of Algorithm BMH. Further, in view of advantage (2), it is reasonable to expect that BMS1 will function more efficiently than BMH, since its average shift length should be greater by at least one. A strategy based on (1) and (2) yields BMS1 in its simplest form, where it is implicitly assumed that the right-to-left processing of $\boldsymbol{p}$ is maintained. But advantage (3) gives us in addition the freedom to change this ordering of the positions of $\boldsymbol{p}$ if we can find one that will on average lead to faster execution. Of course the approach that springs to mind is to process the positions in $\boldsymbol{p}$ in increasing order of the frequencies in the language of the letters at those positions. We now outline this more sophisticated version of BMS1.

The fully general version of BMS1 makes use of a permutation of the positions $1..m$ of $\boldsymbol{p}$ that is stored in an array $\boldsymbol{\pi} = \boldsymbol{\pi}[1..m]$. Although *any* permutation of $1..m$ is possible, the permutation based on letter frequency is defined as follows for every $j \in 1..m$:

$$\boldsymbol{\pi}[j] = j' \iff \boldsymbol{p}[j'] \text{ is the } j\text{th rarest letter that occurs in } \boldsymbol{p},$$

where we shall suppose that ties for jth rarest will be broken by choosing the positions in right-to-left order. Using

$$\boldsymbol{p} = \overset{1}{m}\ \overset{2}{a}\ \overset{3}{r}\ \overset{4}{z}\ \overset{5}{i}\ \overset{6}{p}\ \overset{7}{a}\ \overset{8}{n},$$

for example, we would find

$$\boldsymbol{\pi} = 4\,6\,1\,3\,8\,5\,7\,2.$$

To make use of $\boldsymbol{\pi}$, BMS1 and its associated match routine simply access $\boldsymbol{p}\big[\boldsymbol{\pi}[j]\big]$ in place of $\boldsymbol{p}[j]$. This time the details are straightforward and are left to Exercise 8.3.7. Observe that now the attempted match of $\boldsymbol{p}$ against $\boldsymbol{x}$ can always take place in the natural order $j = 1, 2, \ldots, m$, with left-to-right or right-to-left scans of $\boldsymbol{p}$ handled by the permutations

$$\boldsymbol{\pi} = 1\,2 \cdots n \text{ or } \boldsymbol{\pi} = n\; n-1 \cdots 1,$$

respectively.

For pattern-matching on an alphabet whose frequency distribution is known and moreover highly skewed, Algorithm BMS1 is very attractive, particularly if combined with an efficient skip loop such as the one used in BMFAST. Such a hybrid algorithm will not only be fast, it also remains simple. There is a down side however to BMS1: the indirect addressing $\boldsymbol{p}\big[\boldsymbol{\pi}[j]\big]$ used to access elements of $\boldsymbol{p}$ will increase the time required for all such accesses over that of BM, and so this approach will not be suitable if the frequency distribution of the letters of A is not sufficiently skewed. Also, of course, as indicated in Exercise 8.3.6, at least $\Theta(m)$ additional time will be required to preprocess the array $\boldsymbol{\pi}$, perhaps a significant cost for large values of m.

BM-Sunday(2)

In this subsection we show how freeing BM from the tyranny of the right-to-left scan can be used also to modify $\boldsymbol{\delta_2}$, yielding an algorithm BMS2 that may in some cases be even faster than BMS1.

We observe first that of course the $\boldsymbol{\Delta_1}$ array of BMS1 can be combined with the $\boldsymbol{\delta_2}$ array of standard BM, provided the right-to-left scan of $\boldsymbol{p}$ is maintained. Apart from the preprocessing that computes $\boldsymbol{\Delta_1}$, the only changes required to BM would be a new initialization step

$$i_0 \leftarrow i \leftarrow m$$

together with replacement of the shift instruction by

$$i_0 \leftarrow i \leftarrow \max\big\{i_0 + \mathbf{\Delta_1}\big[\boldsymbol{x}[i_0+1]\big],\ i + \boldsymbol{\delta_2}[j]\big\}$$

As pointed out in [S90], we can perhaps do better in cases where the frequency distribution of the letters of the alphabet is important. There Sunday discusses an array $\mathbf{\Delta_2}$ that generalizes $\boldsymbol{\delta_2}$ by using $\boldsymbol{\pi}[j]$ instead of j. If as above we denote by i_0 the position of $\boldsymbol{x}$ aligned with position m of $\boldsymbol{p}$, then for every $h \in 1..m$, $\boldsymbol{\pi'}[h] = i_0 - \big(m - \boldsymbol{\pi}[h]\big)$ will be the position of $\boldsymbol{x}$ aligned with $\boldsymbol{p}\big[\boldsymbol{\pi}[h]\big]$. If we may by a slight abuse of notation employ the abbreviations

$$\begin{aligned}\boldsymbol{p}\big[\boldsymbol{\pi}[1..h]\big] &= \boldsymbol{p}\big[\boldsymbol{\pi}[1]\big]\boldsymbol{p}\big[\boldsymbol{\pi}[2]\big]\cdots\boldsymbol{p}\big[\boldsymbol{\pi}[h]\big],\\ \boldsymbol{x}\big[\boldsymbol{\pi'}[1..h]\big] &= \boldsymbol{x}\big[\boldsymbol{\pi}[1..h] + (i_0 - m)\big]\\ &= \boldsymbol{x}\big[\boldsymbol{\pi'}[1]\big]\boldsymbol{x}\big[\boldsymbol{\pi'}[2]\big]\cdots\boldsymbol{x}\big[\boldsymbol{\pi'}[h]\big],\end{aligned}$$

for arbitrary $h \in 0..m$, then every partial match of $\boldsymbol{p}$ with $\boldsymbol{x}$ can be expressed in the form

$$\boldsymbol{x}\big[\boldsymbol{\pi'}[1..j-1]\big] = \boldsymbol{p}\big[\boldsymbol{\pi}[1..j-1]\big]$$

for some $j \in 1..m+1$, terminated by the mismatch

$$\boldsymbol{x}\big[\boldsymbol{\pi'}[j]\big] \neq \boldsymbol{p}\big[\boldsymbol{\pi}[j]\big]. \tag{8.2}$$

As with BM, we assume that $\boldsymbol{x}$ and $\boldsymbol{p}$ are padded with special nonmatching letters $\$_1$ and $\$_2$, respectively, so that when $j = m+1$ ($\boldsymbol{p}$ actually matches a substring of $\boldsymbol{x}$), we may suppose that

$$\boldsymbol{x}\big[\boldsymbol{\pi'}[m+1]\big] = \$_1,\ \boldsymbol{p}\big[\boldsymbol{\pi}[m+1]\big] = \$_2,$$

and so even in this case (8.2) holds.

Using this notation and these conventions, it is now possible to describe the computation of $\mathbf{\Delta_2}$ in terms that generalize the Type I and Type II calculations for BM in a straightforward way:

- **(Type I)** If there exists a greatest positive integer $j' < \boldsymbol{\pi}[j]$ and corresponding displacement $d = \boldsymbol{\pi}[j] - j'$ such that

  $$\boldsymbol{p}\big[\boldsymbol{\pi}[1..j-1] - d\big] = \boldsymbol{p}\big[\boldsymbol{\pi}[1..j-1]\big]$$

  while

  $$\boldsymbol{p}[j'] \neq \boldsymbol{p}\big[\boldsymbol{\pi}[j]\big],$$

  then the shift corresponding to a mismatch at position $\boldsymbol{\pi}[j]$ is recorded by the assignment

  $$\mathbf{\Delta_2}[j] \leftarrow d.$$

- **(Type II)** If there exists no such j', we seek instead the longest border $\boldsymbol{p}\big[\boldsymbol{\pi}[1..j']\big]$, $0 \le j' \le m-1$, of the string $\boldsymbol{p}\big[\boldsymbol{\pi}[1..m]\big]$ that satisfies the following conditions: $j' < j$, $d = \boldsymbol{\pi}[m] - \boldsymbol{\pi}[j'] > 0$, and

$$\begin{aligned} d &= \boldsymbol{\pi}[m-1] - \boldsymbol{\pi}[j'-1] \\ &= \boldsymbol{\pi}[m-2] - \boldsymbol{\pi}[j'-2] \\ &\;\;\vdots \\ &= \boldsymbol{\pi}[m-h'+1] - \boldsymbol{\pi}[1]. \end{aligned}$$

  Using the redefined d, again we set

$$\boldsymbol{\Delta_2}[j] \leftarrow d.$$

  Since it is possible that $j' = 0$ $\big(\boldsymbol{p}\big[\boldsymbol{\pi}[1..m]\big]$ has only the empty border$\big)$, we need to define $\boldsymbol{\pi}[0] \equiv 0$ in order to make sense of these conditions in all cases. In fact, as shown in Exercise 8.3.9, whenever the permutation $\boldsymbol{\pi}$ corresponds to increasing frequency of occurrence, it will almost always be true that $j' = 0$; thus in these cases we will make the maximum possible shift $d = \boldsymbol{\pi}[m]$.

We see then that the calculation of $\boldsymbol{\Delta_2}$ can be specified in terms of a permuted pattern $\boldsymbol{p}\big[\boldsymbol{\pi}[1..m]\big]$ in a way that is exactly analogous to the calculation of $\boldsymbol{\delta_2}$ for the original pattern $\boldsymbol{p}[1..m]$. The main new difficulty in this calculation arises from the fact that positions $\boldsymbol{\pi}[h]$ and $\boldsymbol{\pi}[h+1]$ are no longer necessarily adjacent, so that account needs to be taken of the displacement d. Due to this complication and the requirement for indirect addressing, the preprocessing that computes $\boldsymbol{\Delta_2}$ is certainly messier and more time-consuming than that required for $\boldsymbol{\delta_2}$, but retains complexity $\Theta(m)$ and introduces no new ideas. We shall therefore not attempt to describe it in greater detail.

With $\boldsymbol{\Delta_2}$ defined, it is easy to specify a revised algorithm, say BMS2, whose shift instruction is either

$$i_0 \leftarrow i_0 + \max\Big\{\boldsymbol{\Delta_1}\big[\boldsymbol{x}[i_0+1]\big],\ \boldsymbol{\Delta_2}\big[\boldsymbol{\pi}[j]\big]\Big\}$$

or (as proposed in [S90]) simply

$$i_0 \leftarrow i_0 + \boldsymbol{\Delta_2}\big[\boldsymbol{\pi}[j]\big].$$

It appears [HS91] that these rather elaborate versions of BMS2 are competitive only for long patterns, where also the increased cost of preprocessing could become a significant factor; perhaps it is these doubts about the overall efficiency of BMS2 that have led to a much simplified approach [HS91], which we call BMS2*.

Algorithm BMS2* assumes that a skip loop is used with a skip character $\boldsymbol{p}[j]$ chosen either to be the rightmost letter of $\boldsymbol{p}$ ($j = m$) as in BMFAST (Section 8.1) or else based on frequency considerations (see above, **More Effective Skipping**). After the skip has been executed, either the algorithm terminates because $\boldsymbol{p}$ has skipped off the end of $\boldsymbol{x}$ or else $\boldsymbol{p}[j]$ matches $\boldsymbol{x}\big[i_0 - (m-j)\big]$. (We suppose as usual that position m of $\boldsymbol{p}$ is aligned with position i_0 of $\boldsymbol{x}$.) Then after a mismatch is found by the *hctam* routine, Algorithm BMS2* performs a shift

$$i_0 \leftarrow i_0 + \boldsymbol{\Delta}[j],$$

where $\boldsymbol{\Delta}[j]$ is defined as follows:

- if there exists a greatest integer value $j' < j$ such that $\boldsymbol{p}[j'] = \boldsymbol{p}[j]$, then $\boldsymbol{\Delta}[j] = j - j'$;
- otherwise, $\boldsymbol{\Delta}[j] = j$.

BMS2* is therefore a very simple algorithm with simple preprocessing; nevertheless, it seems also to be very efficient in practice [HS91], especially when a guard is used on the match routine and an efficient implementation of the skip loop is used. We leave the details of the algorithm and its preprocessing to Exercise 8.3.11.

Exercises 8.3

1. Prove the correctness of the guard algorithm given on page 212, taking into account the following facts:

   - $\delta_1\left[\boldsymbol{x}[i]\right]$ is used even though it has not been established whether or not $\boldsymbol{x}[i] = \boldsymbol{p}[m]$;
   - $\delta_2[j]$ is used even though it is unknown whether or not

   $$\boldsymbol{x}\left[i - (m - j) + 1..i\right] = \boldsymbol{p}[j + 1..m].$$

2. Researcher Etaoin Shrdlu has computed an array $\boldsymbol{prob}$ in which for every letter $\lambda \in A$, $\boldsymbol{prob}[\lambda]$ specifies the estimated probability of occurrence of λ at an arbitrary position in an arbitrary text on A. Write a preprocessing algorithm that uses $\boldsymbol{prob}$ to compute j_{rarest} and determine your algorithm's complexity.
3. Discuss a modification of the guard algorithm that, in addition to the assignment $j \leftarrow j_{\text{rarest}}$, also assigns $i \leftarrow i - (m - j_{\text{rarest}})$. Will the modification be more or less efficient?
4. Modify Algorithm KMP so as to make use of a guard.
5. Rewrite the BM guard algorithm so that comparisons of $\boldsymbol{p}[j_{\text{rarest}}]$ are confined within a skip loop. Will your skip version execute more quickly than the guard version?
6. Devise an algorithm that uses Professor Shrdlu's $\boldsymbol{prob}$ array to compute the permutation array $\boldsymbol{\pi}[1..m]$ used by Algorithm BMS1. What is your algorithm's time complexity? Perhaps you can suggest to Professor Shrdlu an arrangement of the elements of his $\boldsymbol{prob}$ array that will enable you to reduce these time requirements?
7. Making use of the permutation array $\boldsymbol{\pi}$, write out BMS1 and its associated match routine. What is the worst-case time complexity of BMS1?

8. Modify BMS1 to make use of a skip loop as described in Section 8.1. What difficulty do you find arising that did not occur for BMFAST?
9. Show that when the permutation π gives the letters of $\boldsymbol{p}$ in increasing order of their frequency of use in the language, the string $\boldsymbol{p}[\pi[1..m]]$ has an empty border except when $\boldsymbol{p} = \lambda^m$ for some letter λ. Show more generally that this result holds whenever all occurrences in $\boldsymbol{p}$ of the same letter are adjacent in π. Hence describe for these cases an easier implementation of Type II shifts.
10. Describe the computation of a modified KMP shift array $\boldsymbol{B}'[1..m]$ analogous to $\boldsymbol{\beta}'$ that uses a permutation π of the positions of $\boldsymbol{p}$ rather than the positions themselves. Then rewrite Algorithm KMP and its *match* routine to correspond to $\boldsymbol{B}'$. Do you think the new algorithm will be more efficient in practice or in theory than KMP itself?
11. Assuming that the positions j and j_{rarest} in $\boldsymbol{p}$ of the skip character and least frequent character, respectively, are known, write out Algorithm BMS2* including a guard feature. Also specify the associated *hctam* routine and the computation of $\boldsymbol{\Delta}$. What is the worst-case time complexity of BMS2*?

8.4 BM-Galil

The chief defect of Henry King
Was chewing little bits of string.

— Hilaire BELLOC (1870–1953), *Cautionary Tales*

The preceding sections of this chapter have been largely concerned with rather straightforward (albeit interesting) modifications to BM's $\boldsymbol{\delta_1}$ array. In this section we return to a consideration of the dreaded $\boldsymbol{\delta_2}$.

The reader will recall that in Section 7.2 the possible periodicity of the pattern $\boldsymbol{p}$ was an impediment to establishing an unconditional upper bound of $4n$ on the number of letter comparisons required by Algorithm BM. Here we describe another very simple modification to BM, called Algorithm BMG, and then yet another simple modification to BMG (called, immodestly perhaps, Algorithm BMGS) that permits the bound to be established without any condition imposed on $\boldsymbol{p}$. In effect, we use BMGS to get a proof of a stronger version of Theorem 7.2.4.

The main idea of BMG [G79] is that whenever a match with $\boldsymbol{p}$ is found in $\boldsymbol{x}$, we should make use of our knowledge of the period of $\boldsymbol{p}$ to avoid redundant testing. That is, having found that

$$\boldsymbol{x}[i+1..i+m] = \boldsymbol{p}[1..m], \quad \boldsymbol{x}[i] \neq \boldsymbol{p}[0],$$

we will output $i+1$ and shift $\boldsymbol{p}$ right by

$$\boldsymbol{\delta_2}[0] = m - \boldsymbol{\beta}[m]$$

positions — in other words, by the period of $\boldsymbol{p}$ (Section 1.2). After the shift, we know that the prefix $\boldsymbol{p}[1..\boldsymbol{\beta}[m]]$ matches with the aligned section of the text $\boldsymbol{x}$, and so to establish whether or not there exists a second match with $\boldsymbol{p}$ in $\boldsymbol{x}$, it is necessary only to test the suffix $\boldsymbol{p}[\boldsymbol{\beta}[m]+1..m]$ of length $m-\boldsymbol{\beta}[m]$.

The Galil variant of Algorithm BM is thus extremely simple: if (and only if) a complete match with $\boldsymbol{p}$ has just been found, the value $\boldsymbol{\beta}[m]$ of the longest border of $\boldsymbol{p}$ is used to limit the search conducted by the next invocation of *hctam* to at most the final $m-\boldsymbol{\beta}[m]$ positions of $\boldsymbol{p}$. Recall from Section 7.2 that $\boldsymbol{\beta}[m]$ is available as a byproduct of the calculation of $\boldsymbol{\delta_2}$ (in particular g_{II}) during preprocessing. Algorithm 8.4.1 displays the necessary revisions, most of them concerned with setting a variable *jump* that equals the length of the prefix of $\boldsymbol{p}$ that does not need to be compared with $\boldsymbol{x}$ in the next invocation of *hctam*. The algorithm assumes that *hctam* continues to return $j=0$ when a match with $\boldsymbol{p}$ is found, whether or not *jump* $=0$. The corresponding revisions to *hctam* and proof of correctness of Algorithm BMG are left to Exercises 8.4.1 and 8.4.2.

Algorithm 8.4.1 (BMG)

– *Find all occurrences of* $\boldsymbol{p}$ *in* $\boldsymbol{x}$

$i \leftarrow m$; *jump* $\leftarrow 0$
while $i \le n$ **do**
  $(i, j) \leftarrow$ *hctam*$(i, jump, m)$
  if $j = 0$ **then**
    output $i+1$; *jump* $\leftarrow \boldsymbol{\beta}[m]$
  else
    jump $\leftarrow 0$
  $i \leftarrow i + \max\{\boldsymbol{\delta_1}[\boldsymbol{x}[i]], \boldsymbol{\delta_2}[j]\}$

On most test cases Algorithm BMG will execute more slowly than BM, since the additional instructions required to deal with *jump* will certainly slow down every iteration of the main loop while conferring benefits only when a highly periodic pattern $\boldsymbol{p}$ occurs frequently in $\boldsymbol{x}$. BMG was however useful in a theoretical context: in [G79] it was used to prove a worst-case upper bound of $14n$ letter comparisons for arbitrary pattern $\boldsymbol{p}$, periodic or not. We now show how a further slight extension of BMG yields a straightforward proof of a $4n$ upper bound on letter comparisons.

The extension from BMG to BMGS is a natural and obvious one: instead of using *jump* only when an occurrence of $\boldsymbol{p}$ has been found, use it every time a Type II shift (Figure 7.4) of $\boldsymbol{p}$ has taken place in the preceding attempted match. Of course a Type II shift is easily recognized by virtue of the fact that $\boldsymbol{p}$ is shifted to the right of the position in $\boldsymbol{x}$ at which the preceding mismatch occurred — thus $\boldsymbol{\delta_2}[j] \ge m$. Hence whenever $\boldsymbol{\delta_2}[j] \ge \boldsymbol{\delta_1}[\boldsymbol{x}[i]]$, a correct setting of *jump* for the next attempted match is ensured by

jump $\leftarrow j'$, if $\boldsymbol{\delta_2}[j] \ge m$;
  $\leftarrow 0$, otherwise;

where as shown in Figure 7.4, j' is the length of the longest border of $\boldsymbol{p}[j+1..m]$. Since for a Type II shift, $\delta_2[j] = (m-j) + (m-j')$, we can compute j' from the identity

$$j' = (m-j) - (\boldsymbol{\delta_2}[j] - m).$$

The details of the revisions to BMG (yielding Algorithm BMGS) are left to Exercise 8.4.3. Note that BMGS requires scarcely more overhead than BMG, but may provide benefit (reduced letter comparisons) every time a Type II shift is performed. In fact, BMGS guarantees that, after a Type II shift, the following attempted match performs letter comparisons only on positions in $\boldsymbol{x}$ that have never previously been examined; and in no case does BMGS perform more letter comparisons than BMG (or, for that matter, than BM). Thus in practice we can expect BMGS to be on average faster than BMG. On the theoretical side, use of BMGS leads to a proof of Theorem 7.2.4 that holds for any pattern $\boldsymbol{p}$.

Recall that based on Lemma 7.2.3, we were able to conclude that for *any* pattern $\boldsymbol{p}$, periodic or not, a Type I shift (Figure 7.3) would always yield

$$t + 1 \leq f + 3s, \tag{8.3}$$

where $t = m - j$ is the length of the currently matched suffix of $\boldsymbol{p}$, f is the number of positions being compared for the first time, and s is the length of the shift. We saw also that (8.3) holds for all matches that give rise to a "long shift" ($3s \geq t$). Thus in our effort to improve upon Theorem 7.2.4, we need only consider short Type II shifts (Figure 7.4) of 3-periodic patterns $\boldsymbol{p}$ whose right normal form is

$$\boldsymbol{p} = \boldsymbol{v}'\boldsymbol{v}^{\boldsymbol{k}},\ k \geq 3,$$

where $\boldsymbol{v}$ is not a repetition and where, as in Section 7.2, we let $v = |\boldsymbol{v}|$.

With these observations fresh in our minds, we will now be able to prove the main result of this section:

Theorem 8.4.1 *Algorithm BMGS requires at most $4n$ letter comparisons.*

Proof As we have seen, the result holds for all non-3-periodic patterns whose shifts are either long or of Type I. Suppose then that the period $v \leq m/3$ and that the shift $\boldsymbol{s}$ induced by the current attempted match is a short Type II one. Because $\boldsymbol{s}$ is short, $3s < t$, and because it is Type II, a mismatch of $\boldsymbol{p}[m-t]$ must have been found, leading therefore to a shift of length $s \geq m - t$. From these two conditions on s and t, we conclude that for an attempted match of a 3-periodic pattern that gives rise to a short Type II shift,

$$t \geq 3m/4 \geq 9v/4. \tag{8.4}$$

Now consider the attempted match immediately prior to the current one. If there were no such previous match, or if it were one that gave rise to a Type II shift, the execution of BMGS would ensure that, for the *current* match, the only positions of $\boldsymbol{x}$ being tested would be those being tested for the first time. In other words, for the current match, it would be

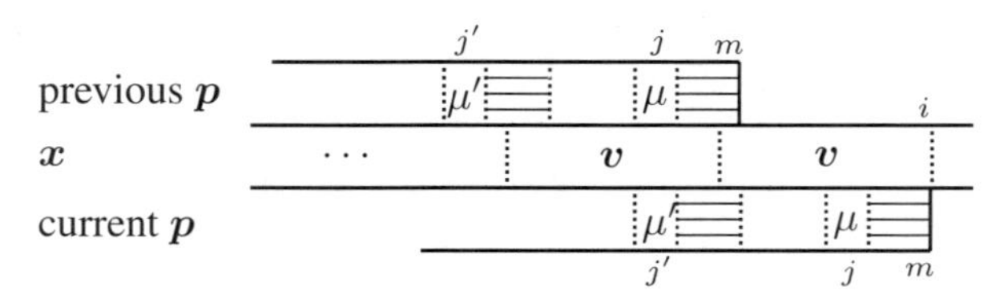

Figure 8.1. Previous shift Type I; current match $t \geq 9v/4$

true that $t = f$, so that (8.3) would hold. Therefore, for BMGS, we need consider only attempted matches preceded by a Type I shift, as illustrated in Figure 8.1.

Suppose then that the previous match gave rise to a shift of length $s_{\text{prev}} = j - j'$ after $t_{\text{prev}} = m - j$ letter comparisons were performed. If we suppose further that $t_{\text{prev}} \geq v$, then there must exist a substring

$$\boldsymbol{u} = \boldsymbol{p}[j' + 1..j' + (m - j)]$$

of length at least v such that $\boldsymbol{u} = \boldsymbol{p}[j + 1..m] = \boldsymbol{u'v}$ for some string $\boldsymbol{u'}$. Thus $\boldsymbol{u}$ has suffix $\boldsymbol{v}$. But since the letter $\boldsymbol{p}[j'] = \mu'$ does not equal μ, the suffix $\boldsymbol{v}$ cannot be aligned with any of the k occurrences of $\boldsymbol{v}$ in $\boldsymbol{p} = \boldsymbol{v'v}^k$, and so necessarily equals a rotation $R_j(\boldsymbol{v})$, $0 < j \leq n-1$, of $\boldsymbol{v}$ (Section 1.4). By Theorem 1.4.2, this is possible only if $\boldsymbol{v}$ is a repetition, a possibility specifically excluded in the definition of $\boldsymbol{v}$. We conclude that $t_{\text{prev}} = |\boldsymbol{u}| < v$.

Turning our attention now to the value of s_{prev}, we observe first that the substring $\boldsymbol{u}$ cannot be a suffix of $\boldsymbol{v}$ — it it were, it could not be preceded on the left by $\mu' \neq \mu$. Hence $\mu'\boldsymbol{u}$ is either a substring of $\boldsymbol{v}$ that occurs to the left of $\boldsymbol{p}[j + 1..m]$ in every occurrence of $\boldsymbol{v}$, or $\mu'\boldsymbol{u}$ is a substring of some rotation of $\boldsymbol{v}$ that in turn occurs as a substring of $\boldsymbol{v}^2$. In both of these cases the rightmost position $j' + (m - j)$ of $\mu'\boldsymbol{u}$ is distance less than v from position m of $\boldsymbol{p}$. That is, the previous shift

$$s_{\text{prev}} = m - \big(j' + (m - j)\big) = j - j' < v.$$

From (8.4) it follows that a substring $\boldsymbol{v}^2$ of $\boldsymbol{x}$ matches a suffix $\boldsymbol{v}^2$ of $\boldsymbol{p}$ at its current alignment (position i in Figure 8.1). Thus, since $s_{\text{prev}} < v$, the substring $\boldsymbol{v}$ exists in $\boldsymbol{x}$, as shown in Figure 8.1, to the left of the previous alignment of $\boldsymbol{p}$. By the correctness of BM (and BMGS), this occurrence of $\boldsymbol{v}$ must already have been matched with $\boldsymbol{p}$, resulting in a shift of $\boldsymbol{p}$ to the left of its previous position in order to attempt the match shown in Figure 8.1. Since no backtracking occurs in BM (or BMGS), this is impossible, and so we conclude that the previous shift could not have been Type I. □

The reader may wish to compare this proof with that of Lemma 7.2.3: essentially the same ideas are involved.

We shall discover in the next section that BMGS can be thought of as a simplified version of a more general algorithm, Turbo-BM, that in fact requires at most $2n$ letter comparisons.

In addition to Turbo-BM, Algorithm BMG has inspired at least one other algorithm [AG86], one that can store information about several previous attempted matches, rather than just a single attempted match, and so is able to match $\boldsymbol{p}$ against $\boldsymbol{x}$ using at most

$2n - m + 1$ letter comparisons. Unfortunately, the BMGAG algorithm in its original form achieves its result only at an additional cost of roughly $9n$ "non-letter" comparisons together with additional preprocessing, and so, in view of Turbo-BM, appears to be competitive neither in a practical nor in a theoretical context.

However, a recent improvement to BMGAG [CL97, CHL02] reduces and simplifies the preprocessing requirement and also reduces the upper bound on letter comparisons to $1.5n$, a bound shown to be sharp. Thus, in terms of letter comparisons, the CL variant of Algorithm BMGAG seems to take the Boyer-Moore algorithm just about as far as it can go.

Exercises 8.4

1. Rewrite the routine *hctam* in accordance with Algorithm BMG.
2. Prove the correctness of Algorithm BMG given that BM is correct.
3. Write out Algorithm BMGS and prove its correctness.
4. With a little ingenuity, it is possible to extend BMGS to avoid redundant letter comparisons also in the case that a Type I shift of $\boldsymbol{p}$ has taken place in the preceding attempted match. The main difference will be that a preceding Type II shift ensures that the previously matched substring is a prefix of $\boldsymbol{p}$, whereas for a preceding Type I shift it can be assumed only to be a substring of $\boldsymbol{p}$. Thus it does not suffice to store only a single value *jump* — the rightmost position of the "jumped" substring must also be specified.

   Write out the revised algorithm BMGS* together with its revised routine *hctam*. It is clear that BMGS* will never require more letter comparisons than are used by BMGS. Do you think BMGS* will be more or less efficient in practice than BMGS?

8.5 Turbo-BM

A man thinks that by mouthing hard words
he understands hard things.

— Herman MELVILLE (1819–1891)

We have seen in the preceding section (Exercise 8.4.4) that an algorithm BMGS* can be designed to avoid repeating letter comparisons that were performed in the preceding attempted match. In this section we show how to extend BMGS* so as to make still further use of the stored information about the preceding partial match — we discover that in some cases this information can also be used to increase the length of the shift of $\boldsymbol{p}$ along $\boldsymbol{x}$. The new algorithm is called Turbo-BM and the new shifts ***turbo-shifts*** [CCGJ94].

As illustrated in Figures 8.2 and 8.3, we suppose that after a partial match of $\boldsymbol{t} = \boldsymbol{p}[j+1..m]$ against $\boldsymbol{x}$, a shift $\boldsymbol{s}$ of Type I or II, respectively, has taken place, followed by a partial match of $\boldsymbol{t'}$ against $\boldsymbol{x}$. As before, we let $t = |\boldsymbol{t}|$, $s = |\boldsymbol{s}|$, $t' = |\boldsymbol{t'}|$. The case $t' = s$ is

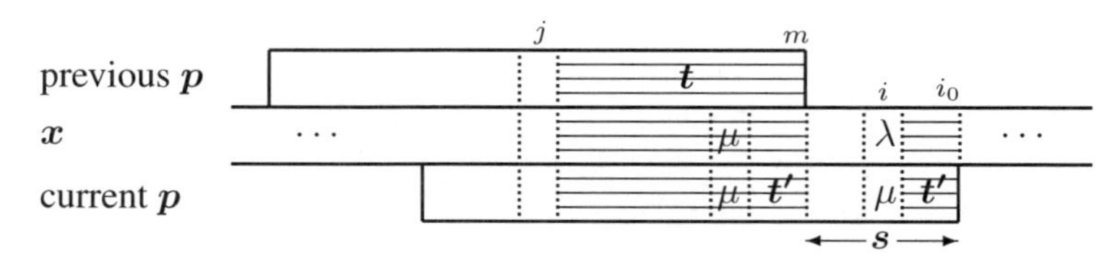

Figure 8.2. Previous shift Type I: $t' < s, t' < t$

taken care of by BMGS*, since then either the suffix $\boldsymbol{ts}$ of $\boldsymbol{p}$ matches $\boldsymbol{x}$ with only $\boldsymbol{p}[1..j']$ remaining to be tested (previous shift of Type I) or else the entire pattern $\boldsymbol{p}$ matches $\boldsymbol{x}$ (previous shift of Type II).

Thus, in order to study turbo-shifts, we consider the case $t' < s$, with a mismatch occurring at

$$\boldsymbol{p}[m - t'] = \mu \neq \lambda = \boldsymbol{x}[i_0 - t']. \tag{8.5}$$

Then the string $\mu\boldsymbol{t'}$ is a suffix of $\boldsymbol{s}$. Confining ourselves further to the case $t' < t$, we see that $\mu\boldsymbol{t'}$ is also a suffix of $\boldsymbol{t}$, where it matches with a substring of $\boldsymbol{x}$, and where in particular μ matches with a position in $\boldsymbol{x}$:

$$\boldsymbol{p}[m - t'] = \boldsymbol{p}\big[(m - s) - t'\big] = \mu = \boldsymbol{x}\big[(i_0 - s) - t'\big]. \tag{8.6}$$

Comparing (8.5) and (8.6), we see that μ and λ are separated in $\boldsymbol{ts}$ by s positions. Recalling (Section 7.2, **Efficiency of BM**) that $\boldsymbol{ts}$ necessarily has period s, we conclude that μ and λ in these positions cannot both match $\boldsymbol{ts}$, and so we can safely shift the unmatched portion of $\boldsymbol{p}$ to the right of position $i - s$ in $\boldsymbol{x}$, where μ occurs.

Figure 8.2 shows that in the case of a previous Type I shift, $\boldsymbol{p}[m - s - t]$ can be shifted right $t - t'$ positions, so that in the next attempted match $\boldsymbol{p}[m]$ will be aligned with $\boldsymbol{x}[i + t]$. This corresponds to the turbo-shift

$$i \leftarrow i + t. \tag{8.7}$$

On the other hand, from Figure 8.3 we see that for a previous Type II shift, $\boldsymbol{p}[1]$ can be shifted right $m - s - t'$ positions, aligning $\boldsymbol{p}[m]$ with $\boldsymbol{x}[i + m - s]$ and corresponding to the turbo-shift

$$i \leftarrow i + (m - s). \tag{8.8}$$

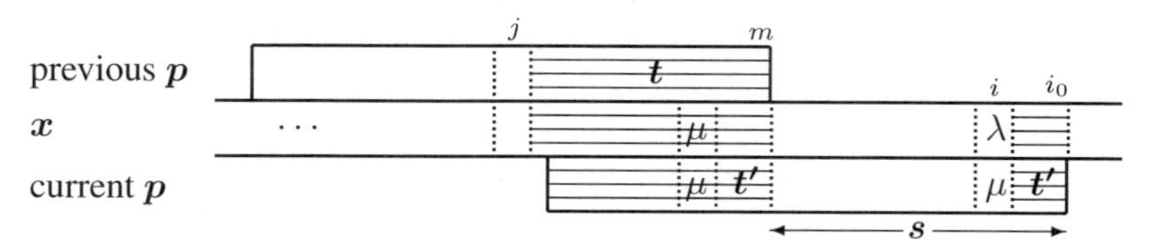

Figure 8.3. Previous shift Type II: $t' < s, t' < t$

Note that for a Type I shift, $t \le m - s$, while for a Type II shift, $m - s \le t$. Thus, since $t = m - j$, the assignments

$$s \leftarrow \boldsymbol{\delta_2}[j] - (m - j)$$
$$jump \leftarrow m - \max\{j, s\}$$

identify, for either type of shift, the number (*jump*) of positions already matched at distance s from the rightmost end of the next alignment of $\boldsymbol{p}$. Both (8.7) and (8.8) will therefore be taken account of in the next attempted match.

Algorithm 8.5.1 (Turbo-BM)

- *Find all occurrences of* $\mathbf{p}$ *in* $\boldsymbol{x}$

$i \leftarrow m$; $jump \leftarrow 0$
while $i \le n$ **do**
  – *For jump* > 0, *if the rightmost* s *positions of* $\mathbf{p}$
  – *match with* $\boldsymbol{x}$, *then skip the next* jump *positions*
  $(i, j) \leftarrow hctam(i, s, jump, m)$
  if $j = 0$ **then output** $i + 1$

  $shift \leftarrow \max\{jump, \boldsymbol{\delta_1}[\boldsymbol{x}[i]]\}$
  $i \leftarrow i + \max\{\boldsymbol{\delta_2}[j], shift\}$
  if $\boldsymbol{\delta_2}[j] < shift$
    then $jump \leftarrow 0$
  else
    $s \leftarrow \boldsymbol{\delta_2}[j] - (m - j)$; $jump \leftarrow m - \max\{j, s\}$

Turbo-BM is displayed as Algorithm 8.5.1. Note that no additional preprocessing is needed beyond that required for Algorithm BM, and that the two values s and *jump* constitute an elegant sufficiency of additional storage. The upper bound of $4n$ letter comparisons satisfied by Algorithm BMGS* of course holds also for Turbo-BM; in fact, as remarked earlier, it can be shown with a bit more work [CCGJ94] that Turbo-BM requires at most $2n$ letter comparisons. In a theoretical context then, the worst-case number of letter comparisons required for Turbo-BM matches those required for Algorithm KMP (Section 7.1), though as observed earlier (Section 7.5), the average-case behaviour will be much better.

[CCGJ94] also describes a further extension of Turbo-BM that uses the Directed Acyclic Word Graph (Section 5.3.1) of $\boldsymbol{p}$ to select even more useful values of s and *jump* for storage. This variant appears to execute more quickly in practice than Turbo-BM, but of course requires considerably more preprocessing effort for construction of the DAWG.

Exercise 8.5

1. Rewrite the routine *hctam* in accordance with Algorithm Turbo-BM. If you write it well, you may be surprised at its simplicity!

8.6 Daughter of KMP Rides Too!

A word is dead
When it is said,
Some say.

— Emily DICKINSON (1830–1886), *No. 1212*

After so many variations on Algorithm BM, finally in this section we consider a variant of KMP (Section 7.1) that reduces significantly (from $2n$ to $1.5n$) the worst-case number of letter comparisons required to compute all occurrences of pattern $\boldsymbol{p}$ in a text $\boldsymbol{x}$. Derived from a suggestion made in [AC91], the algorithm was first presented in a Ph. D. thesis [H93]. We therefore call it Algorithm KMP-Hancart.

The main idea of KMPH stems from an observation made at the beginning of Section 7.1, where we first considered the complexity of KMP. We remarked that comparing a pattern $\boldsymbol{p} = a^{m-1}b$ against a text $\boldsymbol{x} = a^n$ required $2n - m$ letter comparisons, a quantity later shown to be worst possible for KMP. To avoid such cases, Algorithm KMPH splits $\boldsymbol{p}$ into two parts that are dealt with more or less separately:

- a maximum-length prefix of the form $\lambda^{m'-1}$, where $1 < m' < m$ and λ is some letter of the alphabet;
- the nonempty suffix $\boldsymbol{p}[m'..m]$, where $\boldsymbol{p}[m'] \neq \lambda$.

In the case that $\boldsymbol{p} = \lambda^m$, we set $m' \leftarrow 1$, so that the prefix is empty and the suffix $\boldsymbol{p}[m'..m] = \lambda^m$. Of course m' is computed as part of the preprocessing of $\boldsymbol{p}$.

We use a triple

$$(i, j_L, j_R)$$

to keep track of partial matches with $\boldsymbol{x}[i+1..i+m]$ in the two portions (left and right), as shown in Figure 8.4:

- $j_L \in 1..m'$ is the position in the prefix of $\boldsymbol{p}$ such that

$$\boldsymbol{p}[1..j_L - 1] = \boldsymbol{x}\big[i+1..i+(j_L-1)\big];$$

- $j_R \in m'..m+1$ is the position in the suffix of $\boldsymbol{p}$ such that

$$\boldsymbol{p}[m'..j_R - 1] = \boldsymbol{x}\big[i+m'..i+(j_R-1)\big].$$

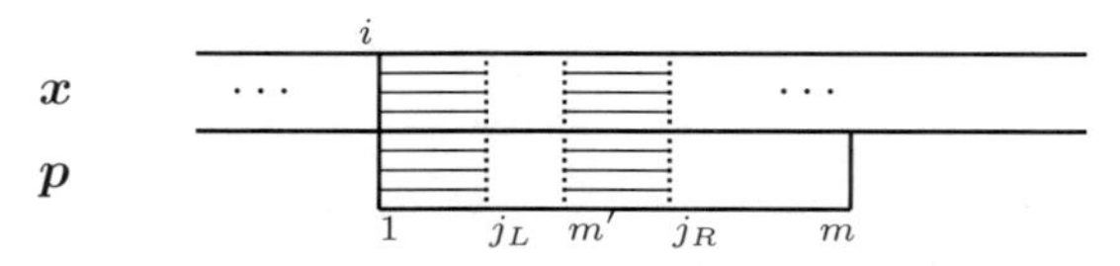

Figure 8.4. Partial matches of $\boldsymbol{p}[1..j_L-1]$ and of $\boldsymbol{p}[m'..j_R-1]$

The initial value of this triple when KMPH starts to compare $\boldsymbol{p}$ with $\boldsymbol{x}$ is $(i, j_L, j_R) = (0, 1, m')$, reflecting the fact that so far no partial match has occurred, neither with the prefix $\boldsymbol{p}[1..m'-1]$ nor with the suffix $\boldsymbol{p}[m'..m]$.

The execution of Algorithm KMPH is similar to that of KMP in that the pattern $\boldsymbol{p}$ is shifted from left to right without backtracking across $\boldsymbol{x}$, so that at each stage a window of length m beginning at position $i+1$ of $\boldsymbol{x}$ is compared with $\boldsymbol{p}$. Also in the KMP style, comparisons take place from left to right in each of the two "subwindows" defined by m'. The details of the processing can be described in terms of updates to the triple $t = (i, j_L, j_R)$, and three cases arise, depending on the current value of j_R:

(1) $j_R = m'$ (no current partial match with suffix $\boldsymbol{p}[m'..m]$)
In this case, we compare $\boldsymbol{p}[j_R]$ with $\boldsymbol{x}[i+j_R]$. If equality is found, the new triple is $t = (i, j_L, j_R+1)$. If not, the new triple is $t = (i+1, j'_L, m')$, where $j'_L = \max\{1, j_L - 1\}$, reflecting a shift of $\boldsymbol{p}$ to the right by one position: after the shift, there will be a partial match in the left subwindow with $\boldsymbol{p}[1..j'_L - 1] = \lambda^{j'_L - 1}$.

(2) $m' < j_R < m+1$ (current match with $\boldsymbol{p}[m'..j_R - 1]$)
Again we compare $\boldsymbol{p}[j_R]$ with $\boldsymbol{x}[i+j_R]$. If equality is found, the new triple, as in case (1), is (i, j_L, j_R+1).

Now suppose that $\boldsymbol{p}[j_R] \neq \boldsymbol{x}[i+j_R]$. This means that we must shift $\boldsymbol{p}$ to the right in order to reestablish (if possible) a partial match between $\boldsymbol{p}$ and $\boldsymbol{x}$. But observe that since

$$\boldsymbol{x}[i+m'] = \boldsymbol{p}[m'] \neq \boldsymbol{p}[j] = \lambda,\ \forall j \in 1..m'-1,$$

this shift cannot be less than m' in length. Further, recall that in Algorithm KMP, either $\boldsymbol{\beta}''[j]-1$ is the length of the longest border of $\boldsymbol{p}[1..j-1]$ such that $\boldsymbol{p}\big[\boldsymbol{\beta}''[j]\big] \neq \boldsymbol{p}[j]$, or $\boldsymbol{\beta}''[j] = 0$ if there is no such border.

Let q be the period of $\boldsymbol{p}[1..j_R - 1]$, and recall from Section 1.2 that therefore $(j_R - 1) - q$ is the length of the longest border of $\boldsymbol{p}[1..j_R - 1]$. Thus $j_R - 1 - q \geq \boldsymbol{\beta}''[j_R] - 1$, from which it follows that

$$j_R - \boldsymbol{\beta}''[j_R] \geq q.$$

But by the definition of m', we know that $q \geq m'$, so that

$$j_R - \boldsymbol{\beta}''[j_R] \geq m'. \tag{8.9}$$

Thus the standard KMP shift of $j_R - \boldsymbol{\beta}''[j_R]$ resulting from a mismatch at position j_R will reposition $\boldsymbol{p}$ correctly for the next attempted match. After the shift, a prefix of $\boldsymbol{p}$ of length $\boldsymbol{\beta}''[j_R] - 1$ will match with $\boldsymbol{x}$. If $\boldsymbol{\beta}''[j_R] - 1 < m'$, the prefix $\boldsymbol{p}\big[1..\boldsymbol{\beta}''[j_R] - 1\big] = \lambda^{\boldsymbol{\beta}''[j_R]-1}$, and the new triple is

$$t = \Big(i + \big(j_R - \boldsymbol{\beta}''[j_R]\big), \max\big\{\boldsymbol{\beta}[j_R], 1\big\}, m'\Big).$$

If on the other hand $\boldsymbol{\beta}''[j_R] - 1 \geq m'$, the new triple is

$$t = \Big(i + \big(j_R - \boldsymbol{\beta}''[j_R]\big), m', \boldsymbol{\beta}[j_R]\Big).$$

(3) $j_R = m+1$ (current match with suffix $\boldsymbol{p}[m'..m]$)
If in addition $j_L = m'$, then a complete match with $\boldsymbol{p} = \boldsymbol{p}[1..m'-1]\boldsymbol{p}[m'..m]$ has been found at position $i+1$ of $\boldsymbol{x}$. After the match has been reported, the new triple may be computed as described in (2) above for the case in which $\boldsymbol{p}[j_R] \neq \boldsymbol{x}[i+j_R]$.
For $j_L < m'$, we compare $\boldsymbol{p}[j_L]$ with $\boldsymbol{x}[i+j_L]$. If equality is found, the new triple is $t = (i, j_L+1, j_R)$; otherwise, the new triple is again computed as in (2).

The details of the processing described above are shown in Algorithm 8.6.1, with implementation of the routine *match* (very similar to the one used in KMP) left to Exercise 8.6.2.

Algorithm 8.6.1 (KMPH)

```
  – Find all occurrences of p in x
(i, j_L, j_R) ← (0, 1, m')
while i ≤ n−m do
    – Try to find a partial match with the suffix p[m'..m]
  j_R ← match(i, j_R, m)
    – If a complete match is found with the suffix,
    – try to find a partial match with the prefix p[1..m'−1]
  if i = m + 1 then
    j_L ← match(i, j_L, m'−1)
    if j_L = m' then output i + 1

    – Recompute the triple t according to cases (1)–(3)
  i ← i + (j_R − β''[j_R])
  if j_R = m' then
    j_L ← max{1, j_L − 1}
  elsif β''[j_R] ≤ m' then
    (j_L, j_R) ← (max{β''[j_R], 1}, m')
  else
    (j_L, j_R) ← (m', β''[j_R])
```

Observe that i increases by at least one in each step of this algorithm, which therefore terminates. Based on the correctness of the updating of the triple t, we claim that

Theorem 8.6.1 *Algorithm 8.6.1 correctly computes all occurrences of a given nonempty pattern* $\boldsymbol{p}$ *in a given string* $\boldsymbol{x}$. □

In order to compute the number of letter comparisons required by Algorithm KMPH, we introduce a quantity C_i, the number of letter comparisons performed in the current execution of the **while** loop. If we let (i, j_L, j_R) be the value of the triple at the beginning of the loop, and let (i, j_L, j_R^*) be its value upon exit from the call to *match*(i, j_R, m), then we may compute/bound C_i as follows:

(a) For $j_R^* = m'$, only one letter comparison has taken place in *match*, and no others will take place in the current **while** loop. Thus $C_i = 1$.

(b) For $j_R^* \in m'+1..m$, $C_i = j_R^* - j_R + 1$.

(c) For $j_R^* = m+1$, a complete match with the suffix $\boldsymbol{p}[m'..m]$ has been found, so that $match(i, j_L, m'-1)$ will be invoked to test the prefix $\boldsymbol{p}[1..m'-1]$. In this case, assuming that *match* executes without making use of a special character $\boldsymbol{p}[m+1] = \$$, $C_i \le (m - j_R + 1) + (m' - j_L)$.

It is convenient now to identify the quantity

$$Q_i = 3i/2 + j_L/2 + j_R, \tag{8.10}$$

whose change ΔQ_i from the beginning to the end of the current **while** loop can be computed/bounded as follows:

(a) For $j_R^* = m'$, observe that $i^* = i+1$ and $j_R^* = j_R$, while j_L will be reduced by at most 1. Thus

$$\Delta Q_i \ge 3(1)/2 - 1/2 = 1.$$

(b′) In the case that $j_R^* > m'$ and $\boldsymbol{\beta}''[j_R^*] \le m'$, $\boldsymbol{p}$ is shifted right by $j_R^* - \boldsymbol{\beta}''[j_R^*]$ positions, while j_L is changed to $\max\{\boldsymbol{\beta}''[j_R^*], 1\}$ and j_R is reduced to m'. Thus

$$\begin{aligned}\Delta Q_i &\ge 3\left(j_R^* - \boldsymbol{\beta}''[j_R^*]\right)/2 + \left(\boldsymbol{\beta}''[j_R^*] - j_L\right)/2 + (m' - j_R) \\ &= \left(j_R^* - \boldsymbol{\beta}''[j_R^*]\right) + (j_R^* - j_L)/2 + (m' - j_R).\end{aligned}$$

(c′) Finally, if $j_R^* > m'$ and $\boldsymbol{\beta}''[j_R^*] > m'$, we find that

$$\begin{aligned}\Delta Q_i &= 3\left(j_R^* - \boldsymbol{\beta}''[j_R^*]\right)/2 + (m' - j_L)/2 + \left(\boldsymbol{\beta}''[j_R^*] - j_R\right) \\ &= (j_R^* - j_R) + \left(j_R^* - \boldsymbol{\beta}''[j_R^*]\right)/2 + (m' - j_L)/2.\end{aligned}$$

We show that the number of letter comparisons in any execution of the **while** loop is bounded above by ΔQ_i:

Lemma 8.6.2 *In the* **while** *loop for i of Algorithm 8.6.1, $C_i \le \Delta Q_i$.*

Proof There are five cases to consider, formed from the three possibilities (a,b,c) for C_i and the three possibilities (a,b′,c′) for ΔQ_i. We denote these cases (a), (bb′), (bc′), (cb′) and (cc′).

(a) Here we have immediately $\Delta Q_i \ge 1 = C_i$.

(bb′) In this case,

$$\begin{aligned}\Delta Q_i - C_i &\ge \left(m' - \boldsymbol{\beta}''[j_R^*] - 1\right) + (j_R^* - j_L)/2 \\ &\ge -1/2,\end{aligned}$$

since $j_R^* > m' \ge j_L$ and $\boldsymbol{\beta}''[j_R^*] \le m'$. But $\Delta Q_i - C_i$ must be an integer, so $\Delta Q_i - C_i \ge 0$.

(bc′) Here

$$\Delta Q_i - C_i \geq \big(j_R^* - \boldsymbol{\beta}''[j_R^*]\big)/2 + (m' - j_L)/2 - 1$$
$$\geq -1/2,$$

since $j_R^* - \boldsymbol{\beta}''[j_R^*] \geq 1$ and $m' - j_L \geq 0$. As in case (bb′), we argue that therefore $\Delta Q_i - C_i \geq 0$.

(cb′) In this case we substitute $j_R^* = m+1$ and compute

$$\Delta Q_i - C_i \geq (m+1)/2 - \boldsymbol{\beta}''[m+1]$$
$$= \big((m+1) - \boldsymbol{\beta}''[m+1]\big)/2 - \boldsymbol{\beta}''[m+1]/2.$$

Now observe that (8.9) holds also in the case that $j_R^* = m+1$ and recall that for (b′), $\boldsymbol{\beta}''[m+1] \leq m'$. Thus $\Delta Q_i - C_i \geq (m' - m')/2 = 0$, as required.

(cc′) Finally we compute

$$\Delta Q_i - C_i \geq \big((m+1) - \boldsymbol{\beta}''[m+1]\big)/2 - (m' - j_L)/2.$$

Applying (8.9) again, this becomes $\Delta Q_i - C_i \geq (m' - m' + j_L)/2 = j_L/2 \geq 0$. □

We are now in a position to prove the main result of this section:

Theorem 8.6.3 *Algorithm 8.6.1 computes all occurrences of a nonempty pattern* $\boldsymbol{p}$ *in a string* $\boldsymbol{x}$ *using* $O(n-m)$ *time and* $\Theta(m)$ *additional space, while performing at most* $\lfloor (3n-m)/2 \rfloor$ *comparisons on elements of* $\boldsymbol{x}$.

Proof Lemma 8.6.2 tells us that, over all executions of the **while** loop in Algorithm 8.6.1, the total number $C = \sum_i C_i$ of letter comparisons performed on elements of $\boldsymbol{x}$ satisfies

$$C \leq \sum_i \Delta Q_i = Q_{\text{final}} - Q_{\text{init}},$$

where Q_{final} is the value of Q at the end of the final execution of the loop, and Q_{init} is its value at the beginning of the first execution. Since the initial triple $t = (0, 1, m')$, we see from (8.10) that

$$Q_{\text{init}} = 1/2 + m',$$

while the final value must satisfy

$$Q_{\text{final}} \leq 3(n-m)/2 + m'/2 + m.$$

Thus

$$C \leq (3n - m)/2 - (m' + 1)/2,$$

yielding the desired result. □

Exercises 8.6

1. Write an $O(m)$-time algorithm to compute m' for pattern $\boldsymbol{p}[1..m]$.
2. Write a *match* routine appropriate for Algorithm KMPH, bearing in mind that the objective now is to minimize letter comparisons rather than to produce the most efficient code.

8.7 Mix Your Own Algorithm

Always and never are two words
you should always remember never to use.

— Wendell A. L. JOHNSON (1906–1965)

Even the uncaring and inattentive reader will have observed that our discussion to date of Son-of-BM and Daughter-of-KMP options has been far from exhaustive, even if attention is confined only to the issue of performance in practice. What happens, for example, when a skip loop is added to Turbo-BM or to KMPH? How does Turbo-BM compare to Algorithm BMH? How do methods based on letter frequency compare with the more sophisticated search strategies based on $\boldsymbol{\delta_2}$? Is Turbo-BM really faster on average than BMGS*? Is it generally true, as suggested in [HS91], that algorithms employing a single (good) shift strategy (say $\boldsymbol{\Delta_1}$ only) are superior to those employing two or more? How strongly is performance affected by implementation details or differing hardware/software platforms?

A really thorough study of exact pattern-matching would surely have dealt (at great length) with these and dozens of other vexing questions. Happily, we may excuse ourselves from embarking on such a Sisyphean task here, because a start has already been made in [HS91], where a taxonomy of exact pattern-matching algorithms is proposed, and where implementations of many important algorithms are presented, compared, and discussed. More recently, other major experimental studies have been published [L95, L00, MM01], while a wealth of similar material is maintained at a website [ChL97]. In this section we present the classification system of [HS91], thus providing a framework within which algorithms to locate $\boldsymbol{p}$ in $\boldsymbol{x}$ can be described in a reasonably precise way according to three main parameters:

- type of skip loop;
- order in which the letters of the pattern are compared;
- basis upon which the pattern is shifted.

Tables 8.1–8.3 give abbreviations and corresponding definitions for many (but again, far from all) of the techniques that have been discussed in this chapter and the previous one. Using these abbreviations, one can characterize a matching strategy in terms of a triple

<skip, match, shift>,

where each parameter in the triple takes one of the values given in the corresponding table. Thus, for example, Algorithm Easy (Section 2.2) is specified by the triple <*none*, *fwd*, +1>, the one that defines BMFAST is <δ_1', *rev*, $\boldsymbol{\delta_1\delta_2}$>, while that of BMS2* is <*lowcost*, *rev*, $\boldsymbol{\Delta}$>.

Note that the total number of different triples, hence the total number of different algorithms, that can be constructed from the parameters given in the three tables is $240 = 4 \times 6 \times 10$. Even though some of the triples are impossible (for example <*none*, *rev*, $\boldsymbol{\beta'}$>), still this classification system suggests that there are a great many possible algorithms to consider. You can truly mix your own!

Exercises 8.7

1. Mix an algorithm of your own from Tables 8.1–8.3 and say what you can about its average-case and worst-case time requirements.
2. Observe that Algorithm KMPH, described in the previous section, is not included in the taxonomy given here. How would you include it?

none	
δ_1'	the BMFAST strategy
rare	skip to next occurrence of a rare letter [H80]
lowcost	skip to lowest cost letter [HS91]

Table 8.1: Skip loops

fwd	left-to-right as in KMP
rev	right-to-left as in BM
π	use a permutation array, especially based on letter frequency
fwd $+ g$	*fwd* with guard (Section 8.3)
rev $+ g$	*rev* with guard
$\pi + g$	π with guard

Table 8.2: Order of match

$+1$	on mismatch, shift pattern right one position
β'	shift according to border of matched substring (KMP)
β''	shift using border and mismatched position (KMP)
$\delta_1\delta_2$	original BM shift
δ_2	BM with δ_1 shift removed
δ_1''	shift using modified δ_1 array (BMH)
Δ_1	Sunday's revised δ_1 shift
Δ_2	Sunday's revised δ_2 based on π
Δ	Sunday's shift based on a skip letter
turbo	shift based on δ_1, δ_2 and turbo-shift

Table 8.3: Shifts

8.8 The Exact Complexity of Exact Pattern-Matching

With women you don't have to talk your head off.
You just say a word and let them fill in from there.

— Satchel PAIGE (c. 1906–1982)

In these two chapters we have studied, primarily from a practical point of view, algorithms to compute all occurrences of a given pattern $\boldsymbol{p} = \boldsymbol{p}[1..m]$ in a given text string $\boldsymbol{x} = \boldsymbol{x}[1..n]$. Based primarily on the results of experimental studies, we have seen that variants of the Boyer-Moore algorithm have an average-case behaviour that is "sublinear": the average number of letter comparisons required is cn, where c is typically 0.3 or less and may for Algorithm BMH be as little as $1/\alpha$, where α is the size of the alphabet. Further, we have seen that at least one BM variant (Algorithm Turbo-BM) requires at most $2n$ letter comparisons in the worst case, while another (Algorithm BMGAG as revised in [CL97]) requires at most $1.5n$. At the same time, a variant of KMP (Algorithm KMPH) requires only $(3n - m)/2$ letter comparisons. In other words, essentially the same worst-case time complexity, at least as measured by letter comparisons, has been achieved by variants both of BM and of KMP. Fundamental theoretical questions arise naturally out of these discoveries:

(1) What is the least number of comparisons of letters of $\boldsymbol{x}$ required in the worst case by any pattern-matching algorithm?

(2) Does there exist an algorithm that achieves this lower bound?

(3) What is the least number of comparisons of letters of $\boldsymbol{x}$ that could be employed in the *average* case by any pattern-matching algorithm?

An initial answer to (1) was given in the same year that Algorithms KMP and BM were published: it was shown in [R77] that for every fixed pattern $\boldsymbol{p} = \boldsymbol{p}[1..m]$ and for every algorithm that correctly determines whether or not $\boldsymbol{p}$ is a substring of arbitrary text strings $\boldsymbol{x} = \boldsymbol{x}[1..n]$, there exists $\boldsymbol{x} = \boldsymbol{x}^*$ in which at least $n - m + 1$ positions must be examined. However, no algorithm was displayed that could guarantee that *at most* $n - m + 1$ positions would need to be examined, and so the question remained open as to whether the lower bound was tight. Further, the bound related only to determining the existence of $\boldsymbol{p}$ in $\boldsymbol{x}$, not to the computation of all occurrences of $\boldsymbol{p}$ in $\boldsymbol{x}$, the problem that we have considered here.

In the intervening quarter-century or so, a great deal of intellect and much energy have been devoted to answering questions (1) and (2), with particularly significant contributions in [CGG90, Co91, GG91, GG92, CH92, Co94]. Variants of these questions that seek an optimal trade-off between time and space efficiency have also been addressed in [GS83, CP91]. In particular, [CH92] presents an algorithm CH that computes all the occurrences of $\boldsymbol{p}$ in $\boldsymbol{x}$ while examining at most

$$\left(1+\frac{8}{3m+3}\right)n \tag{8.11}$$

letters of $\boldsymbol{x}$. Furthermore, Algorithm CH is ***on-line*** in a special sense: the positions of the text are available only in a window of length m that slides monotonely from left to right without backtracking. In terms of letter comparisons, CH is currently the most efficient known algorithm: perhaps its only defect is that it requires an $O(m^2)$-time preprocessing stage.

Later, in [CHPZ95], more precise answers to question (1) were provided that showed the performance of Algorithm CH to be very close to the theoretical lower bound on letter comparisons. For on-line algorithms that employ both pattern-text and text-text comparisons, a lower bound of

$$\left(1+\frac{9}{4m+4}\right)n \;-\; \text{a positive constant} \tag{8.12}$$

comparisons of letters of $\boldsymbol{x}$ was proved for all patterns of length satisfying $m = 36k + 35$, $k \geq 0$. For on-line algorithms restricted to pattern-text comparisons, a slightly larger lower bound was established. Thus every on-line algorithm must perform at least the number of letter comparisons specified by (8.12).

For general algorithms making use of both pattern-text and text-text comparisons, [CHPZ95] proved that at least

$$\left(1+\frac{2}{m+1}\right)n \;-\; \text{a positive constant} \tag{8.13}$$

comparisons of letters of $\boldsymbol{x}$ must be performed in the worst case for all patterns of length $m = 2k + 1$, $k \geq 2$.

In an asymptotic sense (that is, for sufficiently large m and n), this means that the so-called "exact complexity of pattern-matching", at least in terms of worst-case letter comparisons, has been almost established: we see from (8.11)–(8.13) that the best algorithm must perform

$$(1+d/m)n \tag{8.14}$$

letter comparisons in the worst case, where for on-line algorithms $9/4 \leq d \leq 8/3$ and for general algorithms $2 \leq d \leq 8/3$.

We turn finally to the neglected question (3). Since average-case analysis of algorithmic behaviour is generally much more difficult than worst-case analysis, it may seem a little paradoxical that a rather precise answer to this question has been known for some time. In fact [KMP77] describes a simple pattern-matching algorithm that on average requires

$$O\left(\frac{\log_\alpha m}{m}n\right), \tag{8.15}$$

where $\alpha = |A|$ is the alphabet size and the average is taken over all strings on A. Two years later in [Y79] it was shown that, for sufficiently large m and $n > 2m$, the upper bound (8.15) is also a lower bound on the least number of letter comparisons that could be employed in the average-case by any pattern-matching algorithm. [L92] describes a variant of Algorithm BM that also achieves the same average-case bound. Comparing (8.15) with (8.14), we observe that

$$\frac{\log_\alpha m}{m} < \frac{m+d}{m},$$

and so the average-case behaviour of the best possible average-case algorithm is asymptotically better by a logarithmic factor (roughly $\log_\alpha m/m$) than the worst-case behaviour of the best worst-case algorithm, a result that is for once in agreement with intuition.

Exercises 8.8

1. Explain why, even though the expressions (8.12) and (8.13) hold only for certain values of m, they can nevertheless be used as lower bounds over all patterns.
2. Derive (8.14) from (8.11)–(8.13).

Chapter 9

String Distance Algorithms

I need a language such as lovers use,
words of one syllable such as children speak . . .
I have done with phrases.

— Virginia WOOLF (1882–1941), *The Waves*

After two chapterfuls of exact pattern-matching algorithms, we turn now to the study of algorithms that compute *approximate* matches of a given pattern $\boldsymbol{p}$ in a given string $\boldsymbol{x} = \boldsymbol{x}[1..n]$. But we approach our new topic by an indirect route. In order to compute an approximate match of $\boldsymbol{p}$ with a substring $\boldsymbol{u}$ of $\boldsymbol{x}$, we need to define in some way the meaning of "approximate". As discussed in Section 2.2, there are at least three meanings of this term in common use:

- An approximate match between $\boldsymbol{p}$ and $\boldsymbol{u}$ can be measured by "distance" $d(\boldsymbol{p}, \boldsymbol{u})$. In Section 2.2, we defined Hamming distance $d = d_H$, edit distance $d = d_E$, Levenshtein distance $d = d_L$, and weighted distance $d = d_W$. Further, we indicated that other definitions of distance were possible — the inclusion of transpositions, for example.
- The pattern itself perhaps includes metacharacters that represent certain classes of letters rather than specific ones — thus the pattern itself is approximate. We saw then that a pattern could contain "wild-card" metacharacters such as $\bullet$, representing an arbitrary single letter, or $*$, representing an arbitrary string of letters.
- More generally, the pattern could be a regular expression that includes the metacharacters $|$ and *, denoting selection between alternative substrings and repetition of substrings, respectively.

Using the first of these notions of approximate, it becomes necessary, in order to recognize a match, to be able to compute $d(\boldsymbol{p}, \boldsymbol{u})$ for arbitrary substrings $\boldsymbol{u}$ of $\boldsymbol{x}$. Thus in this case the string distance problem arises naturally as a subproblem of approximate pattern-matching.

In this chapter, therefore, as a prelude to approximate pattern-matching, we study algorithms that compute the distance $d(\boldsymbol{x_1}, \boldsymbol{x_2})$ between given strings $\boldsymbol{x_1} = \boldsymbol{x_1}[1..n_1]$ and $\boldsymbol{x_2} = \boldsymbol{x_2}[1..n_2]$. If $\boldsymbol{x_1}$ and $\boldsymbol{x_2}$ are DNA or RNA sequences, then as we shall see computing $d(\boldsymbol{x_1}, \boldsymbol{x_2})$ also determines what is called an ***optimal alignment*** of the two strings. We shall further find that some of the algorithms discussed here have a dual role: they are really *both* string-distance *and* approximate pattern-matching algorithms.

It was shown in [H88] that, for distance defined by arbitrary scoring matrices W, there is an $\Omega(n \log n)$ lower bound on the time complexity of the string distance problem, a bound that applies also to approximate pattern-matching and to approximate repetitions (Section 13.3). Nevertheless, we shall study, in this chapter and the next, several algorithms that make use of special properties of edit or Hamming distance to achieve lower time complexity in many cases.

Here we present four main algorithms in detail, two general-purpose ones that compute string distance $d(\boldsymbol{x_1}, \boldsymbol{x_2})$ directly for arbitrary d, two that compute a (not necessarily unique) longest common subsequence $\text{LCS}(\boldsymbol{x_1}, \boldsymbol{x_2})$ from which, as shown in Section 2.2, the Levenshtein distance $d_L(\boldsymbol{x_1}, \boldsymbol{x_2})$ can immediately be inferred. All of these algorithms are essentially refinements of an original dynamic-programming algorithm that was rediscovered many times by many different researchers.

In the next chapter we shall see how the two general-purpose string distance algorithms can be transformed into approximate pattern-matching algorithms.

9.1 The Basic Recurrence

I give you the end of a golden string;
Only wind it into a ball,
It will lead you in at Heaven's gate,
Built in Jerusalem's wall.

— William BLAKE (1757–1827), *Jerusalem*

As shown in Exercise 9.1.1, there is a straightforward $\Theta(n)$-time algorithm to compute the Hamming distance $d_H(\boldsymbol{x_1}, \boldsymbol{x_2})$ between two strings $\boldsymbol{x_1}$ and $\boldsymbol{x_2}$ of the same length n. We assume therefore throughout this section that $d = d_L$ or d_E or d_W as described in Section 2.2 — Levenshtein or edit or weighted distance, respectively. Further, since the algorithms discussed below will often apply uniformly to each of these three kinds of distance, we will normally denote distance simply by d, which we assume satisfies the conditions (2.2)–(2.5) of a metric. Then we can express the properties of d in terms of edit operations performed on single (nonempty) letters λ and μ:

- (insert: replace ε by λ) $d(\varepsilon, \lambda) > 0$;
- (delete: replace λ by ε) $d(\lambda, \varepsilon) > 0$;
- (substitute: replace λ by μ) $d(\lambda, \mu) > 0$ iff $\lambda \neq \mu$.

Recall that for $d = d_L$ or d_E, $d(\varepsilon, \lambda) = d(\lambda, \varepsilon) = 1$ for all λ; while for $d = d_L$, $d(\lambda, \mu) = 2$, and for $d = d_E$, $d(\lambda, \mu) = 1$.

Observe that each single-letter distance may also be thought of as the ***cost*** of performing the corresponding edit operation; more generally, we think of the distance between two strings as the minimum cost of transforming one string into the other.

The algorithms presented in this chapter are all ***dynamic-programming*** algorithms; that is, they compare the input strings $\boldsymbol{x_1} = \boldsymbol{x_1}[1..n_1]$ and $\boldsymbol{x_2} = \boldsymbol{x_2}[1..n_2]$ from left to right, computing for each pair of positions i and j the minimum cost of transforming $\boldsymbol{x_1}[1..i]$ into $\boldsymbol{x_2}[1..j]$ based on the already-computed minimum costs

- $d(\boldsymbol{x_1}[1..i], \boldsymbol{x_2}[1..j-1])$;
- $d(\boldsymbol{x_1}[1..i-1], \boldsymbol{x_2}[1..j])$;
- $d(\boldsymbol{x_1}[1..i-1], \boldsymbol{x_2}[1..j-1])$.

This basic recurrence is explained below.

It is convenient to introduce a two-dimensional ***cost array*** $\boldsymbol{c} = \boldsymbol{c}[0..n_1, 0..n_2]$ in which

$$\boldsymbol{c}[i, j] = d(\boldsymbol{x_1}[1..i], \boldsymbol{x_2}[1..j])$$

for every $i \in 0..n_1$, $j \in 0..n_2$ with initial values defined as follows:

- $\boldsymbol{c}[0, 0] = 0$, the minimum cost of transforming the empty string into itself;
- $\boldsymbol{c}[0, j] = \sum_{1 \leq h \leq j} d(\varepsilon, \boldsymbol{x_2}[h])$, the minimum cost of inserting the first j letters of $\boldsymbol{x_2}$ into the empty string;
- $\boldsymbol{c}[i, 0] = \sum_{1 \leq h \leq i} d(\boldsymbol{x_1}[h], \varepsilon)$, the minimum cost of deleting the first i letters of $\boldsymbol{x_1}$ so as to form the empty string.

To understand the cost array, consider the example $\boldsymbol{c}[0..5, 0..6]$ corresponding to the computation of the Levenshtein distance $d_L(\boldsymbol{x_1}, \boldsymbol{x_2})$, where $\boldsymbol{x_1} = rests$, $\boldsymbol{x_2} = stress$. As shown in Table 9.1, the desired distance is given by

$$\boldsymbol{c}[5, 6] = d_L(\boldsymbol{x_1}[1..5], \boldsymbol{x_2}[1..6]) = 3.$$

In this example, observe that the minimum cost $\boldsymbol{c}[4, 4] = 4$ corresponding to the transformation $rest \to stre$ is achieved by adding the cost of one deletion (of the t in $rest$) to

	j	**0**	**1**	**2**	**3**	**4**	**5**	**6**
i		ε	s	t	r	e	s	s
0	ε	0	1	2	3	4	5	6
1	r	1	2	3	2	3	4	5
2	e	2	3	4	3	2	3	4
3	s	3	2	3	4	3	2	3
4	t	4	3	2	3	4	3	4
5	s	5	4	3	4	5	4	3

Table 9.1: Cost array for d_L(*rests, stress*)

the minimum cost $\boldsymbol{c}[3,4] = 3$ of the transformation $res \rightarrow stre$. Alternatively, $\boldsymbol{c}[4,4]$ can also be obtained by adding the cost of one insertion (of the e in $stre$) to the minimum cost $\boldsymbol{c}[4,3] = 3$ of the transformation $rest \rightarrow str$. Thus

$$\boldsymbol{c}[4,4] = \boldsymbol{c}[3,4] + 1 = \boldsymbol{c}[4,3] + 1.$$

On the other hand, we see that $\boldsymbol{c}[5,6] = 3$ corresponding to the complete transformation $rests \rightarrow stress$ may be computed from the minimum cost $\boldsymbol{c}[4,5] = 3$ of the transformation $rest \rightarrow stres$ since the cost of transforming the final $s \rightarrow s$ is zero:

$$\boldsymbol{c}[5,6] = \boldsymbol{c}[4,5] + 0.$$

Finally, we observe that in fact the minimum cost can sometimes be obtained in three distinct ways: for the transformation of $rest \rightarrow stress$ we find that

$$\boldsymbol{c}[4,6] = \boldsymbol{c}[3,6] + 1 = \boldsymbol{c}[4,5] + 1 = \boldsymbol{c}[3,5] + 2.$$

These observations are just special cases of the basic recurrence relation that enables any cost array $\boldsymbol{c}$ to be computed from the initial settings $\boldsymbol{c}[0,0]$, $\boldsymbol{c}[0,j]$, $j = 1, 2, \ldots, n_2$, and $\boldsymbol{c}[i,0]$, $i = 1, 2, \ldots, n_1$:

Lemma 9.1.1 *For every* $i \in 1..n_1$, $j \in 1..n_2$,

$$\begin{aligned} \boldsymbol{c}[i,j] = \min\{\ & \boldsymbol{c}\,[i-1,j] + d(\boldsymbol{x_1}[i], \varepsilon), \\ & \boldsymbol{c}\,[i,j-1] + d(\varepsilon, \boldsymbol{x_2}[j]), \\ & \boldsymbol{c}\,[i-1,j-1] + d(\boldsymbol{x_1}[i], \boldsymbol{x_2}[j])\}. \end{aligned}$$

Proof We base the proof on a fact established in Exercise 2.2.13: that inserts and deletes can be performed in any order. In particular, we consider the three possible orderings of inserts and deletes that leave the operations on $\boldsymbol{x_1}[i]$ and/or $\boldsymbol{x_2}[j]$ to be performed last:

- One possibility is that $\boldsymbol{x_1}[i]$ should be deleted at a cost of $d(\boldsymbol{x_1}[i], \varepsilon)$ added to $\boldsymbol{c}[i-1,j]$.
- A second possibility is that $\boldsymbol{x_2}[j]$ should be inserted at a cost of $d(\varepsilon, \boldsymbol{x_2}[j])$ added to $\boldsymbol{c}[i,j-1]$.
- The final possibility is that a substitution could take place — $\boldsymbol{x_1}[i]$ deleted and $\boldsymbol{x_2}[j]$ inserted — at a cost of $d(\boldsymbol{x_1}[i], \boldsymbol{x_2}[j])$ added to $\boldsymbol{c}[i-1,j-1]$. Of course in the case that $\boldsymbol{x_1}[i] = \boldsymbol{x_2}[j]$, this increment would be zero.

Since there are no other orderings of inserts and deletes that leave operations on $\boldsymbol{x_1}[i]$ and $\boldsymbol{x_2}[j]$ to the last, the minimum cost $\boldsymbol{c}[i,j]$ must be the minimum over these three cases. □

The critical factor in the proof of this lemma is that inserts and deletes can be performed in any order. The industrious reader may recall that in working through Exercise 2.2.13, he or she found that this property of the inserts and deletes did *not* depend on the Symmetry Property (2.4) of a metric; that is, it held even if, for some strings $\boldsymbol{x}_1$ and $\boldsymbol{x}_2$,

$$d(\boldsymbol{x}_1, \boldsymbol{x}_2) \neq d(\boldsymbol{x}_2, \boldsymbol{x}_1).$$

Consequently, Lemma 9.1.1 holds even if the Symmetry Property does not. This observation makes it clear that cost need not be quite the same thing as distance: since we have defined cost in terms of transforming $\boldsymbol{x}_1$ into $\boldsymbol{x}_2$, it is possible to imagine that the cost of transforming $\boldsymbol{x}_2$ into $\boldsymbol{x}_1$ may not be the same. This situation in fact arises in applications to molecular biology, where the scoring matrix W may in some cases not be symmetric and where as a result it can happen that $d_W(\boldsymbol{x}_1, \boldsymbol{x}_2) \neq d_W(\boldsymbol{x}_2, \boldsymbol{x}_1)$.

Finally, we remark that the cost array and the recurrence scheme for computing it given by Lemma 9.1.1 are very simple and natural ideas with widespread consequences: not only does the recurrence form the basis of distance (hence LCS) calculations, it is also fundamental to almost all approaches to the calculation of the shortest common superstring of a collection of strings, and to the "best" alignment of a collection of strings. These NP-complete problems are of particular importance in computational biology.

Exercises 9.1

1. In accordance with the claim made in the text, describe a $\Theta(n)$-time algorithm to compute the Hamming distance between two strings of length n.
2. Justify the initial settings of $\boldsymbol{c}[0, 0]$, $\boldsymbol{c}[0, j]$, $\boldsymbol{c}[i, 0]$.

9.2 Wagner-Fischer *et al.*

My words fly up, my thoughts remain below;
Words without thoughts never to heaven go.

— William SHAKESPEARE (1564–1616), *Hamlet*

Algorithm WF computes the cost array, hence the distance between two given strings $\boldsymbol{x}_1[1..n_1]$ and $\boldsymbol{x}_2[1..n_2]$, and is of course an immediate consequence of Lemma 9.1.1, and its time complexity and space complexity will both obviously be $\Theta(n_1 n_2)$. The details are left to Exercise 9.2.1. We observe that this algorithm is in a certain sense on-line (Section 4.1): in the course of computing $d(\boldsymbol{x}_1, \boldsymbol{x}_2)$, the distances between every pair of prefixes of $\boldsymbol{x}_1$ and $\boldsymbol{x}_2$ are also computed. In Section 9.3 we make a simple observation that enables us to reduce the additional space requirement of the Wagner-Fischer algorithm to $\Theta(\min\{n_1, n_2\})$.

As pointed out in [SK83], the cost array algorithm has been independently discovered many times [V68, NW70, VZ70, SC71, S72, RCW73, H73, WF74] in contexts as diverse as speech recognition, automatic spelling correction, and molecular biology. We select

[WF74] to represent this body of work, not because of the cost-array algorithm itself, but rather because of the "trace" algorithm described there that permits both the sequence of edit operations and $\text{LCS}(\boldsymbol{x_1}, \boldsymbol{x_2})$ to be explicitly displayed.

Suppose $\boldsymbol{x_1}[1..n_1]$ and $\boldsymbol{x_2}[1..n_2]$ are given, together with the corresponding cost array $\mathbf{c}[0..n_1, 0..n_2]$. For any position $q = [i, j]$ in $\mathbf{c}$, $1 \leq i \leq n_1$, $1 \leq j \leq n_2$, we define a *move*

$$\nu : q \to q'$$

to be a function that maps q into a preceding position q' of $\mathbf{c}$ according to the following rules:

- **if** $\mathbf{c}[i, j] = \mathbf{c}[i-1, j] + d(\boldsymbol{x_1}[i], \varepsilon),\ q' = [i-1, j]$;
- **elsif** $\mathbf{c}[i, j] = \mathbf{c}[i, j-1] + d(\varepsilon, \boldsymbol{x_2}[j]),\ q' = [i, j-1]$;
- **else** $q' = [i-1, j-1]$.

Thus a move specifies one of the three positions of $\mathbf{c}$ that, with a single edit operation, yields $\mathbf{c}[i, j]$, where the operations are chosen according to the precedence

<delete, insert, substitute> . (9.1)

Since ν reduces at least one of i, j, it follows that there exists some integer $k \in 1..i+j$ such that the composition

$$\nu^k([i, j]) = [i', j']$$

with $\min\{i', j'\} = 0$. In the example of the previous section (Table 9.1), we can trace the sequence

$$\nu([5,6]) = [4,5];\ \nu([4,5]) = [3,5];\ \nu([3,5]) = [2,4];$$

$$\nu([2,4]) = [1,3];\ \nu([1,3]) = [0,2];$$

so that $k = 5$ and $\nu^5([5,6]) = [0,2]$. Now if for each substitution (but not delete or insert) in this sequence of moves, the current value $[i, j]$ is displayed, the result is called a *trace* and denoted by τ. In the example, then, we find

$$\tau([5,6]) = \{[5,6],\ [3,5],\ [2,4],\ [1,3]\}$$

corresponding to the substitutions shown below:

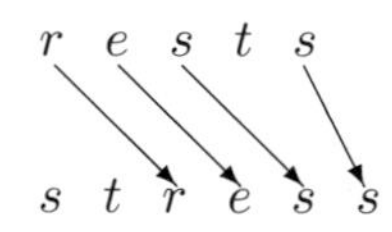

Of course in this example each of the "substitutions" simply replaces a letter by itself, and so has cost zero; we shall see in Exercise 9.2.3 that this is not necessarily always the case — the trace may also contain positions reached by substitutions of nonzero cost.

In general, a trace $\tau([n_1, n_2])$ specifies all the substitutions, including those of zero cost, in a least-cost sequence of edit operations that transforms $\boldsymbol{x_1}$ into $\boldsymbol{x_2}$. Observe that positions i in $\boldsymbol{x_1}$ that do *not* occur in τ determine letters $\boldsymbol{x_1}[i]$ that are deleted during this edit sequence, while non-occurring positions j in $\boldsymbol{x_2}$ correspond to inserted letters $\boldsymbol{x_2}[j]$. Thus a trace can be used in a straightforward way to specify completely a least-cost edit sequence that transforms $\boldsymbol{x_1} \rightarrow \boldsymbol{x_2}$. In our example, the non-occurring positions are $\boldsymbol{x_1}[4] = t$, to be deleted, and $\boldsymbol{x_2}[1..2] = st$, to be inserted, a situation that can also be represented using a space character Δ, as follows:

$$\begin{array}{cccccc} \Delta & \Delta & r & e & s\ t & s \\ s & t & r & e & s\ \Delta & s \end{array}$$

Thus this representation provides an optimal (least-cost) alignment of $\boldsymbol{x_1}$ and $\boldsymbol{x_2}$, as noted earlier a formulation of importance in computational biology. We leave as exercises the detailed description of algorithms both to specify the edit sequence and to compute the trace itself (Exercise 9.2.6).

We now remark that the zero-cost substitutions in a trace determine a common subsequence of $\boldsymbol{x_1}$ and $\boldsymbol{x_2}$ though, as shown in Exercise 9.2.3, not necessarily a *longest* common subsequence $\text{LCS}(\boldsymbol{x_1}, \boldsymbol{x_2})$. But if we confine ourselves to $d = d_L$, so that $d(\lambda, \mu) = 2$ for every pair of letters $\lambda \neq \mu$, it is easy to see that

$$\mathbf{c}[n_1, n_2] = n_1 + n_2 - 2|\text{LCS}(\boldsymbol{x_1}, \boldsymbol{x_2})|, \tag{9.2}$$

an identity that holds for every $\boldsymbol{x_1}[1..n_1]$ and $\boldsymbol{x_2}[1..n_2]$. It follows then that, for $d = d_L$, the length of an LCS is given by

$$|\text{LCS}(\boldsymbol{x_1}, \boldsymbol{x_2})| = (n_1 + n_2 - \mathbf{c}[n_1, n_2])/2. \tag{9.3}$$

Of course, as we see in Exercise 9.2.4, $\text{LCS}(\boldsymbol{x_1}, \boldsymbol{x_2})$ itself is computable in an obvious way from the corresponding trace for $d = d_L$.

As pointed out in [WF74], one interesting application of the cost array is to spelling correction, where $d = d_W$ is used with a scoring matrix W whose entries reflect the fact that neighbouring letters on the standard typewriter keyboard are substituted for each other with greater probability (lower cost) than more widely separated letters. A similar phenomenon occurs in DNA scoring matrices, where often the nucleotide pairs C and G, for example, can be substituted for each other more easily than C and T, or G and A. In both of these applications, another kind of substitution also becomes important: transposition of adjacent symbols. It is shown in [LW75] that if transposition is included as a basic editing operation, the distance between $\boldsymbol{x_1}$ and $\boldsymbol{x_2}$ can be still be computed in $\Theta(n_1 n_2)$ time, provided that the cost of transposition of distinct letters λ and μ is not less than

$$(d(\varepsilon, \lambda) + d(\mu, \varepsilon))/2.$$

Exercises 9.2

1. Specify a $\Theta(n_1 n_2)$-time algorithm to compute the cost array for two given strings $\boldsymbol{x_1}[1..n_1]$ and $\boldsymbol{x_2}[1..n_2]$.
2. Prove that if the Symmetry Property holds for d, then the cost array for transforming $\boldsymbol{x_2}$ into $\boldsymbol{x_1}$ is just the transpose of the cost array for transforming $\boldsymbol{x_1}$ into $\boldsymbol{x_2}$. Verify your result by computing $\boldsymbol{c}$ for $\boldsymbol{x_1} = stress$, $\boldsymbol{x_2} = rests$.
3. For edit distance $d = d_E$, compute the cost array and trace corresponding to $\boldsymbol{x_1} = sowsear$, $\boldsymbol{x_2} = silkpurse$, hence observe that not every position in the trace is necessarily zero cost. Observe also that in this case the common subsequence determined by the trace is *not* an LCS$(\boldsymbol{x_1}, \boldsymbol{x_2})$.
4. For Levenshtein distance $d = d_L$, compute the cost array and trace corresponding to $\boldsymbol{x_1} = sowsear$, $\boldsymbol{x_2} = silkpurse$, hence verify that the common subsequence obtained from the trace in fact equals LCS$(\boldsymbol{x_1}, \boldsymbol{x_2})$.
5. What is the result of permuting the precedence of operations specified in (9.1)? Does a valid trace and corresponding sequence of edit operations result?
6. Given the cost array $\boldsymbol{c}[0..n_1, 0..n_2]$, describe an $O(n_1 + n_2)$-time algorithm to compute the trace. Then describe a $\Theta(n_1 + n_2)$-time algorithm to compute the corresponding edit sequence.
7. Prove (9.2).
8. Describe an $O(\min\{n_1, n_2\})$-time algorithm that computes LCS$(\boldsymbol{x_1}, \boldsymbol{x_2})$ from the trace corresponding to $d = d_L$.
9. Design a scoring matrix W for the English lower case alphabet that takes into account the increased likelihood of substitution of neighbouring letters on the standard typewriter keyboard. For example, you might have

$$W[d,e] = W[d,r] = W[d,s] = W[d,f] = W[d,x] = W[d,c] = 1/2,$$

while $W[d,\lambda] = 1$ for other letters λ. Then recompute the cost array for $\boldsymbol{x_1} = sowsear$ and $\boldsymbol{x_2} = silkpurse$ using $d = d_W$.

9.3 Hirschberg

Apt words have power to suage
The tumours of a troubled mind.

— John MILTON (1608–1674), *Paradise Regained*

Hirschberg [H75] presents a $\Theta(n_1 n_2)$-time algorithm to compute LCS$(\boldsymbol{x_1}, \boldsymbol{x_2})$ directly, without explicit computation of the cost array $\boldsymbol{c}$, and shows how to do so using only $\Theta(\min\{n_1, n_2\})$ additional space. First define, analogous to the cost array, an array $\gamma[0..n_1, 0..n_2]$ in which

$$\gamma[i,j] = |\mathrm{LCS}(\boldsymbol{x_1}[1..i], \boldsymbol{x_2}[1..j])|,$$

for every $i \in 0..n_1$, $j \in 0..n_2$. Thus $\gamma[i,j]$ gives the length of an LCS of $\boldsymbol{x_1}[1..i]$ and $\boldsymbol{x_2}[1..j]$. Of course, since the empty string ε has a zero length LCS with every other string, we know that $\gamma[i,0] = \gamma[0,j] = 0$. For the other elements of γ, it is not difficult to prove an analogue of Lemma 9.1.1:

Lemma 9.3.1 *For Levenshtein distance $d = d_L$ and for every $i \in 1..n_1$, $j \in 1..n_2$,*

- $\gamma[i,j] = \gamma[i-1, j-1] + 1$, *whenever* $\boldsymbol{x_1}[i] = \boldsymbol{x_2}[j]$*;*
- $\gamma[i,j] = \max\{\gamma[i-1,j], \gamma[i,j-1]\}$, *otherwise.*

Proof We make use of the observation that, for $d = d_L$, (9.2) and (9.3) hold when n_1 and n_2 are replaced by $i \in 1..n_1$ and $j \in 1..n_2$, respectively, and then apply Lemma 9.1.1 using $d = d_L$. □

Algorithm 9.3.1 (H1)

- *Compute the length of* LCS$(\boldsymbol{x_1}, \boldsymbol{x_2})$

- *Initialize row zero and column zero of* γ

for $i \leftarrow 0$ **to** n_1 **do**
  $\gamma[i,0] \leftarrow 0$
for $j \leftarrow 1$ **to** n_2 **do**
  $\gamma[0,j] \leftarrow 0$

- *Compute* γ

for $i \leftarrow 1$ **to** n_1 **do**
  for $j \leftarrow 1$ **to** n_2 **do**
    if $\boldsymbol{x_1}[i] = \boldsymbol{x_2}[j]$ **then**
      $\gamma[i,j] \leftarrow \gamma[i-1, j-1] + 1$
    else
      $\gamma[i,j] \leftarrow \max\{\gamma[i-1,j], \gamma[i,j-1]\}$

Algorithm 9.3.1 (H1) provides an implementation of the calculations specified by this result. We display this obvious algorithm not in order to insult the reader's undoubted intelligence, but so that a comparison may be made with an improved version: based on Lemma 9.3.1, Hirschberg makes the key observation that only row $i-1$ of γ is required for the evaluation of row i, and that therefore only two rows of γ need to be stored at any one time. Thus we may define a $2 \times n$ array $\boldsymbol{\gamma}'[0..1, 0..n_2]$ that represents the current $(i\text{th})$ and previous $((i-1)\text{th})$ rows of $\boldsymbol{\gamma}$. We use a ***flipflop variable*** $i' \in 0..1$ to access positions $\boldsymbol{\gamma}'[i', j]$ in the current row, while $1 - i'$ accesses positions $\boldsymbol{\gamma}'[1 - i', j]$ in the previous one. The resulting improved implementation is shown as Algorithm 9.3.2 (H2), now written as a function that computes and returns the final row $\boldsymbol{\gamma}[n_1, 0..n_2]$ of $\boldsymbol{\gamma}$.

Algorithm 9.3.2 (H2)

– *Compute the length of* $\text{LCS}(\boldsymbol{x_1}, \boldsymbol{x_2})$
function $\text{H2}(n_1, \boldsymbol{x_1}, n_2, \boldsymbol{x_2})$

– *Initialize* $\boldsymbol{\gamma}'[1, 0]$ *and row zero of* $\boldsymbol{\gamma}'[0..1, 0..n_2]$
$\boldsymbol{\gamma}'[1, 0] \leftarrow 0$
for $j \leftarrow 0$ **to** n_2 **do**
  $\boldsymbol{\gamma}'[0, j] \leftarrow 0$
– *Initialize the flipflop variable*
$i' \leftarrow 0$

– *Compute* $\boldsymbol{\gamma}'[1-i', j]$
for $i \leftarrow 1$ **to** n_1 **do**
  $i' \leftarrow 1-i'$ – *Flipflop before entering j loop*
  for $j \leftarrow 1$ **to** n_2 **do**
    if $\boldsymbol{x_1}[i] = \boldsymbol{x_2}[j]$ **then**
      $\boldsymbol{\gamma}'[i', j] \leftarrow \boldsymbol{\gamma}'[1-i', j-1] + 1$
    else
      $\boldsymbol{\gamma}'[i', j] \leftarrow \max\{\boldsymbol{\gamma}'[1-i', j], \boldsymbol{\gamma}'[i', j-1]\}$
– $\boldsymbol{\gamma}'[i', j] = |\text{LCS}(\boldsymbol{x_1}, \boldsymbol{x_2}[1..j])|$, $j = 0, 1, \ldots, n_2$
return $\boldsymbol{\gamma}'[i', 0..n_2]$

Note that use of the flipflop variable makes Algorithm H2 nearly as efficient as its predecessor while reducing additional space requirements to $\Theta(n_2)$. Of course, since processing could just as well take place column-wise as row-wise, the additional space requirement can actually be reduced to $\Theta(\min\{n_1, n_2\})$. Similar modifications to the Wagner-Fischer minimum distance algorithm that reduce its space complexity to the same quantity are left to Exercise 9.3.3.

So far, so good. But we have only computed the *length* of $\text{LCS}(\boldsymbol{x_1}, \boldsymbol{x_2})$, not an LCS itself. Nevertheless, it turns out that a recursive divide-and-conquer LCS algorithm can be derived from repeated calls to Algorithm H2. In order to describe this algorithm, we need more notation. Let $\boldsymbol{\Gamma_{ij}}$ be an $\text{LCS}(\boldsymbol{x_1}[1..i], \boldsymbol{x_2}[1..j])$ and let $\boldsymbol{\Gamma^*_{ij}}$ be an $\text{LCS}(\boldsymbol{x_1}[i+1..n_1], \boldsymbol{x_2}[j+1..n_2])$, with

$$\gamma[i, j] = |\boldsymbol{\Gamma_{ij}}|,\ \gamma^*[i, j] = |\boldsymbol{\Gamma^*_{ij}}|$$

and

$$\gamma_i^{\max} = \max_{0 \le j \le n_2} \{\gamma[i, j] + \gamma^*[i, j]\}.$$

We prove below that for every $i \in 0..n_1$, $\gamma_i^{\max} = \gamma[n_1, n_2]$; that is, an LCS of $\boldsymbol{x_1}$ and $\boldsymbol{x_2}$ can be computed by splitting $\boldsymbol{x_1}$ into a prefix $\boldsymbol{x_1}[1..i]$ and a suffix $\boldsymbol{x_1}[i+1..n]$ for any arbitrary choice of i, then performing Algorithm H2 on each part:

$$\text{H2}(i, \boldsymbol{x_1}[1..i], n_2, \boldsymbol{x_2}) \text{ and } \text{H2}(n-i, \boldsymbol{x_1}[i+1..n], n_2, \boldsymbol{x_2})$$

in order to compute

$$\boldsymbol{\gamma}[i, n_2] \text{ and } \boldsymbol{\gamma}^*[i, 0].$$

Of course this result can be applied recursively also to $\boldsymbol{\gamma}[i, n_2]$ and $\boldsymbol{\gamma}^*[i, 0]$. Thus if at each step of the recursion we choose $i = \lfloor n_1/2 \rfloor$, we can after $\log_2 n$ steps reduce the calculation of $\gamma[n_1, n_2]$ to a sum of 0s and 1s, where corresponding to each occurrence of 1 we can test to determine whether or not a letter of the LCS has been found. Before showing the algorithm in detail, we prove

---

Lemma 9.3.2 *For every $i \in 0..n_1$, $\gamma_i^{\max} = \boldsymbol{\gamma}[n_1, n_2]$.*

Proof For arbitrary i, consider $j = j_0$ such that

$$\gamma_i^{\max} = \boldsymbol{\gamma}[i, j_0] + \boldsymbol{\gamma}^*[i, j_0]$$

with corresponding LCSs $\boldsymbol{\Gamma_{ij_0}}$ and $\boldsymbol{\Gamma^*_{ij_0}}$. Then the string $\boldsymbol{\Gamma} = \boldsymbol{\Gamma_{ij_0}\Gamma^*_{ij_0}}$ is a common subsequence of $\boldsymbol{x_1}$ and $\boldsymbol{x_2}$ with length

$$\gamma_i^{\max} \le \boldsymbol{\gamma}[n_1, n_2]. \tag{9.4}$$

Next consider $\boldsymbol{\Gamma_{n_1 n_2}}$, an LCS of $\boldsymbol{x_1}$ and $\boldsymbol{x_2}$. $\boldsymbol{\Gamma_{n_1 n_2}}$ is a subsequence of $\boldsymbol{x_2}$ that may be written in the form $\boldsymbol{\Gamma_1 \Gamma_2}$, where $\boldsymbol{\Gamma_1}$ is a subsequence of $\boldsymbol{x_1}[1..i]$, $\boldsymbol{\Gamma_2}$ is a subsequence of $\boldsymbol{x_1}[i+1..n_1]$, and the value of i is the same as in (9.4). Then there exists $j = j_1$, say, such that $\boldsymbol{\Gamma_1}$ is a subsequence of $\boldsymbol{x_2}[1..j_1]$ and $\boldsymbol{\Gamma_2}$ is a subsequence of $\boldsymbol{x_2}[j_1+1..n_2]$. Hence

$$\begin{aligned}
|\boldsymbol{\Gamma_1}| &\le |\text{LCS}(\boldsymbol{x_1}[1..i], \boldsymbol{x_2}[1..j_1])| \\
&= \boldsymbol{\gamma}[i, j_1]; \\
|\boldsymbol{\Gamma_2}| &\le |\text{LCS}(\boldsymbol{x_1}[i+1..n_1], \boldsymbol{x_2}[j_1+1..n_2])| \\
&= \boldsymbol{\gamma}^*[i, j_1].
\end{aligned}$$

Thus

$$\begin{aligned}
\boldsymbol{\gamma}[n_1, n_2] &= |\boldsymbol{\Gamma_{n_1 n_2}}| \\
&= |\boldsymbol{\Gamma_1}| + |\boldsymbol{\Gamma_2}| \\
&\le \boldsymbol{\gamma}[i, j_1] + \boldsymbol{\gamma}^*[i, j_1] \\
&\le \gamma_i^{\max}.
\end{aligned} \tag{9.5}$$

From (9.4) and (9.5) we conclude that, for any choice of i, $\gamma_i^{\max} = \gamma[n_1, n_2]$, as required. □

In view of this result, we are now in a position to present Algorithm 9.3.3 (H3) and prove its correctness. Again, the algorithm is implemented as a function, this time a recursive one, that returns the desired LCS.

Algorithm 9.3.3 (H3)

– *Compute an* LCS$(\boldsymbol{x_1}, \boldsymbol{x_2})$
function H3$(n_1, \boldsymbol{x_1}, n_2, \boldsymbol{x_2})$

– *Recursion has reduced minimum string length to 0 or 1*
if $\min\{n_1, n_2\} = 0$ **then return** $\boldsymbol{\Gamma} \leftarrow \varepsilon$
elsif $n_1 = 1$ **then**
  if $\exists\, j \in 1..n_2$ such that $\boldsymbol{x_1}[1] = \boldsymbol{x_2}[j]$ **then**
    return $\boldsymbol{\Gamma} \leftarrow \boldsymbol{x_1}[1]$
  else
    return $\boldsymbol{\Gamma} \leftarrow \varepsilon$

– *Recursive step: split* $\boldsymbol{x}$ *in half*
else
  $i \leftarrow \lfloor n_1/2 \rfloor$
  $\gamma'_1[1..n_2] \leftarrow H2(i, \boldsymbol{x_1}[1..i], n_2, \boldsymbol{x_2}[1..n_2])$
  $\gamma'_2[1..n_2] \leftarrow H2(n-i, \boldsymbol{x_1}[i+1..n_1], n_2, \boldsymbol{x_2}[1..n_2])$

  Find $j = j_0$ such that $\gamma'_1[j] + \gamma'_2[j]$ is a maximum

  – *Recursively call the two problems of half the size*
  $\boldsymbol{\Gamma_1} \leftarrow$ H3$(i, \boldsymbol{x_1}[1..i], j_0, \boldsymbol{x_2}[1..j_0])$
  $\boldsymbol{\Gamma_2} \leftarrow$ H3$(n-i, \boldsymbol{x_1}[i+1..n_1], n-j_0, \boldsymbol{x_2}[j_0+1..n_2])$
  – *At the current level of recursion, concatenate*
  – *the two LCSs computed for the two half-problems*
  return $\boldsymbol{\Gamma} \leftarrow \boldsymbol{\Gamma_1}\boldsymbol{\Gamma_2}$

Theorem 9.3.3 *Algorithm 9.3.3 correctly computes* LCS$(\boldsymbol{x_1}, \boldsymbol{x_2})$.

Proof If either n_1 or n_2 is zero, there is only an empty LCS to report; while for $n_1 = 1$ there is a nonempty LCS $\boldsymbol{x_1}[1]$ to report only if $\boldsymbol{x_1}[1]$ equals some letter of $\boldsymbol{x_2}[1..n_2]$.

For $n_1 > 1$ the value of $\gamma[n_1, n_2] = |\text{LCS}(\boldsymbol{x_1}, \boldsymbol{x_2}|$ is correctly computed, according to Lemma 9.3.2, based on the two calls to Algorithm 9.3.2. This calculation determines a position $j_0 \in 0..n_2$ that divides $\boldsymbol{x_2}$ into two parts from which an LCS can be derived. The two recursive calls of Algorithm 9.3.3 correspond to these two parts. When finally the

LCS values $\mathbf{\Gamma_1}$ and $\mathbf{\Gamma_2}$ are returned from the recursion, they need only be concatenated to yield the final result. □

In order to determine the time complexity of Algorithm H3, let $T(n_1, n_2)$ be the time required to execute the algorithm, and observe that, since Algorithm H2 requires $\Theta(n_1 n_2)$ time, there must exist a constant K such that

$$T(n_1, n_2) \leq K n_1 n_2 + T(n_1/2, j_0) + T(n_1/2, n_2 - j_0).$$

(Here an allowance for the $\Theta(n_2)$-time calculation of j_0, as well as for the $O(n_2)$-time determination of j such that $\boldsymbol{x_1}[1] = \boldsymbol{x_2}[j]$, is incorporated into K.) Suppose that for every $i \in 1..n_1 - 1$, $j \in 1..n_2 - 1$, there exists a second constant K' such that

$$T(i, j) < K'ij.$$

Then we find that

$$\begin{aligned} T(n_1, n_2) &\leq K n_1 n_2 + K'(n_1/2)(j_0 + n_2 - j_0) \\ &= (K + K'/2) n_1 n_2, \end{aligned}$$

and so, by choosing $K' > 2K$, we ensure that also

$$T(n_1, n_2) < K' n_1 n_2.$$

Thus, by induction, we conclude that Algorithm H3 requires $\Theta(n_1 n_2)$ time for its execution.

To evaluate the additional space requirement of Algorithm H3, we simply observe that H3 can be implemented so that all recursive calls make use of the same storage for $\boldsymbol{x_1}$ and $\boldsymbol{x_2}$, as well as for the two vectors $\boldsymbol{\gamma_1'}$ and $\boldsymbol{\gamma_2'}$. Hence

Theorem 9.3.4 *Algorithm 9.3.3 computes an LCS of $\boldsymbol{x_1}[1..n_1]$ and $\boldsymbol{x_2}[1..n_2]$ in $\Theta(n_1 n_2)$ time using $\Theta(n_2)$ additional space.* □

There are actually three algorithms due to Hirschberg for the LCS calculation, the one described here [H75] and two others published two years later [H77]. For continuity of presentation we have elected to present the first (and perhaps generally slower) of the three. Both of the other two algorithms are fast when lcs $= |\text{LCS}(\boldsymbol{x_1}, \boldsymbol{x_2})|$ turns out to be small, one of them also when lcs is large. To express this more precisely, let us assume that $n_1 < n_2$, and let $\ell = \text{lcs}(\boldsymbol{x_1}, \boldsymbol{x_2})$. Then Hirschberg's second algorithm computes an LCS in time $O(\ell n_2 + n_2 \log \alpha')$, where $\alpha' \leq \alpha$ is the number of distinct letters in $\boldsymbol{x_2}$; the third algorithm executes in time $O((n_1 + 1 - \ell)\ell \log n_2)$. Thus the second algorithm is efficient when the LCS is short and the alphabet small, but may require $\Theta(n_1 n_2)$ time in the worst case; while the third algorithm is efficient when the LCS is either very short or very long, even though it requires $\Theta(n_1^2 \log n_2)$ worst-case time.

Exercises 9.3

1. Prove Lemma 9.3.1.
2. Describe the changes required to Algorithm H2 so that it will perform its computations column-wise rather than row-wise, and so make use of $\Theta(n_1)$, rather than $\Theta(n_2)$, additional space.
3. Specify an improvement to the Wagner-Fischer algorithm analogous to Algorithm H2.
4. The proof that $T(n_1, n_2) < K'n_1n_2$ for Algorithm H3 is said to follow "by induction". Make the induction explicit.
5. Show by induction that the total number of calls to Algorithm H3 is exactly $2n_1 - 1$.
6. Suppose that k and k' are given fixed real numbers constrained only to satisfy $0 < k < k' < 1$. Consider the class $\mathcal{C}$ of all pairs $(\boldsymbol{x_1}, \boldsymbol{x_2})$ of strings such that

$$kn \leq \text{lcs}(\boldsymbol{x_1}, \boldsymbol{x_2}) \leq k'n,$$

where $n = \min\{|\boldsymbol{x_1}|, |\boldsymbol{x_2}|\}$. Characterize the complexity of Hirschberg's second and third algorithms over the class $\mathcal{C}$.

Explain the significance of your results.

9.4 Hunt-Szymanski

Don't, Sir, accustom yourself to use big words for little matters.

— Samuel JOHNSON (1709–1784), *Boswell's Life of Johnson*

In this section we present another algorithm that directly computes an LCS of two given strings $\boldsymbol{x_1}[1..n_1]$ and $\boldsymbol{x_2}[1..n_2]$ without explicit calculation of the distance $d(\boldsymbol{x_1}, \boldsymbol{x_2})$. Of course, once an LCS has been found, the Levenshtein distance d_L can be determined in constant time using (9.2). Provided that the alphabet is ordered, the new algorithm requires $O((n+r)\log n)$ time and $O(n+r)$ additional space, where

- $n = \max\{n_1, n_2\}$;
- r is the total number of times that $\boldsymbol{x_1}[i] = \boldsymbol{x_2}[j]$ over all $i \in 1..n_1$, $j \in 1..n_2$.

Thus in the worst case (for example, $\boldsymbol{x_1} = \boldsymbol{x_2} = a^n$), Algorithm HS [HS77] could require $\Theta(n^2 \log n)$ time and $\Theta(n^2)$ space. However, in the usual case that $r \in O(n)$ and that the

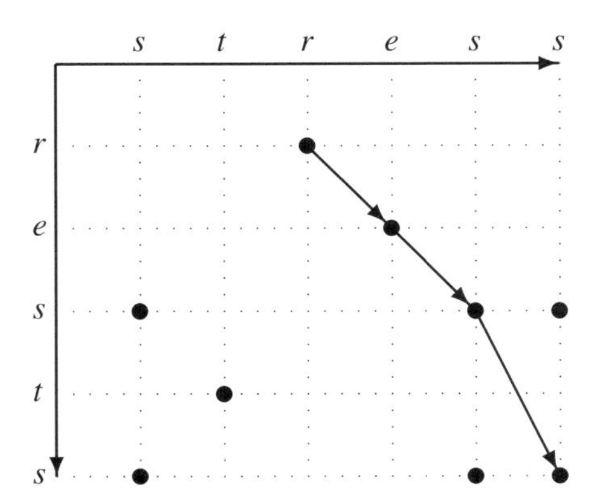

Figure 9.1. The line formed from three diagonal segments spells *ress*

underlying alphabet is ordered, HS needs only $\Theta(n \log n)$ time and $\Theta(n)$ additional space, a big improvement on previous algorithms.

As we have often seen before with other algorithmic innovations, the new algorithm is based on a very simple insight [A90]: an LCS is determined by a "longest" strictly decreasing line in a two-dimensional grid that represents the r occurrences of $\boldsymbol{x_1}[i] = \boldsymbol{x_2}[j]$. Figure 9.1 illustrates this idea for

$$\text{LCS}(\textit{rests,stress}) = \textit{ress},$$

where we see that a continuous line containing the largest number of decreasing diagonal segments must identify an LCS. In fact, observe that *all* the common subsequences of $\boldsymbol{x_1}$ and $\boldsymbol{x_2}$ can easily be read off the grid: for example, *rests* and *stress* also have the common subsequence *sts*. Observe also that even though *rests* and *stress* have a relatively long LCS, the value of r is nevertheless small, satisfying $r < n_1 + n_2 < 2n$.

Of course if we work directly from the $n_1 \times n_2$ grid, we will not be able to improve on the $\Omega(n_1 n_2)$ lower time bound of the preceding algorithms. Instead a two-dimensional array $\boldsymbol{j}$ is defined, in which $\boldsymbol{j}[i,h]$ gives the length of the shortest prefix of $\boldsymbol{x_2}$ that has an LCS of length h with $\boldsymbol{x_1}[1..i]$. In other words, $\boldsymbol{j}[i,h]$ is the least integer j^* such that

$$|\text{LCS}(\boldsymbol{x_1}[1..i], \boldsymbol{x_2}[1..j^*])| = h.$$

For example, for $\boldsymbol{x_1} = \textit{rests}$ and $\boldsymbol{x_2} = \textit{stress}$,

$$\boldsymbol{j}[4,1] = 1,\ \boldsymbol{j}[4,2] = 2,\ \boldsymbol{j}[4,3] = 5,$$

with $\boldsymbol{j}[4,h]$ undefined for $h > 3$, while

$$\boldsymbol{j}[5,1] = 1,\ \boldsymbol{j}[5,2] = 2,\ \boldsymbol{j}[5,3] = 5,\ \boldsymbol{j}[5,4] = 6,$$

with $\boldsymbol{j}[5,h]$ undefined for $h > 4$. More generally, we see that $\boldsymbol{j}[i,h]$ is defined if and only if $h \leq |\text{LCS}(\boldsymbol{x_1}[1..i], \boldsymbol{x_2})| \leq i$; for every undefined position $[i,h]$, we suppose that

$\boldsymbol{j}[i,h] = n_2+1$. We will find it convenient to extend these definitions to $h=0$, assuming that $\boldsymbol{j}[i,0]=0$ for every $i \in 1..n_1$. We also suppose $\boldsymbol{j}[0,0]=0$ with $\boldsymbol{j}[0,h]=n_2+1$ for every $h \in 1..n_1$. Note that elements of $\boldsymbol{j}$ can easily be read off the grid of Figure 9.1.

We now establish some basic properties of $\boldsymbol{j}$ that will be useful in Algorithm HS.

Lemma 9.4.1 *For all integers $i \in 1..n_1$ and $h \in 1..i$,*

$$\boldsymbol{j}[i-1,h-1]+1 \le \boldsymbol{j}[i,h] \le \boldsymbol{j}[i-1,h].$$

Proof The result holds trivially for $i=1$, and so we may suppose $i>1$.

Observe then that if $\boldsymbol{j}[i-1,h]$ is defined, so are $\boldsymbol{j}[i-1,h-1]$ and $\boldsymbol{j}[i,h]$. Observe further that if $\boldsymbol{x_1}[1..i-1]$ and $\boldsymbol{x_2}[1..\boldsymbol{j}[i-1,h]]$ have an LCS of length h, then since $\boldsymbol{j}[i-1,h]$ is a minimum, $\boldsymbol{x_1}[1..i]$ and $\boldsymbol{x_2}[1..\boldsymbol{j}[i-1,h]]$ must have a common subsequence of length at least h. Thus $\boldsymbol{j}[i,h] \le \boldsymbol{j}[i-1,h]$.

Now since $\boldsymbol{j}[i,h]$ is defined, we know that $\boldsymbol{x_1}[1..i]$ and $\boldsymbol{x_2}[1..\boldsymbol{j}[i,h]]$ have a common subsequence of length h. Deleting the rightmost letter from each of these sequences can reduce the length of this subsequence by at most one; thus $\boldsymbol{x_1}[1..i-1]$ and $\boldsymbol{x_2}[1..\boldsymbol{j}[i,h]-1]$ have a common subsequence, hence an LCS, of length at least $h-1$. In other words, $\boldsymbol{j}[i-1,h-1] \le \boldsymbol{j}[i,h]-1$ as required. □

We will use this result to show how to compute $\boldsymbol{j}[i,h]$, given that the values of $\boldsymbol{j}[i-1,h-1]$ and $\boldsymbol{j}[i-1,h]$ are already known. This will in turn provide the basis for the recurrence in Algorithm HS.

Lemma 9.4.2 *For all integers $i \in 1..n_1$ and $h \in 1..|\mathrm{LCS}(\boldsymbol{x_1}[1..i],\boldsymbol{x_2})|$,*

$$\begin{aligned}\boldsymbol{j}[i,h] &= \textit{the smallest integer } j^* \in \boldsymbol{j}[i-1,h-1]+1..\boldsymbol{j}[i-1,h] \\ &\qquad \textit{(if it exists) such that } \boldsymbol{x_1}[i]=\boldsymbol{x_2}[j^*]; \\ &= \boldsymbol{j}[i-1,h], \textit{ if no such } j^* \textit{ exists.}\end{aligned} \tag{9.6}$$

Proof We deal first with the second (and easier) case in which no such j^* exists. Observe that since $\boldsymbol{j}[i,h]$ specifies a *shortest* prefix of $\boldsymbol{x_2}$, it must therefore be true that an LCS of $\boldsymbol{x_1}[1..i]$ and $\boldsymbol{x_2}[1..\boldsymbol{j}[i,h]]$ has $\boldsymbol{x_2}[\boldsymbol{j}[i,h]]$ as its rightmost element. However, since Lemma 9.4.1 tells us that $\boldsymbol{j}[i,h] \in \boldsymbol{j}[i-1,h-1]+1..\boldsymbol{j}[i-1,h]$, and since by hypothesis there exists no j^* in this range such that $\boldsymbol{x_2}[j^*]=\boldsymbol{x_1}[i]$, it therefore must be true that $\boldsymbol{x_1}[i] \ne \boldsymbol{x_2}[\boldsymbol{j}[i,h]]$. Hence the same LCS is a common subsequence of $\boldsymbol{x_1}[1..i-1]$ and $\boldsymbol{x_2}[1..\boldsymbol{j}[i,h]]$, so that $\boldsymbol{j}[i,h] \ge \boldsymbol{j}[i-1,h]$; thus, by Lemma 9.4.1 again, $\boldsymbol{j}[i,h]=\boldsymbol{j}[i-1,h]$, as required.

Now suppose that there does exist a smallest integer $j^* \in \boldsymbol{j}[i-1,h-1]+1..\boldsymbol{j}[i-1,h]$ such that $\boldsymbol{x_1}[i]=\boldsymbol{x_2}[j^*]$. It follows that $\boldsymbol{x_1}[1..i]$ and $\boldsymbol{x_2}[1..j^*]$ have a common subsequence of length h, consisting of $\mathrm{LCS}(\boldsymbol{x_1}[1..i-1],\boldsymbol{x_2}[1..\boldsymbol{j}[i-1,h-1]])$ of length $h-1$ together with $\boldsymbol{x_2}[j^*]$. Thus $\boldsymbol{j}[i,h] \not> j^*$.

We now show that $\boldsymbol{j}[i,h] \not< j^*$. For suppose that in fact $\boldsymbol{j}[i,h] < j^*$. Then since by Lemma 9.4.1 $\boldsymbol{j}[i-1,h-1] < \boldsymbol{j}[i,h] < j^*$, and since j^* is the *least* integer in

$\boldsymbol{j}[i-1,h-1]+1..\boldsymbol{j}[i-1,h]$ such that $\boldsymbol{x_1}[i] = \boldsymbol{x_2}[j^*]$, we conclude that $\boldsymbol{x_1}[i] \neq \boldsymbol{x_2}[\boldsymbol{j}[i,h]]$. Since

$$|\mathrm{LCS}(\boldsymbol{x_1}[1..i], \boldsymbol{x_2}[1..\boldsymbol{j}[i,h]])| = h,$$

it follows that $\boldsymbol{x_1}[1..i-1]$ and $\boldsymbol{x_2}[1..\boldsymbol{j}[i,h]]$ also contain a common subsequence of length h. Thus $\boldsymbol{j}[i-1,h] \leq \boldsymbol{j}[i,h+1]$ and so by Lemma 9.4.1 again, $\boldsymbol{j}[i-1,h] = \boldsymbol{j}[i,h+1]$. But this is impossible since as we have just seen, $\boldsymbol{j}[i-1,h+1] < j^* \leq \boldsymbol{j}[i-1,h]$. The contradiction tells us that $\boldsymbol{j}[i,h] \not< j^*$, hence that $\boldsymbol{j}[i,h] = j^*$, as required. □

Lemma 9.4.2 establishes for the array $\boldsymbol{j}$ essentially the same result established earlier in Lemma 9.3.1 for the array $\boldsymbol{\gamma}$: the values in row i can be computed based only on the values already computed for row $i-1$. In the case of Algorithms H2 and H3, this fact led to a reduction in the space requirement from $\Theta(n_1 n_2)$ to $\Theta(n_2)$; here we shall of course find a similar reduction in space, but in addition it turns out to be feasible also to reduce the time required, at least in the average-case.

Recall that in Algorithm H2 an array $\boldsymbol{\gamma'}[1..2, 1..n_2]$ was introduced to replace $\boldsymbol{\gamma}$; for Algorithm HS it turns out that a one-dimensional array $\boldsymbol{j'}[0..n_2]$ suffices to replace $\boldsymbol{j}$. The calculation is then organized in four main steps:

Step I Compute MATCHLIST

From Lemma 9.4.2 we see that in order to compute $\boldsymbol{j}[i+1, 0..n_1]$ from $\boldsymbol{j}[i, 0..n_1]$ we will need to have available for each value $i \in 1..n_1$ the corresponding positions j in $\boldsymbol{x_2}$ such that $\boldsymbol{x_1}[i] = \boldsymbol{x_2}[j]$. We get this information in a convenient form by computing MATCHLIST$[1..n_1]$, an array in which each position i is a pointer to a list of values j such that $\boldsymbol{x_1}[i] = \boldsymbol{x_2}[j]$. In order to implement Step III properly, we will need to have the values of j corresponding to each i available in descending order.

Note that if $\boldsymbol{x_1}$ and $\boldsymbol{x_2}$ are defined on an ordered alphabet, MATCHLIST can be computed in $O(n \log n + r)$ time and $\Theta(n_1 + r)$ space by using an efficient sort or search tree algorithm, where as above $n = \max\{n_1, n_2\}$. However, if the alphabet is general (not ordered), the computation of MATCHLIST could require $\Theta(n^2)$ time in the worst case. The details are left to Exercise 9.4.1. See also Problem 4.1 (Section 4.1).

Step II Initialize $\boldsymbol{j'}$

Initially we set $\boldsymbol{j'}[0] \leftarrow 0$ while for $h \in 1..n_2$ we indicate that $\boldsymbol{j'}[h]$ is undefined by setting $\boldsymbol{j'}[h] \leftarrow n_2 + 1$.

Step III Compute $\boldsymbol{j}[i,h]$ from $\boldsymbol{j}[i-1,h]$, $i = 1, 2, \ldots, n_1$

This calculation is carried out by replacing the values in the previous array $\boldsymbol{j'}[0..n_1]$ corresponding to $i-1$ by those corresponding to i. Recall from Lemma 9.4.2 that $\boldsymbol{j}[i,h] = \boldsymbol{j}[i-1,h]$ — that is, $\boldsymbol{j'}[h]$ is unchanged — unless there exists $j^* \in \boldsymbol{j'}[h-1]+1..\boldsymbol{j'}[h]$ such that $\boldsymbol{x_1}[i] = \boldsymbol{x_2}[j^*]$. Thus for the current value of i and each value $j^* \in MATCHLIST[i]$, the array $\boldsymbol{j'}$ needs to be searched in order to determine h^* such that

$$\boldsymbol{j'}[h^*-1] + 1 \leq j^* \leq \boldsymbol{j'}[h^*].$$

Then, in accordance with Lemma 9.4.2, $\boldsymbol{j'}$ is updated as follows:

$$\boldsymbol{j'}[h^*] \leftarrow j^*. \tag{9.7}$$

Observe here that having the entries $j^* \in MATCHLIST[i]$ made available in *descending* order ensures that in each range $\boldsymbol{j'}[h^* - 1] + 1..\boldsymbol{j'}[h^*]$ the smallest value j^* is the final one to replace $\boldsymbol{j'}[h^*]$.

Every assignment (9.7) that strictly reduces $\boldsymbol{j'}[h^*]$ corresponds to one of the r points on the grid of Figure 9.1, where $\boldsymbol{x_1}[i] = \boldsymbol{x_2}[j^*]$. At every such assignment a leaf node

$$(i, j^*, \text{LINK}[h^* - 1]$$

of the tree is created, where LINK$[h^* - 1]$ is a pointer to a node previously created for an assignment (9.7) on $\boldsymbol{j'}[h^* - 1]$. (For $h^* = 1$, LINK$[h^* - 1] =$ NULL.) At the same time as the leaf node is created, a pointer to it is stored in LINK$[h^*]$. Thus after all positions $i \in 1..n_1$ have been processed, the largest value h such that $\boldsymbol{j'}[h] < n_2 + 1$ is the length of LCS$(\boldsymbol{x_1}, \boldsymbol{x_2})$, and LINK$[h]$ points to a path of length h from a leaf node of the tree to the root.

Step IV Recover LCS$(\boldsymbol{x_1}, \boldsymbol{x_2})$ in reverse order
First a node in the tree is located corresponding to the largest value of h such that $\boldsymbol{j'}[h] < n_2 + 1$. Then the path from that node to the root of the tree is output: this is an LCS (in reverse order) of $\boldsymbol{x_1}$ and $\boldsymbol{x_2}$.

The details of these steps are shown in Algorithm 9.4.1.

We can gain an intuitive understanding of the operation of Algorithm HS by considering our previous example $\boldsymbol{x_1} =$ *rests*, $\boldsymbol{x_2} =$ *stress*. Then as we see from Figure 9.1:

$$\begin{aligned}
\text{MATCHLIST}[1] &= <3> \\
\text{MATCHLIST}[2] &= <4> \\
\text{MATCHLIST}[3] &= <6, 5, 1> \\
\text{MATCHLIST}[4] &= <2> \\
\text{MATCHLIST}[5] &= <6, 5, 1>
\end{aligned}$$

The values of $\boldsymbol{j'}$ after each iteration are

$$\begin{aligned}
\text{initialize } \boldsymbol{j'}: &\quad 0\,7\,7\,7\,7\,7 \\
i = 1: &\quad 0\,\underline{\mathbf{3}}\,7\,7\,7\,7 \\
i = 2: &\quad 0\,3\,\underline{\mathbf{4}}\,7\,7\,7 \\
i = 3: &\quad 0\,1\,4\,\underline{\mathbf{5}}\,7\,7 \\
i = 4: &\quad 0\,1\,2\,5\,7\,7 \\
i = 5: &\quad 0\,1\,2\,5\,\underline{\mathbf{6}}\,7
\end{aligned}$$

and the corresponding LCS is defined in reverse order by

$$(i = 5, j^* = 6),\ (i = 3, j^* = 5),\ (i = 2, j^* = 4),\ (i = 1, j^* = 3)$$

as indicated by the underlined values.

Algorithm 9.4.1 (HS)

– *Compute an* $\mathrm{LCS}(\boldsymbol{x_1}, \boldsymbol{x_2})$

– *Step I: compute* MATCHLIST
$\forall\, i \in 1..n$ compute MATCHLIST$[i]$, a pointer to a list of monotone decreasing values j such that $\boldsymbol{x_1}[i] = \boldsymbol{x_2}[j]$

– *Step II: Initialize*

```
j'[0] ← 0
for i ← 1 to n1 do
   j'[i] ← n2 + 1
LINK[0] ← NULL
```

– *Step III: Compute* $\boldsymbol{j}[i,h]$ *from* $\boldsymbol{j}[i-1,h]$, $1 \le i \le n_1$

```
for i ← 1 to n1 do
   ∀ j* ∈ MATCHLIST[i] do
      find h* such that j'[h*−1] + 1 ≤ j* ≤ j'[h*]
      if j* < j'[h*] then
         j'[h*] ← j*
         LINK[h*] ← newnode(i, j*, LINK[h*−1])
```

– *Step IV: Recover* $\mathrm{LCS}(\boldsymbol{x_1}, \boldsymbol{x_2})$ *in reverse order*

```
find the largest h such that j'[h] < n2 + 1
PTR ← LINK[h]
while PTR ≠ NULL do
   output the (i, j) pair in the node pointed to by PTR
   advance PTR using LINK[h−1] from this node
```

Theorem 9.4.3 *Given two strings* $\boldsymbol{x_1}[1..n_1]$ *and* $\boldsymbol{x_2}[1..n_2]$ *on an ordered alphabet, Algorithm 9.4.1 correctly computes* $\mathrm{LCS}(\boldsymbol{x_1}, \boldsymbol{x_2})$ *in* $O((n+r)\log n)$ *time and* $O(n+r)$ *additional space, where* $n = \max\{n_1, n_2\}$ *and* r *is the total number of distinct pairs* (i,j) *such that* $\boldsymbol{x_1}[i] = \boldsymbol{x_2}[j]$.

Proof We claim that the correctness of Algorithm 9.4.1 follows from Lemmas 9.4.1 and 9.4.2, together with the above discussion.

To analyze the time and space requirements, we consider each step separately. As we have already seen, Step I requires $O(n \log n + r)$ time and $\Theta(n_1 + r)$ additional space when the alphabet is ordered, while Step II requires $O(n_1)$ time and $\Theta(n_1)$ space for the storage of $\boldsymbol{j'}$. In Step III the **for** loop requires $\Theta(n_1)$ time plus

(a) the total time required to process r entries in MATCHLIST;

(b) the total time required to locate h^* in $\boldsymbol{j'}$;

(c) the total time required to perform at most r updates to $\boldsymbol{j}'$ and to create at most r nodes in the tree.

For (a) the time required is $\Theta(r)$; for (b) binary search can be used at a total cost of $O(r \log n_1)$ time; while for (c) $O(r)$ time will be needed. Thus Step III requires a total of $O(n_1 + r \log n_1)$ time and $O(r)$ additional space. In Step IV $O(n_1)$ time is needed to find position h in $\boldsymbol{j}'$, together with $O(r)$ time to output the LCS. Summing the time and space requirements for the four steps, we get the desired result. □

Finally we remark that Apostolico and Guerra [AG87] have developed a variant of Algorithm HS that executes in time $O(n_1 \log n_2 + t \log(2n_1n_2/t))$, where t is the number of "dominant" matches between letters of $\boldsymbol{x_1}$ and $\boldsymbol{x_2}$. Since $t \leq r$, the maximum time required by Algorithm AG is $\Theta(n_1n_2)$, an improvement by a $\log n$ factor on Algorithm HS, while the minimum is $\Theta(n_1 \log n_2)$.

Exercises 9.4

1. Describe an algorithm to compute MATCHLIST in Algorithm HS assuming

   (a) the alphabet is ordered;

   (b) the alphabet is not ordered.

   In each case specify your algorithm's asymptotic complexity in both the worst and best cases, and give examples of strings $\boldsymbol{x_1}$ and $\boldsymbol{x_2}$ that achieve these bounds.

2. Observe that for Algorithm HS, as for the other LCS algorithms, the roles of $\boldsymbol{x_1}$ and $\boldsymbol{x_2}$ may be interchanged. Hence show that the time and space bounds given in Theorem 9.4.3 can be slightly sharpened.

9.5 Ukkonen-Myers

"Clang, clang, clang" went the trolley,
"Ding, ding, ding" went the bell,
"Zing, zing, zing" went my heart-strings,
From the moment I saw him I fell.

— Hugh MARTIN (1914–) and Ralph BLANE (1914–1995)
— Judy GARLAND (1922–1969), *Meet Me in St Louis*

In this section we return essentially to the methodology of Algorithm WF: iterative processing of the cost array $\boldsymbol{c} = \boldsymbol{c}[0..n_1, 0..n_2]$ based on the recurrence relation of Lemma 9.1.1.

Now however we model $\boldsymbol{c}$ as a directed acyclic graph G with a single source vertex corresponding to position $[0,0]$ of $\boldsymbol{c}$ and sinks corresponding to positions $[n_1,j]$ and $[i,n_2]$, $1 \le i \le n_1, 1 \le j \le n_2$, including in particular $[n_1,n_2]$. We again use Lemma 9.1.1 to find a least-cost ("shortest") path from $\boldsymbol{c}[0,0]$ to $\boldsymbol{c}[n_1,n_2]$, but this time we use the magnitude of the distance $d = d(\boldsymbol{x_1},\boldsymbol{x_2})$ itself as a means of avoiding vertices and arcs (or edges) of G that cannot possibly affect the calculation. The result is an algorithm, called here Algorithm UM in honour of its two independent discoverers [U85a, M86], that computes d in $O(nd)$ time, where $n = \min\{n_1,n_2\}$. Thus for values of d that are not too large with respect to n (a common case in distance calculations), the new algorithm is very attractive.

Further, essentially the same algorithm can be employed in those cases, also common, where the value d represents an upper bound (or threshold) on the value of $d(\boldsymbol{x_1},\boldsymbol{x_2})$ that is of interest; in the modified algorithm, either $d(\boldsymbol{x_1},\boldsymbol{x_2}) < d$ or the value ∞ is returned.

As already indicated, the vertices of the graph $G = (V,E)$ are the positions $[i,j]$ of $\boldsymbol{c}$, for every $i \in 0..n_1$, $j \in 0..n_2$. For every vertex $[i,j] \in V$ such that $i+j > 0$, there are at most three possible inarcs $([i',j'],[i,j])$:

$$\begin{aligned} ([i-1,j],[i,j]) \in E \quad &\text{iff} \quad \boldsymbol{c}[i,j] - \boldsymbol{c}[i-1,j] = d(\boldsymbol{x_1}[i],\varepsilon); \\ ([i,j-1],[i,j]) \in E \quad &\text{iff} \quad \boldsymbol{c}[i,j] - \boldsymbol{c}[i,j-1] = d(\varepsilon,\boldsymbol{x_2}[j]); \\ ([i-1,j-1],[i,j]) \in E \quad &\text{iff} \quad \boldsymbol{c}[i,j] - \boldsymbol{c}[i-1,j-1] = d(\boldsymbol{x_1}[i],\boldsymbol{x_2}[j]). \end{aligned} \tag{9.8}$$

We associate with each inarc the corresponding weight $\boldsymbol{c}[i,j] - \boldsymbol{c}[i',j']$, and we observe that by Lemma 9.1.1, there exists at least one inarc to every vertex $[i,j] \in V$ if and only if $[i,j] \neq [0,0]$.

We call G the ***dependency graph*** of $\boldsymbol{c}$. Note that

- $|E| \le 3(|V|-1)$;
- G is a subgraph of the graph whose structure reflects exactly the structure of the array $\boldsymbol{c}$ (in G arcs that do not correspond to the minimum cost are omitted);
- there exists a directed path from $[0,0]$ to every other vertex $[i,j] \in V$;
- further, the sum of the weights on every directed path in G from $[0,0]$ to $[i,j]$ equals the cost of a minimum-cost edit sequence that transforms $\boldsymbol{x_1}[1..i]$ into $\boldsymbol{x_2}[1..j]$;
- G contains a single source vertex $[0,0]$ and at most n_1+n_2-1 sinks, all of the latter corresponding to row n_1 or column n_2 of $\boldsymbol{c}$.

Figure 9.2 displays the dependency graph for our Levenshtein distance example $\boldsymbol{x_1} = \textit{rests}$, $\boldsymbol{x_2} = \textit{stress}$.

We now show in stages how vertices and arcs can effectively be pruned from G — exactly those vertices and arcs that cannot lie on the minimum-cost path from $[0,0]$ to $[n_1,n_2]$. We shall see that omitting these arcs and vertices from our calculations enables us to compute the path in time $\Theta(nd)$.

Observe first that, for nonnegative weights, $\boldsymbol{c}[i,j] \ge \boldsymbol{c}[i',j']$ for every vertex $[i',j']$ that lies on a path in G from $[0,0]$ to $[i,j]$. Thus no vertex $[i',j']$ for which $\boldsymbol{c}[i',j'] > \boldsymbol{c}[i,j]$ can lie on a minimum-cost path to $[n_1,n_2]$, and so all such vertices and all paths that lead only to such vertices may be removed from G. Conversely, all vertices and arcs that do not satisfy these conditions must lie on a path to $[n_1,n_2]$ and therefore *must* be part of some valid minimum edit sequence for $d(\boldsymbol{x_1},\boldsymbol{x_2})$. As shown for our example in Figure 9.3, this very simple observation makes a dramatic difference to the size of G.

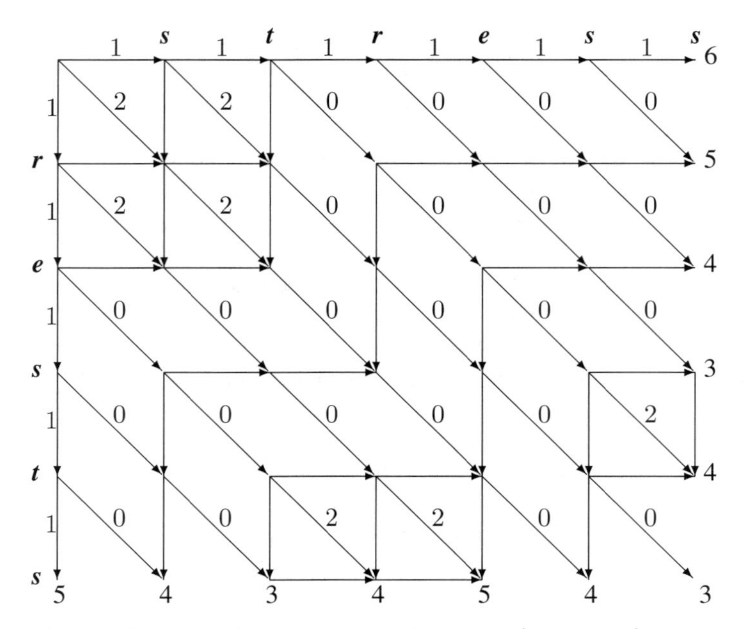

Figure 9.2. Original dependency graph G for $d_L(\textit{rests}, \textit{stress})$

In fact, looking more closely at Figure 9.3, we see that in this case, where all insertions and deletions (***indels***) cost exactly one, no vertex can contribute to the minimum distance if it is more than $\lceil d/2 \rceil = 2$ "steps" away from the diagonal line beginning at $[0, 1]$ and ending at $[n_1, n_2] = [5, 6]$. More precisely, a contributing vertex may be at most $\lfloor (n_2 - n_1 + d)/2 \rfloor$ steps *below* this line, or at most $\lfloor (n_1 - n_2 + d)/2 \rfloor$ steps *above* it. Since the diagonal itself

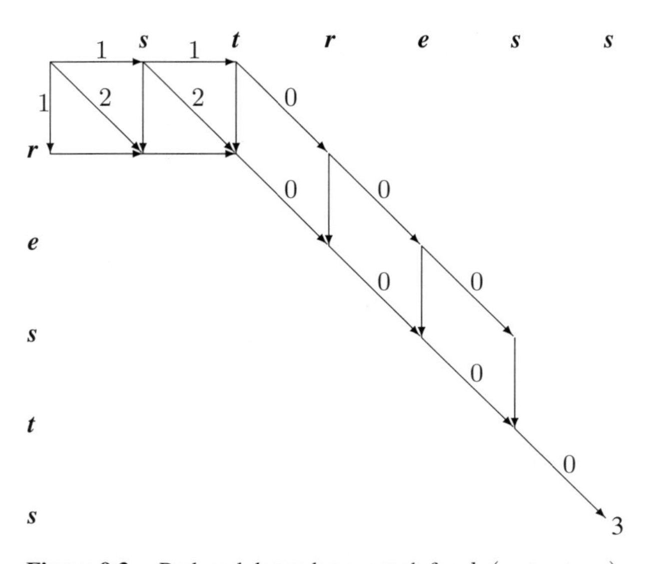

Figure 9.3. Reduced dependency graph for $d_L(\textit{rests}, \textit{stress})$

contains at most n_1 arcs, this means that the number of vertices (hence the number of arcs) that need to be considered is $O(n_1 d)$, and so the problem "should" be solvable in $O(n_1 d)$ time and space.

The remainder of this section simply expresses these observations in a precise mathematical form, while at the same time generalizing their scope to distance functions with arbitrary positive indel values. This development will lead us in a natural way to a straightforward algorithm. In order to simplify calculation, it will be convenient to assume throughout the remainder of this section that $n_2 \geq n_1$; as we have seen, for distance functions d satisfying the conditions of a metric, this is not a restriction, since $d(\boldsymbol{x_1}, \boldsymbol{x_2}) = d(\boldsymbol{x_2}, \boldsymbol{x_1})$ for every pair of strings $\boldsymbol{x_1}$, $\boldsymbol{x_2}$.

Let d^* be the minimum cost of any indel over all letters in the alphabet A; then

$$d^* = \min_{\lambda \in A}\{d(\lambda, \varepsilon), d(\varepsilon, \lambda)\} > 0.$$

Further, let $\Delta_{i,j} \in -n_1..n_2$ be the left-to-right diagonal on which the vertex $[i, j]$ lies, and let $\Delta_{i,j}$ take the value $j - i$. Then of course every vertex $[i', j']$ on diagonal $\Delta_{i,j}$ satisfies $j' - i' = j - i$. Now consider a directed path π in G from a vertex $[i', j']$ to another vertex $[i, j]$. If $\Delta_{i,j} - \Delta_{i',j'} \geq 0$, then π contains at least $\Delta_{i,j} - \Delta_{i',j'}$ deletions; while if $\Delta_{i,j} - \Delta_{i',j'} \leq 0$, then π contains at least $\Delta_{i',j'} - \Delta_{i,j}$ insertions. Thus, in view of (9.8),

$$\begin{aligned} \boldsymbol{c}[i, j] - \boldsymbol{c}[i', j'] &\geq |\Delta_{i,j} - \Delta_{i',j'}|d^* \\ &= |(j - i) - (j' - i')|d^*. \end{aligned} \tag{9.9}$$

In particular, setting $[i', j'] = [0, 0]$, we see that $\boldsymbol{c}[i, j] \geq |j - i|d^*$ for every vertex $[i, j]$ on any path from $[0, 0]$ to $[n_1, n_2]$. Since $\boldsymbol{c}[i, j] \leq \boldsymbol{c}[n_1, n_2]$ for every such vertex $[i, j]$, and since the distance $d = d(\boldsymbol{x_1}, \boldsymbol{x_2}) = \boldsymbol{c}[n_1, n_2]$, it follows that

$$d/d^* \geq |j - i|, \tag{9.10}$$

and so, in order to compute $\boldsymbol{c}[n_1, n_2]$, it suffices to consider diagonals $\Delta_{i,j}$ satisfying

$$|\Delta_{i,j}| \leq \lfloor d/d^* \rfloor.$$

For our example $\boldsymbol{x_1}$ = *rests*, $\boldsymbol{x_2}$ = *stress*, $d^* = 1$, this would reduce the diagonals to be considered to the set $\{-3, -2, -1, 0, 1, 2, 3\}$, still almost twice as many as occur in the slender graph of Figure 9.3.

But there is more to say. Consider any path from $[0, 0]$ to $[n_1, n_2]$ in the dependency graph G that passes through an intermediate vertex $[i, j]$. Applying (9.9) twice, first to the path $[i, j] - [n_1, n_2]$, then to the path $[0, 0] - [i, j]$, and recalling that $\boldsymbol{c}[0, 0] = 0$, we conclude that

$$d/d^* \geq |j - i| + |(n_2 - n_1) - (j - i)|, \tag{9.11}$$

a refinement of (9.10). We consider two cases:

- $j \le i$ (This is the case in which $\Delta_{i,j} \le 0$.)
(9.11) now becomes

$$d/d^* \ge (n_2 - n_1) + 2(i - j),$$

since we suppose $n_2 \ge n_1$. In view of the fact that $i - j$ is a nonnegative integer, we may rewrite this equation in the form

$$q \ge i - j \ge 0, \tag{9.12}$$

where

$$q = \left\lfloor (d/d^* - (n_2 - n_1))/2 \right\rfloor, \tag{9.13}$$

true for every intermediate vertex $[i, j]$ on a path in G from $[0, 0]$ to $[n_1, n_2]$.

- $j \ge i$ (This is the case in which $\Delta_{i,j} \ge 0$.)
Here two subcases arise, depending on whether or not $n_2 - n_1 \le j - i$. If we suppose $n_2 - n_1 > j - i$, (9.11) becomes

$$d/d^* \ge n_2 - n_1,$$

which in view of (9.13) we may rewrite as

$$q \ge 0. \tag{9.14}$$

On the other hand, for $n_2 - n_1 \le j - i$, we find that

$$d/d^* \ge 2(j - i) - (n_2 - n_1),$$

which may be rewritten as

$$q + (n_2 - n_1) \ge j - i \ge 0. \tag{9.15}$$

The results of our investigation are encapsulated in equations (9.12), (9.14) and (9.15). For the moment we may ignore (9.14) which imposes only the trivial condition that d/d^* must be at least equal to the difference in length between $\boldsymbol{x_1}$ and $\boldsymbol{x_2}$. But from (9.12) and (9.15) we infer that

$$-q \le j - i \le q + (n_2 - n_1) \tag{9.16}$$

for any intermediate vertex $[i, j]$ lying on any directed path in G from $[0, 0]$ to $[n_1, n_2]$, where q is given by (9.13). Since $\Delta_{i,j} = j - i$, (9.16) constrains the intermediate vertices $[i, j]$ to lie between diagonals $-q$ and $q + (n_2 - n_1)$ of G. If we apply (9.16) to our example, we find for $\boldsymbol{x_1} =$ *rests*, $\boldsymbol{x_2} =$ *stress*, $d^* = 1$, that $q = 1$, so that (9.16) becomes

$$-1 \le j - i \le 2,$$

in accordance with Figure 9.3 where the vertices occur on diagonals $\{-1, 0, 1, 2\}$.

In order to make algorithmic use of (9.16), we consider first the problem of determining whether or not $d(\boldsymbol{x_1}, \boldsymbol{x_2}) \leq D$, for some given positive integer D. Using (9.13) we can compute a trial value

$$Q = \left\lfloor (D/d^* - (n_2 - n_1))/2 \right\rfloor$$

and then go on to compute a minimum sum of weights on paths from $[0, 0]$ to $[n_1, n_2]$ that lie between the diagonals specified by (9.16):

$$-Q \leq \Delta_{i,j} \leq Q + (n_2 - n_1).$$

If the minimum sum over these paths leads to a value at vertex $[n_1, n_2]$ that is greater than D, we can conclude that not all the vertices $[i, j]$ were included in the sum that were required to ensure a minimum value $\mathbf{c}[n_1, n_2]$. In this case, in order to include outlying diagonals and the vertices that lie upon them, we need to increase the value of Q, hence the value of D. If on the other hand we compute a value at $[n_1, n_2]$ that is not greater than D, then we can be sure that the correct value $\mathbf{c}[n_1, n_2] \leq D$ has been found. We display below the function *trial_distance*(D) that implements this calculation: the returned value is $\mathbf{c}[n_1, n_2]$ in all cases, correct if $\mathbf{c}[n_1, n_2] \leq D$, otherwise possibly incorrect.

– *Compute* $\mathbf{c}[n_1, n_2]$ *using the range of diagonals*
– *determined by the argument* D
function *trial_distance*(D)

– *Using (9.13), compute* Q*, the range of diagonals*
– *corresponding to D; observe that since* $D/d^* > n_2 - n_1$,
– *the trivial condition (9.14) is satisfied*
$Q \leftarrow \left\lfloor (D/d^* - (n_2 - n_1))/2 \right\rfloor$

– *Compute* $\mathbf{c}[n_1, n_2]$ *using positions* $[i, j]$ *of* $\mathbf{c}$
– *that lie between diagonals* $-Q$ *and* $Q + (n_2 - n_1)$ *(9.16)*
for $i \leftarrow 1$ **to** n_1 **do**
  for $i \leftarrow \max\{1, i - Q\}$ **to** $\min\{n_1, i + Q + (n_2 - n_1)\}$ **do**
    $\mathbf{c}[i, j] \leftarrow \min\{\mathbf{c}[i-1, j] + d(\boldsymbol{x_1}[i], \varepsilon),$
      $\mathbf{c}[i, j-1] + d(\varepsilon, \boldsymbol{x_2}[j]),$
      $\mathbf{c}[i-1, j-1] + d(\boldsymbol{x_1}[i], \boldsymbol{x_2}[j])\}$
return $\mathbf{c}[n_1, n_2]$

All that is left now for Algorithm UM (presented as Algorithm 9.5.1) to do is to perform some routine initializations on the $\mathbf{c}$ array and then to pass appropriate values of D to *trial_distance*. For reasons that will appear below when we discuss the complexity of Algorithm UM, D is initially taken to be one, then successively doubled until $\mathbf{c}[n_1, n_2] \leq D$. Based on the preceding discussion, we claim that Algorithm UM is correct.

To determine the complexity of our new algorithm, consider first the function *trial_distance*, in particular the double **for** loop: for each input parameter D and for each value of i, the assignment statement will be executed at most $2Q + (n_2 - n_1) + 1$ times, where by (9.13)

$$2Q + (n_2 - n_1) + 1 \leq D/d^* + 1.$$

Thus the total time required for each execution of *trial_distance* is $O(n_1 D)$. Altogether this function will be executed $h + 1$ times using values $D = 1, 2, ..., 2^h$, where h is the least nonnegative integer such that $d = d(\boldsymbol{x_1}, \boldsymbol{x_2}) \leq 2^h$. Thus over all invocations of *trial_distance*, the total time required will be

$$\begin{aligned} O((1 + 2 + \ldots + 2^h)n_1) &= O(2^{h+1} n_1) \\ &= O(4 n_1 d) \\ &= O(n_1 d). \end{aligned}$$

We have proved

Theorem 9.5.1 *For two given strings $\boldsymbol{x_1} = \boldsymbol{x_1}[1..n_1]$ and $\boldsymbol{x_2} = \boldsymbol{x_2}[1..n_2]$, $n_2 \geq n_1$, Algorithm 9.5.1 correctly computes $d = d(\boldsymbol{x_1}, \boldsymbol{x_2})$ in $O(n_1 d)$ time and $\Theta(n_1 n_2)$ space.* □

Algorithm 9.5.1 (UM)

- *Compute* $d(\boldsymbol{x_1}, \boldsymbol{x_2}) = \mathbf{c}[n_1, n_2]$

- *To simplify matters, initialize row zero*
- *and column zero of* $\mathbf{c} = \mathbf{c}[0..n_1, 0..n_2]$

$\mathbf{c}[0, 0] \leftarrow 0$
for $i \leftarrow 1$ **to** n_1 **do**
  $\mathbf{c}[i, 0] \leftarrow \mathbf{c}[i-1, 0] + d(\boldsymbol{x_1}[i], \varepsilon)$
for $j \leftarrow 1$ **to** n_2 **do**
  $\mathbf{c}[0, j] \leftarrow \mathbf{c}[0, j-1] + d(\varepsilon, \boldsymbol{x_2}[j])$

- *Initialize D to its minimum possible nonzero value*

$D \leftarrow (n_2 - n_1 + 1)d^*$

- *Keep doubling D until* $D \geq$ *trial_distance*(D)

while $D <$ *trial_distance*(D) **do**
  $D \leftarrow 2D$
output $\mathbf{c}[n_1, n_2]$

As shown in Exercise 9.5.2, the space bound for Algorithm UM can actually be reduced to $O(n_1 d)$. As shown in Exercise 9.5.3, a straightforward modification to UM allows $d(\boldsymbol{x_1}, \boldsymbol{x_2})$ to be computed with the same time and space requirements if and only if $d(\boldsymbol{x_1}, \boldsymbol{x_2}) < d$, where now d is a given threshold of the calculation.

For the special cases of edit and Levenshtein distance, an "incremental" version of UM has been proposed [LMS98] that permits $d(\boldsymbol{x_1}, \lambda\boldsymbol{x_2})$ and its associated cost array to be computed from $d(\boldsymbol{x_1}, \boldsymbol{x_2})$ and its cost array in $O(d)$ time, where d is again a given threshold and λ is an arbitrary letter. If no threshold can be specified, then the time requirement increases to $O(n_1 + n_2)$. Thus, since $d(\boldsymbol{x_1}, \boldsymbol{x_2}\lambda)$ can be routinely computed from the cost array for $d(\boldsymbol{x_1}, \boldsymbol{x_2})$, Algorithm LMS can be used to perform distance calculations built up from $d(\varepsilon, \varepsilon)$ by appending or prepending letters one-by-one to either argument in any order. When a threshold d can be specified on these calculations, the overall time requirement is $O(d(n_1 + n_2))$, equivalent to that of Algorithm UM. However, the incremental approach of LMS allows it to be applied to solve other approximate pattern-matching problems more efficiently — in particular, problems (P1) and (P2) that arise in Section 13.4. The details of Algorithm LMS are complicated and technical: we do not discuss them further here. See [LMS98].

A much simpler application of the same incremental idea has recently been discovered [KP00] that requires $O(n_1 + n_2)$ time, whether or not a threshold is specified, to compute either $d(\boldsymbol{x_1}, \lambda\boldsymbol{x_2})$ or $d(\boldsymbol{x_1}, \boldsymbol{x_2}\lambda)$. This method would confer benefits if no threshold on the distance could be specified, or if the threshold $d \in \Theta(n_1)$ were large.

Finally, we remark that, just as with Algorithm WF, when $d = d_L$ the array $\boldsymbol{c}$ that represents the dependency graph G can be traced backwards, from vertex $[n_1, n_2]$ to vertex $[0, 0]$, in order to recapture an LCS of $\boldsymbol{x_1}$ and $\boldsymbol{x_2}$.

Exercises 9.5

1. It is claimed in Theorem 9.5.1 that Algorithm UM executes in $O(n_1 d)$ time, even though the initialization of $\boldsymbol{c}$ requires $\Omega(n_2)$ time. To remove any doubt about the correctness of this theorem, show that $n_2 \in O(n_1 d)$.
2. Theorem 9.5.1 also claims that Algorithm UM requires $\Theta(n_1 n_2)$ space for its execution. Describe an implementation of this algorithm that reduces its space requirement to $O(n_1 d)$.
3. Show how to modify UM to calculate $d(\boldsymbol{x_1}, \boldsymbol{x_2})$ if and only if the distance is bounded above by a given threshold d.

9.6 Summary

Wit has truth in it;
wisecracking is simply calisthenics with words.

— Dorothy PARKER (1893–1967), *Paris Review (1956)*

In this section we have discussed four main algorithms that compute either the distance $d(\boldsymbol{x_1}, \boldsymbol{x_2})$ or an LCS$(\boldsymbol{x_1}, \boldsymbol{x_2})$ for given strings $\boldsymbol{x_1} = \boldsymbol{x_1}[1..n_1]$, $\boldsymbol{x_2} = \boldsymbol{x_2}[1..n_2]$:

- Algorithm WF computes $d(\boldsymbol{x_1}, \boldsymbol{x_2})$ for any distance function d in $\Theta(n_1 n_2)$ time and $\Theta(n_1 n_2)$ additional space. When $d = d_L$, the Levenshtein distance, a corresponding LCS$(\boldsymbol{x_1}, \boldsymbol{x_2})$ can be computed as a byproduct of the distance calculation.
- Algorithm H3 computes an LCS$(\boldsymbol{x_1}, \boldsymbol{x_2})$, hence the Levenshtein distance $d_L(\boldsymbol{x_1}, \boldsymbol{x_2})$, also in $\Theta(n_1 n_2)$ time, but using only $\Theta(n)$ additional space, where $n = \min\{n_1, n_2\}$.
- Algorithm HS computes LCS$(\boldsymbol{x_1}, \boldsymbol{x_2})$ in time that is often $\Theta(n \log n)$ and using additional space that is often $\Theta(n)$, where $n = \max\{n_1, n_2\}$. Only in cases where $\boldsymbol{x_1}$ and $\boldsymbol{x_2}$ have altogether more than $\Theta(n)$ matching letters (for example, when alphabet size is small with respect to string length), will the complexity of Algorithm HS increase beyond these bounds.
- Algorithm UM computes the distance $d = d(\boldsymbol{x_1}, \boldsymbol{x_2})$ in $O(dn)$ time and $\Theta(n_1 n_2)$ additional space, where now $n = \min\{n_1, n_2\}$. As for Algorithm WF, an LCS$(\boldsymbol{x_1}, \boldsymbol{x_2})$ can be efficiently computed as a byproduct when the distance computed is d_L, the Levenshtein distance. Thus in cases where a threshold d can be placed on the distance between the given strings that is not too large ($d \in O(\log n)$ or even $d \in O(n/\log n)$, for example), UM becomes the algorithm of choice. Such cases are common: normally distance calculations are of interest only if the distance is relatively small with respect to string length.

  For edit distance d_E, it is shown in [M86] that if one of the input strings is randomly chosen, UM executes in *expected* time $O(n + k^2)$, and in [M86, LV88] that the use of suffix trees reduces *worst-case* time to $O(n \log \alpha + k^2)$, where α is alphabet size. Also, fast incremental versions of Algorithm UM are available for edit and Levenshtein distance [LMS98, KP00].

There are other algorithms for these and closely related problems. [MP80] shows how the "Four Russians" technique [ADKF70] can be applied to break up the cost array $\boldsymbol{c}$ into overlapping square submatrices so as to achieve $O(n^2/\log n)$ time for the distance calculation. However, this algorithm becomes practical only for very long strings ($n \geq 300,000$) that, in view of Algorithm UM, are separated by a relatively large distance. Algorithms have also been proposed for the "heaviest" common subsequence [JV92] and the longest "increasing" subsequence [S61]. Several recent algorithms [CIP01, CIPR01, IP01] apply variants of the bit-vector approach [D68, AD86, BYG92], discussed in Section 7.4 and Chapter 10, to the calculation of an LCS; in particular, [CIP01] shows how to implement the algorithms of Hirschberg and Hunt-Szymanksi using bit vectors. Generally, these algorithms require $O(n_1 n_2/w)$ time, where w is the length of the computer word, and appear to be very fast in practice. Also fast in practice is the algorithm of Allison and Dix [AD86] that uses bit vectors to compute the *length* of the LCS.

Chapter 10

Approximate Pattern-Matching

Take it from me, marriage isn't just a word — it's a sentence.

— King VIDOR (1895–1982), *The Crawl (1928 film)*

The number and variety of approximate pattern-matching algorithms proposed since the early 1980s come close to equalling those proposed for the exact case (see Chapters 7 and 8). At time of writing, the more efficient and effective of these algorithms can all be said to derive from two main ideas, both of which we have already encountered:

(1) In Section 7.4 Algorithm DBG was introduced for exact pattern-matching; it made use of a bit vector $\boldsymbol{s} = \boldsymbol{s}[1..m]$ to describe the "state" of the computation, together with a bit matrix $\boldsymbol{t} = \boldsymbol{t}[1..\alpha, 1..m]$ that encodes the elements of the pattern $\boldsymbol{p} = \boldsymbol{p}[1..m]$ in terms of the letters $1..\alpha$ of an indexed alphabet. It then turns out that elementary bit operations on $\boldsymbol{s}$ and $\boldsymbol{t}$ can be efficiently used to recompute the state vector $\boldsymbol{s}$ for the next matched letter $\boldsymbol{x}[i+1]$ in the text string $\boldsymbol{x} = \boldsymbol{x}[1..n]$.

In Sections 10.2–10.4 we shall look at three applications of this idea to approximate pattern-matching.

(2) In Section 9.1 we introduced a two-dimensional cost array $\boldsymbol{c}$ whose entries were defined by the identity

$$\boldsymbol{c}[i,j] = d\big(\boldsymbol{x_1}[1..i], \boldsymbol{x_2}[1..j]\big),$$

the distance between prefixes of strings $\boldsymbol{x_1} = \boldsymbol{x_1}[1..n_1]$, $\boldsymbol{x_2} = \boldsymbol{x_2}[1..n_2]$. A basic recurrence relation on neighbouring elements of $\boldsymbol{c}$ then gave rise directly to two distance algorithms, called WF and UM, and indirectly to two LCS algorithms, H3 and HS.

Here we shall first see in Section 10.1 how a slight redefinition of $\boldsymbol{c}$ leads to a straightforward adaptation of Algorithm WF for approximate pattern-matching. Then Sections 10.3 and 10.4 deal with further variations in the use and implementation of the array $\boldsymbol{c}$, especially its interaction with the bit-vector approach of DBG. Finally, Section 10.5 provides a brief overview of the theoretical time complexities achieved by recent approximate pattern-matching algorithms based on edit/Levenshtein distance.

[Nav01] is a thorough up-to-date overview of approximate pattern-matching algorithms, including sketches of the methodology of several variants not considered here; it includes also the results of experiments designed to evaluate the algorithms' relative efficiency.

10.1 A General Distance-Based Algorithm

Or bid the soul of Orpheus sing
Such notes, as warbled to the string,
Drew iron tears down Pluto's cheek.

— John MILTON (1608–1674), *Il Penseroso*

We began the first section of Chapter 9 with a discussion of the cost array $\boldsymbol{c}$ and its properties; we begin the first section of this chapter in the same way. Here we redefine the cost array $\boldsymbol{c} = \boldsymbol{c}[0..n_1, 0..n_2]$ of Chapter 9 as $\boldsymbol{c} = \boldsymbol{c}[0..n, 1..m]$ with elements

$$\boldsymbol{c}[i,j] = \min_{0 \le i' \le i} d\big(\boldsymbol{x}[i'..i], \boldsymbol{p}[1..j]\big), \tag{10.1}$$

valid for every $i \in 0..n$, $j \in 1..m$, where as before d is any distance function satisfying the conditions (2.2)–(2.5) for a metric, including weighted distances specified by a scoring matrix (Section 2.2). The minimum over i' needs to be computed in (10.1) because as shown in Exercise 10.1.1, when the distance is nonzero, it is not necessarily true that $i - i' + 1 = j$ — it is possible that $|\boldsymbol{x}[i'..i]| \neq |\boldsymbol{p}[1..j]|$. Thus $\boldsymbol{c}[i,j]$ is now the distance between $\boldsymbol{p}[1..j]$ and the "nearest" suffix of $\boldsymbol{x}[1..i]$ and, in particular, $\boldsymbol{c}[i,m]$ gives the minimum distance between $\boldsymbol{p}$ and suffixes of $\boldsymbol{x}$ ending at position i. Therefore, for any given nonnegative integer k, those values of i for which $\boldsymbol{c}[i,m] \le k$ specify those substrings of the text string $\boldsymbol{x}$ that are distance at most k from the pattern $\boldsymbol{p}$.

Using (10.1), we see that the initial values for row 0 coincide with those given in Section 9.1:

- $\boldsymbol{c}[0,j] = d(\varepsilon, \boldsymbol{p}[1..j])$ for every $j \in 1..m$ — the cost of inserting the prefix $\boldsymbol{p}[1..j]$.

However, the initial values in column 1 are different from those of Section 9.1 due to the fact that we are computing minimum distance to substrings $\boldsymbol{x}[i'..i]$ rather than distance to a single prefix $\boldsymbol{x}[1..i]$. Of course if $\boldsymbol{x}[i] = \boldsymbol{p}[1]$, then we set

$$\boldsymbol{c}[i,1] \leftarrow 0, \tag{10.2}$$

but if not, then we must compute

$$c[i,1] \leftarrow \min\left\{c[i-1,1] + d(x[i],\varepsilon), d(x[i],p[1])\right\}. \tag{10.3}$$

Observe that in fact these initializations constitute the *only* difference: since for any distance calculation, indels may be arbitrarily ordered (Exercise 2.2.14), it follows that, once the boundary values in row 0 and column 1 have been correctly computed, the basic recurrence relation of Lemma 9.1.1 holds in this case also. Thus we may compute each element $c[i,j]$, $i \in 1..n$, $j \in 2..m$, using

$$\begin{aligned} c[i,j] \leftarrow \min\Big\{ & c[i-1,j] + d(x[i],\varepsilon), \\ & c[i,j-1] + d(\varepsilon, p[j]), \\ & c[i-1,j-1] + d(x[i],p[j])\Big\}, \end{aligned} \tag{10.4}$$

just as before.

We use the example of Chapter 9 to illustrate the differences that may arise between the two c arrays. Let c_{dist} be the array that specifies distance, and let c_{best} be the array that specifies the best approximate match over all suffixes ending at position i. Then for Levenshtein distance between *rests* and *stress* we have c_{dist} as shown in Table 10.1.

On the other hand, if we try to find the best approximate matches of prefixes of the pattern p = *stress* in the text string x = *rests*, we get the altered cost array shown in Table 10.2. Observe that for smaller values of j the differences between the two arrays can be quite pronounced, due of course to the changed boundary values in column 1.

In general, given a nonnegative integer k, we can inspect the entries in column j of c_{best} to determine positions i at which there exists a substring $u = x[i'..i]$ such that

$$d(u, p[1..j]) \le k.$$

We say then that a ***k*-approximate** match of $p[1..j]$ ***occurs*** at position i of x. For example, taking $j = 5$ and $k = 2$ in Table 10.2, we can conclude that at positions $i = 3$ and $i = 5$, 2-approximate matches of $p[1..5]$ = *stres* occur; similarly, for $j = 3$ and $k = 1$, we find that at position $i = 4$ a 1-approximate match of $p[1..3]$ = *str* occurs. Note that the matching substring u is not necessarily unique.

	j	**0**	**1**	**2**	**3**	**4**	**5**	**6**
i		ε	*s*	*t*	*r*	*e*	*s*	*s*
0	ε	0	1	2	3	4	5	6
1	*r*	1	2	3	2	3	4	5
2	*e*	2	3	4	3	2	3	4
3	*s*	3	2	3	4	3	2	3
4	*t*	4	3	2	3	4	3	4
5	*s*	5	4	3	4	5	4	3

Table 10.1: c_{dist} for *rests* and *stress* using d_L

	j	**1**	**2**	**3**	**4**	**5**	**6**
i		s	t	r	e	s	s
0	ε	1	2	3	4	5	6
1	r	2	3	2	3	4	5
2	e	2	3	3	2	3	4
3	s	0	1	2	3	2	3
4	t	1	0	1	2	3	4
5	s	0	1	2	3	2	3

Table 10.2: c_{best} for $\boldsymbol{x} = rests$ and $\boldsymbol{p} = stress$ using d_L

It is clear then that Algorithm WF (Section 9.2), making use of the revised initializations (10.2) and (10.3), can also be applied to approximate pattern-matching. But as observed earlier, the length $i - i' + 1$ of the matching substring $\boldsymbol{u}$ may not equal the length m of the pattern. Thus if it is required to specify at least one matching pattern $\boldsymbol{u} = \boldsymbol{x}[i'..i]$, we must in addition provide a value for i'. An easy way to achieve this is provided by the sequence of moves ν described in Section 9.2: since these moves specify edit operations that transform $\boldsymbol{u}$ into $\boldsymbol{p}$, hence $\boldsymbol{p}$ into $\boldsymbol{u}$, they also determine the initial position i' of $\boldsymbol{u}$. As shown in Exercise 10.1.3, the number of moves is at most $2|\boldsymbol{p}| = 2m$, while the number of occurrences i of k-approximate matches is at most n, so the time required to compute all positions i' is $O(mn)$, no greater than the $\Theta(mn)$ time required to compute $\boldsymbol{c}$. Thus

Theorem 10.1.1 *Given a text $\boldsymbol{x}[1..n]$, a pattern $\mathbf{p}[1..m]$ and a nonnegative integer k, a modified Algorithm WF computes all of the k-approximate occurrences of $\mathbf{p}$ in $\boldsymbol{x}$ using $\Theta(mn)$ time and $\Theta(mn)$ space.* □

Of course, as suggested earlier, essentially the same algorithm can also be used to compute all the k-approximate occurrences of any prefix $\boldsymbol{p}[1..j]$, $1 \leq j \leq m$, in $\boldsymbol{x}$.

The algorithm described in this section is both time- and space-consuming: all the elements of the cost array $\boldsymbol{c}$ need to be computed. However, as already shown in Exercise 9.3.3 and shown again in Exercise 10.1.4, the consumption of space is easily reduced by observing that the values in row i depend only on those in row $i - 1$. Moreover, the algorithm has the merit of being general-purpose: it can be used with any distance function d that satisfies the axioms (2.2)–(2.5) for a metric. In the following sections we shall see how imposing various kinds of restrictions on the domain of application of approximate pattern-matching algorithms enables us to use techniques that significantly reduce time complexity.

Exercises 10.1

1. For Levenshtein distance d_L, give an example of a pattern $\boldsymbol{p}[1..m]$ and a text $\boldsymbol{x}$ such that in (10.1) $\boldsymbol{c}[i, m]$ corresponds to a choice i' for which

(a) $i+i'-1<m$;

(b) $i+i'-1=m$;

(c) $i+i'-1>m$.

Then show that these three cases can occur whenever it is true that the cost of some substitution exceeds that of some indel.

2. Prove that for edit distance d_E, it may without loss of generality be assumed that in (10.1) $i+i'-1 \le m$.
3. Show that the maximum number of moves ν that transform a matching substring $\boldsymbol{u}$ into a pattern $\boldsymbol{p}$ is $2|\boldsymbol{p}|$.
4. Specify the algorithm, WF* say, of Theorem 10.1.1 that computes every integer i and a corresponding integer i' such that $d\big(\boldsymbol{x}[i'..i], \boldsymbol{p}\big) \le k$. Show that your algorithm can be implemented using $\Theta\big(\min\{mn, m^2\}\big)$ space. Show further that if output of i' is not required, then only $\Theta(m)$ space need be used.

10.2 An Algorithm for k-Mismatches

Man's word is God in man.

— Alfred, Lord TENNYSON (1809–1892), *Idylles of the King*

In this section we show how an algorithm already presented — Algorithm DBG of Section 7.4 — can easily be modified to compute k-approximate matches under Hamming distance (Section 2.2). In fact, this will be only the first of several applications of the DBG idea to various kinds of approximate pattern-matching, and so we will introduce it in a somewhat more general context than we have done earlier. Nevertheless, the specific application dealt with in this section is Hamming distance — what is called the ***k*-mismatches** problem.

We suppose that an integer $k \ge 0$ is given, along with a pattern $\boldsymbol{p} = \boldsymbol{p}[1..m]$ and a text string $\boldsymbol{x} = \boldsymbol{x}[1..n]$, both defined on an alphabet A that, as for the original DBG algorithm, we require to be finite, indexed, and known in advance. Thus we may assume (see Exercise 4.1.3) that $A = \{1, 2, \ldots, \alpha\}$.

The original DBG algorithm made use of a ***state vector*** $\boldsymbol{s} = \boldsymbol{s}[1..m]$ in which the value $\boldsymbol{s}[j]$ in position j always specified whether or not $\boldsymbol{p}[1..j]$ matched the current substring $\boldsymbol{x}[i-j+1..i]$ of $\boldsymbol{x}$. Since in Section 7.4 we were discussing exact matches, it sufficed that $\boldsymbol{s}$ should be a vector of bits, with $\boldsymbol{s}[j]=0$ corresponding to a match, $\boldsymbol{s}[j]=1$ to a mismatch. In the context of k-approximate pattern-matching under Hamming distance, however, we will need to count the *number* of mismatched positions between $\boldsymbol{p}[1..j]$ and the current substring of $\boldsymbol{x}$. Since we may need to record anything from 0 to m such mismatches, it therefore becomes necessary to have available as many as $\lceil \log_2(m+1) \rceil$ bits corresponding to each position in $\boldsymbol{s}$ rather than just the single bit of DBG.

Let B be the number of bits of storage required for each position in $\boldsymbol{s}$, and for the time being (we shall see later how to reduce it) let $B = \lceil \log_2(m+1) \rceil$. Instead of the original DBG invariant (7.13), then, we seek to maintain

$$\boldsymbol{s}[j] = d_H\big(\boldsymbol{x}[i-j+1..i], \boldsymbol{p}[1..j]\big) \tag{10.5}$$

for every $j \in 1..m$ as i successively takes the values $1, 2, \ldots, n$.

Recall that Algorithm DBG made use of an array $\boldsymbol{t} = \boldsymbol{t}[1..\alpha, 1..m]$, where for every letter $h \in 1..\alpha$ and every position $j \in 1..m$,

$$\begin{aligned} \boldsymbol{t}[h,j] &= 0, \text{ if } \boldsymbol{p}[j] = h; \\ &= 1, \text{ otherwise.} \end{aligned} \tag{10.6}$$

We define exactly the same array here also, but now we suppose that each element of $\boldsymbol{t}$ consists of, not just a single bit, but rather $B \geq 1$ bits. Thus each $\boldsymbol{t}[h,j]$ consists of 0 or 1 preceded by $B-1$ leading zeros. As we shall see, the reason for using this extra space is to permit sums $\boldsymbol{s} + \boldsymbol{t}[h, 1..m]$ to be easily and efficiently computed, the analogue of the logical OR $\boldsymbol{s} \vee \boldsymbol{t}[h, 1..m]$ used for Algorithm DBG. As shown in Exercise 10.2.2, the array $\boldsymbol{t}$ can be efficiently computed in a preprocessing step.

Using our new definitions (10.5) and (10.6), we can now state and prove a result closely analogous to Lemma 7.4.1, with $\vee$ replaced by ordinary addition $+$.

Lemma 10.2.1 *Let $\boldsymbol{s}^{(i)}$ be the string $\boldsymbol{s}$ that defines the state corresponding to position i in the text $\boldsymbol{x}$. Define the initial state $\boldsymbol{s}^{(0)}[0..m] = 0^B 0^B \cdots 0^B$. Then for every $i \in 1..n$ and every $j \in 1..m$,*

$$\boldsymbol{s}^{(i)}[j] = \boldsymbol{s}^{(i-1)}[j-1] \; + \; \boldsymbol{t}\big[\boldsymbol{x}[i], j\big]. \tag{10.7}$$

Proof The proof follows exactly the pattern of the proof of Lemma 7.4.1 and is left to Exercise 10.2.1. □

We remark that in fact all lemmas that take the form of Lemmas 10.2.1 or 7.4.1 will hold provided that the operation defined ($+$ or $\vee$ or whatever) ensures that every position $j-1$ of the previous state is correctly updated to yield position j of the current state. Thus Lemmas 10.2.1 and 7.4.1 may be thought of as special cases of a more general lemma in which the operation, say ***op***, though not specified, nevertheless satisfies this property. We shall suppose in future applications of the DBG approach that such a general lemma holds.

In this application we find that the similarity of Lemma 10.2.1 to Lemma 7.4.1 is reflected in similarity of the Approximate DBG algorithm (ADBG) described here to Algorithm DBG as described in Section 7.4. At each step the computation

$$\boldsymbol{s} \leftarrow \textit{rightshift}(\boldsymbol{s}, B) + \boldsymbol{t}[\boldsymbol{x}[i], 1..m] \tag{10.8}$$

is performed, where

- first the state vector $\boldsymbol{s}$ is shifted right by B bits (with B zeros introduced into element $\boldsymbol{s}[0]$), reflecting the transition from state $j-1$ to state j for every $j \in 1..m$;
- then the m positions $1..m$ of the shifted vector $\boldsymbol{s}$ are updated by adding the corresponding m positions of $\boldsymbol{t}$ determined by the current letter $\boldsymbol{x}[i]$ in $\boldsymbol{x}$.

The details of Algorithm ADBG are very similar to those of Algorithm DBG and are left to Exercise 10.2.3.

In order to evaluate the complexity of our new algorithm, we first observe that the space requirement for $\boldsymbol{s}[1..m]$ is mB bits while that for $\boldsymbol{t}$ is αmB bits. As for Algorithm DBG, we suppose that the computer word length is w bits; thus $\boldsymbol{s}[1..m]$ requires $\lceil mB/w \rceil$ words of storage, and $\boldsymbol{t}$ requires $\alpha\lceil mB/w \rceil$. We suppose further that arithmetic operations (such as *rightshift* and $+$) on individual computer words can be performed in constant time. Thus for each value of i, (10.8) will be executed in $\Theta(\lceil mB/w \rceil)$ time, and we can state formally the analogue of Theorem 7.4.3:

Theorem 10.2.2 *Given a text $\boldsymbol{x}[1..n]$, a pattern $\mathbf{p}[1..m]$ and a nonnegative integer k, Algorithm ADBG computes all k-approximate occurrences of $\mathbf{p}$ in $\boldsymbol{x}$ under Hamming distance in time $\Theta(\lceil mB/w \rceil n)$ using $\Theta(\alpha\lceil mB/w \rceil)$ additional space.* □

Since w is a constant, and since we have been obliged to choose $B \in \Theta(\log m)$, we see that ADBG is "essentially" a $\Theta(mn \log m)$ algorithm, not as efficient as we might have wished. Of course if the pattern length m is small, so that the quantity $\lceil mB/w \rceil$ is constrained to be a small integer, the claim could be made that in practice ADBG will behave like a $\Theta(n)$ algorithm. It turns out that we can increase the likelihood that mB/w will indeed be small by reducing the size of B to $\Theta(\log k)$.

We make two simple observations: first, it may be supposed that $k < m$, since for $k \geq m$ the k-mismatches problem is trivial; second, values $\boldsymbol{s}[j] > k$ do not need to be explicitly computed — when $\boldsymbol{s}[j]$ first exceeds k, it suffices simply to record that fact, since a mismatch with $\boldsymbol{p} = \boldsymbol{p}[1..m]$ is already assured no matter what further matches or mismatches occur with $\boldsymbol{p}[j+1..m]$. These observations suggest that the required information could be provided by storing only $\lceil \log_2(k+1) \rceil$ bits plus one overflow bit in each of the $m+1$ positions of $\boldsymbol{s}$ — hence also in each of the positions in the array $\boldsymbol{t}$. Thus a modified ADBG algorithm would satisfy Theorem 10.2.2 with $B = \lceil \log_2(k+1) \rceil + 1$, so that the asymptotic time requirement reduces to $\Theta(mn \log k)$. In fact, since as remarked in Section 7.4 $w \in \Omega(\log n)$, we may express the time requirement as $O(\lceil m \log k / \log n \rceil n)$. Note that now for $k = 0$, $B = 1$, and so in this case ADBG behaves just like Algorithm DBG, as one would anticipate.

To understand the computational mechanisms of the modified ADBG algorithm, we consider an example [BYG92], illustrated in Table 10.3. We look for all 2-approximate matches under Hamming distance of $\boldsymbol{p} = ababc$ in $\boldsymbol{x} = abdabababc$. Since $k = 2$, we need $B = 3$ bits for each position in $\boldsymbol{s}$, providing space for a mismatch count of up to 3 plus an overflow bit set to 1 when the number of mismatches exceeds 3. We suppose that every time the overflow bit becomes 1, it is stored in another vector, here called $\boldsymbol{s}'$. The vertical arrows in Table 10.3 point to the two positions in $\boldsymbol{x}$ at which a 2-approximate match of $\boldsymbol{p}$ terminates — more precisely, to the two positions at which $\boldsymbol{s}[5] + 2^2\boldsymbol{s}'[5] \leq 2$.

x	a	b	d	a	b	a	b	a	b	c
$t[x[i]]$	01011	10101	11111	01011	10101	01011	10101	01011	10101	11110
s	01011	10202	12131	02220	10323	02003	10301	02001	10301	12100
s'	01111	00111	00011	00001	00000	00010	00001	00010	00001	00010
								↑		↑

Table 10.3: Modified ADBG matches $p = ababc$ with $x = abdabababc$, $k = 2$

In general, the overflow bit in $s[j]$, $j = 1, 2, \ldots, m$, needs to be inspected at each step: a zero bit can be ignored, but a 1, indicating that an overflow has just occurred at position j, needs to be recorded in the overflow vector s'. After inspection, the overflow bit in $s[j]$ is reset to zero. Thus after this processing has been completed, we will always know that the current number of mismatches at each position j of p is at least

$$s[j] + 2^{B-1}s'[j].$$

Recall that the bitwise logical AND operator yields the result 1 if and only if both operands are zero. Then this processing can be implemented using the operation

$$op = \text{AND}$$

with a mask

$$0^B10^B1\cdots0^B1$$

to pick up overflow bits from s, and the mask's complement

$$1^B01^B0\cdots1^B0$$

to reset overflow bits to zero. As indicated in Table 10.3, s and s' will be shifted right together at each step, so that positions j in each vector are always synchronized. This enables us to use the test

$$s[m] + 2^{B-1}s'[m] \leq k?$$

to determine whether or not $p[1..m]$ has at most k mismatches with the current position in x.

The details of the modified ADBG are left as Exercise 10.2.3. Note that the storage requirements for s, s' and all masks are bounded by $O(\lceil mB/w \rceil)$ as required by Theorem 10.2.2.

Finally, to gain some insight into space and time requirements for Algorithm ADBG, we consider a pair of examples. Suppose that the computer word length $w = 32$. For an approximation constant $k \leq 7$, a value $B = 4$ suffices, so that for every pattern of length $m \leq 8$, $\lceil mB/w \rceil = 1$ — in these cases all the entries in $s[1..8]$ (and therefore also in $s'[1..8]$) fit into a single word, and we can claim that Algorithm ADBG executes in $\Theta(n)$ time. Alternatively, if we know that $k \leq 3$, we can achieve $B = 3$ and hence $\lceil mB/w \rceil = 1$ for every $m \leq 10$. For a more detailed analysis, consult Exercise 10.2.4.

Other Approaches

The first practical algorithm for the k-mismatches problem seems to have been proposed by Landau and Vishkin [LV85, LV86a]. Their algorithm makes use of a two-dimensional "pattern-mismatch" array determined during a preprocessing phase in order to be able to compute a two-dimensional "text-mismatch array" that specifies the k-approximate matches of $\boldsymbol{p}$ in $\boldsymbol{x}$. The time including preprocessing required by this algorithm is $O(k(n+m \log m))$: as shown in Exercise 10.2.5, this worst-case upper bound may be expected to be greater than the ADBG upper bound over a wide range of cases, and it appears that, in practice, ADBG is faster by an order of magnitude [BYG92]. The worst-case time of the LV algorithm was improved to $O(kn + m \log m)$ by Galil and Giancarlo [GG86] using suffix trees, but the revised algorithm is actually slower in practice [BYG92]. Another algorithm that depends on bit operations is described in [BYP92]: a count of the length of partial matches with $\boldsymbol{p}$ is maintained for every position i in $\boldsymbol{x}$. For an alphabet of size α with each letter equally likely to occur, this algorithm executes in *expected* time $\Theta(mn/\alpha)$ and so could provide some advantage over ADBG in cases where α is large ($\alpha \in \Omega(m)$ or $\alpha \in \Omega(n)$).

The Boyer-Moore approach (Section 7.2) has also been applied with some success to the k-mismatches problem. In [TU93] a generalized version of Algorithm BMH (Section 8.2) is described that computes all k-approximate occurrences of $\boldsymbol{p}$ in $\boldsymbol{x}$ under Hamming distance in expected time

$$O\left(k\left(\frac{1}{m-k}+\frac{k}{\alpha}\right)n\right),$$

thus possibly competitive with ADBG in cases where k is small with respect to m and α, and m is large with respect to word length w. Perhaps of greater interest is the proposal of El-Mabrouk and Crochemore [EC96] to augment Algorithm ADBG with Boyer-Moore exact pattern-matching capabilities so that when a partial match between $\boldsymbol{p}$ and $\boldsymbol{x}$ is found, fewer bit operations are performed. Especially for larger values of m and smaller values of k, this approach provides some speed-up in practice over ADBG.

Overall, it appears that the DBG approach and its variants provide the best results currently available for the k-mismatches problem.

Exercises 10.2

1. Prove Lemma 10.2.1.
2. Describe an algorithm that computes the array $\boldsymbol{t}$ in time $\Theta(\alpha mB/w)$.
3. Specify Algorithm ADBG, in both its original ($B = \lceil \log_2(m+1) \rceil$) and improved ($B = \lceil \log_2(k+1) \rceil + 1$) forms.
4. For each of the word lengths $w = 32$ and $w = 64$, construct a table that shows for $k = 0, 1, 3, 7, 15, 31, 63$ and for $\lceil mB/w \rceil \in 1..10$ the maximum corresponding pattern length m that can be processed by Algorithm ADBG.

5. For positive integers $n \geq m \geq k \geq 2$, show that

$$mn \log_2 k / \log_2 n < k(n + m \log_2 m)$$

provided $m \log_2 m/n \leq 2$. Hence draw conclusions about the asymptotic efficiency of Algorithm ADBG compared to that of Landau and Vishkin.

10.3 Algorithms for k-Differences

Like untuned golden strings all women are ...

— Christopher MARLOWE (1564–1593), *Hero and Leander*

In this section we consider approximate pattern-matching algorithms based on edit distance — algorithms to solve the so-called ***k-differences*** problem. Recall that for edit distance the costs of all single-letter indels and substitutions are positive and equal:

$$d_E(\varepsilon, \lambda) = d_E(\lambda, \varepsilon) = d_E(\lambda, \mu) > 0 \tag{10.9}$$

for every letter λ in the alphabet A and every letter $\mu \neq \lambda$. Of course, for $\mu = \lambda$, it follows from (2.3) that $d_E(\lambda, \lambda) = 0$. And since single-letter operation costs are equal, we may assume without loss of generality that the cost is 1. In view of (10.2), (10.3) and Lemma 9.1.1, these restrictions mean that we may suppose the cost array $\boldsymbol{c}$ for edit distance to take the form shown in Table 10.4, where by Lemma 9.1.1 the entries δ_{i1} in column 1 take the values

$$\begin{aligned} \delta_{i1} &= 0, \text{ if } \boldsymbol{x}[i] = \boldsymbol{p}[1]; \\ &= 1, \text{ otherwise.} \end{aligned} \tag{10.10}$$

We can now establish an important result, first proved in a more general form in [MP80], then later in this form in [U85b]:

	j	**0**	**1**	**2**	•	•	m
i		ε	$\boldsymbol{p}[1]$	$\boldsymbol{p}[2]$	·	·	$\boldsymbol{p}[m]$
0	ε	0	1	2	·	·	m
1	$\boldsymbol{x}[1]$	0	δ_{11}	·	·	·	·
2	$\boldsymbol{x}[2]$	0	δ_{21}	·	·	·	·
•	·	·	·	·	·	·	·
•	·	·	·	·	·	·	·
n	$\boldsymbol{x}[n]$	0	δ_{n1}	·	·	·	·

Table 10.4: Cost array for edit distance

Lemma 10.3.1 *In any cost array $\mathbf{c} = \mathbf{c}[0..n, 0..m]$ computed using edit distance:*

$$\begin{array}{ll} (a) & \mathbf{c}[i,j] - \mathbf{c}[i,j-1] \in \{-1,0,1\}; \\ (b) & \mathbf{c}[i,j] - \mathbf{c}[i-1,j] \in \{-1,0,1\}; \\ (c) & \mathbf{c}[i,j] - \mathbf{c}[i-1,j-1] \in \{0,1\}; \end{array}$$

for every $i \in 1..n$, $j \in 1..m$.

Proof Observe first that by Lemma 9.1.1 and (10.9), no entry $\mathbf{c}[i,j]$ can be greater by more than 1 than any of the entries

$$\mathbf{c}[i-1,j],\ \mathbf{c}[i,j-1],\ \mathbf{c}[i-1,j-1].$$

Since all the entries in $\mathbf{c}$ are integers, we therefore need only prove that the specified differences cannot be less than -1 (cases (a) and (b)) or less than 0 (case (c)).

We prove (a) by induction, noting from Table 10.4 that the result holds for $i = 1$ and supposing therefore that it holds for some $i \geq 1$ and all j. Suppose further that for $i+1$ and some least $j \geq 1$, (a) does not hold, so that

$$\mathbf{c}[i+1,j] - \mathbf{c}[i+1,j-1] < -1.$$

But then $\mathbf{c}[i+1,j-1] > \mathbf{c}[i+1,j] + 1$, and hence

$$\begin{aligned} \mathbf{c}[i+1,j] &= \min\{\mathbf{c}[i,j],\ \mathbf{c}[i,j-1],\ \mathbf{c}[i+1,j-1]\} + 1 \\ &= \min\{\mathbf{c}[i,j],\ \mathbf{c}[i,j-1]\} + 1, \end{aligned}$$

so that

$$\mathbf{c}[i+1,j-1] > \min\{\mathbf{c}[i,j],\ \mathbf{c}[i,j-1]\} + 2.$$

But since we know that for edit distance

$$\mathbf{c}[i,j-1] \geq \mathbf{c}[i+1,j-1] - 1, \tag{10.11}$$

it must therefore be true that

$$\mathbf{c}[i+1,j-1] > \mathbf{c}[i,j] + 2. \tag{10.12}$$

Then

$$\begin{aligned} \mathbf{c}[i,j-1] - \mathbf{c}[i,j] &\geq \mathbf{c}[i+1,j-1] - 1 - \mathbf{c}[i,j], \text{ by (10.11)} \\ &> 1, \text{ by (10.12)} \end{aligned}$$

contrary to Lemma 9.1.1 for edit distance. This contradiction tells us that there is no least j for which (a) does not hold, as required.

Condition (b) follows from (a) by virtue of the symmetric roles played by i and j in the calculation (10.4). The proof of (c) is left as Exercise 10.3.1. □

Armed with these insights, we are now in a position to consider two k-differences algorithms. The first, called Algorithm U, is due to Ukkonen [U85b]. It combines Lemma 10.3.1 with the approach of Algorithm UM (Section 9.5) to compute all k-approximate matches of $\boldsymbol{p}$ in $\boldsymbol{x}$ in expected time $O(kn)$. The second, Algorithm M, is a new algorithm due to Myers [M99] that initially uses the bit-vector approach of Algorithm DBG (Section 7.4) in the context of Lemma 10.3.1 to develop a k-differences algorithm with $\Theta(mn/w)$ worst-case execution time (independent of k); the ideas of Algorithm U are then employed to modify Algorithm M so that $O(kn/w)$ average-case execution time is achieved for the k-differences problem. Algorithm WM, a third algorithm for k-differences that also uses the DBG approach, will be presented in Section 10.4: even though WM requires $\Theta(kmn/w)$ time in the worst case, its range of application extends over a much wider spectrum of distance types than is covered by Algorithms U and M.

10.3.1 Ukkonen's Algorithm

"There are strings," said Mr. Tappertit,
". . . in the human heart that had better not be wibrated."

— Charles DICKENS (1812–1890), *Barnaby Rudge*

The Ukkonen-Myers string-distance algorithm (Section 9.5) achieved worst-case time $O(dn)$, where d was the actual distance between the given strings, by reducing to $O(d)$ the number of diagonals in the cost array $\mathbf{c}$ that needed to be considered in the calculation. A similar strategy is used by the Ukkonen algorithm described here [U85b] to solve the k-differences problem in average-case time $O(kn)$. The basis of the algorithm is Lemma 10.3.1(c), the fact that diagonal elements in $\mathbf{c}$ under edit distance form a nondecreasing sequence. Thus if we record the rightmost position j' in row i for which $\mathbf{c}[i,j'] \leq k$, we will then be assured that every position $j = j'+2, j'+3, \ldots, m$ in row $i+1$ satisfies $\mathbf{c}[i+1,j] > k$. So the rightmost $m-j'-1$ positions in row $i+1$ of $\mathbf{c}$ do not need to be computed, since they cannot determine a k-approximate match for $\boldsymbol{p}$. More generally, we see that in all rows of $\mathbf{c}$ greater than i, we may ignore all diagonals that occur to the right of diagonal $\Delta_{i,j'}$. This simple strategy is the basis of Algorithm 10.3.1, shown below.

Algorithm 10.3.1 (U)

– *Compute all k-approximate matches of $\mathbf{p}[1..m]$ in $\boldsymbol{x}[1..n]$*

initialize row 0 and column 0 of $\mathbf{c}$ (see Table 10.4)

– *In row 0, the rightmost occurrence of k is in column $j' = k$*
$j' \leftarrow k$
for $i \leftarrow 1$ **to** n **do**

$j'' \leftarrow 0$ – *ensures correct execution when* $k = 0$
for $j \leftarrow 1$ **to** $\min\{j'+1, m\}$ **do**
compute $\boldsymbol{c}[i,j]$ using (10.4)
if $\boldsymbol{c}[i,j] \le k$ **then** $j'' \leftarrow j$

– *Output an approximate match if found*
if $j'' = m$ **then output** i
– *Save rightmost position* j' *such that* $\boldsymbol{c}[i,j'] \le k$
$j' \leftarrow j''$

The time complexity of Algorithm U is clearly proportional to the sum of the n values assumed by the variable j' during the algorithm's execution. In the worst case, of course, $\boldsymbol{c}[i,m] \le k$ for every row $i \in 1..n$, so that $j' = m$ for every row and Algorithm U therefore requires $\Theta(mn)$ time. Indeed, even if $\boldsymbol{c}[i,m/2] \le k$ for $\Theta(n)$ rows, for example, Algorithm U will still require $\Theta(mn)$ time.

In order to determine the average-case complexity, we need to estimate the expected value of j' for each row i of $\boldsymbol{c}$ corresponding to the prefixes $\boldsymbol{x}[1..i]$ of $\boldsymbol{x}$. We can estimate $\exp[j']$ by computing the expected edit distance from a prefix $\boldsymbol{p}[1..j]$ of a given pattern $\boldsymbol{p}$ to a randomly-chosen string $\boldsymbol{x}$ of length at most j. [U85b] states that it "should be quite obvious" that $\exp[j'] \in O(k)$, a sentiment in which all right-thinking lexicologists will undoubtedly concur. Alas, it appears that it is *not* obvious, and in fact the result was not proved till 1992, by Chang and Lampe [ChL92].

Their proof begins with two lemmas, the first a simple property of edit distance, the second rather technical.

Lemma 10.3.2 *For all integers* $j \in 1..m$ *and* $i \in j..n$,

$$\boldsymbol{c}[i,j] \ge d_E\big(\boldsymbol{x}[i-j+1..i], \boldsymbol{p}[1..j]\big)/2.$$

Proof Let $d_E\big(\boldsymbol{x}[i-j+1..i], \boldsymbol{p}[1..j]\big) = d$ and recall that

$$\boldsymbol{c}[i,j] = d_E\big(\boldsymbol{x}[i'..i], \boldsymbol{p}[1..j]\big) = d' \le d$$

for some $i' \in i-j+1..i$. By the triangle inequality (2.5),

$$d - d' \le d_E\big(\boldsymbol{x}[i-j-1..i], \boldsymbol{x}[i'..i]\big). \tag{10.13}$$

Now we make the following observation:

> The edit distance between any two strings of lengths n_1 and $n_2 \ge n_1$, respectively, is at least $n_2 - n_1$, a minimum that is attained when the first string is a suffix of the second.

We therefore conclude that

$$d_E\big(\boldsymbol{x}[i-j-1..i], \boldsymbol{x}[i'..i]\big) \le d_E\big(\boldsymbol{p}[1..j], \boldsymbol{x}[i'..i]\big) = d'. \tag{10.14}$$

Combining (10.13) and (10.14), the result follows. □

Lemma 10.3.3 *Let $p_{j,r}$ be the probability that two random strings of length $j > 0$ have a common subsequence of length $\lceil rj \rceil$, $r < 1$. Then for given j and every choice of r satisfying $7/8 \le j < 1$, there exist positive constants s and $t < 1$ such that*

$$p_{j,r} < st^j/j.$$

Proof Omitted: see [ChL92]. □

In order to make use of these two lemmas, we begin by defining the quantity $j'' = 2k/(1-r)$, where $7/8 \le r < 1$ as required by Lemma 10.3.3. Observe that $j - 2k \ge rj$ for all integers $j \ge j''$. Since moreover $j'' > 2$ we can bound $\exp[j']$ as follows:

$$\exp[j'] < (j''-1) + \sum_{j \ge j''} \Big(j \times \text{prob}\big[\boldsymbol{c}[i,j] \le k\big] \Big). \tag{10.15}$$

In this expression, the condition $\boldsymbol{c}[i,j] \le k$ implies by Lemma 10.3.2 that

$$d_E\big(\boldsymbol{x}[i-j+1..i], \boldsymbol{p}[1..j]\big) \le 2k.$$

However, as shown in Exercise 10.3.7,

$$d_E\big(\boldsymbol{x}[i-j+1..i], \boldsymbol{p}[1..j]\big) \ge j - \big|\text{LCS}\big((\boldsymbol{x}[i-j+1..i], \boldsymbol{p}[1..j]\big)\big|, \tag{10.16}$$

and so we conclude that $\boldsymbol{c}[i,j] \le k$ implies

$$\text{LCS}\big(\boldsymbol{x}[i-j+1..i], \boldsymbol{p}[1..j]\big) \ge j - 2k,$$

where as we have seen, for $j \ge j''$, $j - 2k \ge rj$. Then for $j \ge j''$,

$$\begin{aligned} \text{prob}[\boldsymbol{c}[i,j] \le k] &\le \text{prob}[\text{LCS}(\boldsymbol{x}[i-j+1..i], \boldsymbol{p}[1..j]) \ge rj] \\ &\le \text{prob}[\text{LCS}(\boldsymbol{x}[i-j+1..i], \boldsymbol{p}[1..j]) \ge \lceil rj \rceil] \\ &< st^j/j \end{aligned} \tag{10.17}$$

by Lemma 10.3.3, for appropriate choices of positive constants s and $t < 1$. Substituting (10.17) in (10.15), we find

$$\begin{aligned} \exp[j'] &< (j''-1) + \sum_{j \ge j''} (jst^j/j) \\ &= (j''-1) + O(1) \\ &\in O(k). \end{aligned}$$

Thus Algorithm U requires $O(kn)$ time on average. Since, just as for Algorithm WF, we can implement Algorithm U using storage for only two rows of $\boldsymbol{c}$, we have thus established

Theorem 10.3.4 *Algorithm 10.3.1 solves the k-differences problem for a pattern $\boldsymbol{p}[1..m]$ and a text string $\boldsymbol{x}[1..n]$ in average-case time $O(kn)$ using $\Theta(m)$ additional space.* □

Based on the proof given here, the $O(kn)$ average-case upper bound is not entirely satisfactory: even though $j'' \in O(k)$, still $j'' \geq 16k$, so the constant of proportionality could as far as we know be quite large. In practice, however, Algorithm U seems to perform quite efficiently.

Three years after its original publication Algorithm U was modified to execute in worst-case time $O(k^2n)$ and $O(n)$ space using suffix trees [LV88], then later to execute in worst-case time $O(kn)$ and $O(m^2)$ space [LV89, GP90]. In the 1990s further variants have been proposed [ChL92, UW93, WMM96] culminating in a modified version of Algorithm M [M99], the method to be considered next. As indicated in Exercises 10.3.4 and 10.3.5, Algorithm U can be redesigned to deal with Levenshtein, rather than edit, distance.

10.3.2 Myers' Algorithm

Like two doomed ships that pass in storm
We had crossed each other's way;
But we made no sign, we said no word,
We had no word to say.

— Oscar WILDE (1854–1900), *The Ballad of Reading Gaol*

In this subsection we present an algorithm that applies bit vectors in an ingenious way based on the special properties of edit distance (Lemma 10.3.1) in order to achieve very high efficiency — in fact, worst-case processing time that is independent of the approximation constant k. As with all bit-vector algorithms, we suppose that the alphabet is indexed — a requirement normally satisfied in practice — and so may be supposed to take the form $A = \{1, 2, \ldots, \alpha\}$.

The Difference Matrices Δ and Δ'

The first step in describing Algorithm M is to show that for edit distance the cost matrix $\boldsymbol{c}$ can be replaced by a difference matrix $\boldsymbol{\Delta}$ — actually by two difference matrices $\boldsymbol{\Delta}$ and $\boldsymbol{\Delta}'$. We begin by defining row vectors $\boldsymbol{\Delta}_i = \boldsymbol{\Delta}_i[1..m]$ for ***horizontal differences***, $i = 1, 2, \ldots, n$, where

$$\boldsymbol{\Delta}_i[j] = \boldsymbol{c}[i,j] - \boldsymbol{c}[i,j-1], \tag{10.18}$$

for every $j \in 1..m$; and row vectors $\boldsymbol{\Delta}'_i = \boldsymbol{\Delta}'_i[1..m]$ for ***vertical differences***, $i = 1, 2, \ldots, n$, where

$$\boldsymbol{\Delta}'_i[j] = \boldsymbol{c}[i,j] - \boldsymbol{c}[i-1,j], \tag{10.19}$$

for every $j \in 1..m$. Recall from Table 10.4 that for edit distance both row 0 and column 0 of $\boldsymbol{c}$ are known:

$$\boldsymbol{c}[0,j] = j, \text{ for every } j \in 0..m; \tag{10.20}$$
$$\boldsymbol{c}[i,0] = 0, \text{ for every } i \in 0..n. \tag{10.21}$$

Thus (10.18) together with (10.21) can be used to define a ***difference matrix*** $\boldsymbol{\Delta} = \boldsymbol{\Delta}[1..n, 1..m]$, where

$$\boldsymbol{\Delta}[i,j] = \boldsymbol{\Delta_i}[j], \tag{10.22}$$

$i = 1, 2, \ldots, n$, $j = 1, 2, \ldots, m$. Similarly we use (10.19) together with (10.20) to define

$$\boldsymbol{\Delta'}[i,j] = \boldsymbol{\Delta'_i}[i]. \tag{10.23}$$

It will be convenient later on to extend these definitions to include row 0 and column 0 of both $\boldsymbol{\Delta}$ and $\boldsymbol{\Delta'}$:

$$\boldsymbol{\Delta_0}[j] = \boldsymbol{\Delta'_0}[j] = 0, \quad \forall j \in 1..m; \tag{10.24}$$
$$\boldsymbol{\Delta_i}[0] = \boldsymbol{\Delta'_i}[0] = 0, \quad \forall i \in 1..n. \tag{10.25}$$

What actually happens in Algorithm M is that the row vectors (10.18) and (10.19) are successively computed by an iteration that takes

$$\boldsymbol{\Delta_{i-1}} \to \boldsymbol{\Delta_i}, \;\; \boldsymbol{\Delta'_{i-1}} \to \boldsymbol{\Delta'_i},$$

$i = 1, 2, \ldots, n$, thereby ultimately computing all the elements (10.22) and (10.23) of $\boldsymbol{\Delta}$ and $\boldsymbol{\Delta'}$. Of course our objective is to determine $\boldsymbol{c}[i,m]$ for every value of i in order to be able to decide whether or not $\boldsymbol{c}[i,m] \le k$. Observe however from (10.19) that

$$\boldsymbol{c}[i,m] \;=\; \boldsymbol{c}[0,m] + \sum_{i'=1}^{i} \boldsymbol{\Delta'_{i'}}[j] \;=\; m + \sum_{i'=1}^{i} \boldsymbol{\Delta'_{i'}}[j]. \tag{10.26}$$

Thus we can replace the calculation of $\boldsymbol{c}$ by the calculation of the difference matrices $\boldsymbol{\Delta}$ and $\boldsymbol{\Delta'}$; in fact, more precisely, by the calculation of the row vectors $\boldsymbol{\Delta_i}$ and $\boldsymbol{\Delta'_i}$.

The advantage gained from substituting the $\boldsymbol{\Delta}/\boldsymbol{\Delta'}$ problem for the $\boldsymbol{c}$ problem is that the elements of $\boldsymbol{\Delta}$ and $\boldsymbol{\Delta'}$ can, by virtue of Lemma 10.3.1(a)–(b), take only values in the set $\{-1, 0, 1\}$. If we can find a way to compute these values efficiently and as a byproduct compute (10.26), we might then be able to take full advantage of the special properties of edit distance embedded in Lemma 10.3.1 to reduce overall time for the solution of the k-differences problem. This is exactly what Algorithm M achieves.

To gain an understanding of the relationship between the cost matrix $\boldsymbol{c}$ for edit distance and the corresponding difference matrices $\boldsymbol{\Delta}$ and $\boldsymbol{\Delta'}$, compare Tables 10.5 and 10.6. The cost matrix for the same example using Levenshtein distance can be found in Table 10.2.

	j	**0**	**1**	**2**	**3**	**4**	**5**	**6**
i		ε	s	t	r	e	s	s
0	ε	0	1	2	3	4	5	6
1	r	0	1	2	2	3	4	5
2	e	0	1	2	3	2	3	4
3	s	0	0	1	2	3	2	3
4	t	0	1	0	1	2	3	3
5	s	0	0	1	1	2	2	3

Table 10.5: Cost matrix $\boldsymbol{c}$ for $\boldsymbol{x} = rests$ and $\boldsymbol{p} = stress$ using d_E

	j	**1**	**2**	**3**	**4**	**5**	**6**
i		s	t	r	e	s	s
1	r	1	1	0	1	1	1
2	e	1	1	1	−1	1	1
3	s	0	1	1	1	−1	1
4	t	1	−1	1	1	1	0
5	s	0	1	0	1	0	1

$\boldsymbol{\Delta}$ (horizontal differences)

	j	**1**	**2**	**3**	**4**	**5**	**6**
i		s	t	r	e	s	s
1	r	0	0	0	−1	−1	−1
2	e	0	0	0	1	−1	−1
3	s	0	−1	−1	−1	1	−1
4	t	0	1	−1	−1	−1	0
5	s	0	−1	1	0	0	0

$\boldsymbol{\Delta}'$ (vertical differences)

Table 10.6: Difference matrices for $\boldsymbol{x} = rests$ and $\boldsymbol{p} = stress$

A Two-Stage Recurrence

The next step is to see how to compute the row vectors $\boldsymbol{\Delta_i}$ and $\boldsymbol{\Delta'_i}$. We first consider $\boldsymbol{\Delta_i}$, applying (10.18) and the basic recurrence (10.4) to find that for $i \in 1..n$, $j \in 1..m$,

$$\begin{aligned}\boldsymbol{\Delta_i}[j] &= \boldsymbol{c}[i,j] - \boldsymbol{c}[i,j-1] \\ &= \min\Big\{\boldsymbol{c}[i-1,j]+1,\ \boldsymbol{c}[i,j-1]+1,\ \boldsymbol{c}[i-1,j-1]+d\big(\boldsymbol{x}[i],\boldsymbol{p}[j]\big)\Big\} \\ &\qquad - \boldsymbol{c}[i,j-1].\end{aligned}$$

As for Algorithms DBG and ADBG, we may assume that $d(\boldsymbol{x}[i],\boldsymbol{p}[j])$ is available by accessing position $[\boldsymbol{x}[i],j]$ of the precomputed bit array $\boldsymbol{t} = \boldsymbol{t}[1..\alpha,1..m]$ (10.6) whose elements are computable in $\Theta(\alpha\lceil m/w\rceil + m)$ time (Theorem 7.4.4). For brevity, let us abbreviate $t_{ij} = \boldsymbol{t}[\boldsymbol{x}[i],j]$. Then after some manipulation based on (10.18), (10.19), (10.24) and (10.25), we find a formula for horizontal differences:

$$\boldsymbol{\Delta_i}[j] = \min\big\{\boldsymbol{\Delta_{i-1}}[j],\ \boldsymbol{\Delta'_i}[j-1],\ t_{ij}-1\big\} + \big(1 - \boldsymbol{\Delta'_i}[j-1]\big), \tag{10.27}$$

valid for every $i \in 1..n$, $j \in 1..m$. Similarly, for vertical differences:

$$\boldsymbol{\Delta'_i}[j] = \min\big\{\boldsymbol{\Delta_{i-1}}[j],\ \boldsymbol{\Delta'_i}[j-1],\ t_{ij}-1\big\} + \big(1 - \boldsymbol{\Delta_{i-1}}[j]\big). \tag{10.28}$$

Thus each of $\boldsymbol{\Delta}[i,j]$ and $\boldsymbol{\Delta'}[i,j]$ is computed based only on $\boldsymbol{\Delta}[i-1,j]$ and $\boldsymbol{\Delta'}[i,j-1]$ together with the precomputed t_{ij}. The equations (10.27) and (10.28) constitute a two-stage recurrence that forms the basis of Algorithm M.

A Bitwise Formulation

In order to express the calculations (10.27) and (10.28) in terms of bit operations, we must first express the variables themselves in terms of bits. To do so, we introduce four new bit vectors that encode the plus (P) and minus (M) values of the vectors $\boldsymbol{\Delta_i}$ and $\boldsymbol{\Delta'_i}$. Specifically, for $\boldsymbol{\Delta_i}$ we define bit vectors $\boldsymbol{P_i}[1..m]$ and $\boldsymbol{M_i}[1..m]$, where for every $j \in 1..m$,

$$\boldsymbol{P_i}[j] = 1 \text{ iff } \boldsymbol{\Delta_i}[j] = 1, \ \ \boldsymbol{M_i}[j] = 1 \text{ iff } \boldsymbol{\Delta_i}[j] = -1;$$

while for $\boldsymbol{\Delta'_i}$ we define bit vectors $\boldsymbol{P'_i}[1..m]$ and $\boldsymbol{M'_i}[1..m]$, where

$$\boldsymbol{P'_i}[j] = 1 \text{ iff } \boldsymbol{\Delta'_i}[j] = 1, \ \ \boldsymbol{M'_i}[j] = 1 \text{ iff } \boldsymbol{\Delta'_i}[j] = -1.$$

Since $\boldsymbol{\Delta_i}[j] = 0$ iff $\boldsymbol{P_i}[j] = \boldsymbol{M_i}[j] = 0$, and $\boldsymbol{\Delta'_i}[j] = 0$ iff $\boldsymbol{P'_i}[j] = \boldsymbol{M'_i}[j] = 0$, this effectively enables us to replace the calculation of $\boldsymbol{\Delta_i}$ and $\boldsymbol{\Delta'_i}$ by the calculation of the bit vectors $\boldsymbol{P_i}$, $\boldsymbol{M_i}$, $\boldsymbol{P'_i}$, $\boldsymbol{M'_i}$.

Consider first the horizontal row vectors (10.27). We claim that a computation based on this identity can be specified in terms of operations on single bits as follows:

$$\begin{aligned} \boldsymbol{K}[j] &\leftarrow \overline{t_{ij}} \ \vee \ \boldsymbol{M_{i-1}}[j] \\ \boldsymbol{P_i}[j] &\leftarrow \boldsymbol{M'_i}[j-1] \ \vee \ \overline{(\boldsymbol{K}[j] \ \vee \ \boldsymbol{P'_i}[j-1])} \\ \boldsymbol{M_i}[j] &\leftarrow \boldsymbol{P'_i}[j-1] \ \wedge \ \boldsymbol{K}[j] \end{aligned} \tag{10.29}$$

where

- $\boldsymbol{K} = \boldsymbol{K}[1..m]$ is a bit vector introduced for convenience of calculation;
- $\vee$ denotes bitwise logical OR (result is 0 if and only if *both* operands are 0);
- $\wedge$ denotes bitwise logical AND (result is 0 if and only if *any* operand is 0);
- $\overline{<\text{exp}>}$ denotes the bitwise complement of the expression $<$exp$>$.

In order to check this claim, first observe that in (10.27) there are only three variables on the righthand side, two of which $\big(\boldsymbol{\Delta_{i-1}}[j]$ and $\boldsymbol{\Delta'_i}[j-1]\big)$ take values in $\{-1, 0, 1\}$ while the third (t_{ij}) is a single bit. It follows that there are only $3 \times 3 \times 2 = 18$ distinct calculations performed by (10.27). Consider for example one of these:

$$\boldsymbol{\Delta_{i-1}}[j] = 1, \ \boldsymbol{\Delta'_i}[j-1] = -1, \ t_{ij} = 1$$

with corresponding values in (10.29):

$$\boldsymbol{M_{i-1}}[j] = 0, \ \boldsymbol{M'_i}[j-1] = 1, \ \boldsymbol{P'_i}[j-1] = 0, \ \overline{t_{ij}} = 0.$$

Applying (10.27) we find $\boldsymbol{\Delta_i}[j] = 1$ while from (10.29) we find $\boldsymbol{K}[j] = 0$, $\boldsymbol{P_i}[j] = 1$, $\boldsymbol{M_i}[j] = 0$, as expected. We leave verification of the remaining 17 cases to Exercise 10.3.8!

By symmetry we may write down the corresponding reformulated version for the vertical row vectors (10.28):

$$\begin{aligned} \boldsymbol{K'}[j] &\leftarrow \overline{t_{ij}} \ \vee \ \boldsymbol{M'_i}[j-1] \\ \boldsymbol{P'_i}[j] &\leftarrow \boldsymbol{M_{i-1}}[j] \ \vee \ \overline{(\boldsymbol{K'}[j] \ \vee \ \boldsymbol{P_{i-1}}[j])} \\ \boldsymbol{M'_i}[j] &\leftarrow \boldsymbol{P_{i-1}}[j] \ \wedge \ \boldsymbol{K'}[j], \end{aligned} \tag{10.30}$$

where again $\boldsymbol{K'} = \boldsymbol{K'}[1..m]$ is a bit vector introduced for convenience.

Cell Logic

If we suppose that $\boldsymbol{K}[j]$ has been calculated, we can think of (10.29) as a calculation of the horizontal difference for cell $[i, j]$ of $\boldsymbol{c}$ based on knowledge of the vertical difference for cell $[i, j-1]$. Similarly, if the value of $\boldsymbol{K'}[j]$ is given, we can regard (10.30) as computing the vertical difference for $[i, j]$ based on the horizontal difference for cell $[i-1, j]$. These dependencies are illustrated in Figure 10.1, where we show $\boldsymbol{\Delta}$ and $\boldsymbol{\Delta'}$ merged, with each cell containing both a horizontal (h) and vertical (v) difference.

Thus, setting aside for the moment the calculation of K and K', we can make use of the initial values in row 0 and column 0 to compute the elements of $\boldsymbol{\Delta}$ and $\boldsymbol{\Delta'}$ row by row. At the ith step:

- the elements of $\boldsymbol{\Delta_{i-1}}[j]$ are used to compute $\boldsymbol{\Delta'_i}[j]$ (horizontal to vertical);
- the elements of $\boldsymbol{\Delta'_i}[j-1]$ are used to compute $\boldsymbol{\Delta_i}[j]$ (vertical to horizontal).

At the same time, by initializing a variable $cost \leftarrow m$, we can use (10.26) to compute the value of $\boldsymbol{c}[i, m]$ at step i:

$$cost \leftarrow cost + \boldsymbol{\Delta'_{i-1}}[m]. \tag{10.31}$$

Exactly as with Algorithms DBG and ADBG, a critically important consequence of the use of bit vectors is that the bit operations do not need to be performed individually: standard operations available on computer words can be performed so as to process w bits with a single constant-time instruction, where w is the word length, usually 32 or 64 bits. In this formulation, use of a standard *rightshift* operation enables positions $j-1$ to be accessed. Thus individual positions in the bit vectors do not need to be addressed, so that references to j disappear. Further, we assume that for $m > w$, each bit vector is stored in $\lceil m/w \rceil$ adjacent words, on each of which a standard operation needs to be performed just once. Then we can

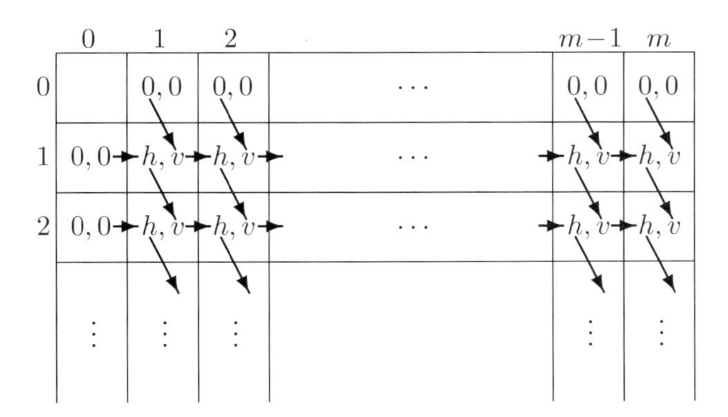

Figure 10.1. Computing horizontal differences from vertical ones, and *vice versa*

try to combine and reformulate (10.29), (10.30) and (10.31) as follows, making use now of the notation $t_i = \boldsymbol{t}[\boldsymbol{x}[i], 1..m]$:

$$\begin{array}{lll} (1) & \boldsymbol{K'} & \leftarrow \overline{t_i} \vee \textit{rightshift}(\boldsymbol{M}'_i, 1) \\ (2) & \boldsymbol{P}'_i & \leftarrow \boldsymbol{M}_{i-1} \vee \overline{(\boldsymbol{K'} \vee \boldsymbol{P}_{i-1})} \\ (3) & \boldsymbol{M}'_i & \leftarrow \boldsymbol{P}_{i-1} \wedge \boldsymbol{K'} \\ (4) & \textit{cost} & \leftarrow \textit{cost} + \boldsymbol{P}'_i[m] - \boldsymbol{M}'_i[m] \\ (5) & \boldsymbol{K} & \leftarrow \overline{t_i} \vee \boldsymbol{M}_{i-1} \\ (6) & \boldsymbol{P}_i & \leftarrow \textit{rightshift}(\boldsymbol{M}'_i, 1) \vee \overline{(\boldsymbol{K} \vee \textit{rightshift}(\boldsymbol{P}'_i, 1))} \\ (7) & \boldsymbol{M}_i & \leftarrow \textit{rightshift}(\boldsymbol{P}'_i, 1) \wedge \boldsymbol{K} \end{array} \tag{10.32}$$

The second and third lines of (10.32) compute the current values of $\boldsymbol{P'}$ and $\boldsymbol{M'}$ from $\boldsymbol{K'}$ together with the previous values of $\boldsymbol{P}$ and $\boldsymbol{M}$; this enables *cost* to be computed in the fourth line; then in lines (6) and (7) the current values of $\boldsymbol{P}$ and $\boldsymbol{M}$ are computed from $\boldsymbol{K}$ together with the shifted current values of $\boldsymbol{P'}$ and $\boldsymbol{M'}$. Clearly this calculation can be carried out using storage only for single variables $\boldsymbol{K}, \boldsymbol{P}, \boldsymbol{M}, \boldsymbol{K'}, \boldsymbol{P'}, \boldsymbol{M'}$, each a bit vector of length m, while the time required for the execution of lines (1)–(3) and (5)–(7) will be $\Theta(\lceil m/w \rceil)$. Line (4) executes in constant time. Thus the space requirement is $\Theta(\lceil m/w \rceil)$ words and the overall time requirement for n steps is $\Theta(\lceil m/w \rceil n)$. It is assumed in the *rightshift* operation that a 0 bit is shifted into position $j = 1$.

Unfortunately the calculation (10.32) does not quite work! The difficulty lies with the calculation of $\boldsymbol{K'}$ in line (1): it depends on the availability of the bit vector $\boldsymbol{M}'_i$ that is not actually calculated until line (3). A rather technical and lengthy argument can be invoked [M99] to persuade us that $\boldsymbol{K'}$ can instead be computed as follows:

$$\boldsymbol{K'} \leftarrow \overline{t_i} \vee \Big(\big((\overline{t_i} \wedge \boldsymbol{P}_{i-1}) + \boldsymbol{P}_{i-1}\big) \;\hat{}\; \boldsymbol{P}_{i-1}\Big), \tag{10.33}$$

where ^ denotes bitwise exclusive OR (result is 0 if and only if both operands are the same). We omit the proof of (10.33) but use it nevertheless to replace line (1) of (10.32), thus completing the specification of the main steps of Algorithm M. We leave the details of the algorithm to Exercise 10.3.10 and state formally the main result of this subsection:

Theorem 10.3.5 *As described in (10.32) and (10.33), Algorithm M correctly computes all k-approximate matches under edit distance of a pattern $\boldsymbol{p} = \boldsymbol{p}[1..m]$ in a text string $\boldsymbol{x} = \boldsymbol{x}[1..n]$ in time $\Theta(\lceil m/w \rceil n)$ using $\Theta(\lceil m/w \rceil)$ additional space.* □

Since we may assume, as remarked in Section 7.4, that $w \in \Omega(\log n)$, the upper bounds of Theorem 10.3.5 may also be expressed in the alternative forms $O(kn/\log n)$ and $O(mn/\log n)$ respectively.

Enhancement

In the preceding subsection, we presented Algorithm U that achieves $O(kn)$ expected time performance by taking advantage of the fact (Lemma 10.3.1(c)) that diagonal entries in the cost array $\boldsymbol{c}$ are nondecreasing. Essentially, Algorithm U records the rightmost column j' in each row i at which $\boldsymbol{c}[i, j'] \leq k$, then limits the processing in row $i+1$ to $j'+1$ columns. Since the expected value of j' is $O(k)$, the expected time required to process each row is also $O(k)$.

In [M99] it is shown that the same strategy can be applied to modify Algorithm M, thus limiting to $O(k)$ the expected number of bits that need to be processed in each row. If this enhanced algorithm is called Algorithm M*, we have then

Theorem 10.3.6 *Algorithm M* computes all k-approximate matches under edit distance of a pattern $\boldsymbol{p} = \boldsymbol{p}[1..m]$ in a text string $\boldsymbol{x} = \boldsymbol{x}[1..n]$ in expected time $\Theta(\lceil k/w \rceil n)$ and worst-case time $O(\lceil m/w \rceil n)$.* □

Efficiency Considerations

Apart from straightforward processing of the cost matrix $\boldsymbol{c}$ (such as the algorithm described in Section 10.1), the earliest nontrivial k-differences algorithm appears to be due to Landau and Vishkin [LV86b], also the first proposers of a practical k-mismatches algorithm. Their algorithm has been enhanced, notably by Chang and Lawler [ChL90, ChL94], to yield so-called "filter" algorithms that execute in $O(kn+m)$ worst-case time and as little as $O\left(\frac{k \log m}{m} n\right)$ average-case time for values of $k < m/(\log m + O(1))$. In these cases such algorithms can be competitive with Algorithm M*.

Another approach to filtering has been suggested by Ukkonen [U92b] based on alternative definitions of distance [EH88, U92b] that have an important property: they give rise to efficient k-approximation algorithms that identify a set of substrings of $\boldsymbol{x}$ guaranteed to include all the solutions to the k-differences problem. Once identified, the elements of the set can be efficiently processed to eliminate substrings that are not solutions of the k-differences problem.

In [M99] experiments are described in which Algorithm M* is compared with other more recent k-differences algorithms — in particular, [ChL92, WMM96, BYN96] where [ChL92] is derived from Algorithm U and [WMM96, BYN96] are bit-vector approaches whose techniques extend those described in [BYG92] and [WM92]. (In fact, the interesting algorithm of [BYN96] is described below, in Subsection 11.2.4, as a leading algorithm for multiple approximate pattern-matching.) Over a wide range of test cases, M* is found to execute faster, thus confirming in practice the theoretical improvement in asymptotic complexity stated by Theorem 10.3.6.

For approximate pattern-matching restricted to edit distance, Algorithm M* appears to be the current algorithm of choice. In the next section, however, we study an algorithm that, although not as fast either in theory or in practice as Algorithm M* for edit distance, nevertheless has a much wider range of application.

Exercises 10.3

1. Prove Lemma 10.3.1(c).

2. Prove the correctness of Algorithm U.
3. Try to characterize cases in which Algorithm U requires $\Theta(mn)$ time for its execution.
4. State and prove an analogue of Lemma 10.3.1 for Levenshtein distance d_L.
5. In view of the preceding exercise, design an analogue of Algorithm U based on Levenshtein distance.
6. Does Lemma 10.3.2 hold also for other kinds of distance? Extend its scope as far as you can and then prove your result, making clear the assumptions on which it is based.
7. Prove (10.16).
8. Prove that equations (10.27) and (10.28) are correct.
9. Verify the equivalence of (10.27) and (10.29).
10. Specify Algorithm M.
11. Discuss the problems that arise in recasting Algorithm M to handle Levenshtein distance instead of edit distance.

10.4 A Fast and Flexible Algorithm — Wu and Manber

Three merry boys, and three merry boys,
And three merry boys are we,
As ever did sing in a hempen string
Under the gallows tree.

— John FLETCHER (1579–1625), *Rollo, Duke of Normandy*

In Section 10.2 we described a straightforward generalization of Algorithm DBG (Section 7.4) that provided an efficient solution of the k-mismatches problem. Here we describe another equally straightforward, though technically more complicated, extension of the DBG approach that provides an efficient solution to approximate pattern-matching problems over a wide range of distance types, including in particular the k-differences problem discussed in Section 10.3 as well as wild-card and regular expression matching (Section 2.2). As we shall see, for most of these distance types, Algorithm WM executes in worst-case time $O(k\lceil m/w\rceil n)$, where w is the length of the computer word [WM92] and $k \geq 1$ the given approximation bound. Although not as fast for k-differences as Algorithm M*, WM is however much more flexible.

Like most other approximate pattern-matching algorithms, Algorithm WM originates in the recurrence relation (10.4) for the nonnegative cost matrix $\boldsymbol{c} = \boldsymbol{c}[0..n, 1..m]$ introduced in Section 9.1:

$$\boldsymbol{c}[i,j] \leftarrow \min\Big\{ \boldsymbol{c}\,[i-1,j] + d\big(\boldsymbol{x}[i],\varepsilon\big), \\ \boldsymbol{c}\,[i,j-1] + d\big(\varepsilon,\boldsymbol{p}[j]\big), \\ \boldsymbol{c}\,[i-1,j-1] + d\big(\boldsymbol{x}[i],\boldsymbol{p}[j]\big)\Big\}$$

for every $i \in 1..n, j \in 2..m$. This relation expresses $\boldsymbol{c}[i,j]$ as a minimum over three adjacent entries in $\boldsymbol{c}$ together with a delete (of $\boldsymbol{x}[i]$), an insert (of $\boldsymbol{p}[j]$), and a substitution ($\boldsymbol{x}[i]$ for $\boldsymbol{p}[j]$), respectively. In Section 9.1 we studied Algorithm WF that uses (10.4) to compute the elements of $\boldsymbol{c}$ in an obvious manner. It is the key insight of Algorithm WM that the DBG bit-vector approach can be used to implement this calculation of $\boldsymbol{c}$.

We make use of a bit vector $\boldsymbol{s} = \boldsymbol{s}[1..m]$ and corresponding bit array $\boldsymbol{t} = \boldsymbol{t}[1..\alpha, 1..m]$ exactly as they were defined in Section 7.4. Thus for given i and every $j \in 1..m$, $\boldsymbol{s}[j] = 0$ iff $\boldsymbol{x}[i-j+1..i] = \boldsymbol{p}[1..j]$, while $\boldsymbol{t}[h,j] = 0$ iff $\boldsymbol{p}[j] = h$, where $h \in 1..\alpha$ is a letter of the alphabet. Of course the new algorithm described here will also be subject to the same restrictions that Algorithm DBG is: the alphabet A must be finite, known in advance, and indexed (Section 4.1), allowing us to assume that it takes the form $\{1, 2, \ldots, \alpha\}$.

Instead of just a single bit vector $\boldsymbol{s}$, however, Algorithm WM employs $k+1$ bit vectors, one corresponding to each integer $q \in 0..k$:

$$\begin{aligned} \boldsymbol{s_q}[j] &= 0, \ \text{ if } \boldsymbol{c}[i,j] \le q; \\ &= 1, \ \text{ otherwise.} \end{aligned} \tag{10.34}$$

Observe that $\boldsymbol{s_0} = \boldsymbol{s}$ as defined above; observe also that for fixed i and every integer $q \in 0..k-1$,

$$\boldsymbol{s_q}[j] = 0 \ \Rightarrow \ \boldsymbol{s_{q+1}}[j] = 0. \tag{10.35}$$

To simplify matters a little, we introduce the notation

$$I = d\big(\varepsilon, \boldsymbol{x}[i]\big), \ \ D = d\big(\boldsymbol{p}[j], \varepsilon\big), \ \ S = d\big(\boldsymbol{x}[i], \boldsymbol{p}[j]\big),$$

where now we make the assumption that I, D and S are nonnegative integers, with $S = 0$ if and only if $\boldsymbol{t}\big[\boldsymbol{x}[i], j\big] = 0$. We may then rewrite (10.4) as

$$\boldsymbol{c}[i,j] = \min\big\{\boldsymbol{c}[i-1,j] + I, \ \boldsymbol{c}[i,j-1] + D, \ \boldsymbol{c}[i-1,j-1] + S\big\}. \tag{10.36}$$

At the same time, we introduce the notation $\boldsymbol{s}_{\boldsymbol{q}}^{(i)}$, $q = 0, 1, \ldots, k$, to denote the $k+1$ bit vectors corresponding to row i of $\boldsymbol{c}$. What Algorithm WM actually does is to compute all the $\boldsymbol{s}_{\boldsymbol{q}}^{(i)}$, $q = 0, 1, \ldots, k$, from all the $\boldsymbol{s}_{\boldsymbol{q}}^{(i-1)}$, for every $i = 1, 2, \ldots, n$. As we shall see, the calculation takes place *in situ*, just like the original DBG calculation, and so storage is required only for $k+1$ bit vectors rather than for $(k+1)n$ of them.

For some value $q \in 0..k$, consider $\boldsymbol{s}_{\boldsymbol{q}}^{(i)}[j]$. We wish to determine whether

$$\boldsymbol{s}_{\boldsymbol{q}}^{(i)}[j] = 0 \text{ or } 1;$$

that is, whether

$$\boldsymbol{c}[i,j] \le q \text{ or } > q.$$

Observe from (10.36) that $\boldsymbol{s}_q^{(i)}[j] = 0$ if and only if at least one of the following conditions is satisfied:

$$\begin{array}{lll} (1) & \boldsymbol{c}[i-1,j] \le q-I & \Leftrightarrow \; \boldsymbol{s}_{q-I}^{(i-1)}[j] = 0; \\ (2) & \boldsymbol{c}[i,j-1] \le q-D & \Leftrightarrow \; \boldsymbol{s}_{q-D}^{(i)}[j-1] = 0; \\ (3) & \boldsymbol{c}[i-1,j-1] \le q-S & \Leftrightarrow \; \boldsymbol{s}_{q-S}^{(i-1)}[j-1] = 0. \end{array}$$

Note that (3) holds also in the case that $S = 0$ (that is, in the case that $\boldsymbol{t}[\boldsymbol{x}[i],j] = 0$). Thus the calculation of $\boldsymbol{s}_q^{(i)}[j]$ may be specified as follows:

$$\begin{aligned} \boldsymbol{s}_q^{(i)}[j] \leftarrow & \Big(\boldsymbol{s}_{q-I}^{(i-1)}[j] \;\wedge\; \boldsymbol{s}_{q-D}^{(i)}[j-1] \;\wedge\; \boldsymbol{s}_{q-S}^{(i-1)}[j-1]\Big) \\ & \wedge \Big(\boldsymbol{s}_q^{(i-1)}[j-1] \;\vee\; \boldsymbol{t}\big[\boldsymbol{x}[i],j\big]\Big), \end{aligned} \tag{10.37}$$

where $\wedge$ denotes bitwise logical AND and $\vee$ denotes bitwise logical OR as defined in Section 10.3 for the computation (10.29).

Recall from Section 7.4 that a bit vector $\boldsymbol{s}^{(i)}$ is initially formed from $\boldsymbol{s}^{(i-1)}$ by a right shift of one bit, where we suppose that 0 is always shifted into the leftmost position, so that initially $\boldsymbol{s}^{(i)}[1] = 0$. Otherwise, for $j \in 2..m$, the right shift moves $\boldsymbol{s}^{(i-1)}[j-1]$ into $\boldsymbol{s}^{(i)}[j]$. Using the notation of Section 7.4, then, and adopting the convention that every $\boldsymbol{s}_q^{(0)}[0..m]$ is initialized to 01^m, we may express the calculation (10.37) as follows:

$$\begin{aligned} \boldsymbol{s}_q^{(i)} \leftarrow \boldsymbol{s}_{q-I}^{(i-1)} & \;\wedge\; \mathit{rightshift}\big(\boldsymbol{s}_{q-D}^{(i)},1\big) \;\wedge\; \mathit{rightshift}\big(\boldsymbol{s}_{q-S}^{(i-1)},1\big) \\ & \wedge \Big(\mathit{rightshift}\big(\boldsymbol{s}_q^{(i-1)},1\big) \;\vee\; \boldsymbol{t}\big[\boldsymbol{x}[i],1..m\big]\Big), \end{aligned} \tag{10.38}$$

for $i = 1,2,\ldots,n$. After each execution of (10.38), we need only inspect the current value of $\boldsymbol{s}_{\boldsymbol{k}}[m] = \boldsymbol{s}_{\boldsymbol{k}}^{(i)}[m]$ in order to determine using (10.34) whether or not $\boldsymbol{c}[i,m] \le k$; that is, whether or not there is a k-approximate match of $\boldsymbol{p}$ terminating at position i of $\boldsymbol{x}$. We leave as an (easy) exercise (10.4.1) the embedding of (10.38) into Algorithm WM, directly analogous to Algorithm DBG.

Note that the calculation (10.38) depends upon fixed known integer values of $I \le q$, $D \le q$ and $S \le q$. Thus for $I = D = S = 1$ (10.38) computes edit distance, while for $I = D = 1$ and $S = 2$, Levenshtein distance is computed. Table 10.7 shows the array $\boldsymbol{t}$ and the successive values of $\boldsymbol{s}_0^{(i)}$ and $\boldsymbol{s}_1^{(i)}$ for 1-approximate matching under Levenshtein distance of $\boldsymbol{p} = \mathit{str}$ in $\boldsymbol{x} = \mathit{rests}$, an example whose $\boldsymbol{c}$-array is shown in Table 10.2. Other variations on the type of distance employed are possible, provided only that I, S and D are integers fixed in advance.

Algorithm WM must execute (10.38) $(k+1)n$ times in order to locate all k-approximate matches of $\boldsymbol{p}$ in $\boldsymbol{x}$. Each execution requires three right shifts, three AND operations and one OR operation on each word of each $\boldsymbol{s}_{\boldsymbol{q}}$, $q = 1,2,\ldots,k$ — hence constant time per word. Thus as before denoting by w the number of bits in a computer word, and making due allowance for the preprocessing and storage of the array $\boldsymbol{t}$ (Theorem 7.4.4), we may state the analogue of Theorem 7.4.3:

	s	t	r
e	1	1	1
r	1	1	0
s	0	1	1
t	1	0	1

Bit array $\boldsymbol{t}$

	s_0	s_1
r	111	111
e	111	111
s	011	001
t	001	00$\underline{0}$
s	011	001

Bit vectors

Table 10.7: Approximate pattern-matching under Levenshtein distance: $A = \{e, r, s, t\}$, $\boldsymbol{p} = str$, $\boldsymbol{x} = rests$, $k = 1$

Theorem 10.4.1 *Algorithm WM computes all k-approximate matches of* $\boldsymbol{p}[1..m]$ *in* $\boldsymbol{x}[1..n]$ *in time* $\Theta(k\lceil m/w\rceil n)$ *using* $\Theta((\alpha + k)\lceil m/w\rceil)$ *additional space.* □

As with all algorithms employing the DBG methodology, we may claim for small or even bounded values of m that the $\lceil m/w\rceil$ term may be absorbed into the constant of proportionality implied by the asymptotic complexity notation, and that therefore the time requirement may be thought of as being only $\Theta(kn)$. Further examination of the validity in practice of this claim is left to Exercise 10.4.4. Recall also the remark made at the end of Section 7.4 that we may reasonably suppose that $w \in \Omega(\log n)$: the WM time bound then approximates to $O(kmn/\log n)$.

So far in this section we have described an efficient and remarkably flexible algorithm for approximate pattern-matching. The algorithm deals easily with all definitions of distance based on fixed known integer costs for insertions, deletions, and substitutions. It is thus considerably more flexible than the k-mismatches and k-differences algorithms discussed in Sections 10.2 and 10.3, though not suited to variable or non-integer costs such as those that commonly occur in scoring matrices (Section 2.2). On the other hand, as we discuss below, Algorithm WM can, with little or no sacrifice of efficiency, be easily extended in several ways that make it very attractive as a general-purpose pattern-matching algorithm. In fact, Algorithm WM has been implemented in its full generality as a Unix package *agrep*, supported and maintained by Internet WorkShop at

`http://webglimpse.net`

Matching Sets of Letters

In many pattern-matching applications it is desirable to be able to specify patterns in which a match at some position j will be recognized if the matching letter is drawn from some specified set. For example, we might want to search for occurrences of "ref" followed by an integer in the range 1–7, or for occurrences of "ion" preceded by any one of "c", "s", or "t". Thus we could specify patterns $\boldsymbol{p} = ref[1\text{–}7]$ and $\boldsymbol{p} = [c, s, t]ion$, respectively, identifying sets of letters by the metacharacters [ ]. Algorithm WM can handle such cases simply by taking sets into account in the preprocessing that computes the bit array $\boldsymbol{t}$. All that would be necessary to deal correctly with these examples would be to set

$$\boldsymbol{t}[1,4] \leftarrow \boldsymbol{t}[2,4] \leftarrow \cdots \leftarrow \boldsymbol{t}[7,4] \leftarrow 0$$

in the first example to ensure that a match with 1–7 will be recognized in the fourth position of $\boldsymbol{p}$, and to set

$$\boldsymbol{t}[c,1] \leftarrow \boldsymbol{t}[s,1] \leftarrow \boldsymbol{t}[t,1] \leftarrow 0$$

in the second to ensure that a match with one of $[c, s, t]$ is recognized in $\boldsymbol{p}[1]$. No other change to WM is required.

Matching Wild-Card Characters

Recall from Section 2.2 that there are two wild-card symbols:

- $\bullet$ matches any single letter of A;
- $*$ matches any string of A^*.

Thus $\bullet$ is just a special case of a set of letters, and can be handled by the preprocessing just described above.

For a wild-card string, however, a small change is required to the algorithm itself. We suppose that the pattern is

$$\boldsymbol{p} = \boldsymbol{p}[1..j_1] * \boldsymbol{p}[j_1+1..j_2] * \cdots * \boldsymbol{p}[j_{r-1}+1..j_r] * \boldsymbol{p}[j_r+1..m],$$

with metacharacters $*$ located immediately after positions $j_1, j_2, \ldots, j_r$ of $\boldsymbol{p}$, but not themselves counted as letters in $\boldsymbol{p}$. Therefore, to specify the location of the metacharacters, we introduce a new bit vector $\boldsymbol{t}^* = \boldsymbol{t}^*[1..m]$ in which $\boldsymbol{t}^*[j] = 0$ if and only if $j \in \{j_1, j_2, \ldots, j_r\}$.

Observe that if for some $h \in 1..r$, $i \in 1..n$, and $q \in 0..k$, $\boldsymbol{s}_q^{(i-1)}[j_h] = 0$, then there exists a q-approximate match of $\boldsymbol{p}[1..j_h]$ (including consideration of any occurrences of $*$ to the left of position j_h) with a substring $\boldsymbol{x}[i'..i-1]$ of $\boldsymbol{x}$ terminating at position $i-1$. Further, because of the $*$ that follows $\boldsymbol{p}[1..j_h]$, *every* substring $\boldsymbol{x}[i'..i'']$, $i'' > i-1$, is also a q-approximate match with $\boldsymbol{p}[1..j_h]$. In other words, if it is true for some position $i-1$ in $\boldsymbol{x}$ that $\boldsymbol{s}_q^{(i-1)}[j_h] = 0$, then for every $i'' > i-1$ it should also be true that $\boldsymbol{s}_q^{(i'')}[j_h] = 0$. We can ensure that this condition is satisfied by executing after every assignment (10.38) the following:

$$\boldsymbol{s}_q^{(i)} \leftarrow \boldsymbol{s}_q^{(i)} \wedge \left(\boldsymbol{s}_q^{(i-1)} \vee \boldsymbol{t}^*\right). \tag{10.39}$$

In any case in which $\boldsymbol{s}_q^{(i)} \leftarrow 1$ as a result of (10.38), assignment (10.39) will set $\boldsymbol{s}_q^{(i)}[j] \leftarrow 0$ if and only if $\boldsymbol{t}^*[j] = 0$ *and* $\boldsymbol{s}_q^{(i)}[j] = 0$ for the preceding position $i-1$ in $\boldsymbol{x}$. Thus (10.39) ensures that zero values corresponding to positions $j_1, j_2, \ldots, j_r$ in $\boldsymbol{p}$ are preserved for all subsequent positions in $\boldsymbol{x}$, but otherwise has no effect.

Note that although (10.39) adds an extra AND and an OR to the processing required for each q and each i, nevertheless the asymptotic complexity of the algorithm is unaffected.

Combining Exact and Approximate Matching

It can happen that an exact match is appropriate for some parts of the pattern, while for other parts an approximate match is more suitable. For instance, we may wish to search for a DNA fragment CGATGAGAGCGC in which the first four letters (CGAT) and the last three letters (CGC) are required to match exactly, while for the middle five letters (GAGAG) only a 2-approximate match is necessary. Effectively what this means is that for certain positions (j =1–4 and j =10–12) in the pattern $\boldsymbol{p}$, no insertions, deletions or substitutions are allowed — so that if the match is not exact we must set $\boldsymbol{s_q}[j] \leftarrow 1$ for those positions j and for every $q \in 0..k$. We can accomplish this very easily just by introducing a new bit vector $\boldsymbol{e} = \boldsymbol{e}[1..m]$ in which $\boldsymbol{e}[j] = 1$ if and only if j is a position in $\boldsymbol{p}$ at which an exact match is required. Thus in our example we would set $\boldsymbol{e} \leftarrow 111100000111$. We use $\boldsymbol{e}$ to force $\boldsymbol{s_q^{(i)}}[j]$ to take the value 1 unless $\boldsymbol{s_q^{(i-1)}}[j-1] = 0$ and $\boldsymbol{x}[i] = \boldsymbol{p}[j]$, the condition for exact matching; we accomplish this by modifying (10.38) as follows:

$$\boldsymbol{s_q^{(i)}} \leftarrow \Big(\Big(\boldsymbol{s_{q-I}^{(i-1)}} \wedge \mathit{rightshift}\big(\boldsymbol{s_{q-D}^{(i)}}, 1\big) \wedge \mathit{rightshift}\big(\boldsymbol{s_{q-S}^{(i-1)}}, 1\big) \Big) \vee \boldsymbol{e} \Big)$$
$$\wedge \Big(\mathit{rightshift}\big(\boldsymbol{s_q^{(i-1)}}, 1\big) \vee \boldsymbol{t}\big[\boldsymbol{x}[i], 1..m\big] \Big). \qquad (10.40)$$

Observe that since (10.40) adds only a single OR operation to (10.38), the asymptotic complexity again remains the same.

Matching with Minimum k

It may be of interest to determine the *least* value k^* for which a given pattern (say a DNA fragment $\boldsymbol{p} =$ CGATGAGAGCGC) has one or more k^*-approximate matches in a given text (say a strand of several thousand base-pairs). In order to compute such a k^*, we can first search for an exact match ($k = 0$), perhaps using the original Algorithm DBG; if no exact match is found, Algorithm WM can be executed for

$$k = 1, 3, \ldots, 2^B - 1,$$

until for some value $B \geq 1$ at least one k^*-approximate match of $\boldsymbol{p}$ is found in $\boldsymbol{x}$, for some $k^* \in 2^{B-1}..2^B - 1$. Each execution of WM requires time proportional to the value of k used, so that the total time required by the B executions of WM will be proportional to

$$\Sigma = 1 + 3 + \cdots + (2^B - 1) < 2^{B+1}.$$

As we have seen, the actual value k^* found by the final execution of WM satisfies $k^* \geq 2^{B-1}$. Thus $\Sigma / k^* < 4$ and so the time required to determine k^* is $O\big(4k^* \lceil m/w \rceil n\big)$, essentially unchanged.

Further Extensions

We shall see in the next chapter that Algorithm WM can be extended to match regular expressions, and that it also can be modified to deal efficiently with multiple patterns. Of course these extensions, together with those already already described above, can be compounded: for example, WM can be used to search for multiple patterns $\boldsymbol{p}$, each of which makes use of wild-card symbols and combines exact and approximate matching, and for which a minimum value of k is required. Given that the pattern $\boldsymbol{p}$ is not extremely long, WM is indeed a flexible algorithm!

Enhancements to Algorithm WM have been proposed in [W94]. In particular, we shall discover in Section 11.2.4 that the algorithm described in [BYN96] forms the basis of a very effective approach to multiple approximate pattern-matching.

Exercises 10.4

1. Using (10.38), specify Algorithm WM and prove its correctness. Be sure to deal appropriately with cases where one of I, D, S exceeds q. And don't neglect to explain why only $\boldsymbol{s_k}[m]$ needs to be examined for each value of i, rather than all of the bit vectors $\boldsymbol{s_q}[m]$, $q = 1, 2, \ldots, k$.
2. Specify changes to (10.38) that permit Algorithm WM to compute Hamming distance. Discuss the space and time complexity of the modified algorithm in comparison to those of Algorithm ADBG.
3. For each of $k = 1$ and $k = 2$, compute the equivalent of Table 10.7 using edit distance.
4. For each of the word lengths $w = 32$ and $w = 64$, construct a table that shows the maximum pattern length m that can be handled by Algorithm WM for each $k \in 1..10$ and for each upper bound $B \in 1..10$ on the number of words allocated to each bit vector $\boldsymbol{s_q}$.

10.5 The Complexity of Approximate Pattern-Matching

True words are not beautiful;
Beautiful words are not true.

— LAO Tzu (c. 604–531 BC), *The Way of Lao Tzu*

Just as for exact pattern-matching (Section 8.8), so also in the approximate case, there has been much work done to establish theoretical limits on the complexity of the problem itself. In view of the fact that for a small approximation threshold k the pattern and the text would be very close to an exact match, it was conjectured that for small k it should be possible to find algorithms with linear worst-case execution time and sublinear average-case execution time — that is, with time requirements close to those attained by the fastest exact pattern-matching algorithms.

There are several algorithms for the k-differences problem whose asymptotic performance comes close to meeting these requirements. In Table 10.8 we give execution times for two algorithms already studied (M and M*), one that will be investigated in the next chapter (BYN), and two others of recent interest. Observe that

Algorithm	Worst-case time bound	Average-case time bound
M (Subsection 10.3.2)	$O((m/\log n)n)$	$O((m/\log n)n)$
M* (Subsection 10.3.2)	$O((m/\log n)n)$	$O((k/\log n)n)$
BYN (Subsection 11.2.4)	$O\left(\frac{k(m-k)}{\log n}n\right)$	
Chang and Lawler [ChL90]	$\Theta(kn+m)$	$O\left(\frac{k \log m}{m}n\right)$
Cole and Hariharan [CH98]	$O\left(\left(\frac{k^4}{m}+1\right)n+m\right)$	

Table 10.8: Complexity of approximate pattern-matching algorithms

- For values of m small with respect to n ($m \in O(\log n)$), Algorithms M and M* achieve worst-case times linear or sublinear in n. Further, Algorithm M* executes in average time $O(n)$ whenever $k \in O(\log n)$, a much weaker condition that as conjectured prescribes the value of k only.
- For Algorithm BYN, the worst-case time bound will generally be somewhat greater than that of Algorithms M and M*. A variant of BYN executes in average time $O(n)$ for certain medium-sized values of k/m, but does not achieve the consistent linear average-case bound attained by M*.
- The Chang-Lawler algorithm is guaranteed to execute in $O(n)$ average-case time for $k < m/\big(\log m + O(1)\big)$, whereas the corresponding condition for M* is $k \in O(\log n)$. Thus, if

$$m \in \Theta(\log m \log n), \tag{10.41}$$

  these upper bounds on k will be roughly equivalent, with some advantage, at least theoretically, to M* when m is smaller than indicated in (10.41), and to Chang-Lawler when m is greater.
- The algorithm of Cole-Hariharan is linear in n provided $k \in O(m^{1/4})$. Thus, this algorithm, like that of Chang-Lawler, guarantees linear behaviour in the worst case based only on an upper bound on k, in accordance with conjecture. In [CH98] it is further conjectured that the least upper bound attainable may be $k \in O(m^{1/3})$.

For a more detailed discussion of the comparative efficiency of approximate pattern-matching algorithms, we again refer the interested reader to [Nav01].

Chapter 11

Regular Expressions and Multiple Patterns

Proper words in proper places
make the true definition of a style.

— Jonathan SWIFT (1667–1745)
Letter to a Young Clergyman, 9 January 1720

I am aware that many of my colleagues will be shocked and appalled that it is only in Chapter 11, almost 300 pages into the book, that I have finally got around to introducing the finite automaton, an idea central to computer science in general, and furthermore perhaps particularly to the processing of strings.

In my defence I can only say that in a book focussed on algorithms, I have tried to adopt a minimalist approach: to explain the ideas involved in the algorithms in as economical a way as possible, with a minimum of conceptual overhead. Rightly or wrongly, it seemed to me that the algorithms introduced up to this point could be made satisfactorily clear without recourse to the idea of a finite automaton.

Now however the situation changes: the matching of both regular expressions and multiple patterns is inextricably tied, both historically and conceptually, to the use of finite automata. In this chapter, then, we will deal with several different varieties of FA, each of them directly related to a problem to be solved. To make this material intelligible, we give here a brief and very informal tutorial on the FA. For a more careful and structured development, consult [LP98].

An FA can most simply be thought of as a labelled directed graph, in which

- vertices represent "states" — states of a machine, states of a computation;
- vertex labels are just the identifiers of states;
- arcs denote transitions from one state to another;
- arc labels are letters or strings whose input or occurrence is thought of as causing the corresponding transitions.

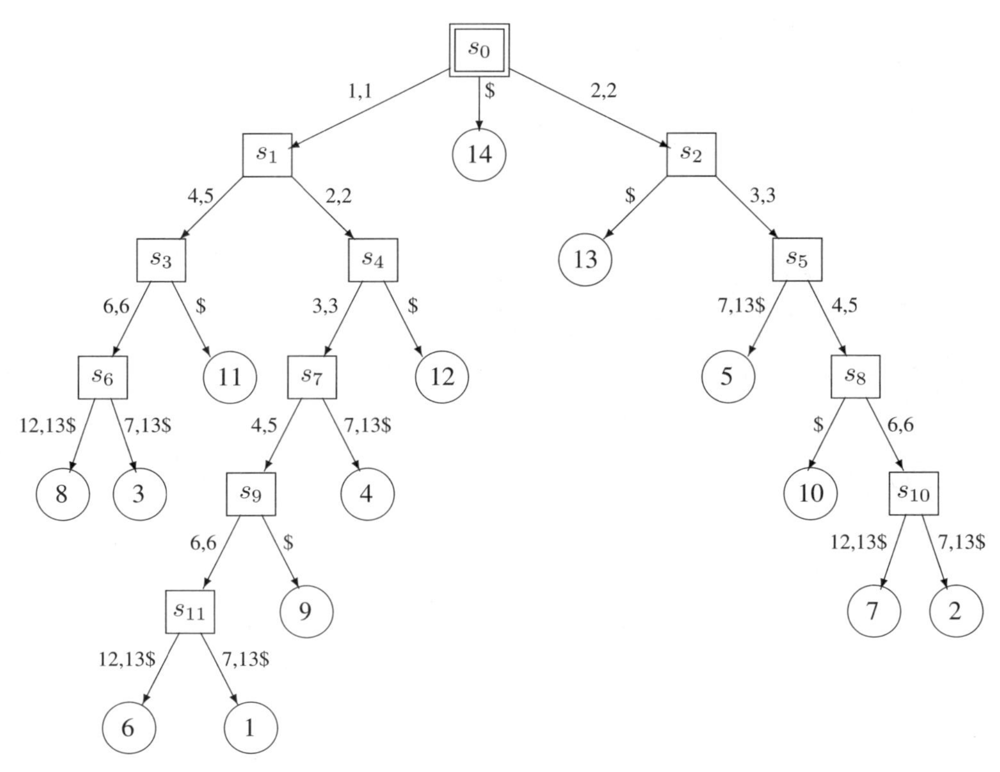

Figure 11.1. The suffix FA for $\boldsymbol{f_6} = abaababaabaab\$$

Reflecting on this definition, we realize that we have already encountered an FA in this book: the suffix tree! To make this clear, we reproduce here as Figure 11.1 the suffix tree given earlier as Figure 5.2 with links removed, nonterminal vertices labelled as states $s_1, s_2, \ldots, s_{11}$, and downward edges replaced by directed arcs.

If now we imagine inputting a string $\boldsymbol{p} = aab\$$ at vertex (state) s_0, the first letter $\boldsymbol{p}[1] = \boldsymbol{f_6}[1] = a$ will cause a transition to state s_1, where the substring $\boldsymbol{p}[2..3] = \boldsymbol{f_6}[4..5] = ab$ will trigger a transition to state s_3, where finally $\boldsymbol{p}[4] = \$$ will cause a transition to terminal state 11, telling us that $\boldsymbol{f_6}[11..14] = \boldsymbol{p}$. It is evident that any pattern $\boldsymbol{p}$ input to our suffix automaton will reach a terminal state t if and only if $\boldsymbol{p}$ is in fact a suffix $\boldsymbol{f_7}[t..14]$ of $abaababaabaab\$$ — a pattern $\boldsymbol{p}$ that is not such a suffix will inevitably get "stuck" at one of the internal vertices, since no current letter or substring will match the label of any of the downward arcs from that vertex.

Of course the suffix tree of any string $\boldsymbol{x}$ (or indeed of any collection of strings) can be interpreted and searched in this fashion: we see that the idea of an FA essentially provides a graphical way of representing and thinking about the course of a search calculation. The various kinds of FA that we will be studying in this chapter will all represent searches based on the same mechanisms that we have just described for the Suffix FA.

It should however be pointed out that the Suffix FA solves a problem that is in a sense inverse to the one discussed in this chapter: the Suffix FA deals easily with matching a single pattern against multiple texts simply by incorporating all the texts in a single suffix tree;

as we shall see, to match multiple patterns against a single text, we shall on the other hand (again rather easily) construct an FA that incorporates all the patterns.

In Section 11.1 we first consider a non-deterministic FA (NDFA) for the recognition of regular expression matches; we then show how to construct a deterministic FA (DFA) from an NDFA to provide a more efficient solution of the same problem. In the last subsection of Section 11.1 we consider another extension of Algorithm WM (Section 10.4), this time to k-approximate regular-expression matching, one that models the execution of an NDFA using bit vectors. Thus we come to understand that Algorithm DBG (Section 7.4) and its numerous extensions and derivatives can in general be thought of as modelling the execution of an FA.

Section 11.2 begins with FA solutions for exact multiple pattern-matching based first on the KMP algorithm (Section 7.1), next on the BM algorithm (Section 7.2). Then two approaches to multiple approximate pattern-matching are presented, one derived yet again from the ubiquitous WM algorithm, the other due to Baeza-Yates and Navarro.

11.1 Regular Expression Algorithms

Our eye-beams twisted, and did thread
Our eyes, upon one double string.

— John DONNE (1572–1631), *The Extasy*

In this section we present three algorithms, two of them "classical" — in fact, developed before 1970 — one more recent, to compute all the substrings of a text string $\boldsymbol{x}$ that match a pattern $\boldsymbol{p}$ expressed as a regular expression. In a sense, as discussed in Section 2.2, regular expressions constitute an extremity of the idea of an "approximate match".

11.1.1 Non-Deterministic FA

False words are not only evil in themselves,
but they infect the soul with evil.

— PLATO (c. 428–348 BC), *Phaedo*

If we are to use an FA corresponding to a given regular expression $\boldsymbol{p}$, we must first have a means of constructing it. We describe a method due to Thompson [T68] for constructing such an FA, based on four rules recursively applied. Actually we construct a ***nondeterministic finite automaton*** (NDFA), so called because it may happen, as we shall see, that from some vertex there is more than one transition (outarc) labelled by the empty string ε: thus the outarc to be followed from the vertex is not uniquely determined. As we shall also see, an NDFA is similar to the Suffix FA described above in having a single ***start state*** or ***source*** but differs in that it contains only one terminal vertex, called an ***accepting state*** or ***sink***. The form of NDFA described here will have its arcs labelled either by a single letter or by the empty string, and so is in this respect analogous to an uncompacted suffix tree (Section 2.1).

We represent NDFA($\boldsymbol{p}$) corresponding to a regular expression $\boldsymbol{p}$ by

where i and j are the labels of the start and accepting states, respectively, and the dotted line denotes one or more paths from i to j.

Given a regular expression $\boldsymbol{p}$ defined on an alphabet $A = \{\lambda_1, \lambda_2, \ldots, \lambda_\alpha\}$, we begin by computing

$$\boldsymbol{p}^* \leftarrow (\lambda_1|\lambda_2|\cdots|\lambda_\alpha)^*\boldsymbol{p}$$

in order to permit matching of an input string $\boldsymbol{x}$ with $\boldsymbol{p}$ beginning at any position in $\boldsymbol{x}$ — thus matching $\boldsymbol{p}$ beginning from any position in $\boldsymbol{x}$ is equivalent to matching $\boldsymbol{p}^*$ from the start of $\boldsymbol{x}$. Then we imagine that $\boldsymbol{p}^*$ is scanned position-by-position from left to right, with the following rules applied, in order to construct NDFA($\boldsymbol{p}$):

(R1) For a nonmetacharacter (ordinary letter) $\lambda \in A$, construct NDFA(λ) as follows:

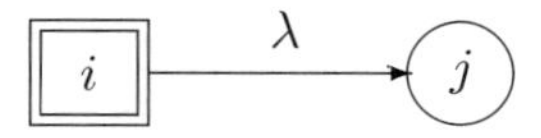

NDFA(λ) clearly accepts the letter λ and no other string.

(R2) Suppose that NDFA($\boldsymbol{q_1}$) and NDFA($\boldsymbol{q_2}$) are specified for regular expressions $\boldsymbol{q_1}$ and $\boldsymbol{q_2}$:

Then NDFA($\boldsymbol{q_1 q_2}$) is constructed as follows:

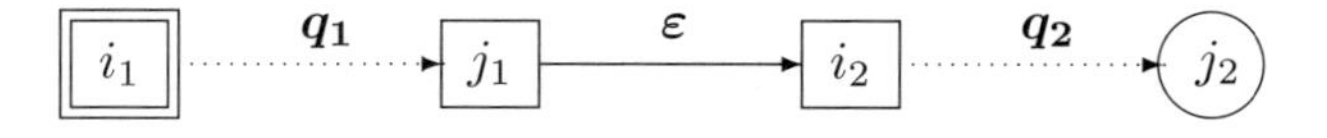

Here the empty label ε denotes a transition that occurs without any input, so that the accepting state j_1 of NDFA($\boldsymbol{q_1}$) is effectively identified with the start state i_2 of NDFA($\boldsymbol{q_2}$): both become normal internal states of NDFA($\boldsymbol{q_1 q_2}$). Thus NDFA($\boldsymbol{q_1 q_2}$) can accept only the string $\boldsymbol{q_1}\varepsilon\boldsymbol{q_2} = \boldsymbol{q_1 q_2}$.

(R3) Suppose that for regular expressions $\boldsymbol{q_1}$ and $\boldsymbol{q_2}$, NDFA($\boldsymbol{q_1}$) and NDFA($\boldsymbol{q_2}$) are specified. Then NDFA($\boldsymbol{q_1}|\boldsymbol{q_2}$) is constructed as follows:

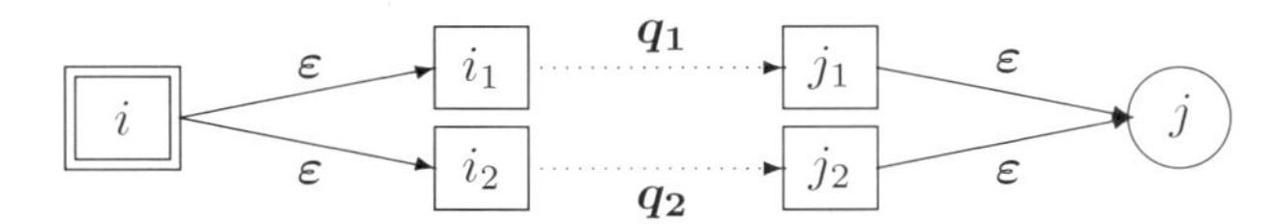

Here both the start and accepting states of NDFA($\boldsymbol{q_1}$) and NDFA($\boldsymbol{q_2}$) become normal internal states of NDFA($\boldsymbol{q_1}|\boldsymbol{q_2}$). Note that the empty string ε labels both outarcs of the new start state i as well as both inarcs of the new accepting state j. Thus NDFA($\boldsymbol{q_1}|\boldsymbol{q_2}$) will accept either $\boldsymbol{q_1}$ or $\boldsymbol{q_2}$ but no other string.

(R4) Suppose that NDFA($\boldsymbol{q_1}$) is specified for a regular expression $\boldsymbol{q_1}$. Then NDFA($\boldsymbol{q_1^*}$) is constructed as follows:

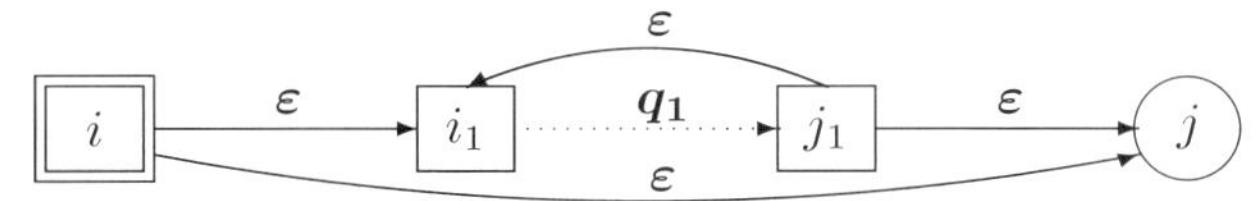

By virtue of the empty transitions from j_1 to i_1 and from j_1 to j, NDFA($\boldsymbol{q_1^*}$) will accept $\boldsymbol{q_1}$ or any repetition $\boldsymbol{q_1^k}$, $k \geq 2$; by virtue of the empty transition from i to j, NDFA($\boldsymbol{q_1^*}$) will also accept the empty string. No other strings will be accepted. Again the start state i_1 and the accepting state j_1 of NDFA($\boldsymbol{q_1}$) become normal internal states of NDFA($\boldsymbol{q_1^*}$).

These rules tell us how to process occurrences of individual letters, with concatenations of letters and of regular expressions, and with occurrences of the metacharacters $|$ and *. But they do not provide for metacharacters) and (, nor do they tell us how to handle nesting of operators, as in expressions such as

$$p = (\boldsymbol{q_1} \mid (\boldsymbol{q_2^*}|\boldsymbol{q_3}))^* \mid (\boldsymbol{q_4}|\boldsymbol{q_5})^*, \tag{11.1}$$

where the $\boldsymbol{q_i}$, $i = 1, 2, \ldots, 5$, are regular expressions (possibly just ordinary strings).

Assuming that letters of the alphabet can be recognized in constant time, an expression such as (11.1) can be recognized and parsed in time proportional to the number of characters in it into a ***syntax tree***. Figure 11.2 shows the tree corresponding to (11.1), in which we observe that each of the subexpressions $\boldsymbol{q_i}$ occurs as a leaf node.

A syntax tree can then in turn be traversed and transformed into NDFA($\boldsymbol{p}$), also in time proportional to $|\boldsymbol{p}|$. For further discussion of these matters, consult standard references such as [HU79].

We make the following observations about NDFA($\boldsymbol{p}$):

- As shown in Exercise 11.1.2, if m is the number of occurrences of letters in $\boldsymbol{p}$, then $|\boldsymbol{p}| \leq 6m$.
- Since each of rules (R1)–(R4) yields an NDFA with exactly one start state and one accepting state, it follows that NDFA($\boldsymbol{p}$) also has exactly one start state and one accepting state.
- For similar reasons, each nonaccepting vertex has either exactly one outarc labelled by a letter or at most two outarcs labelled by ε; also, each nonstart vertex has either exactly one inarc labelled by a letter or at most two inarcs labelled by ε. For our algorithm, the critical facts are the following:

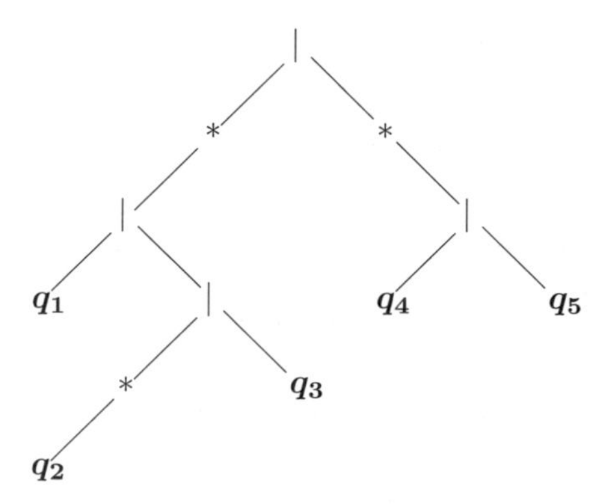

Figure 11.2. The syntax tree for (11.1)

- A nonaccepting vertex has a unique nonempty outarc if and only if it has no empty outarc.
- Every nonempty outarc leads to a vertex that is either the accepting state or a vertex with empty outarcs.

■ Let NDFA$(\boldsymbol{p}) = (V, A)$, where V is the set of vertices and A is the set of arcs. Then, as shown in Exercise 11.1.3,

$$|V| \leq 8m, \quad |A| \leq 13m, \tag{11.2}$$

where as above m is the number of individual letters in $\boldsymbol{p}$. Thus NDFA$(\boldsymbol{p})$ can be computed in a preprocessing phase in $\Theta(m)$ time.

Figure 11.3 shows a typical NDFA for the pattern (regular expression) $\boldsymbol{p} = (a|b)^*ab^*a$, where the prefix $(a|b)^*$ is prepended, as noted above, to ensure that a match is found with the letter occurring at any position of a given string $\boldsymbol{x}$ on $\{a, b\}$.

Having thus convinced ourselves that NDFA$(\boldsymbol{p})$ can be computed in time linear in the number of letters in the regular expression $\boldsymbol{p}$, we turn now to a consideration of how to use it. Essentially what we want to do is to input a text string $\boldsymbol{x} = \boldsymbol{x}[1..n]$ letter by letter, computing at each position in $\boldsymbol{x}$ the possible transitions from the currently activated set of

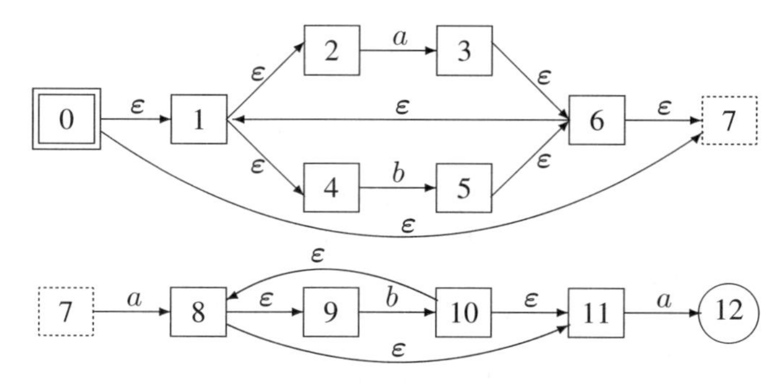

Figure 11.3. NDFA$((a|b)^*ab^*a)$

states (vertices) in NDFA($\boldsymbol{p}$), and reporting a match if and only if the computed transitions take us to the accepting state. For this purpose it is convenient to have available a function that, given a set S of states and a letter $\lambda \in A$ (the current input letter), computes a new set of states that can be "reached" by following an outarc labelled λ from a state in S. More precisely, we say that a state j in NDFA($\boldsymbol{p}$) is ***λ-reachable*** from a state i in NDFA($\boldsymbol{p}$) if and only if

- state i has an outarc labelled λ;
- there exists a directed path in NDFA($\boldsymbol{p}$) from i to j whose arcs are labelled $\lambda\varepsilon^{\boldsymbol{k}}$ for some integer $k \geq 0$;
- state j has no outarc labelled ε.

Then, given a set S of states in NDFA($\boldsymbol{p}$) and $\lambda \in A$, we define *trans*(S, λ) to be the set of states λ-reachable from the states of S. We extend this definition in order to accommodate two special cases:

- In the case that $S = \emptyset$, the empty set, *trans*(S, λ) is interpreted to mean *trans*$(\{0\}, \lambda)$, where 0 is the label of the start state of NDFA($\boldsymbol{p}$).
- We interpret *trans*(S, ε) to mean the set of all states reachable from the states of S by following all possible outarcs labelled ε until states are reached that have no outarcs labelled ε.

Thus in Figure 11.3 we would find

$$\begin{aligned} \textit{trans}(\emptyset, a) = \textit{trans}(\{3\}, a) = \textit{trans}(\{12\}, \varepsilon) = \emptyset, \\ \textit{trans}(\{2\}, a) = \textit{trans}(\emptyset, \varepsilon) = \textit{trans}(\{5\}, \varepsilon) = \{2, 4, 7\}. \end{aligned}$$

Algorithm 11.1.1 makes use of the function *trans* to compute every position i in a given string $\boldsymbol{x}$ at which a substring $\boldsymbol{x}[i'..i]$ matches a given nonempty regular expression $\boldsymbol{p}$. We leave to Exercise 11.1.8 the proof of the algorithm's correctness, based on a correct implementation of *trans* and the properties of an NDFA.

Algorithm 11.1.1 (NDFA)

– *Identify all substrings of $\boldsymbol{x}$ that match a given nonempty regular expression $\boldsymbol{p}$*

construct NDFA($\boldsymbol{p}$) with accepting state ω

$S \leftarrow$ *trans*$(\emptyset, \varepsilon)$
for $i \leftarrow 1$ **to** n **do**
  $S \leftarrow$ *trans*$(S, \boldsymbol{x}[i])$
  if $\omega \in S$ **then**
    output i
  $S \leftarrow$ *trans*(S, ε)
  if $\omega \in S$ **then**
    output i

In order to determine the complexity of Algorithm NDFA, we clearly need first to determine the complexity of *trans*: this function is executed a total of $2n + 1$ times, alternately using the arguments ε and $\boldsymbol{x}[i]$. We can imagine that *trans* is implemented using two queues requiring $O(|V|)$ space and a bit map requiring $\Theta(|V|)$ space, where $|V| \in \Theta(m)$ is the number of states (vertices) in NDFA$(\boldsymbol{p})$ and m is the number of occurrences of letters in $\boldsymbol{p}$. Given a set of distinct states of the NDFA in one of the queues, we can compute at most two transitions for each state and store the states resulting from these transitions in the other queue. We use the bit map to ensure that the states stored in the other queue are also distinct. Since at most $|V|$ states can be stored, and since there are at most two transitions for each state, each execution of *trans* requires $O(|V|) = O(m)$ time. As we have seen, NDFA$(\boldsymbol{p})$ can itself be computed in $\Theta(m)$ time. We have therefore the main result of this subsection:

Theorem 11.1.1 *Given a nonempty regular expression* $\mathbf{p}$ *containing* m *occurrences of letters and a text string* $\boldsymbol{x} = \boldsymbol{x}[1..n]$, *all the substrings in* $\boldsymbol{x}$ *that match* $\mathbf{p}$ *are computed by Algorithm 11.1.1 in* $O(mn)$ *time and* $\Theta(m)$ *additional space.* □

The "Four Russians" technique [ADKF70] can be used to reduce the time requirement for this approach to $O(mn/\log n)$ at a cost of using $O(mn/\log n)$ additional space [M88].

11.1.2 Deterministic FA

There is a Southern proverb:
fine words butter no parsnips.

— *English proverb, borrowed from Latin*

In this subsection we see how to construct a deterministic FA (DFA) for a given regular expression $\boldsymbol{p}$ — that is, a finite automaton with no empty transitions and exactly one transition from each state corresponding to each letter of the alphabet A. We shall see that DFA$(\boldsymbol{p})$ can be used to compute all matches with $\boldsymbol{p}$ in a given string $\boldsymbol{x}$ more efficiently in many cases than is possible using NDFA$(\boldsymbol{p})$. As for many other algorithms, we will find it convenient here to assume that A is indexed (Section 4.1).

For a given regular expression $\boldsymbol{p}$, the construction of DFA$(\boldsymbol{p})$ is based on an obvious idea: eliminate empty transitions (and hence nondeterminism) from NDFA$(\boldsymbol{p})$ by coalescing vertices that at each stage can be reached by empty transitions.

Consider for example the NDFA of Figure 11.3. Beginning with the start state 0, we follow all possible empty outarcs until we arrive at vertices that have nonempty outarcs — in this case, vertices $2, 4, 7$ of the NDFA. Thus the start vertex of DFA$((a|b)^*ab^*a)$ is

$$\boxed{2, 4, 7}$$

Next we consider all states reachable from $2, 4, 7$ by making a single a-transition in the NDFA followed again by all possible empty transitions. These states will be $2, 4, 7, 9, 11$ and so we encapsulate them in a new vertex, while at the same time recording the a-transition:

b
2, 4, 7
a
2, 4, 7, 9, 11

Here we have also recorded the fact that a b-transition from vertices $2, 4, 7$ simply brings us back to the same set of vertices.

It is easy to continue this process to complete DFA$((a|b)^*ab^*a))$ as shown below:

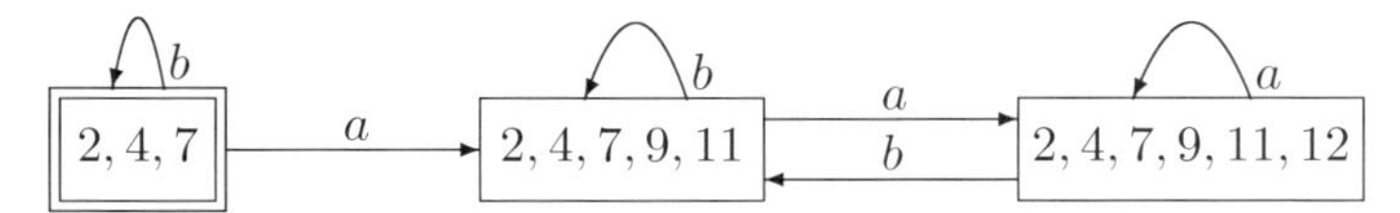

Of course this construction can be carried out on any NDFA to yield a corresponding DFA in which each vertex represents a unique set of states that are reachable based on a single specified input letter from a previously-defined set of states. The construction process terminates when no outarc from any vertex requires the introduction of a previously-undefined set of states. Any vertex of the DFA that includes among its labels the accepting state of the NDFA is called an ***accepting state*** of the DFA. It is not difficult to show that the construction process described here must terminate and must yield a DFA with at least one accepting state. Indeed, based on our example, it is tempting to suppose that for any regular expression $\boldsymbol{p}$,

- the number of vertices in DFA$(\boldsymbol{p})$ is $O(m)$, where m is the number of occurrences of letters in $\boldsymbol{p}$;
- there exists exactly one accepting state of DFA$(\boldsymbol{p})$.

However, as shown in Exercise 11.1.10, both these conjectures are false! The regular expression

$$\boldsymbol{p} = (a|b)^*a(a|b)^{(m-3)/2}$$

describes all strings on $\{a, b\}$ in which an occurrence of a is followed by exactly $m' = (m-3)/2$ occurrences of any letter (including the letter a). Thus for $m' = 2$, $\boldsymbol{p}$ would have exactly one match with $\boldsymbol{x} = abbab$ but two with $\boldsymbol{x} = abaab$ and three with $\boldsymbol{x} = aaabb$: in order to count the number of occurrences of letters following the most recent occurrence of a, it becomes necessary to create $\Theta(2^{m/2})$ distinct DFA vertices, many of which include the accepting state of the NDFA, and which are therefore accepting states of the DFA. This is essentially the worst case: as shown in Exercise 11.1.13, the number of vertices in a DFA can be as much as $\Theta(2^m)$.

The construction process described here is not hard to understand, but its implementation in a computer is not such an easy matter. In particular, each new transition in the DFA determines a vertex whose label consists of integers drawn from a set of $m + 1$ integers: we need to know whether or not the current set of integers has occurred before — that is, whether or not the current label was previously created — and if so, where it is located in memory. We can give a vertex with labels $\{s_1, s_2, \ldots, s_t\}$, $1 \le t \le m+1$, a unique integer identifier such as

$$S = 2^{s_1} + 2^{s_2} + \cdots + 2^{s_t},$$

essentially a bit map of the labels that requires $O(m)$ bits of storage. Then at each stage we need to know whether or not there already exists, among the $O(2^m)$ previously-created vertices, one with identifier S.

We may suppose that S takes values in the range $1..2^{m+1}$ — if not, we can renumber the labels of the NDFA in time proportional to m so that this property holds. We can then make use of an array $I[1..2^{m+1}]$ in which each element $I[S]$ is either `NULL` or a pointer to the vertex of the DFA whose identifier is S. We see that I requires $\Theta(2^m)$ storage and that each element $I[S]$ can be tested and updated in constant time, so that searches for $O(2^m)$ identifiers require only $O(2^m)$ time. Observe however that the storage required for each vertex of the DFA is proportional to α, the alphabet size, since each vertex may have as many as α distinct outarcs. Thus the processing related to each vertex is $O(\alpha)$. Hence

Theorem 11.1.2 *Given a regular expression* $\boldsymbol{p}$ *containing* m *occurrences of letters,* DFA$(\boldsymbol{p})$ *can be constructed in* $O(2^m\alpha)$ *time using* $O(2^m\alpha)$ *space.* □

As shown in [MY60], a DFA can also be constructed directly from a syntax tree, but the complexity bounds given in Theorem 11.1.2 are unaffected.

Once DFA$(\boldsymbol{p})$ has been constructed, it is a trivial matter to search it to find all substrings of a given text string $\boldsymbol{x} = \boldsymbol{x}[1..n]$ that match $\boldsymbol{p}$. The DFA algorithm will make use of a much-simplified *trans* function to perform transitions from vertex to vertex in the DFA, outputting the current position in $\boldsymbol{x}$ whenever an accepting vertex is reached. For an indexed alphabet, each transition can be computed in constant time; otherwise, as for suffix trees, $O(\log \alpha)$ time may be required. We therefore state

Theorem 11.1.3 *Given* DFA$(\boldsymbol{p})$ *corresponding to a nonempty regular expression* $\boldsymbol{p}$ *and a text string* $\boldsymbol{x} = \boldsymbol{x}[1..n]$*, both defined on an alphabet* A*, all the substrings in* $\boldsymbol{x}$ *that match* $\boldsymbol{p}$ *can be computed in time* $O(n \log \alpha)$ *(*$O(n)$ *if* A *is indexed) and constant space.* □

For an indexed alphabet, the efficiency of Algorithm DFA can be improved by a constant factor by defining a ***transition matrix*** $T = T[1..m, 1..\alpha]$, where M is the number of vertices in DFA$(\boldsymbol{p})$. This requires that the vertices be relabelled $1..M$ instead of using the identifiers S introduced above for DFA construction. Then each element $T[j, h]$ is the label of the vertex of DFA$(\boldsymbol{p})$ that is reached from the vertex labelled j by a transition resulting from the input of the letter $h \in 1..\alpha$.

Even with a nonindexed alphabet, the availability of a DFA certainly enables us to locate regular expression matches extremely quickly; especially in cases where multiple text strings are to be matched against the same regular expression, the overhead required for DFA construction assumes greatly reduced significance.

Overall, however, the results of our investigation so far are a little discouraging: all that we seem to be able to say is that the use of a DFA for regular expression matching can require a total of $O(2^m\alpha+n\log\alpha)$ time and $O(2^m\alpha)$ space, a very doubtful improvement for a single match over the corresponding $O(mn)$ and $O(m)$ bounds for an NDFA. The good news is that the worst case of $O(2^m)$ vertices is pathological: in most cases (such as our example), only $O(m)$ DFA vertices need to be created, and so the time and space bounds for the DFA approach can often be reduced to $O(m\alpha+n)$ and $O(m\alpha)$, respectively. Furthermore, there are various implementation techniques [ASU86] that can be applied to reduce space and time requirements. One of these is "lazy transition evaluation" that computes transitions on a just-in-time basis — according to [A90], this approach reduces total observed time for the construction and use of DFA($\boldsymbol{p}$) to $O(m+n)$, even in the worst case.

11.1.3 Algorithm WM Revisited

Words are wise men's counters,
they do but reckon with them,
but they are the money of fools ...

— *Thomas HOBBES (1588–1679), Leviathan*

In this subsection we outline an extension to the bit-vector techniques of Algorithm WM (Section 10.4) that allows us to efficiently compute all k-approximate matches in a text string $\boldsymbol{x}$ to a given regular expression $\boldsymbol{p}$. The method is based on a model of the execution of NDFA($\boldsymbol{p}$) in which each state is represented by a single bit: if the vertices of NDFA($\boldsymbol{p}$) are labelled $0..m'$, then the bit vector contains $m'+1$ bits. Recall from Subsection 11.1.1 that if $\boldsymbol{p}$ contains m occurrences of letters, then $m' \in \Theta(m)$. Of course the method is subject to the same restrictions that Algorithm WM is: the alphabet must be finite, known in advance, and indexed, thus taking the form $A=\{1,2,\ldots,\alpha\}$.

In this new modification of WM, called WM*, we construct bit vectors $\boldsymbol{s_q}[1..m']$, $q=0,1,\ldots,k$, exactly analogous to the vectors (10.34); the only difference is that now the position j in $\boldsymbol{s_q}$ denotes the label j in NDFA($\boldsymbol{p}$) rather than position j in $\boldsymbol{p}$ itself. Of course, if an exact match is sought, so that $k=0$, there will be only a single bit vector $\boldsymbol{s_0}$.

Similarly, we redefine the bit array $\boldsymbol{t}=\boldsymbol{t}[1..\alpha,0..m']$ with $\boldsymbol{t}[h,j]=0$ if and only if the outarc (transition) from the vertex labelled j to the vertex labelled $j+1$ is labelled with letter h. Then for any position i in the given text string $\boldsymbol{x}$, the OR calculation

$$\boldsymbol{s_q^{(i-1)}}[j-1] \;\vee\; \boldsymbol{t}\big[\boldsymbol{x}[i],j\big] \tag{11.3}$$

included in (10.37) will execute correctly for vertices of NDFA($\boldsymbol{p}$) that are appropriately labelled. Referring to Figure 11.3, we see that (11.3) is correct for vertices $2,3$ and letter a, for $4,5$ and b, and so on. Further, we see that we can always ensure correctness for nonempty transitions simply by assigning consecutive labels to the end vertices of nonempty arcs.

However, to ensure the correctness of the calculation for all vertices, we need to deal with two other cases:

(1) Corresponding to every occurrence of the $|$ operator in $\boldsymbol{p}$, there must exist at least one pair $(j, j+1)$ of consecutive labels in NDFA$(\boldsymbol{p})$ that has *no* transition between labels (Rule 3); for example, the pair $(3, 4)$ in Figure 11.3. Such cases can be easily handled by introducing a mask that suppresses the transition from j to $j+1$.

(2) There will be many vertex pairs (j_1, j_2) that are joined by an empty arc. With some care in vertex labelling, we can ensure that many of these vertex pairs are consecutive, so that $j_2 = j_1 + 1$. For example, the simple labelling algorithm of Figure 11.3 ensures that *every* nonaccepting vertex j has an outarc (empty or nonempty) to vertex $j+1$. But inevitably, for every occurrence of either $|$ or * in $\boldsymbol{p}$, there will be outarcs (j_1, j_2) for which $j_2 \neq j_1 + 1$, simply as a result of the application of Rules 3 and 4 — for example, the arcs $(1, 4)$, $(3, 6)$, $(8, 11)$ and $(10, 8)$ in Figure 11.3.

We describe below a mechanism for dealing with empty arcs, whether between consecutive vertices or not.

The meaning of the existence of an empty arc (j_1, j_2) is simply that any approximate match corresponding to vertex j_1 holds identically for j_2. Recall from Subsection 11.1.1 that there are only two types of vertex in an NDFA:

- **(Type I)** Those with one or two empty outarcs.
- **(Type II)** Those with either a single nonempty outarc or no outarc.

Thus what we really want to do is to replace all Type I vertices j_1 by corresponding Type II vertices j_2. In other words, we wish to follow empty outarcs from vertices of Type I until we reach a vertex of Type II, something that the construction rules for an NDFA guarantee we can always do. We can conceive of this process as a function

$$f : J_1 \to J_2$$

that maps a set J_1 of Type I vertices into another set J_2 of corresponding Type II vertices. Since the elements of J_1 and J_2 are distinct integers in the range $0..m'$, f effectively maps some bit vector of length $m'+1$ into another one.

In order to accomplish this mapping efficiently, Algorithm WM* makes use of additional preprocessing to construct arrays that describe *all possible* transitions $J_1 \to J_2$. There is one array corresponding to each byte (eight consecutive bits) in any $\boldsymbol{s_q}$. Thus there are $\lceil m'/8 \rceil$ arrays T_r say, $r = 0, 1, \ldots, \lceil m'/8 \rceil - 1$, corresponding respectively to the positions of $\boldsymbol{s_q}$ (vertices of NDFA$(\boldsymbol{p})$)

$$8r..8r+7 = 0..7,\ 8..15, \ldots,\ 8(\lceil m'/8 \rceil - 1)..m'.$$

Thought of as binary integers, these bytes take values b in the range $0..255$. For each value of $b \in 0..255$, the values of certain positions

$$j_1 \in 0..7$$

in b will be zero, indicating that vertex

$$j_1 + 8r$$

is a currently activated vertex of NDFA$(\boldsymbol{p})$: in other words, a q-approximate match between some substring of the text string $\boldsymbol{x}$ and the regular expression $\boldsymbol{p}$ is valid up to vertex $j_1 + 8r$ of NDFA$(\boldsymbol{p})$. Then $T_r[b]$ is defined to be a bit vector of length $m' + 1$ in which position $j_2 \in 0..m' + 1$ takes the value 0 if and only if vertex j_2 of NDFA$(\boldsymbol{p})$ is a Type II vertex reachable by a sequence of empty arcs from one of the vertices $j_1 + 8r$. Effectively, each T_r is a two-dimensional array containing $2^8(m' + 1)$ bits that can be constructed in $\Theta(m')$ time. Then the total preprocessing time for all $m'/8$ arrays T_r is $\Theta((m')^2/8)$.

If we denote by $s_{q,r}$ the value of the rth byte of $\boldsymbol{s_q}$, we can make use of T_r by computing

$$T_r[s_{q,r}] \wedge \boldsymbol{s_q} \tag{11.4}$$

for every $r \in 0..\lceil m'/8 \rceil$, where $\wedge$ denotes bitwise logical AND. Then the bitwise result of (11.4) will be 0 unless both operands are 1, so that any currently activated vertex $j_1 + 8r$ remains activated, while in addition all vertices j_2 of Type II reachable from $j_1 + 8r$ also become activated. Since the AND operation can be executed a word at a time, the $\lceil m'/8 \rceil$ executions of (11.4) require $\Theta((m')^2/8w))$ time at each step of the algorithm, where as before w is the computer word length. Thus overall (11.4) will be executed kn times, in conjunction with the usual WM processing (10.37); a match is reported whenever as a result $\boldsymbol{s_k}[m'] = 0$.

According to [WM92], Algorithm WM* can actually be implemented in

$$O(k\lceil 2m'/\log_2 n \rceil \lceil m'/w \rceil n)$$

time. For values of m' that are not too large and values of n that are large enough, this means that WM* can handle approximate regular expression matching in close to $O(kn)$ time with $\Theta((m')^2)$ preprocessing.

Exercises 11.1

1. For the regular expression (11.1), draw NDFA$(\boldsymbol{p})$.
2. Suppose that a regular expression $\boldsymbol{p}$ contains m occurrences of letters. Show that

   - at most $2m$ parentheses
   - at most m occurrences of $|$
   - at most $2m$ occurrences of *

   may be required in the worst case to specify $\boldsymbol{p}$ unambiguously. Hence show that the number $|\boldsymbol{p}|$ of characters in $\boldsymbol{p}$ can be assumed to satisfy $|\boldsymbol{p}| \leq 6m$.
3. Based on the preceding exercise, prove the inequalities (11.2). Can you suggest modifications to rules (R1)–(R4) to reduce these upper bounds?
4. With reference to Figure 11.3, we computed *trans*$(\{5\}, \varepsilon) = \{2, 4, 7\}$. Was it correct to omit states 1 and 6 in this result?

5. We observe from Figure 11.3 that NDFA($\boldsymbol{p}$) may contain directed cycles. Show that, nevertheless, for any nonstart state s, $s \notin trans(s, \varepsilon)$. Discuss the case when s is the start state.
6. Write efficient pseudocode for the function $trans(S, \lambda)$ and prove its correctness. Be careful when handling the case where the second argument is ε!
7. Show by example that if $(\lambda_1|\lambda_2|\cdots|\lambda_\alpha)^*$ were *not* prepended to the regular expression $\boldsymbol{p}$, substrings of $\boldsymbol{x}$ that actually match $\boldsymbol{p}$ could be missed by Algorithm 11.1.1 in its implementation of NDFA($\boldsymbol{p}$).
8. Prove the correctness of Algorithm 11.1.1.
9. Prove that in a syntax tree corresponding to a regular expression, a node is an internal node if and only if its label is a metacharacter.
10. Construct the syntax tree, the NDFA and the DFA for

    - $\boldsymbol{p} = (a|b)^*aba$;
    - $\boldsymbol{p} = ab^{m-1}$;
    - $\boldsymbol{p} = a(a|b)(a|b)$;
    - $\boldsymbol{p} = a(a|b|c)(a|b|c)$.

    Based on the last two of these exercises, compute the number of vertices in DFA($\boldsymbol{p}$), where

    $$\boldsymbol{p} = \lambda_1(\lambda_1|\lambda_2|\cdots|\lambda_\alpha)^{(m-1)/\alpha},$$

    and as usual m is the number of occurrences of letters in $\boldsymbol{p}$, α the alphabet size. How many of these are accepting vertices of the DFA?
11. Prove that the construction of a DFA from an NDFA is a process that must terminate.
12. Prove that every DFA has at least one accepting state.
13. Observe that, apart from the accepting vertex, the only labels of the NDFA that can occur in vertices of the corresponding DFA are those of vertices that have nonempty outarcs. Hence show that the maximum number of vertices in a DFA is 2^{m+1}.
14. Specify an efficient algorithm to convert an NDFA into a DFA.
15. Specify Algorithm DFA that uses DFA($\boldsymbol{p}$) to compute in $\Theta(n)$ time all matches of substrings of a text string $\boldsymbol{x} = \boldsymbol{x}[1..n]$ with the regular expression $\boldsymbol{p}$. For an alphabet that is

    (a) ordered but not indexed,

    (b) indexed,

    include a modified function *trans* that, given the identifier S of a vertex and a letter λ, computes the next vertex $trans(S, \lambda)$.
16. Design an algorithm to perform the preprocessing for Algorithm WM*.

11.2 Multiple Pattern Algorithms

It's good food and not fine words
that keeps me alive.

— Jean-Baptiste POQUELIN (MOLIÈRE) (1622–1673)
Les Femmes Savantes

In this section we consider the problem of matching multiple patterns

$$P = \{\boldsymbol{p_1}, \boldsymbol{p_2}, \ldots, \boldsymbol{p_r}\}$$

against a given text string $\boldsymbol{x} = \boldsymbol{x}[1..n]$. Since P may be written in the form $\boldsymbol{p_1}|\boldsymbol{p_2}|\cdots|\boldsymbol{p_r}$, we see that, for exact matching, this problem is actually a special case of regular expression matching and can therefore be solved using the algorithms of the previous section. We find in this section however that we can design algorithms specific to multiple patterns that are more efficient.

Algorithms for exact pattern-matching are outlined in Subsections 11.2.1 and 11.2.2, straightforward extensions of Algorithms KMP (Section 7.1) and BM (Section 7.2), respectively. In the final two subsections we deal with multiple approximate pattern-matching: Subsection 11.2.3 outlines yet another extension to Algorithm WM (Section 10.4), while Subsection 11.2.4 describes a new approach due to Baeza-Yates and Navarro. The new extension to WM is of course applicable over a wide range of distance types, while Algorithm BYN is applicable to edit, Levenshtein and Hamming distance.

11.2.1 Aho-Corasick FA: KMP Revisited

Where seldom is heard a discouraging word
And the skies are not cloudy all day.

— Brewster HIGLEY (1820–1911) and Daniel E. KELLEY (1845–?)
Home on the Range

Now that we understand the idea of an FA, we can easily see that the KMP algorithm can be conceived in FA terms: we look first for a match on the next input symbol, but if that match fails, then we look for a match with the symbol following the longest border at the current input position. In other words, we simply design an automaton to follow the KMP border array rule.

The nature of the Aho-Corasick FA [AC75] will be made clear by an example. Suppose first that there is just a single pattern $\boldsymbol{p} = abaababa$ against which we wish to match a given text string $\boldsymbol{x} = \boldsymbol{x}[1..n]$. The border array (Section 1.3) of $\boldsymbol{p}$ is as follows:

$$\begin{array}{rccccccc} & 1 & 2 & 3 & 4 & 5 & 6 & 7 \\ \boldsymbol{p} = & a & b & a & a & b & a & b \\ \boldsymbol{\beta} = & 0 & 0 & 1 & 1 & 2 & 3 & 2 \end{array}$$

Then the corresponding ACFA($\boldsymbol{p}$) is easy to design:

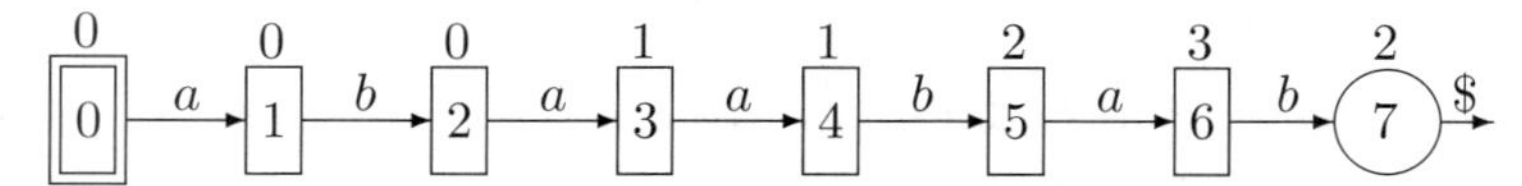

Here the labels *inside* the vertices are just the positions in $\boldsymbol{p}$, while the labels on the arcs are the elements of $\boldsymbol{p}$ with which a match is required in order to proceed to the next vertex (state). In fact, recalling the introduction of suffix trees in Section 2.1, we see that ACFA($\boldsymbol{p}$) is nothing but the uncompacted trie for $\boldsymbol{p}$. Added to the trie are the labels *above* the vertices, which are exactly the corresponding entries in the border array; each such label specifies the vertex (state) of the FA that the automaton should return to in the event of a *failure* of the input character to match with the label of the outarc. (So at last we understand the reason for referring to the border as the "failure function"!) Thus these labels really stand for backward arcs from the current one to a preceding one determined by the entry in the border array. To enable mismatches at vertex 0 to be handled in a standard way, we also give vertex 0 a "backward" arc to itself.

As for suffix trees, we adopt the convention that outarcs from terminal vertices of ACFA($\boldsymbol{p}$) are labelled with a special letter \$ that matches no input letter. This ensures correct behaviour of the matching algorithm after a match has been found.

In general, since the border array can be computed in linear time (Theorems 1.3.2 and 1.3.3), it is clear that for any single pattern $\boldsymbol{p} = \boldsymbol{p}[1..m]$, ACFA($\boldsymbol{p}$) can be constructed in $\Theta(m)$ time using $\Theta(m)$ space. Then Algorithm 11.2.1 makes use of ACFA($\boldsymbol{p}$) in a straightforward way to compute all the occurrences of $\boldsymbol{p}$ in $\boldsymbol{x}$. Algorithm ACFA can be thought of as a variant of Algorithm KMP, but the reader should observe also that its structure mirrors that of Algorithm 1.3.1 for the calculation of the border array.

Algorithm 11.2.1 (ACFA)

– *Compute all positions i in $\boldsymbol{x}$ such that a substring $\boldsymbol{x}[i'..i]$ traces a path from state* 0 *of an ACFA to an accepting state*

$s \leftarrow 0$
for $i \leftarrow 1$ **to** n **do**
  while $s > 0$ **and not** $match(s, \boldsymbol{x}[i])$ **do**
    $s \leftarrow \boldsymbol{\beta}[s]$
  if $match(s, \boldsymbol{x}[i])$ **then**
    $s \leftarrow s + inc(s, \boldsymbol{x}[i])$
    if $accept(s)$ **then output** (i, s)

The execution of Algorithm ACFA depends on four calculations:

$\boldsymbol{\beta}$
The border array will of course be precomputed during the construction of the NDFA so that $\boldsymbol{\beta}[s]$ is available for every state s in constant time. As noted above, we define $\boldsymbol{\beta}[0] = 0$.

inc
In case of a match, the current value of s is incremented according to the letter $\boldsymbol{x}[i]$ with which the match occurs. In our example the increment is always one, but as we shall see, for multiple patterns this is not always true. The increment will be stored with the outarc from s corresponding to $\boldsymbol{x}[i]$ and so will always be available in constant time once the match has been identified.

accept
Each state will contain a single bit indicating whether or not it is an accepting state.

match
This function will be true if there is a match of the label on an outarc from s with $\boldsymbol{x}[i]$. In our example of a single pattern, this function can always be evaluated in constant time by a single letter comparison, but as we shall see, in the case of multiple patterns, there is more to talk about.

Let us extend the above example to a set P of three patterns:

$$P = \{\boldsymbol{p_1}, \boldsymbol{p_2}, \boldsymbol{p_3}\} = \{abaabab, abab, ababcabab\}. \quad (11.5)$$

Clearly we can use the same methodology to construct the uncompacted trie for (11.5), and so we get ACFA(P), as shown in Figure 11.4. Note that, if m is the total length of the patterns in P, the number of (forward and backward) outarcs in ACFA(P) cannot exceed $2m + 1$.

We see from this example that adjustments need to be made to the normal border array calculation in order to take into account the offsets that occur when a position in one of the patterns is renumbered as a state label in ACFA(P), where P denotes a set of patterns. For instance, positions 4 in $abab$ and $ababcabab$ become state 8, position 9 in $ababcabab$ becomes state 13, and $\boldsymbol{\beta}[13] = 8$. We leave the details of these adjustments to Exercise 11.2.2. Also included in the same exercise is an exploration of what happens when one pattern happens to be a border of another, as in (11.5).

Given that $\boldsymbol{\beta}$ is correctly computed for ACFA(P), it is not difficult to check that Algorithm 11.2.1 will correctly identify all substrings of $\boldsymbol{x}$ that match any one of the patterns in P: the **while** loop ensures that whenever a mismatch occurs with some prefix $\boldsymbol{p}[1..j]$ of some pattern $\boldsymbol{p}$, then possible matches with every border of $\boldsymbol{p}[1..j-1]$ are considered in

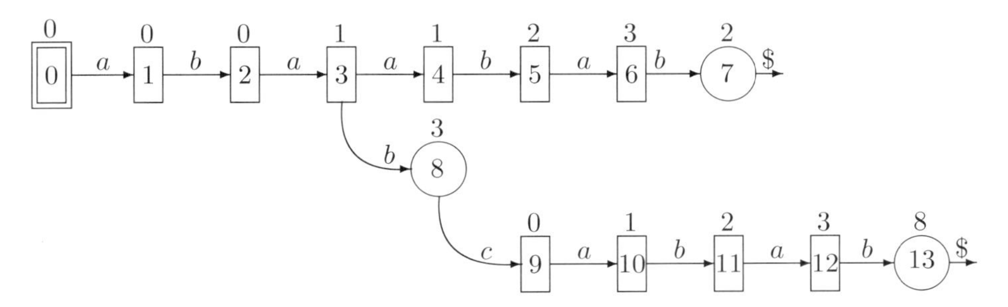

Figure 11.4. ACFA(P) for $P = \{abaabab, abab, ababcabab\}$

decreasing order of border length, just as in Algorithm KMP. Indeed, correctness of Algorithm ACFA for multiple patterns is really a direct consequence of the correctness of KMP. Furthermore, if the function *match* can be computed in constant time, then the execution time of Algorithm 11.2.1 will be proportional to n. This fact follows from the observation that the total number of iterations of the **while** loop, summed over the n steps of the algorithm, cannot exceed $n-1$ — exactly the same observation that was made in Section 1.3 for the **while** loop of the border array calculation! Thus, provided that *match* executes in constant time, the linearity of Algorithm 11.2.1 is established by the same argument that proves the linearity of Algorithm Border Array (Theorem 1.3.3).

What then can be said about the execution time of *match*? In general, ACFA(P) is just a (slightly embellished) uncompacted trie on a set

$$P = \{\boldsymbol{p_1}, \boldsymbol{p_2}, \ldots, \boldsymbol{p_r}\}$$

of distinct strings of lengths $m_1, m_2, \ldots, m_r$, respectively, and total length $m = m_1 + m_2 + \cdots + m_r$. On an alphabet of size α, it could happen that one or more vertices of ACFA(P) might have $O(\alpha)$ outarcs; in order to select the correct outarc, a data structure of size $\Theta(\alpha)$ would need to be stored at each vertex, from which the correct outarc could be selected in constant time if the alphabet were indexed, $O(\log \alpha)$ time if it were ordered. The situation is therefore just as outlined in Digression 2.1.1 for suffix trees.

Even when the patterns are long and their number relatively small, the selection of outarcs during the search process can be time-consuming. Suppose that $r = \alpha$ and pattern $\boldsymbol{p_h}$ has prefix λ_h^2, $h = 1, 2, \ldots, \alpha$. Then vertex 0 of ACFA(P) has α forward outarcs. Let $\boldsymbol{x} = \boldsymbol{x}[1..n]$ be a string free of single-letter squares in which each letter of the alphabet occurs n/α times. Using ACFA(P) to search $\boldsymbol{x}$ for matches with any of the patterns P will result in vertex 0 being visited n times; each visit will require one outarc out of a total of α to be selected. If the alphabet is not indexed and a balanced search tree or ordered array needs to be searched, the processing at vertex 0 will therefore require $O(n \log \alpha)$ time.

We summarize the above discussion as follows:

Theorem 11.2.1 *Let $P = \{\boldsymbol{p_1}, \boldsymbol{p_2}, \ldots, \boldsymbol{p_r}\}$ be a set of patterns of total length m and $\boldsymbol{x} = \boldsymbol{x}[1..n]$ a given text string, all defined on an alphabet A of size α. Then* ACFA(P) *requires $\Theta(\alpha m)$ storage and*

(a) *for indexed A, $\Theta(m)$ time is required to construct* ACFA(P) *and $\Theta(n)$ time is required to execute Algorithm 11.2.1;*

(b) *for A ordered but not indexed, $O(m \log \alpha)$ time is required to construct* ACFA(P) *and $O(n \log \alpha)$ time is required for Algorithm 11.2.1.*

□

Thus use of the Aho-Corasick automaton enables a text string $\boldsymbol{x}[1..n]$ to be matched against a set of patterns of total length m in $\Theta(m+n)$ time when the alphabet is indexed, $O((m+n)\log \alpha)$ time when it is ordered. In other words, Algorithm KMP extends to multiple pattern-matching with little or no decrease in efficiency.

Algorithm ACFA has been extended using DAWGs (Section 5.3) so as to require only $2n$ letter comparisons in the worst case [CCGL99]; the modified algorithm appears to be most useful when the alphabet is small and the shortest pattern is long.

11.2.2 Commentz-Walter FA: BM Revisited

It was in the barbarous gothic times when words had a meaning; in those days writers expressed thoughts.

— Anatole FRANCE (1844–1924), *La Vie Littéraire II*

Just as Algorithm ACFA extends KMP to multiple patterns, so Algorithm CWFA extends BM [CW79]. CWFA makes use of essentially the same $\boldsymbol{\delta_1}$ and $\boldsymbol{\delta_2}$ arrays employed by BM (Section 7.2); the difference is that the shifts specified by these arrays are minima over positions in *all* the patterns in a set P of patterns, not just over the positions in a single pattern. To make this clearer, consider CWFA(P) for the three patterns (11.5) introduced in the previous subsection. As shown in Figure 11.5, the states of the Commentz-Walter FA correspond to the positions of the strings in reverse order; the FA is of course constructed this way because Algorithm BM conducts pattern-matching by making comparisons with letters of the text from right to left in the pattern. Recall that at each mismatch in BM, a new value of i is calculated and then $\boldsymbol{x}[i]$ is compared with the rightmost character in the pattern — this corresponds to returning to vertex 0 in Figure 11.5 after every mismatch.

Suppose a text string

$$\boldsymbol{x} = \cdots b\underline{bab}\cdots$$

were input to the CWFA(P) of Figure 11.5 and a partial match found with the underlined letters, say $\boldsymbol{x}[i'+1..i'+3] = bab$. Then on the outarc from vertex 3 a mismatch occurs with $\boldsymbol{x}[i'] = b \neq a$. For this mismatch, the standard BM algorithm would compute $\boldsymbol{\delta_2}$-shifts individually for each of the three strings in P yielding new values

$$i = i'+5,\ i'+8,\ i'+9$$

corresponding to

$$abab,\ abaabab,\ ababcabab, \tag{11.6}$$

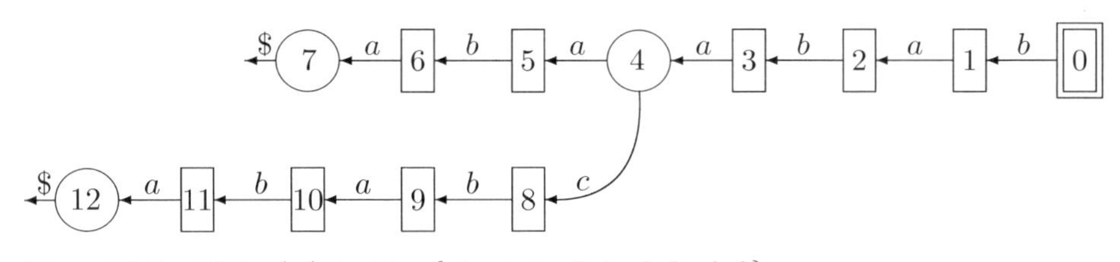

Figure 11.5. CWFA(P) for $P = \{abaabab, abab, ababcabab\}$

respectively. The shift value used by Algorithm CWFA would be $i' + 5$, the minimum of the three.

Similarly, input of a letter c at vertex 0 would determine $\boldsymbol{\delta_1}$-shifts yielding

$$i = i' + 4,\ i' + 6,\ i' + 4,$$

again corresponding to the strings (11.6). The minimum value $i = i' + 4$ would be chosen.

More generally, suppose $P = \{\boldsymbol{p_1}, \boldsymbol{p_2}, \ldots, \boldsymbol{p_r}\}$, a set of r patterns of lengths $m_1, m_2, \ldots, m_r$, respectively, and total length $m = m_1 + m_2 + \cdots + m_r$. We may think of $\boldsymbol{\delta_1}$ and $\boldsymbol{\delta_2}$ values being computed for each pattern in P, then the minimum over the r patterns being determined. As in the above example, this means finding the minimum for each state s over the at most r patterns that have a common prefix with the path to s. Since both $\boldsymbol{\delta_1}$ and $\boldsymbol{\delta_2}$ can be computed for any single pattern in time proportional to its length (Exercise 7.2.4 and Theorem 7.2.2), it follows that these shifts can be computed for P in $O(m)$ time. Though straightforward, the details are somewhat technical and are omitted.

We replace the BM routine

$$(i, j) \leftarrow hctam(i, m)$$

by a new routine

$$(i, s) \leftarrow hctam(i, 0)$$

that decrements i along a path of outarcs from vertex 0 until a mismatch occurs at some vertex s (as it eventually must, because of the \$ outarcs from terminal vertices). We also now delegate to *hctam* the task of outputting (i, s) whenever an accepting state s is reached ($accept(s) = 0$). Thus Algorithm 11.2.2 is formally almost identical to BM.

Algorithm 11.2.2 (CWFA)

– *Compute all positions i in $\boldsymbol{x}$ such that*
a substring $\boldsymbol{x}[i..i']$ traces a path from state 0
of CWFA(P), $P = \{\boldsymbol{p_1}, \boldsymbol{p_2}, \ldots, \boldsymbol{p_r}\}$,
to an accepting state

$i \leftarrow \min\{m_1, m_2, \ldots, m_r\}$
while $i \le n$ **do**
  $(i, s) \leftarrow hctam(i, 0)$
  $i \leftarrow i + \max\left\{\boldsymbol{\delta_1}\left[\boldsymbol{x}[i], s\right], \boldsymbol{\delta_2}[s]\right\}$

Observe that since every $\boldsymbol{\delta_2}$-shift with respect to any pattern $\boldsymbol{p}$ is positive, therefore i increases by at least one at each step, and so Algorithm 11.2.2 terminates.

Like ACFA(P), the automaton CWFA(P) requires $\Theta(\alpha m)$ storage and can be constructed in $\Theta(m)$ time if the alphabet is indexed, $O(m \log \alpha)$ time if it is ordered. But, like BM, Algorithm CWFA has difficulty with periodic patterns and so can require as much as $\Theta(mn)$ time in the worst case. According to [A90], Algorithm CWFA is faster on average than ACFA for fewer patterns (smaller values of r), but somewhat slower for larger values of r.

11.2.3 Approximate Patterns: WM Revisited Again!

... all words,
and no performance!

— Philip MASSINGER (1583–1640), *The Parliament of Love*

We turn now to the problem of identifying all occurrences in a given text string $\boldsymbol{x} = \boldsymbol{x}[1..n]$ of k-approximate matches with a set $P = \{\boldsymbol{p}_1, \boldsymbol{p}_2, \ldots, \boldsymbol{p}_r\}$ of patterns of lengths $m_1, m_2, \ldots, m_r$ respectively. In this subsection we outline a straightforward extension of Algorithm WM that deals efficiently with this problem [WM92].

The main idea of the modified algorithm is to process the concatenated pattern

$$\boldsymbol{p} = \boldsymbol{p}_1\boldsymbol{p}_2 \cdots \boldsymbol{p}_r$$

of length $m = m_1 + m_2 + \cdots + m_r$ as if it were a single pattern, then to correct for possible errors on the boundaries between patterns.

Recall that for each $q \in 0..k$ and $i \in 1..n$, WM computes bit vectors $\boldsymbol{s}_q^{(i)}$, where for every $j \in 1..m$,

$$\boldsymbol{s}_q^{(i)}[j] = 0 \;\Leftrightarrow\; \boldsymbol{c}[i,j] \leq q$$

and $\boldsymbol{c}[i,j]$ is the cost matrix entry (Section 10.1) for a substring $\boldsymbol{x}[i'..i]$ of the text and a prefix $\boldsymbol{p}[1..j]$ of the pattern.

WM further makes use of the bit array $\boldsymbol{t}$ in which for every $h \in 1..\alpha$ and $j \in 1..m$,

$$\boldsymbol{t}[h,j] = 0 \;\Leftrightarrow\; \boldsymbol{p}[j] = h.$$

In order to identify boundaries between the patterns, the extended algorithm introduces a mask

$$M = 01^{m_1-1}01^{m_2-1}\cdots 01^{m_r-1}$$

of length m and uses it to compute

$$\boldsymbol{t}'[h,1..m] = \boldsymbol{t}[h,1..m] \,\vee\, M,$$

where as before (Section 10.4), $\vee$ denotes the bitwise OR operation. Then an entry $\boldsymbol{t}'[h,j] = 0$ if and only if j is the first position of one of the r patterns and $\boldsymbol{p}[j] = h$.

The first bit vector $\boldsymbol{s}_0^{(i)}[j]$ is computed in the normal way according to (10.37):

$$\boldsymbol{s}_q^{(i)}[j] \leftarrow \left(\boldsymbol{s}_{q-\boldsymbol{I}}^{(i-1)}[j] \,\wedge\, \boldsymbol{s}_{q-\boldsymbol{D}}^{(i)}[j-1] \,\wedge\, \boldsymbol{s}_{q-\boldsymbol{S}}^{(i-1)}[j-1]\right)$$
$$\wedge \left(\boldsymbol{s}_q^{(i-1)}[j-1] \,\vee\, \boldsymbol{t}\big[\boldsymbol{x}[i],j\big]\right).$$

For $q = 0$, this calculation will be correct for all positions j except those positions j_0 at the beginning of each pattern for which values will have been incorrectly set corresponding to values in the preceding pattern. To compensate for this, we need to set each $s_0^{(i)}[j_0] \leftarrow 0$ if $x[i] = p[j_0]$, $s_0^{(i)}[j_0] \leftarrow 1$ otherwise. We may accomplish this by computing

$$s_0^{(i)}[j] \wedge t'\left[x[i], j\right]$$

for every $j \in 1..m$, where $\wedge$ denotes the bitwise AND: the bits in non-start positions are unchanged, and the bit in a start position can change (from 1 to 0) only if $t'\left[x[i], j_0\right] = 0$.

Turning now to the calculation of $s_1^{(i)}[j]$ for $q = 1$, we see that we must set

$$s_1^{(i)}[j_0] \leftarrow 0$$

for every start position j_0 of a pattern in P in order to allow for $q = 1$ errors to occur at the beginning of the pattern. This is analogous to the original WM algorithm's shift of 0 into $s_1^{(i)}[1]$ as part of the *rightshift* procedure (10.38). Thus, after (10.37) has been executed, we compute

$$s_1^{(i)}[j] \leftarrow s_1^{(i)}[j] \wedge M_1,$$

for every $j \in 1..m$, where $M_1 = M$, in order to set the first bit for each pattern to zero, while at the same time preserving the values of the other bits. In general, we must compute

$$s_q^{(i)}[j] \leftarrow s_q^{(i)}[j] \wedge M_q,$$

where

$$M_q = 0^q 1^{m_1 - q} 0^q 1^{m_2 - q} \cdots 0^q 1^{m_r - q}.$$

The additional bit operations required to correct the $s_q^{(i)}$ can be performed in constant time, and since they are all logical operations, they can be performed on words, in the style of (10.38), rather than just on bits as we have presented them here. Thus the complexity of Algorithm WM is not affected by the extensions, and we can therefore state the analogue of Theorem 10.4.1:

Theorem 11.2.2 *Suppose a text string $x[1..n]$ and a set $P = \{p_1, p_2, \cdots, p_r\}$ of patterns are given, all defined on an indexed alphabet of size α. Let $m = |p_1| + |p_2| + \cdots + |p_r|$. Then Algorithm WM can be extended to compute all k-approximate occurrences in x of every pattern in P, using $\Theta(k\lceil m/w \rceil n)$ time and $\Theta((\alpha + k)\lceil m/w \rceil)$ additional space, where w is the computer word length.* □

In [MM96] an algorithm is described that combines bit map and hashing techniques to compute matches with multiple approximate patterns; the method is apparently fast, but it is confined to edit distance and to approximation threshold $k = 1$ only.

11.2.4 Approximate Patterns: (Baeza-Yates)-Navarro

Oaths are but words, and words but wind.

— Samuel BUTLER (1612–1680), *Hudibras*

We conclude this section and this chapter with a description of another bit-vector algorithm originally proposed for single approximate patterns [BYN96], then later extended to multiple approximate patterns [BYN97]. As applied to multiple patterns, Algorithm BYN is an example of a "filter algorithm", a concept mentioned earlier in Subsection 10.3.2: essentially, a very fast technique is used to identify a *superset* of the substrings of the text that approximately match at least one of the patterns, then another fast technique is used to filter any spurious solutions out of the superset. Like all bit vector algorithms, BYN depends on the use of an indexed alphabet $1, 2, \ldots, \alpha$ (Section 4.1).

We first describe the algorithm that applies to single patterns under edit distance (the k-differences problem discussed in Section 10.3), then go on to show how it can be extended to multiple pattern-matching. It is important to remark that the BYN methodology is not however confined to edit distance. As we shall see, the algorithm is expressed in terms of a finite automaton that incorporates insertion, deletion and substitution: Levenshtein distance and Hamming distance can therefore be handled merely by eliminating substitution and insertion/deletion respectively.

BYNFA($\boldsymbol{p}, k$)

The main feature of the single-pattern algorithm is a novel design of a finite automaton corresponding to a given pattern $\boldsymbol{p}$ and approximation threshold k — here called BYNFA($\boldsymbol{p}, k$). As shown in Figure 11.6 for the pattern $\boldsymbol{p}$ = *this* with $k = 2$, the BYNFA may be thought of as a rectangular grid of $k+1$ rows $h \in 0..k$ and $m+1$ columns $j \in 0..m$, where of course we may assume $m \geq k$. Row h of the grid corresponds to h differences (under edit distance) between a substring of the input text string and the pattern, while column j corresponds to an approximate match up to position j of the pattern. Each intersection (h, j) in the grid is thought of as a state of the computation: if state (h, j) is ***active***, then there is an h-approximate match between $\boldsymbol{p}[1..j]$ and a substring of the text string $\boldsymbol{x}$ ending at the current position i. Initially, therefore, all states in column $j = 0$ are active, but

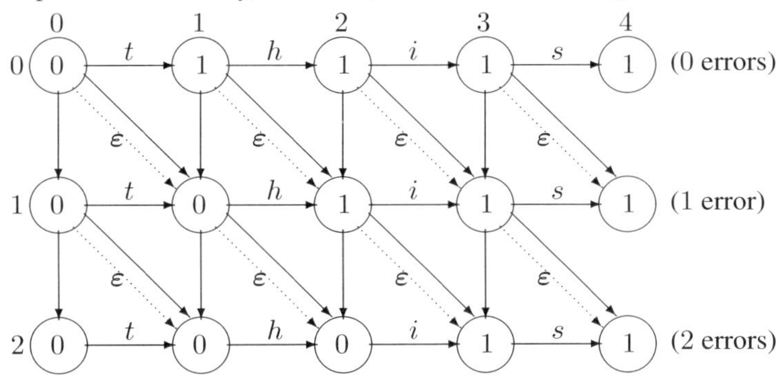

Figure 11.6. BYNFA(*this*, 2)

more generally all states (h, j), $0 \leq j \leq h \leq k$, will be active to allow for up to $k = 2$ mismatches between the first k positions of $\boldsymbol{p}$ and an empty prefix of the given text string $\boldsymbol{x}$. Initially active states are labelled 0 in Figure 11.6.

As in the other FAs that we have studied, the arcs denote transitions from one state to another based on input of the current (ith) letter of $\boldsymbol{x}$. States are active or inactive according to the following rules:

- initially active states remain active to permit a match to begin at any position in the text;
- every state reached by a transition from an active state based on the current letter itself becomes active;
- every noninitial state not reached by a transition based on the current letter becomes inactive.

In a BYNFA the transitions themselves are of four types, one that maintains the current error threshold h, three that increase it by one:

(1) A horizontal transition $(h, j-1) \rightarrow (h, j)$ represents a match between $\boldsymbol{x}[i]$ and $\boldsymbol{p}[j]$, the letter that labels the transition; h is unchanged because there is no additional difference introduced by $\boldsymbol{x}[i]$.

(2) A vertical transition $(h-1, j) \rightarrow (h, j)$ stands for the possibility that the current text letter $\boldsymbol{x}[i]$ could be inserted in the pattern at position j to force a match: the position j in $\boldsymbol{p}$ is unchanged, but the current error threshold increases.

(3) A solid diagonal transition $(h-1, j-1) \rightarrow (h, j)$ represents a substitution for $\boldsymbol{p}[j]$ by $\boldsymbol{x}[i]$: thus the positions in $\boldsymbol{p}$ and $\boldsymbol{x}$ are both advanced by one at the cost of increasing h by one.

(4) A dotted diagonal transition (marked by ε) stands for the possibility that the current letter $\boldsymbol{p}[j]$ can be deleted from $\boldsymbol{p}$ in order to force a match; this is an empty or automatic transition since the letter is deleted from the pattern without any advance being required in the text.

Observe that the empty diagonal transitions ensure that if (h, j) is active, then so is $(h + q, j + q)$ for every integer $q \in 0..k-h$. Thus in particular (k, m) becomes active whenever any $(k-q, m-q)$ becomes active, $0 \leq q \leq k$, and so we can use (h, m) active as the signal that a substring of $\boldsymbol{x}$ ending at the current position i is an h-approximate match with $\boldsymbol{p}[1..j]$.

To gain some intuition about the operation of the finite automaton, consider the input of $\boldsymbol{x}$ = *hash* and $\boldsymbol{x}$ = *whist* to BYNFA(*this*, 2). Tables 11.1 and 11.2 show the results, including the states, other than the initial ones, that become active as each letter is processed. Observe that (i, h) is output if and only if there exists a least integer h such that state (h, m) becomes active as a result of the input of $\boldsymbol{x}[i]$.

Compact Representations of BYNFA

In general, it is clear that the operation of BYNFA$(\boldsymbol{p}, k)$ depends on keeping track of state changes. In order to accomplish this, we first introduce a bit matrix $A = A[0..k, 0..m]$ in which for every $h \in 0..k$ and every $j \in 0..m$:

i	$\boldsymbol{x}[i]$	**Active states**	**Match** (i, h)
1	h	$(1,2),(2,3)$	
2	a	$(2,3)$	
3	s	$(2,4)$	$(3,2)$
4	h	$(1,2),(2,3)$	

Table 11.1: Matching $\boldsymbol{x} = $ *hash* using BYNFA$($*this*$, 2)$

i	$\boldsymbol{x}[i]$	**Active states**	**Match** (i, h)
1	w		
2	h	$(1,2),(2,3)$	
3	i	$(1,3),(2,3),(2,4)$	$(3,2)$
4	s	$(1,4),(2,4)$	$(4,1)$
5	t	$(0,1),(1,2),(2,3)$	

Table 11.2: Matching $\boldsymbol{x} = $ *whist* using BYNFA$($*this*$, 2)$

$$\begin{aligned} A[h,j] &= 0, \text{ iff the current state } (h,j) \text{ is active;} \\ &= 1, \text{ otherwise.} \end{aligned}$$

Initially, $A[h,j] = 0$ if and only if $0 \le j \le h \le k$. We denote this initial state of BYNFA as $A^{(0)}$ and subsequent states as $A^{(i)}$, $i = 1, 2, \ldots, n$, corresponding to the input of $\boldsymbol{x}[1], \boldsymbol{x}[2], \ldots, \boldsymbol{x}[n]$. Then for each i and all $0 \le h < j \le m$, the states of BYNFA are updated from those for $i-1$ by

$$\begin{aligned} A^{(i)}[h,j] \leftarrow & \Big(A^{(i-1)}[h,j-1] \,\vee\, \boldsymbol{t}\big[\boldsymbol{x}[i],j\big]\Big) \\ & \wedge\, A^{(i-1)}[h-1,j] \,\wedge\, A^{(i-1)}[h-1,j-1] \,\wedge\, A^{(i)}[h-1,j-1], \end{aligned} \tag{11.7}$$

where

- the bit array $\boldsymbol{t} = \boldsymbol{t}[1..\alpha, 1..m]$ is precomputed exactly as described for Algorithms DBG (Section 7.4) and WM (Subsection 11.2.3) — for every letter $r \in 1..\alpha$ and every $j \in 1..m$, $\boldsymbol{t}[r,j] = 0$ iff $\boldsymbol{p}[j] = r$;
- $\vee$ denotes bitwise logical OR (result is 0 iff *both* operands are 0);
- $\wedge$ denotes bitwise logical AND (result is 0 iff *any* operand is 0).

Note that in order to update $A^{(i)}[h,j]$ correctly from the preceding term $A^{(i)}[h-1,j-1]$ in the same diagonal, the values of h and j must be considered in ascending order.

By expressing the update of all the states of BYNFA in terms of logical operations that can be performed in parallel on computer words, we have satisfied the central requirement of a bit-vector algorithm. Use of the array A would require storage of $(k+1)(m+1)$ bits for the finite automaton, and, as shown in Exercise 11.2.9, computation of $(k+1)(2m-k)/2 \in \Theta(mk)$ bits corresponding to every position i in $\boldsymbol{x}$. It turns out that the existence of empty transitions enables us to express the essential information about the states in a somewhat more convenient way: since by virtue of the term $A^{(i)}[h-1,j-1]$, the active states propagate automatically down each diagonal, we need only store the row index of the least active state in each diagonal in order to completely specify all active (and inactive) states.

Observe in fact that, due to the effect of empty transitions in every diagonal, no state (h, j), $j > k + h$, can become active unless state (k, m) has already become active. Consequently, it suffices to consider changes in diagonals originating from states $(0, j)$, $j = 1, 2, \ldots, m - k$. This simplified model allows us to report all positions i in $\boldsymbol{x}$ at which an h-approximate match occurs, $h \leq k$, but does not allow us to specify the corresponding value of h, as in Table 11.2.

We therefore define a single-dimensional array $D = D[0..m-k+1]$ in which each position $j \in 0..m-k+1$ corresponds to the diagonal beginning at state $(0, j)$ in BYNFA$(\boldsymbol{p}, k)$. The value of each $D[j]$ is an integer in the range $0..k+1$: if $D[j] = k+1$, no entry in the diagonal is active, while otherwise $D[j]$ is the index of the least active row in diagonal j. Let $D^{(0)}$ be the initial state of the diagonal array D; then $D^{(0)}[0] = 0$, while for every $j \in 1..m-k$, $D^{(0)}[j] = k+1$. To ensure that (11.8) is correct for $j = m-k$, we set $D^{(0)}[m-k+1] = 0$. Denoting by $D^{(i)}$ the diagonal array corresponding to position i of $\boldsymbol{x}$, we can easily derive a recurrence relation for $D^{(i)}[j]$, $j \in 1..m-k$, from (11.7):

$$D^{(i)}[j] \leftarrow \min\left\{D^{(i-1)}[j+1]+1,\ D^{(i-1)}[j]+1,\ g(i-1, j-1)\right\}, \tag{11.8}$$

where $g(i-1, j-1)$ is the minimum row index h_{min} in diagonal $j-1$ such that

- $D^{(i-1)}[j-1] \leq h_{\text{min}} \leq k\}$ (in other words, $A[h_{\text{min}}, j-1]$ was active for $\boldsymbol{x}[i-1]$), and
- $\boldsymbol{t}[\boldsymbol{x}[i], j-1+h_{\text{min}}] = 0$ (in other words, the current letter $\boldsymbol{x}[i]$ equals the letter $\boldsymbol{p}[j-1+h_{\text{min}}]$ in the pattern);

or else, if no such h_{min} exists, then $g(i-1, j-1) = k+1$. The first term in (11.8) describes a vertical transition (insertion of $\boldsymbol{x}[i]$) in BYNFA$(\boldsymbol{p}, k)$, the second a diagonal transition (substitution of $\boldsymbol{x}$[i] for $\boldsymbol{p}$[j]), and the third a horizontal transition (based on a match between $\boldsymbol{x}[i]$ and some letter of the pattern originating at an active position in diagonal $j-1$). Thus (11.8) provides a means of identifying all active states corresponding to each diagonal $j \in 1..m-k$ resulting from the input of $\boldsymbol{x}[i]$, based on knowledge of the least active state in each of the diagonals $j+1, j, j-1$ corresponding to $i-1$. The only calculations required are ordinary addition, retrieval of a bit from $\boldsymbol{t}$, and computation of a minimum.

Observe that for $h \in 0..k$, there is an h-approximate match of $\boldsymbol{p}$ with a substring of $\boldsymbol{x}$ ending at position i if and only if $D^{(i)}[m-k] \leq k$.

The Single-Pattern BYN Algorithm

The usual computer representation of each least active row index $D[j]$, $1 \leq j \leq m-k$, would of course be binary, and would therefore require $\lceil \log_2(k+2) \rceil$ bits, yielding a total for $D[1..m-k]$ of $\lceil \log_2(k+2) \rceil (m-k)$ bits overall. Unfortunately, this representation does not lead to an efficient algorithm. For this reason, BYN employs a ***unary*** representation of each $D[j]$:

$$0^{k+1-D[j]}1^{D[j]}.$$

Thus $D[j] = 0$ is represented by 0^{k+1}, $D[j] = 2$ by $0^{k-1}1^2$, $D[j] = k+1$ by 1^{k+1}, and so on. With a single zero separator bit introduced at the left of each $D[j]$, the complete representation of $D[1..m-k]$ becomes

$$D = 0\,0^{k+1-D[1]}1^{D[1]}\;0\,0^{k+1-D[2]}1^{D[2]}\cdots 0\,0^{k+1-D[m-k]}1^{D[m-k]},$$

a total of $(k+2)(m-k)$ bits, roughly the number required for storage of A.

In this formulation the required operations on elements $D[j]$ reduce to easily-implemented bit operations:

(O1) $D[j]+1:$ *leftshift*$\big(D[j],1\big)\ \vee\ 0^k1$;

(O2) $\min\big\{D[j_1],D[j_2]\big\}:\ D[j_1]\ \wedge\ D[j_2]$;

(O3) accessing $D[j\pm 1]:$ *leftshift*$\big(D[j+1],k+2\big)$ or *rightshift*$\big(D[j-1],k+2\big)$.

In order to implement the function g, we need first to define an inverted form of each row of the bit array $\boldsymbol{t}$:

$$\boldsymbol{t}'[r] = \boldsymbol{t}[r,m]\boldsymbol{t}[r,m-1]\cdots\boldsymbol{t}[r,1],$$

for every letter $r \in 1..\alpha$. Then a mask for $D[1..m-k]$ is defined by

$$T'[r] = 0\ \textit{rightshift}(\boldsymbol{t}'[r],0)\ 0\ \textit{rightshift}(\boldsymbol{t}'[r],1)\cdots 0\ \textit{rightshift}(\boldsymbol{t}'[r],m-k-1),$$

again for every $r \in 1..\alpha$. Making the usual assumption that all shifts, whether right or left, bring zeros into positions in the word that bits are shifted out of, we may now represent $m-k$ calculations (11.8), $j = 1,2,\ldots,m-k$, by a word-based operation on $D = D[1..m-k]$:

$$\begin{aligned} D = {} & (\textit{leftshift}(D,k+3)\ \vee\ (0^{k+1}1)^{m-k-1}01^{k+1}) \\ & \wedge\ (\textit{leftshift}(D,1)\ \vee\ (0^{k+1}1)^{m-k}) \\ & \wedge\ \textit{rightshift}((D' + (0^{k+1}1)^{m-k})\ \wedge\ D',1) \\ & \wedge\ D^{(0)}[1..m-k], \end{aligned} \tag{11.9}$$

where

$$D^{(0)}[1..m-k] = (0\,1^{k+1})^{m-k}$$

and

$$D' = \textit{rightshift}(D,k+2)\ \vee\ T[\boldsymbol{x}[i]].$$

The first three lines of (11.9) correspond one by one to the first three lines of (11.8), making use of the bit operations (O1)–(O3) specified above, and $D^{(0)}$ is introduced in the fourth line to ensure that any overflow into the separator bits is removed.

Of course the current index i is output if and only if, after execution of (11.9),

$$D\ \wedge\ 0^{(k+2)(m-k-1)}\,0\,10^k = 0;$$

that is, if and only if $D^{(i)}[m-k] \leq k$. Whenever i is output, we ensure that no false occurrences of a match are subsequently reported by clearing $D[m-k]$:

$$D \leftarrow D \vee 0^{(k+2)(m-k-1)}\, 0\, 1^{k+1}.$$

The earnest reader may rightly feel that (11.9) embodies sufficient complexities for the moment, and so we leave the exact specification of Algorithm BYN to Exercise 11.2.10. In [BYN96] a form of skip loop (Section 8.1) is proposed as a means of speeding up the algorithm: the text $\boldsymbol{x}$ is scanned to locate occurrences of any one of the first $k+1$ letters of the pattern, and the automaton is started only from those occurrences.

So far we have described the simulation of the BYNFA as if the computer word length w were sufficient to contain D; that is, we have implicitly assumed that $w \geq (k+2)(m-k)$. When this assumption is valid, each of the n steps of Algorithm BYN requires only constant time for the operations on a single word, and so the algorithm executes in $\Theta(n)$ time overall.

Partitioning the Automaton

In the general case that $w < (k+2)(m-k)$, the partitioning of the problem is a more complex matter than it has been for other bit-mapping algorithms (Section 10.4), due to the two-dimensional nature of BYNFA. Three cases may be identified:

(1) $w \geq k+2$

In this case $\lceil w/(k+2) \rceil$ out of the total of $m-k$ diagonals can be accomodated in a single word, forming a "subautomaton", and so altogether $\lceil (m-k)/\lceil w/(k+2) \rceil \rceil$ words are required for the complete BYNFA. Since each $D[j]$ depends both on $D[j-1]$ and $D[j+1]$, the algorithm is complicated by the necessity for the leftmost and rightmost diagonal in each subautomaton to communicate with its neighbouring subautomaton, but as discussed in [BYN96] and shown in Exercise 11.2.11, each word can still be processed in constant time. Thus the overall time required for the algorithm is $O(k(m-k)n/w)$.

(2) $w < k+2$ but $w \geq 2(m-k)$

When a single diagonal exceeds the storage capacity of a computer word, we can try to partition the BYNFA into vertical components, one above another, where each component contains a section of each of the $m-k$ diagonals. Then, since the portion of each diagonal that occurs in each component must have a separator bit, there are only $w-(m-k)$ bits available for storage of at least one bit from each of $m-k$ diagonals. Thus for vertical partitioning to work, we require $w \geq 2(m-k)$, and then each word will contain

$$\lfloor (w-(m-k))/(m-k) \rfloor = \lfloor w/(m-k) \rfloor - 1$$

out of a total of $k+1$ nonseparator bits in each diagonal. Hence altogether $\lceil (k+1)/(\lfloor w/(m-k) \rfloor - 1) \rceil$ words are required to represent the complete BYNFA. Again the algorithm is complicated by the need for communication, but each word can still be processed in constant time, so that the overall time requirement remains $O(k(m-k)n/w)$.

(3) $w < k+2$ and $w < 2(m-k)$

Here a combination of partitions (1) and (2) once more results in the use of $\Theta(k(m-k)/w)$ space and $O(k(m-k)n/w)$ time. For details see [BYN96].

Recalling from Digression 1.3.1 that $w \in \Omega(\log n)$, we have then

Theorem 11.2.3 *Given a single pattern $\boldsymbol{p} = \boldsymbol{p}[1..m]$ and a nonnegative integer $k < m$, Algorithm BYN computes all k-approximate matches under edit distance in a text string $\boldsymbol{x} = \boldsymbol{x}[1..n]$ using $\Theta(k(m-k)/w)$ additional space and $O(k(m-k)n/\log n)$ time, where w is the computer word length.* □

In [BYN96] a number of more sophisticated strategies are discussed that are designed to reduce running time, and a more detailed analysis of time and space requirements is used to show that the worst-case bounds of Theorem 11.2.3 are pessimistic. A modified algorithm is described that executes in $O(n)$ time in the average-case, provided that k/m assumes certain medium-sized values. Still, for approximate matching of single patterns, both the worst-case and average-case time bounds for BYN appear generally to be greater than those achieved by Algorithm M* (Subsection 10.3.2).

The Multiple-Pattern BYN Algorithm

The main idea of the multiple-pattern version of Algorithm BYN is to use the single-pattern version as a filter to compute a superset of the set of k-approximate matches of $\boldsymbol{x}$ with a collection $P = \{\boldsymbol{p_1}, \boldsymbol{p_2}, \ldots, \boldsymbol{p_r}\}$ of patterns, then use efficient techniques to refine the superset into the required solution set. In our discussion of the multiple-pattern BYN algorithm, we make the simplifying assumption that each pattern is of the same length m/r, where m is the total length of all the patterns: if this is initially not true, we can extend each pattern in P with nonmatching letters until each pattern is the same length as the one that initially was longest.

The technique used to compute the superset is effectively superposition of the patterns, so that the first transition in BYNFA(P, k) recognizes a match not just with the first letter of a single pattern but with any one of

$$\boldsymbol{p_1}[1], \boldsymbol{p_2}[1], \ldots, \boldsymbol{p_r}[1];$$

and more generally the jth transition recognizes a match with any one of

$$\boldsymbol{p_1}[j], \boldsymbol{p_2}[j], \ldots, \boldsymbol{p_r}[j].$$

Suppose for example we extend BYNFA(*this*, 2) illustrated in Figure 11.6 to

$$\text{BYNFA}\big(\{\textit{this,worm}\}, 2\big).$$

The resulting FA is shown in Figure 11.7.

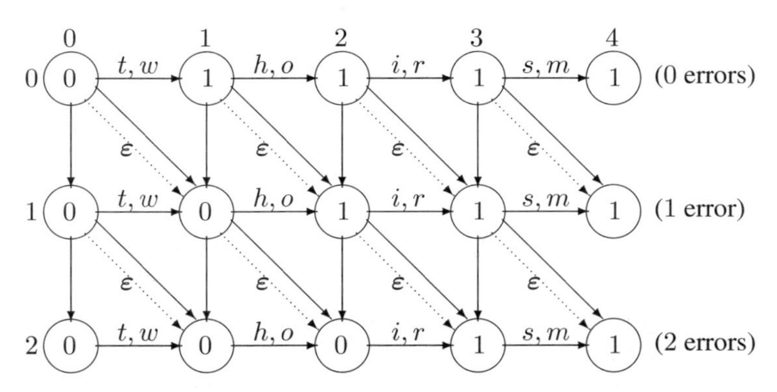

Figure 11.7. BYNFA$(\{this,worm\}, 2)$

If a string $\boldsymbol{x} = what$ were input to this FA, a 2-approximate match would be found for $i = 2$ even though $d_E(what,this) = d_E(what,worm) = 3$; on the other hand, matches would correctly be found for $i = 4$ in $\boldsymbol{x} = wham$ (edit distance 2 from *worm*) as well as for $i = 3$ and $i = 4$ in $\boldsymbol{x} = whim$ (edit distance 2 from both *this* and *worm*).

For arbitrary $\boldsymbol{x}$ this approach clearly will identify all positions in $\boldsymbol{x}$ that can possibly be part of a k-approximate match with at least one element of P. Furthermore, it is easy to implement: in order to recognize a match with multiple letters on each transition, it suffices to redefine the bit array $\boldsymbol{t}[1..\alpha, 1..m]$ so that $\boldsymbol{t}[h, j] = 0$ for *every* letter h that occurs in position j of some pattern of P. The preprocessing required for this change is described in Section 10.4 under **Matching Sets of Letters**: constant time is required for each new update to $\boldsymbol{t}$ and so in accordance with Theorem 7.4.4, $\boldsymbol{t}$ can be computed in time $\Theta(\alpha\lceil m/w\rceil + m)$, where as above m is the sum of the lengths of the r patterns in P. Then after preprocessing, in accordance with Theorem 11.2.3, the superset of solutions to the multiple approximate pattern-matching problem can be computed in $O(k(m/r - k)n/\log n)$ time.

Let $S = S(P, k, \boldsymbol{x}) = \{i_1, i_2, \ldots, i_q\}$ denote the superset computed by Algorithm BYN corresponding to a set P of patterns, an approximation threshold k, and a given text string $\boldsymbol{x}$. The elements of S are positions in $\boldsymbol{x}$ at which a match has been found by BYNFA(P, k).

In order to filter out elements of S that do not identify k-approximate matches with at least one individual pattern in P, the substring of $\boldsymbol{x}$ determined by each i_s, $s = 1, 2, \ldots, q$, needs to be compared with each $\boldsymbol{p_t}$, $t = 1, 2, \ldots, r$; if no k-approximate match under edit distance is found with any $\boldsymbol{p_t}$, then i_s is discarded. Each of these comparisons is essentially an edit distance calculation that can be performed using Algorithm UM (Section 9.5) in time $O(km/r)$; thus over q elements of S the total time requirement will be $O(kmq)$, and since certainly $q \leq n$, the filtering can therefore be done in time $O(kmn)$.

The effectiveness of this approach compared with the "rerevisited" Algorithm WM (Subsection 11.2.3) depends both on the size of q and on the number q' of elements of S actually discarded by the filtering process: if q is bounded above by a small integer, then the time required to execute Algorithm UM reduces to $O(km)$; if moreover q' is small, then most of the executions of Algorithm UM will be devoted to the necessary task of verifying elements of S rather than to discarding the excess. [BYN97] discusses strategies for reducing the size of q', especially executing Algorithm BYN on partitioned patterns grouped as much as possible to contain frequent identical subpatterns.

Exercises 11.2

1. In Algorithm ACFA the *match* function may need to be called twice for many values of i. Rewrite the algorithm so as to eliminate this inefficiency.
2. As shown in Figure 11.4, when multiple patterns P occur, many state values s for an individual pattern, and hence in some cases also the border values $\beta[s]$, may need to be offset from their natural values within the pattern. Explain how you would ensure that these offsets were correctly computed during the construction of ACFA(P).
3. Given a set of patterns $P = \{\boldsymbol{p}_1, \boldsymbol{p}_2, \ldots, \boldsymbol{p}_r\}$ of lengths $m_1, m_2, \ldots, m_r$, respectively, where $m = m_1 + m_2 + \cdots + m_r$, design efficient algorithms to construct an efficient implementation of ACFA(P) under *each* of the following assumptions:

   (a) the alphabet is indexed;

   (b) the alphabet is ordered but not indexed.

   In each case characterize the time complexity of your algorithm and the search efficiency of the corresponding ACFA.
4. For each of the cases identified in the previous exercise, design an appropriate function *match* and characterize its time requirement.
5. Suppose that in a set $P = \{\boldsymbol{p}_1, \boldsymbol{p}_2, \ldots\}$ of patterns, $\boldsymbol{p}_1$ is a border of $\boldsymbol{p}_2$. As we observed in Figure 11.4, the accepting vertex corresponding to $\boldsymbol{p}_1$ will therefore not be a terminal vertex of ACFA($\boldsymbol{p}$). Show that in this case, if Algorithm ACFA reports a match with $\boldsymbol{p}_2$ at position i of $\boldsymbol{x}$, then no match will be reported with $\boldsymbol{p}_1$ at position i. Discuss the wisdom or otherwise of this approach, then show how to modify the algorithm to output matches with $\boldsymbol{p}_1$ also.
6. Recall that in Algorithm KMP (Section 7.1) a slightly modified array β' was precomputed. Then we realized that in fact a more efficient array β'' could be precomputed to specify for each position j the longest border of $\boldsymbol{p}[1..j-1]$ *not* followed by the letter $\boldsymbol{p}[j]$.

   Show that the same improvement can be be made to the calculation of β in ACFA($\boldsymbol{p}$), and write pseudocode for a linear-time algorithm to compute a modified border array, $\hat{\beta}$ say.
7. Write pseudocode for the routine *hctam* of Algorithm CWFA.
8. (Research Project!) For your favourite variant of Algorithm BM (Section 7.2), develop an FA suitable for multiple pattern-matching, together with associated shift procedures and a pattern-matching algorithm. Characterize the asymptotic complexity of your routines and compare them with Algorithms ACFA and CWFA.
9. Show that the array A defined for BYNFA has exactly $(k+1)(2m-k)/2$ entries that are not initially active.

10. Assuming the computer word length w is sufficiently long to hold $D = D[1..m-k]$, specify an implementation of the single-pattern form of Algorithm BYN.

11. Now assuming that the computer word length w satisfies $w < (k+2)(m-k)$ and $w \geq k+2$, modify Algorithm BYN to execute in $O(k(m-k)n)$ time.

12. Show how to modify Algorithm BYN so that it will output not only positions i that identify an h-approximate match, $h \in 0..k$, but also output the corresponding value of h. Is the complexity of the algorithm increased by this modification?

13. In the text a form of skip loop is briefly described, in which the automaton is started only at positions i in $\boldsymbol{x}$ such that $\boldsymbol{x}[i]$ equals one of the letters of $\boldsymbol{p}[1..k+1]$. Prove that at least one of the letters of $\boldsymbol{p}[1..k+1]$ must be present in any substring of $\boldsymbol{x}$ that forms a k-approximate match with $\boldsymbol{p}$.

14. Specify an algorithm that precomputes a Boolean array $B[1..\alpha]$ that tells whether or not each letter $r \in 1..\alpha$ is one of the first $k+1$ letters of $\boldsymbol{p}$. What is the complexity of your algorithm?

15. Show that whenever $m \leq 2(\sqrt{w}-1)$, BYNFA$(\boldsymbol{p}[1..m], k)$ can be stored in a single computer word of length w, no matter what the value of k. Then compute the maximum allowable values of m for $w = 32$ and $w = 64$.

Part IV

Computing Generic Patterns

Part IV

My father still reads the dictionary every day.
He says your life depends on your power to master words.

— Arthur SCARGILL (1938–) *The Sunday Times, 10 January 1982*

Our algorithmic voyage concludes with a study of methods for computing generic patterns — that is, patterns described by stated properties rather than by a specification or characterization of the letters contained in them. For example, we recognize both aa and $bcbc$ as repetitions because they share a common property; we don't care at all what particular letters they contain.

We shall find in Part IV that generic patterns generally relate to periodicity properties of strings, though there seems to be no reason why properties quite unrelated to periodicity should not also be "generic". In Chapter 12 we first (in Section 12.1) study "classical" algorithms for computing all the repetitions in a given string, then we go on in Section 12.2 to look at recent insights that permit a more compact representation and thus more efficient calculation of repetitions encoded as runs (Definition 2.3.3).

Chapter 13 deals with various relaxed forms of periodicity. The first of these is "quasiperiodicity": a periodic string contains concatenated occurrences of a generator $\boldsymbol{u}$, but in a quasiperiodic string occurrences of $\boldsymbol{u}$ may also overlap each other. In accordance with the discussion in Definition 2.3.3, we say that a quasiperiodic string is "covered" by $\boldsymbol{u}$, and in Section 13.1 we describe an algorithm that computes all the covers of every prefix of a string in linear time. Sections 13.2 and 13.3 generalize the idea of a repetition to that of a "repeat", where occurrences of a repeating substring need not be concatenated, but may also be separated from each other. We discuss algorithms to generate all repeats, both exact and approximate, and then finally we extend our ideas to include "approximate periodicity".

Chapter 12

Repetitions (Periodicity)

Thanks to words, we have been able to rise above the brutes; and thanks to words, we have often sunk to the level of demons.

— Aldous HUXLEY (1894–1963), *Adonis and the Alphabet*

In this chapter we describe algorithms that compute the repetitions (tandem repeats) in a given string $\boldsymbol{x} = \boldsymbol{x}[1..n]$. Before embarking on this material, the reader should study the beginning of Section 2.3 where the basic terminology is introduced and the encodings (i, p^*, r^*) of repetitions and (i, p^*, r^*, t) of runs are defined.

In the late 1960s it was shown [LS69] that there were $O(n \log n)$ squares in any string $\boldsymbol{x}$, provided that the generators of the squares were constrained to not themselves be repetitions. In the early 1980s three algorithms appeared [C81, AP83, ML84] that computed all the repetitions in $\boldsymbol{x}$ in $O(n \log n)$ time. In the first two sections of this chapter we present two of these algorithms, but before doing so we show that the upper bound is best possible for strings on an unordered alphabet, a result established in [ML84].

Theorem 12.0.1 *Let $\boldsymbol{x} = \boldsymbol{x}[1..n]$ be a string defined on an unordered alphabet. Any algorithm that decides on the basis of letter comparisons whether or not $\boldsymbol{x}$ contains a repetition requires $\Omega(n \log n)$ time in the worst case.*

Proof Recall from Section 4.1 that symbols from a general alphabet can be compared only to determine equality or inequality. Suppose that in fact $\boldsymbol{x}$ is a string in which each of the n letters is distinct. Assuming for simplicity of presentation that n is even, we show that $\Omega(n \log n)$ letter comparisons are required in order to discover that $\boldsymbol{x}$ contains no squares, hence no repetitions.

Suppose that for some $1 \leq i < j \leq n$, we compare $\boldsymbol{x}[i]$ with $\boldsymbol{x}[j]$, discovering that $\boldsymbol{x}[i] \neq \boldsymbol{x}[j]$. The only squares in $\boldsymbol{x}$ that can be eliminated by this comparison are those that require $\boldsymbol{x}[i] = \boldsymbol{x}[j]$. If we let $k = j - i$, there are therefore at most k such squares eliminated, each of length $2k$:

$$\boldsymbol{x}[i-k+1..j], \boldsymbol{x}[i-k+2..j+1], \ldots, \boldsymbol{x}[i..j+k-1]. \tag{12.1}$$

Note that in fact there could be fewer than k squares eliminated, since the list of eliminated squares resulting from the knowledge that $\boldsymbol{x}[i+1] \neq \boldsymbol{x}[j+1]$ is

$$\boldsymbol{x}[i-k+2..j+1], \boldsymbol{x}[i-k+3..j+2], \ldots, \boldsymbol{x}[i+1..j+k],$$

containing only one square not already eliminated in (12.1).

For each integer $k \in 1..n/2$, there are (as remarked in Section 1.2) altogether $n-2k+1$ different possible substrings of length $2k$, hence $n - 2k + 1$ different possible squares of length $2k$; hence to eliminate all possible squares of length $2k$ we must make at least $(n-2k+1)/k$ letter comparisons.

Observe however that by eliminating all possible squares of length $2k$, we get no information about the existence or nonexistence of any squares of length $2k'$, where $k' \neq k$. This is because, in order to determine whether or not any substring of length $2k'$ is a square, we must compare pairs of letters separated by k' positions. Thus the elimination of squares of different lengths $2k$, $k = 1, 2, \ldots, n/2$, must be treated as independent activities. Consequently, to ensure that $\boldsymbol{x}$ is square-free, at least

$$\sum_{k=1}^{n/2}(n-2k+1)/k \;\geq\; (n+1)H_{n/2} - n$$

letter comparisons must be made, where H_i is the ith harmonic number. Since $H_i > \ln i + \gamma$ [K73a], where $\gamma \approx 0.577$ is Euler's constant, we find that the required number of letter comparisons exceeds

$$\begin{aligned}(n+1)\big(\ln n - (\ln 2 - \gamma)\big) - n &> (n+1)(\ln n - 1.13)\\ &\in \Omega(n \log n).\end{aligned}$$

□

This result is a fundamental one for the calculation of repetitions, and it leads directly to other fundamental questions. For example:

- Does the $n \log n$ lower bound hold also in the case that the alphabet is not general?

  In [C84] Crochemore describes an algorithm that, using suffix trees (Section 5.2), determines in time $O(n \log \alpha)$ whether or not a given string $\boldsymbol{x}[1..n]$ on an ordered alphabet of size α is square-free. In [ML85] Main and Lorentz describe an algorithm with similar properties that however avoids the use of suffix trees. If in fact $\alpha \in \Theta(n)$, then the time bound for these algorithms is no improvement on Theorem 12.0.1; on the other hand, if α is a constant known in advance, then so is $\log \alpha$, and thus these algorithms can be represented as requiring only $O(n)$ time. See Subsection 5.2.5.

- For a string $\boldsymbol{x}[1..n]$ on an ordered alphabet, can the repetitions themselves be computed in $O(n)$ time — or at least in less than $\Theta(n \log n)$ time?

  It seemed for some time that there was little hope that this question could be answered in the affirmative, since in view of Crochemore's result ([C81] and Section 3.4), Fibostrings of length f_n contain $\Theta(f_n \log f_n)$ repetitions encoded in triples (i, p^*, r^*) that essentially describe the normal form (Section 1.2) of each maximal repeating substring. Thus it seemed that just to output the repetitions could require $\Theta(n \log n)$ time for a string of length n. But more recently the idea of a run (Section 2.3) encoded as a 4-tuple (i, p^*, r^*, t) has been introduced [M89] and used by various authors [G97, IMS97, KK98]. Further, it was shown in [KK98, KK00] that the number of runs in any string is linear in the string length, thus opening up the possibility that, at least for strings on a known ordered alphabet, the repetitions, encoded as runs, might be computed in linear time. In fact, [KK00] describes an algorithm that makes use of suffix trees (thus assuming an ordered alphabet) to compute all runs in $\boldsymbol{x}$; if the alphabet is moreover assumed to be not too large and either fixed or indexed (Section 4.1), the potential $\log n$ factor hidden in suffix tree construction (Subsection 5.2.5) can again be ignored, thus yielding a $\Theta(n)$ algorithm. We discuss Algorithm KK in Section 12.2. Whether or not there exists a $\Theta(n)$-time all-runs algorithm for ordered alphabets that does not depend on the use of suffix trees remains an open question.

12.1 All Repetitions

Yes, I had two strings to my bow;
both golden ones, agad! and both cracked.

— Henry FIELDING (1707–1754), *Love in Several Masques*

In this section we describe two very different algorithms that achieve the same purpose with comparable complexity: an $O(n \log n)$-time algorithm [C81] and a $\Theta(n \log n)$-time algorithm [ML84] that both compute all the repetitions in a given string $\boldsymbol{x} = \boldsymbol{x}[1..n]$.

12.1.1 Crochemore

Man shall not live on bread alone,
but on every word that proceeds
out of the mouth of God.

— *MATTHEW 4:4*

As a preliminary to explaining Algorithm C, we first describe a naïve algorithm that achieves the same result at each of its stages that C achieves: a specification of all repeating substrings in $\boldsymbol{x}$ of length ℓ, where $\ell = 1, 2, \ldots, n-1$.

Consider the Fibostring

$$\begin{array}{rccccccccccccc} & {\scriptstyle 1} & {\scriptstyle 2} & {\scriptstyle 3} & {\scriptstyle 4} & {\scriptstyle 5} & {\scriptstyle 6} & {\scriptstyle 7} & {\scriptstyle 8} & {\scriptstyle 9} & {\scriptstyle 10} & {\scriptstyle 11} & {\scriptstyle 12} & {\scriptstyle 13} \\ \boldsymbol{f_6} = & a & b & a & a & b & a & b & a & a & b & a & a & b, \end{array}$$

and suppose that at stage (level) $\ell = 1$, we compute sequences of positions in $\boldsymbol{f_6}$ at which equal letters occur:

$$\begin{array}{lcc} \ell = 1 & <1,3,4,6,8,9,11,12> & <2,5,7,10,13> \\ & a & b \end{array}$$

Recall (Section 4.1) that on an unordered alphabet, this calculation may require as much as $\Theta(n^2)$ time, but only $O(n \log n)$ and $\Theta(n)$ time on ordered and indexed alphabets, respectively.

Now for $\ell = 2$ we wish to identify all repeating substrings of length 2. Since clearly all repeating substrings of length $\ell \geq 2$ will have a common prefix of length $\ell - 1$, we see that the level ℓ calculation must yield sequences that are *subsequences* of those computed for $\ell - 1$. In other words, the decomposition at level $\ell \geq 2$ is a ***refinement*** of the decomposition at level $\ell - 1$, as shown in Table 12.1.

$\ell = 2$	$<1,4,6,9,12>$	$<3,8,11>$	$<2,5,7,10>$	$<13>$	
	ab	*aa*	*ba*	*b*\$	
$\ell = 3$	$<1,4,6,9>$	$<12>$	$<3,8,11>$	$<2,7,10>$	$<5>$
	aba	*ab*\$	*aab*	*baa*	*bab*
$\ell = 4$	$<1,6,9>$	$<4>$	$<3,8>$	$<11>$	$<2,7,10>$
	abaa	*abab*	*aaba*	*aab*\$	*baab*
$\ell = 5$	$<1,6,9>$	$<3>$	$<8>$	$<2,7>$	$<10>$
	abaab	*aabab*	*aabaa*	*baaba*	*baab*\$
$\ell = 6$	$<1,6>$	$<9>$	$<2>$	$<7>$	
	abaaba	*abaab*\$	*baabab*	*baabaa*	
$\ell = 7$	$<1>$	$<6>$			
	abaabab	*abaabaa*			

Table 12.1: The Crochemore refinement of $\boldsymbol{f}_6 = abaababaabaab$

Note: In this display singletons have been dropped after their first occurrence, since no further refinement is possible.

It is not difficult to imagine an approach to this calculation that would in the worst case require $\Theta(n^2)$ time: a straightforward refinement of each sequence reported for ℓ into subsequences for $\ell + 1$. For such a refinement the string $\boldsymbol{x} = (ab)^{n/2}$ for example would at each level ℓ require $n - \ell + 1$ comparisons, hence $\Theta(n^2)$ comparisons overall. But there are two insights, applications of an idea introduced in [H71], that allow this time complexity to be reduced:

(1) The refinement of each sequence can be accomplished indirectly rather than using the direct approach suggested above.

At each level ℓ, the direct refinement of any sequence $c_j^{(\ell)}$ takes place based on the individual letters that occur in the first position to the right of the substring corresponding to $c_j^{(\ell)}$. More precisely, if $c_j^{(\ell)} = <p_1, p_2, \ldots, p_r>$, the positions

$$\boldsymbol{x}[p_1 + \ell], \boldsymbol{x}[p_2 + \ell], \ldots, \boldsymbol{x}[p_r + \ell]$$

need to be inspected; if any pair is equal, say

$$\boldsymbol{x}[p_{h_1} + \ell] = \boldsymbol{x}[p_{h_2} + \ell],$$

then p_{h_1} and p_{h_2} are placed in the same sequence at level $\ell + 1$.

Thus, for example, in the refinement of $\boldsymbol{f_6}$ displayed above, for $\ell = 3$, the sequence $<1, 4, 6, 9>$ is refined into separate sequences for $\ell = 4$, $<1, 6, 9>$ and $<4>$, because

$$\boldsymbol{f_6}[1 + 3] = \boldsymbol{f_6}[6 + 3] = \boldsymbol{f_6}[9 + 3] \neq \boldsymbol{f_6}[4 + 3].$$

But the refinement can also be handled indirectly by focussing, not on the sequence to be refined, but rather on the sequences *with respect to which* other sequences will be refined.

Consider again $\ell = 3$ and the sequence

$$c_1^{(3)} = <p_1, p_2, p_3, p_4> = <1, 4, 6, 9>$$

relating to substring aba in the above example. For each $p_h > 1$, we know that

$$\boldsymbol{f_6}[p_h - 1..p_h + \ell - 1]$$

specifies a substring of $\boldsymbol{f_6}$ of length $\ell + 1$ that belongs to some sequence $c_{j'}^{(\ell+1)}$ at level $\ell + 1$. In fact, since $c_j^{(\ell)}$ corresponds to a unique substring of $\boldsymbol{f_6}$, each such sequence $c_{j'}^{(\ell+1)}$ must be formed from exactly those positions $p_{h_1}, p_{h_2}, \ldots, p_{h_k}$ in $c_j^{(\ell)}$ that constitute an equivalence class:

$$\boldsymbol{f_6}[p_{h_1} - 1] = \boldsymbol{f_6}[p_{h_2} - 1] = \cdots = \boldsymbol{f_6}[p_{h_k} - 1].$$

In our example, $c_j^{(\ell)} = c_1^{(3)} = <1, 4, 6, 9>$, and since

$$\boldsymbol{f_6}[4 - 1] = \boldsymbol{f_6}[9 - 1] = a,$$

the level 3 sequence $<3, 8, 11>$ is refined into $<3, 8>$ and $<11>$ at level 4. Further, since

$$\boldsymbol{f_6}[5 - 1] = b,$$

the level 3 sequence $<5>$ is trivially refined into itself at level 4. Thus refinement of $<3, 8, 11>$ and $<5>$ is achieved *with respect to* $<1, 4, 6, 9>$.

Similarly, refinement with respect to sequence $<2, 7, 10>$ at level 3 yields

$$\boldsymbol{f_6}[2-1] = \boldsymbol{f_6}[7-1] = \boldsymbol{f_6}[10-1] = a,$$

and so $<1, 6, 9>$ necessarily forms a sequence at level 4, a refinement of $<1, 4, 6, 9>$ at level 3.

We find then that refinement at level $\ell + 1$ can be accomplished indirectly by considering the position one to the left of each starting position in each sequence at level ℓ.

(2) But the indirect refinement does not need to be carried out with respect to *every* sequence at level ℓ: in every set of sequences with the same parent sequence in level $\ell - 1$, we can always avoid using exactly one of them to perform refinements of sequences at level ℓ.

To see this, consider again the refinement of sequence $<1, 4, 6, 9>$ at level 3 into $<1, 6, 9>$ and $<4>$ at level 4. Sequence $<1, 4, 6, 9>$ corresponds to a string with proper suffix ba, represented at level 2 by $<2, 5, 7, 10>$; thus any refinement of $<1, 4, 6, 9>$ at level 4 is derived from the corresponding refinement of $<2, 5, 7, 10>$ at level 3. In fact, we see that $<1, 6, 9>$ is derived from subsequence $<2, 7, 10>$ (suffix baa), while $<4>$ is derived from subsequence $<5>$ (suffix bab). This observation leads us to understand that in order to refine $<1, 4, 6, 9>$ at level 4, it is sufficient to make use of *all but one* of the refinements of $<2, 5, 7, 10>$ at level 3 — the final refinement can only be what is left over from $<1, 4, 6, 9>$ after the other refinements have been computed. In this example we have two choices: we can either derive $<1, 6, 9>$ from $<2, 7, 10>$ leaving $<4>$ as the remainder, or we can derive $<4>$ from $<5>$ leaving $<1, 6, 9>$ as the remainder. Of course the second approach requires less processing, since only a single calculation needs to be made.

Motivated by these insights, we therefore introduce the following definition.

Definition 12.1.1 *In every refinement of a sequence* $c_j^{(\ell)}$ *into subsequences*

$$\left\langle c_1^{(\ell+1)}, c_2^{(\ell+1)}, \ldots, c_q^{(\ell+1)} \right\rangle,$$

$q \geq 1$, *we call one subsequence of largest cardinality* ***big****; the remaining* $q - 1$ *subsequences are called* ***small****. For* $\ell = 1$, *every sequence is* ***small****.*

Note that if $q = 1$, there will be no small subsequence — in particular, whenever $c_j^{(\ell)}$ contains a single element, there will be no small subsequence.

The basic idea of Algorithm C can then be succinctly stated as follows: to carry out the refinement of sequences at each level ℓ *only* with respect to sequences that are small at that level. It has already been remarked that this strategy is sufficient to ensure that the new refinement at level $\ell + 1$ is complete: for every sequence at level $\ell - 1$, exactly one of its subsequences at level ℓ is big, and it is only that big subsequence that is not used in the new refinement — whatever positions are left over in the original sequences after all the small sequences have been used for refinement simply become elements of subsequences at the next level. But the strategy also has an impact on the efficiency of the refinement process, as the next result shows:

Lemma 12.1.2 *Suppose that refinement of the sequences corresponding to a given string $\boldsymbol{x} = \boldsymbol{x}[1..n]$ is performed for $\ell = 1, 2, \ldots, \ell^*$, where $\ell^* \in 1..n-1$ is the least level at which every sequence contains a single position. Then each position i in $\boldsymbol{x}$ is contained in a small sequence $O(\log n)$ times.*

Proof Observe that if a sequence $c_j^{(\ell)}$ is refined into

$$\langle c_1^{(\ell+1)}, c_2^{(\ell+1)}, \ldots, c_q^{(\ell+1)} \rangle,$$

every small subsequence $c_{j'}^{(\ell+1)}$ must satisfy

$$|c_{j'}^{(\ell+1)}| \leq |c_j^{(\ell)}|/2.$$

In other words, every small subsequence, $\ell \geq 2$, is at most half the size of its parent sequence. Since for $\ell = 1$ an initial small sequence can contain at most n positions, it follows that no position can occur in more than $\lceil \log_2(n+1) \rceil$ small sequences. $\square$

Since there are n positions in $\boldsymbol{x}$, Lemma 12.1.2 tells us that there are overall $O(n \log n)$ positions in small sequences at all levels. Thus if the processing at each level ℓ can be performed in time proportional to the number of elements in the small sequences at that level, the entire refinement process can be carried out in $O(n \log n)$ time. Algorithm 12.1.1 presents an outline of this approach that is seen to be very simple; what has given Algorithm C its reputation as the mother of all algorithms is the complexity of the data structures required to achieve $O(n \log n)$ time. Having made clear the main idea, we are now in a better position to present those data structures.

Algorithm 12.1.1 (C)

– *Compute all the repetitions in a string $\boldsymbol{x} = \boldsymbol{x}[1..n]$*

$\ell \leftarrow 1$
compute sequences at level 1
    and label them all *small*
while level ℓ contains a small sequence
    output repetitions of period ℓ (if any)
    refine sequences at level ℓ into subsequences at level $\ell + 1$
        using only positions in small sequences
    $\ell \leftarrow \ell + 1$
    compute small sequences at level ℓ

We explain Algorithm C in more detail by classifying its data structures according to the main function with which each one is associated:

(F1) Recording the current sequence for each position in $\boldsymbol{x}$

array SEQ$[1..n]$:	SEQ$[i]$ gives the index of the current sequence to which position i in $\boldsymbol{x}$ belongs
array SEQLIST$[1..2n]$:	SEQLIST$[j]$ points to a doubly-linked list of positions in increasing order that belong to the sequence of index j
array SEQSIZE$[1..2n]$:	SEQSIZE$[j]$ gives the number of positions in the sequence of index j — that is, the number of positions in the list pointed to by SEQLIST$[j]$
INDEXSTACK:	a stack of currently unused sequence indices initialized to contain indices $1, 2, \ldots, 2n$

Whenever a new sequence needs to be formed, whether in level 1 or because the new sequence is required as part of the refinement of an existing sequence, a sequence index is popped from INDEXSTACK; whenever a sequence becomes empty, as a result of its refinement into subsequences, the sequence index is returned to INDEXSTACK. Whenever a position i is added to a sequence of index j, the assignments

$$\text{SEQ}[i] \leftarrow j; \ \text{SEQSIZE}[j] \leftarrow \text{SEQSIZE}[j] + 1$$

are performed, and i is added at the end of SEQLIST$[j]$; whenever a position i is removed from a sequence of index j, as a result of its assignment to a subsequence of j, the assignment SEQSIZE$[j] \leftarrow$ SEQSIZE$[j] - 1$ is performed, and i is removed from its position in SEQLIST$[j]$.

As shown in Exercise 12.1.4, no more than $2n$ sequence indices need to be in use at any one time; thus the array sizes $[1..2n]$ are sufficient. Observe also that since positions i are always stored in SEQLIST in increasing order, and consequently always removed from SEQLIST in increasing order, both storage and removal can be performed in constant time. We see then that all updates required during the refinement process to record the changes in assignments of positions to sequences can be performed in constant time per update.

(F2) Managing small sequences

SMALL:	a queue of indices j of small sequences arranged so that (for $\ell > 1$) all the small sequences occur together that are subsequences of the same sequence at the previous level

QUEUE: a queue of pairs (i, j), where i is a position in the small sequence of index j that is to be used for refinement of the current level; for each j, the positions i are contiguous and maintained in ascending order

LASTSMALL$[1..2n]$: LASTSMALL$[j'] = j > 0$ if and only if j is the index of the small sequence most recently used to refine the sequence of index j'

For $\ell = 1$ the queue SMALL just consists of each of the α' sequence indices that are initially computed corresponding to each letter that actually occurs in $\boldsymbol{x}$; for $\ell > 1$, SMALL is recreated by inspecting the subsequences of those sequences that have been refined (split) in level ℓ, and by adding to SMALL all but the single "big" subsequence — see the discussion of SPLIT and SUBSEQ under (F3). For each entry j in SPLIT, this processing can be carried out in a single scan of the list pointed to by SUBSEQ$[j]$. Thus SMALL is updated in time proportional to the number of entries in it.

QUEUE is created by removing the entries from SMALL at the beginning of the processing for every level $\ell > 1$. For every sequence j found in SMALL, the positions $i > 1$ found in SEQLIST$[j]$ are extracted, and entries (i, j) are added to QUEUE.

The main processing of each level $\ell > 1$ takes place as entries (i, j) are removed one by one from QUEUE. For each i removed, the sequence $j' = SEQ[i-1]$ is the one that will be refined; therefore, if LASTSMALL$[j'] \neq j$, j' was most recently refined with respect to some other class, and so a new sequence index is required. Accordingly, we set LASTSMALL$[j'] \leftarrow j$, pop the next sequence number from INDEXSTACK and add it to the list pointed to by SUBSEQ$[j']$ (see (F3)).

We find that the processing associated with each entry in QUEUE requires only constant time.

(F3) Managing subsequences

SPLITFLAG$[1..2n]$: SPLITFLAG$[j] = 1$ if and only if the sequence of index j is going to be refined in the current level (using sequences stored in SMALL — see (F2))

SPLIT: a list of sequences j that are going to be refined in the current level — those for which SPLITFLAG$[j] = 1$

SUBSEQ$[1..2n]$: when sequence j is going to be refined in the current level, SUBSEQ$[j]$ points to a list of indices of subsequences of sequence j — the first entry in this list is j itself

All of these data structures are initialized for each level $\ell > 1$ at the time that QUEUE is formed from SMALL (see (F2)). In particular, when (i, j) is added to

QUEUE, we know that i will be used to refine the sequence j' in which position $i-1$ is currently found ($j' = \text{SEQ}[i-1]$); therefore, if $\text{SPLITFLAG}[j'] = 0$, j' is at this time stored in the list SPLIT of sequences to be refined at level ℓ, while the following assignments are made:

$$\text{SPLITFLAG}[j'] \leftarrow 1;\ \text{SUBSEQ}[j'] \leftarrow <j'>;\ \text{LASTSMALL}[j'] \leftarrow 0.$$

We see then that putting each entry in QUEUE requires only constant time, and we have already seen in (F2) that removal of each entry from QUEUE also gives rise only to constant-time processing. Since the entries in QUEUE are exactly the positions in small sequences, we see that subsequences and small sequences can be managed in time proportional to the number of small sequences — by Lemma 12.1.2, in $O(n \log n)$ time.

At the completion of the processing for each level $\ell > 1$, the sequences j in SPLIT are removed one by one: if $\text{SEQSIZE}[j] = 0$, the sequence has been emptied and so its index is no longer required; accordingly j is pushed onto INDEXSTACK and removed from the list pointed to by $\text{SUBSEQ}[j]$.

(F4) Computing repetitions

$\text{GAP}[1..n]$:	$\text{GAP}[i]$ is the positive difference between position i in $\boldsymbol{x}$ and the next larger position in the same sequence at the current level; if there is no larger position, $\text{GAP}[i] = \infty$
$\text{GAPLIST}[1..n]$:	$\text{GAPLIST}[g]$ points to a doubly-linked list of positions i for which $\text{GAP}[i] = g$

For $\ell = 1$, GAP and GAPLIST are both initialized in a straightforward way by processing each list pointed to by $\text{SEQLIST}[j]$, $j = 1, 2, \ldots, \alpha'$, computing $\text{GAP}[i]$ for each position i in $\text{SEQLIST}[j]$, then adding i to the list pointed to by $\text{GAPLIST}\big[\text{GAP}[i]\big]$.

For $\ell > 1$, GAP and GAPLIST are updated when entry (i, j) is removed from QUEUE, leading to a refinement of sequence $j' = SEQ[i-1]$ in which position $i-1$ in $\text{SEQLIST}[j']$ will be moved to the end of $\text{SEQLIST}[j'']$. Since the lists pointed to by both SEQLIST and GAPLIST are doubly-linked, the recalculation of gaps required by deletion of $i-1$ from one list and its insertion in another can also be done in a straightforward way.

We observe that after the sequences for level ℓ have been created, every entry in $\text{GAPLIST}[\ell]$ describes two identical substrings of $\boldsymbol{x}$ of length ℓ whose first letters occur distance ℓ apart — in other words, a square. GAP and GAPLIST can then be used as shown in the following processing fragment to output the normal form of the maximal repetitions for each level ℓ in time proportional to the number of squares:

while $\text{GAPLIST}[\ell]$ **not** empty **do**
  $i_0 \leftarrow$ next position in $\text{GAPLIST}[\ell]$
  $i \leftarrow i_0;\ r \leftarrow 1$
  repeat

```
            delete i from GAPLIST[ℓ]
            i ← i + ℓ; r ← r + 1
        until GAP[i] ≠ ℓ
        output (i0, ℓ, r)
```

Suppose that a substring $\boldsymbol{u^2}$ occurs in $\boldsymbol{x}$, where $|\boldsymbol{u}| = \ell$ and $\boldsymbol{u}$ is itself a repetition $\boldsymbol{v^k}$, for some integer $k \geq 2$. Then $\boldsymbol{u^2} = \boldsymbol{v^{2k}}$ contains $k+1$ occurrences of $\boldsymbol{u}$ with gap $\ell/k < \ell$, so that $\boldsymbol{u}$ will never be selected for output as generator of a repetition. Since as remarked earlier there are at most $O(n \log n)$ squares in $\boldsymbol{x}$ whose generators are not themselves repetitions [LS69], the repetitions can therefore be output in $O(n \log n)$ time.

This completes our description of Algorithm C. In its implementation of the functions (F1)–(F4), Algorithm C is necessarily long and rather technical, and we do not present all the details here. The interested reader is referred to [C81]. We have tried nevertheless to provide the reader with a good understanding of how the complex data structures of the algorithm can be used to refine each level and to compute the repetitions, and to do so in $O(n \log n)$ time. Recalling that level 1 of the refinement can be computed in $O(n \log n)$ time when the alphabet is ordered, we state this important result formally as follows:

Theorem 12.1.3 *Algorithm 12.1.1 computes all the repetitions in a given string $\boldsymbol{x} = \boldsymbol{x}[1..n]$ on an ordered alphabet using $O(n \log n)$ time and $\Theta(n)$ additional space.*

Proof Observe that the time bound depends on the truth of three separate assertions:

- that the alphabet is ordered;
- Lemma 12.1.2;
- that the number of squares processed is $O(n \log n)$ [LS69].

□

It is important to remark that for the calculation of the repetitions, at most $\lfloor n/2 \rfloor$ levels need to be considered by Algorithm C, since the period ℓ of every repetition in $\boldsymbol{x}$ must satisfy $\ell \leq \lfloor n/2 \rfloor$. However, the $O(n \log n)$ upper bound on the time requirement holds even if the refinement of levels continues until there are only single-element lists remaining in SEQLIST — this could require levels $\ell > \lfloor n/2 \rfloor$ to be processed in cases such as $\boldsymbol{x} = abaababaababa$ where $\boldsymbol{x}$ contains long overlapping substrings. We observe then that Algorithm C can be easily extended to compute all repeats in $\boldsymbol{x}$ with a time requirement of $O(n \log n + Q)$, where Q is the size of the output. We return to this point in Section 13.2, where we show that all the repeating substrings in $\boldsymbol{x}$ can be represented in $Q \in O(n)$ space.

In fact, we observe further that Algorithm C in its extended form is essentially an algorithm for suffix tree construction. This is easily seen if we represent in tree form (Figure 12.1) the refinement of $\boldsymbol{f_6}$ used as an example above, labelling each edge with the letter on which the refinement is based and creating a terminal node for each position whenever it becomes a singleton in the Crochemore refinement.

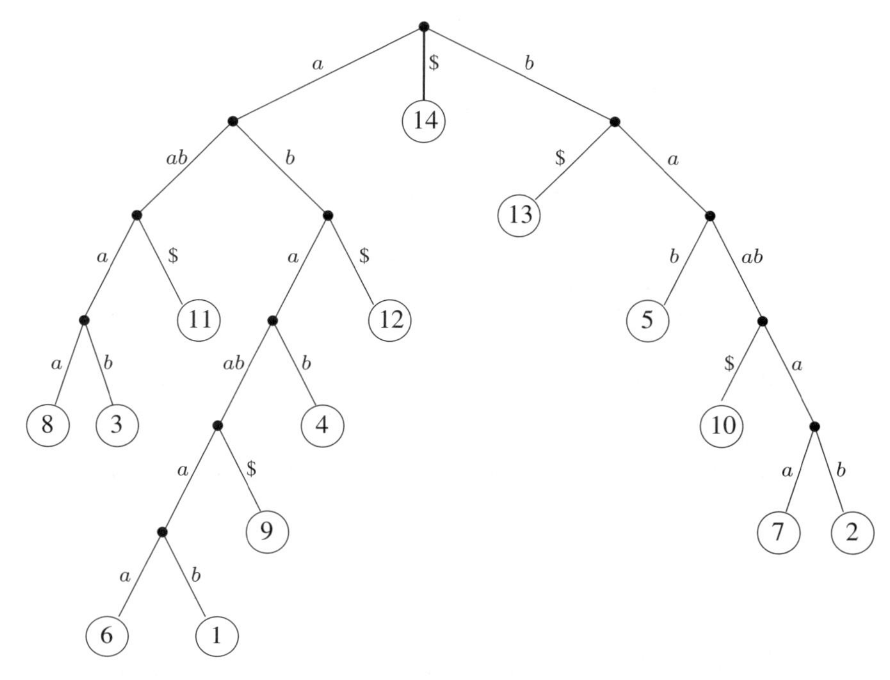

Figure 12.1. The Crochemore refinement tree for $\boldsymbol{f_6} = abaababaabaab\$$

Comparing Figure 12.1 with the suffix tree of $\boldsymbol{f_6}$ shown in Figure 5.2, we see that in order to transform the former into the latter, only a trivial relabelling of some edges is required; in particular, we need only ensure that all edges that lead to terminal nodes represent suffixes. Thus we discover that the extended version of Algorithm C computes all the common prefixes of the suffixes of $\boldsymbol{x}[1..n]$, exactly the information displayed in the suffix tree $T_{\boldsymbol{x}}$.

Unfortunately, Algorithm C also shares another of the properties of suffix tree construction algorithms: it requires a great deal of space for its execution. Indeed, by virtue of its copious use of arrays, queues and doubly-linked lists, general implementations of Algorithm C have required at least $80n$ bytes of storage, though this amount has recently been cut in half by a new approach [FSX02]. It was suggested in Subsection 5.2.5 that a general suffix tree implementation would require upward of $10n$ bytes — this then is a lower bound on the space required by any algorithm for suffix tree construction.

12.1.2 Main and Lorentz

I am not yet so lost in lexicography as to forget
that words are the daughters of earth,
and that things are the sons of Heaven.

— Samuel JOHNSON (1709–1784), *Preface to the Dictionary*

The approach taken by Algorithm ML is entirely different from that of Algorithm C. Interestingly, ML executes on an unordered alphabet, but still achieves a $\Theta(n \log n)$ time bound: it requires no knowledge of the lexicographical ordering of any letters or substrings of $\boldsymbol{x}$, information that is crucial to the first step of Algorithm C. ML is a divide-and-conquer algorithm [AHU74] that computes all the squares, hence all the repetitions, in $\boldsymbol{x} = \boldsymbol{x}[1..n]$ by considering them in three distinct classes:

(C1) squares $\boldsymbol{w}^2$ that begin and end in $\boldsymbol{x}[1..\lfloor n/2 \rfloor]$: $\boldsymbol{w}^2 = \boldsymbol{x}[i..i+2|\boldsymbol{w}|-1]$, where $i+2|\boldsymbol{w}|-1 \le \lfloor n/2 \rfloor$;

(C2) squares $\boldsymbol{w}^2$ that begin and end in $\boldsymbol{x}[\lfloor n/2 \rfloor + 1..n]$: $\boldsymbol{w}^2 = \boldsymbol{x}[i..i+2|\boldsymbol{w}|-1]$, where $i > \lfloor n/2 \rfloor$;

(C3) ***new*** squares $\boldsymbol{w}^2$ that begin in $\boldsymbol{x}[1..\lfloor n/2 \rfloor]$ and end in $\boldsymbol{x}[\lfloor n/2 \rfloor + 1..n]$: $\boldsymbol{w}^2 = \boldsymbol{x}[i..i+2|\boldsymbol{w}|-1]$, where $1 \le i \le \lfloor n/2 \rfloor$ and $\lfloor n/2 \rfloor < i+2|\boldsymbol{w}|-1 \le n$.

Let $\boldsymbol{u} = \boldsymbol{x}[1..\lfloor n/2 \rfloor]$, $v = \boldsymbol{x}[\lfloor n/2 \rfloor + 1..n]$. The basic idea of Algorithm ML is to compute the squares in $\boldsymbol{x} = \boldsymbol{uv}$ on the basis of an efficient computation of the new squares in class (C3): in fact, recursive splitting of $\boldsymbol{x}$ into halves reduces the calculation of *all* squares to those of class (C3). This observation leads immediately to a straightforward recursive algorithm, shown here as Algorithm 12.1.2.

Algorithm 12.1.2 (ML)

– *Compute all the squares in a string* $\boldsymbol{x} = \boldsymbol{x}[1..n]$

procedure ML($\boldsymbol{x}$)
if $|\boldsymbol{x}| > 1$ **then**
  – *Compute the class (C3) squares*
  C3($\boldsymbol{x}[1..\lfloor n/2 \rfloor], \boldsymbol{x}[\lfloor n/2 \rfloor + 1..n]$)
  – *Recursive steps*
  ML($\boldsymbol{x}[1..\lfloor n/2 \rfloor]$)
  ML($\boldsymbol{x}[\lfloor n/2 \rfloor + 1..n]$)

We shall discover below that for arbitrary strings $\boldsymbol{u}$ and $\boldsymbol{v}$, procedure C3($\boldsymbol{u}, \boldsymbol{v}$) can be implemented in $\Theta(|\boldsymbol{uv}|)$ time. If we let t_n be the maximum time required to execute procedure ML on any string of length n, we see immediately that for $n > 1$, there exists a constant k_2 such that

$$t_n \le k_2 n + 2t_{\lceil n/2 \rceil}, \tag{12.2}$$

while for $0 \le n \le 1$, there exists a constant k_1 such that

$$t_1 \le k_1. \tag{12.3}$$

As shown in Exercise 12.1.5, we can solve (12.2) and (12.3) to find that $t_n \in O(n \log n)$. But since the time required for procedure C3 is *exactly* proportional to $|\boldsymbol{uv}|$, we see also that there exist constants k_1' and k_2' such that

$$t'_1 \geq k'_1;\ t'_n \geq k'_2 n + 2t'_{\lceil n/2 \rceil},\ n > 1, \tag{12.4}$$

where now t'_n is defined to be the *minimum* time required to execute ML on any string of length n. Again we can solve the recurrence relation (12.4) to show that $t'_n \in \Omega(n \log n)$. Thus we have

Theorem 12.1.4 *Based on a linear time requirement for procedure C3, Algorithm 12.1.2 computes all the squares in a given string $\boldsymbol{x} = \boldsymbol{x}[1..n]$ in $\Theta(n \log n)$ time.* □

In order to explain procedure C3$(\boldsymbol{u}, \boldsymbol{v})$, we first observe that the new squares $\boldsymbol{w}^2$ in $\boldsymbol{uv}$ are of two kinds:

- ***newright:*** the second occurrence of $\boldsymbol{w}$ lies entirely in $\boldsymbol{v}$;
- ***newleft:*** a nonempty prefix of the second occurrence of $\boldsymbol{w}$ lies in $\boldsymbol{u}$.

For example, $\boldsymbol{u} = aba$ and $\boldsymbol{v} = ba$ give rise to a newright square $\boldsymbol{w}^2 = (ba)^2$ and a newleft square $\boldsymbol{w}^2 = (ab)^2$. We describe here the processing required to compute newright squares and leave as Exercise 12.1.10 the symmetric processing needed for newleft squares. Procedure C3 is implemented then just as a combination of newright and newleft.

Let $\boldsymbol{u} = \boldsymbol{u}[1..n_1]$, $\boldsymbol{v} = \boldsymbol{v}[1..n_2]$. In order to find all newright squares in $\boldsymbol{uv}$, two arrays need to be computed, both of them bearing some similarity to the border array (Section 1.3). To define these arrays, recall (Subsection 5.2.1) that lcp$(\boldsymbol{x_1}, \boldsymbol{x_2})$ is the length of the longest common prefix of $\boldsymbol{x_1}$ and $\boldsymbol{x_2}$, and let us use lcs$(\boldsymbol{x_1}, \boldsymbol{x_2})$ to denote the length of the longest common suffix of $\boldsymbol{x_1}$ and $\boldsymbol{x_2}$ — not it is hoped to be confused with the LCS (longest common *subsequence*) of Chapter 9! The new arrays can then be succinctly defined as follows:

- **LP**$[2..n_2 + 1]$: for each position $i = 2, 3, \ldots, n_2$ in $\boldsymbol{v}$, LP$[i]$ = lcp$(\boldsymbol{v}[i..n_2], \boldsymbol{v})$, while for $i = n_2 + 1$, LP$[i] = 0$;
- **LS**$[1..n_2]$: for each position $i = 1, 2, \ldots, n_2$ in $\boldsymbol{v}$, LS$[i]$ = lcs$(\boldsymbol{v}[1..i], \boldsymbol{u})$.

For example, given $\boldsymbol{v} = abaababa$, we find

$$\text{LP}[2..9] = 01303010;$$

if in addition $\boldsymbol{u} = abaab$, then

$$\text{LS}[1..8] = 02005020.$$

Roughly speaking, LP identifies maximum-length recurrences of prefixes of $\boldsymbol{v}$ within $\boldsymbol{v}$; LS identifies maximum-length occurrences of suffixes of $\boldsymbol{u}$ within $\boldsymbol{v}$.

As might be expected of arrays that are so similar to the border array, both LP and LS turn out to be computable in time $\Theta(n_2)$. We describe later how this can be done, but first we show how these arrays can be used to compute the newright squares of $\boldsymbol{u}$ and $\boldsymbol{v}$. For this we need

Lemma 12.1.5 *Let $\boldsymbol{u}[1..n_1]$ and $\boldsymbol{v}[1..n_2]$ be given nonempty strings, and let $p \le n_2$ and $i \in p..2p-1$ be positive integers. Then there is a square of length $2p$ in $\boldsymbol{uv}$ ending at position i of $\boldsymbol{v}$ if and only if*

$$2p - \mathrm{LS}[p] \le i \le p + \mathrm{LP}[p+1].$$

Proof For any square $\boldsymbol{w}^2$ of length $2p$ in $\boldsymbol{uv}$ ending at position i, the first occurrence of $\boldsymbol{w}$ must begin at $\boldsymbol{u}[n_1 - 2p + i + 1]$ and the second occurrence must begin at $\boldsymbol{v}[i - p + 1]$. Thus such a square occurs if and only if

$$\boldsymbol{u}[n_1 - 2p + i + 1..n_1]\boldsymbol{v}[1..i-p] = \boldsymbol{v}[i-p+1..i];$$

that is, if and only if

(a) $\boldsymbol{u}[n_1 - 2p + i + 1..n_1] = \boldsymbol{v}[i-p+1..p]$, and

(b) $\boldsymbol{v}[1..i-p] = \boldsymbol{v}[p+1..i]$.

Note that if $i = p$, so that the square is formed from a suffix $\boldsymbol{w}$ of $\boldsymbol{u}$ and a prefix $\boldsymbol{w}$ of $\boldsymbol{v}$, condition (b) reduces to the statement that $\varepsilon = \varepsilon$, and so becomes unnecessary.

Condition (a) is exactly the statement that the string $\boldsymbol{v}[i-p+1..p]$ of length $2p-i$ is a suffix of $\boldsymbol{u}$, true if and only if $\mathrm{LS}[p] \ge 2p - i$ — that is, if and only if

$$2p - \mathrm{LS}[p] \le i.$$

Similarly, as shown in Exercise 12.1.7, condition (b) is equivalent to

$$i \le p + \mathrm{LP}[p+1],$$

completing the proof. □

A newright square can possibly occur only for the values of p and i specified by Lemma 12.1.5. Furthermore, for each allowable value of $p \in 1..n_2$, the corresponding allowable values of i can only lie in a range specified by the endpoints

$$2p - \mathrm{LS}[p] \;..\; p + \mathrm{LP}[p+1].$$

Therefore, provided that the arrays LP and LS have been precomputed, the actual values of i can be specified in constant time by outputting these endpoints whenever

$$\mathrm{LP}[p+1] + \mathrm{LS}[p] \ge p.$$

Hence, over all allowable values of p, the newright squares can be computed in $\Theta(n_2)$ time.

We leave to Exercise 12.1.8 the pseudocode statement of the simple algorithm that uses the arrays LP and LS to perform this calculation.

In order to deal with the newleft squares, we need analogous arrays:

- **LP**$'[2..n_1]$: for each position $i = 2, 3, \ldots, n_1$ in $\boldsymbol{u}$, $\text{LP}'[i] = \text{lcp}(\boldsymbol{u}[i..n_1], \boldsymbol{v})$;
- **LS**$'[1..n_1 - 1]$: for each position $i = 1, 2, \ldots, n_1 - 1$ in $\boldsymbol{u}$, $\text{LS}'[i] = \text{lcs}(\boldsymbol{u}[1..i], \boldsymbol{u})$.

As shown in Exercise 12.1.9, the new arrays LP$'$ and LS$'$ can be used to prove an analogue of Lemma 12.1.5, and thus immediately to compute all the newleft squares of $\boldsymbol{uv}$ in $\Theta(n_1)$ time.

As observed earlier, procedure C3 makes use of the newright and newleft calculations in a straightforward way to compute all the new squares in class (C3) in time $\Theta(n_1 + n_2) = \Theta(|\boldsymbol{uv}|)$. Thus if the arrays LP$[2..n_2 + 1]$ and LS$[1..n_2]$ for newright can be computed in $\Theta(n_2)$ time, while the analogous arrays LP$'[2..n_1]$ and LS$'[1..n_1 - 1]$ for newleft can be computed in $\Theta(n_1)$ time, then by Theorem 12.0.1 Algorithm ML will achieve $\Theta(n \log n)$ running time. We now show how the required arrays can be efficiently calculated.

Computing LP, LS, LP$'$, LS$'$

For a given string $\boldsymbol{v}[1..n_2]$ and every integer $i \in 1..n_2 + 1$, recall that LP$[i]$ is the length of the longest prefix of $\boldsymbol{v}[i..n_2]$ that matches a prefix of $\boldsymbol{v}$ itself. We imagine computing LP from left to right for increasing values of i, so that if we are currently computing LP$[i]$, we may assume that every LP$[k]$, $2 \leq k < i$, has already been computed. Suppose that for some value of $i \geq 3$, there exists an integer $k \in 2..i-1$ such that $k + \text{LP}[k] > i$. As shown in Figure 12.2, this means that there exists a substring

$$\boldsymbol{w} = \boldsymbol{v}\left[i..k + \text{LP}[k] - 1\right]$$

of length $\ell = k + \text{LP}[k] - i$ that matches a suffix $\boldsymbol{v}\left[i-k+1..\text{LP}[k]\right]$ of the prefix $\boldsymbol{v}\left[1..\text{LP}[k]\right]$ corresponding to k.

Now consider LP$[i - k + 1]$, the length of the longest substring occurring at $i - k + 1$ that matches a prefix of $\boldsymbol{v}$. If

$$\text{LP}[i - k + 1] < |\boldsymbol{w}| = \ell,$$

then it must be true that

$$\text{LP}[i] = \text{LP}[i - k + 1]. \tag{12.5}$$

On the other hand, if

$$\text{LP}[i - k + 1] \geq \ell,$$

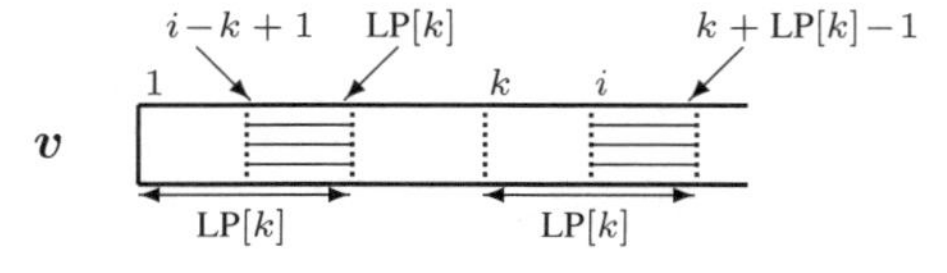

Figure 12.2. Computing LP$[i]$ when $k + \text{LP}[k] > i$

then

$$\mathrm{LP}[i] \geq \ell \tag{12.6}$$

and so $\boldsymbol{w}$ is a prefix of the longest substring occurring at i that matches a prefix of $\boldsymbol{v}$. Thus if there exists k such that $k + \mathrm{LP}[k] > i$, either by (12.5) $\mathrm{LP}[i]$ is determined, or else by (12.6) a lower bound on $\mathrm{LP}[i]$ is established.

Of course if there exists no integer $k \in 2..i-1$ such that $k + \mathrm{LP}[k] > i$, then the only lower bound that can be established is

$$\mathrm{LP}[i] \geq 0. \tag{12.7}$$

The three cases (12.5)–(12.7) are the ones that need to be considered in the algorithm that computes LP. As shown in Algorithm 12.1.3, once a lower bound on $\mathrm{LP}[i]$ is determined, cases (12.6) and (12.7) are treated in the same way: a letter-by-letter match computes the exact value of $\mathrm{LP}[i]$ beyond the lower bound. Observe that the algorithm maintains the value of $k < i$ that maximizes $k + \mathrm{LP}[k]$ — this will ensure the largest possible lower bound for $\mathrm{LP}[i]$. It is convenient, in order to correctly compute the terminating positions in LP, to append to $\boldsymbol{v}$ a single letter $\$_2$ guaranteed not to match with any other letter.

Algorithm 12.1.3 (calcLP)

– *Compute the array LP$[2..n_2+1]$ corresponding to $\boldsymbol{v}[1..n_2]\$_2$*
– *First compute LP$[2]$ (case (12.7)) and initialize k*
$\mathrm{LP}[2] \leftarrow 0$
while $\boldsymbol{v}\big[\mathrm{LP}[2]+1\big] = \boldsymbol{v}\big[\mathrm{LP}[2]+2\big]$ **do** $\mathrm{LP}[2] \leftarrow \mathrm{LP}[2]+1$
$k \leftarrow 2$

for $i \leftarrow 3$ **to** n_2+1 **do**
  $\ell \leftarrow k + \mathrm{LP}[k] - i$ – *possibly negative!*
  if $\mathrm{LP}[i-k+1] < \ell$ **then**
    – *Case (12.5)*
    $\mathrm{LP}[i] \leftarrow \mathrm{LP}[i-k+1]$
  else
    – *Cases (12.6) and (12.7) — just like $i = 2$!*
    $\mathrm{LP}[i] \leftarrow \max\{0, \ell\}$
    while $\boldsymbol{v}\big[\mathrm{LP}[i]+1\big] = \boldsymbol{v}\big[\mathrm{LP}[i]+i\big]$ **do** $\mathrm{LP}[i] \leftarrow \mathrm{LP}[i]+1$
    $k \leftarrow i$

Based on the above discussion, we can perhaps have confidence in the correctness of Algorithm 12.1.3: it certainly implements the three cases (12.5)–(12.7), and does so even for $\boldsymbol{v}[n_2+1] = \$_2$. To estimate the time complexity of the algorithm, observe that the instructions in the **for** loop all require constant time, with the exception of the **while** loop. Thus the time requirement is $\Theta(n_2)$ plus the number of times that $\mathrm{LP}[i]$ is incremented in the **while** loop. But each increment to $\mathrm{LP}[i]$ increases by one the value of $k + \mathrm{LP}[k]$ that in subsequent iterations will be the minimum position in $\boldsymbol{v}$ that will be subject to a letter

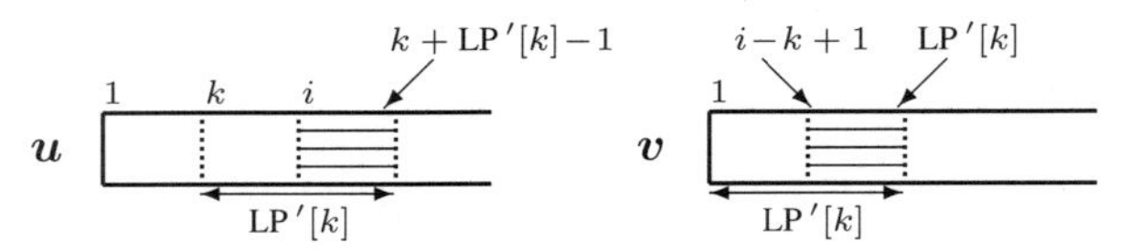

Figure 12.3. Computing $LP'[i]$ when $k + LP'[k] > i$

comparison. Since k is initialized to 2, we see therefore that the instruction $LP[i] \leftarrow LP[i]+1$ can be executed at most $n_2 - 1$ times over *both* the **while** loops, and so the time requirement of calcLP is $\Theta(n_2)$.

It is not difficult to extend the approach of Algorithm calcLP to compute $LP'[2..n_1]$; that is, to compute the length of the longest prefix of $\boldsymbol{u}[i..n_1]$ (rather than of $\boldsymbol{v}[i..n_2]$) that matches a prefix of $\boldsymbol{v}$. We again compute from left to right, so that as before for every $i \geq 3$ there exists an integer $k \in 2..i-1$ that maximizes $k + LP'[k]$. As shown in Figure 12.3, this means that for $k + LP'[k] > i$ there exists a substring

$$\boldsymbol{w} = \boldsymbol{u}\big[i..k + LP'[k] - 1\big]$$

of length $\ell' = k + LP'[k] - i$ that matches $\boldsymbol{v}\big[i - k + 1..LP'[k]\big]$.

We argue as for LP that three cases arise, analogous to (12.5)–(12.7): assuming that $LP[2..n_2 + 1]$ has already been calculated, it remains true that if $LP[i-k+1] < \ell'$, then

$$LP'[i] = LP[i-k+1], \tag{12.8}$$

while if not, then

$$LP'[i] \geq \ell'. \tag{12.9}$$

Of course, if $k + LP'[k] \leq i$, we can assume only that

$$LP'[i] \geq 0. \tag{12.10}$$

Implementing the three cases (12.8)–(12.10) leads to Algorithm calcLP′, almost identical to Algorithm calcLP, shown as Algorithm 12.1.4. By exactly the same argument used to establish the time complexity of calcLP, we establish a $\Theta(n_1)$ time requirement for calcLP′.

Algorithm 12.1.4 (calcLP′)

– *Compute the array* $LP'[2..n_1]$ *corresponding to* $\boldsymbol{u}[1..n_1]\$_1$
– *and* $\boldsymbol{v}[1..n_2]\$_2$*, assuming* $LP[2..n_2+1]$ *already computed*

– *First compute* $LP'[2]$ *and initialize* k
$LP'[2] \leftarrow 0$
while $\boldsymbol{v}\big[LP'[2]+1\big] = \boldsymbol{u}\big[LP'[2]+2\big]$ **do** $LP'[2] \leftarrow LP'[2]+1$
$k \leftarrow 2$

for $i \leftarrow 3$ **to** n_1 **do**
  $\ell' \leftarrow k + LP'[k] - i$ – *possibly negative!*

– *Note use of LP rather than LP′*
if LP$[i-k+1] < \ell'$ **then**
  LP$'[i] \leftarrow$ LP$[i-k+1]$
else
  LP$'[i] \leftarrow \max\{0, \ell'\}$
  while $\boldsymbol{v}\big[\text{LP}'[i]+1\big] = \boldsymbol{u}\big[\text{LP}'[i]+i\big]$ **do** LP$'[i] \leftarrow$ LP$'[i]+1$
  $k \leftarrow i$

The specification of the corresponding algorithms, calcLS and calcLS′, is left to Exercise 12.1.14.

Summary

Making use of the linear-time algorithms that compute LP, LP′, LS and LS′, we can finally state unconditionally that procedure C3$(\boldsymbol{u}, \boldsymbol{v})$ executes in $\Theta(|\boldsymbol{uv}|)$ time every time it is invoked, hence by Theorem 12.1.4 that Algorithm 12.1.2 requires $\Theta(n \log n)$ time. Recall that Algorithm 12.1.1 requires only $O(n \log n)$ time. On the other hand, since it is based entirely on the use of simple array structures, Algorithm 12.1.2 has a very low space requirement, certainly less than the $40n$ bytes consumed by Algorithm 12.1.1, even in its most recent space-efficient form [FSX02].

It is interesting to compare the output produced by Algorithms C and ML. Algorithm C needs to report every repetition individually in normal form, making use of the (i, p^*, r^*) encoding; recall that although Algorithm C inspects each square individually, it possesses the ability to combine adjacent squares until a maximal exponent $r \geq 2$ is obtained. On the other hand, Algorithm ML outputs squares only, but in view of Lemma 12.1.5 and its analogue for LP′ and LS′, it outputs them in blocks that correspond closely to the "runs" introduced in Section 2.3. Consider for example the string

$$\begin{array}{ccccccccccccccc} & & 1 & 2 & 3 & 4 & 5 & 6 & 7 & 8 & 9 & 10 & 11 & 12 & 13 \\ \boldsymbol{x} & = & a & b & a & a & b & a & b & a & a & b & a & b & a. \end{array}$$

For repetitions of period $p = 5$, Algorithm C would output

$$(1, 5, 2),\ (2, 5, 2),\ (3, 5, 2),\ (4, 5, 2),$$

corresponding to the square $(abaab)^2$ and its three rotations $(baaba)^2$, $(aabab)^2$, and $(ababa)^2$. Algorithm ML on the other hand would in this case need to output only the triple

$$(p, \text{first}, \text{last}) = (5, 10, 13),$$

indicating that squares of period 5 end at positions 10–13 inclusive.

We see then that ML foreshadows the idea of a run that was introduced in [M89] and used there to reduce the asymptotic complexity of the repetitions problem. We explore these matters further in the next section.

Exercises 12.1

1. Prove the following "refinement" of Lemma 12.1.2: no position in $\boldsymbol{x}$ can occur in more than $\lceil \log_2(n-\alpha'+2) \rceil$ small sequences, where α' is the number of distinct letters that occur in $\boldsymbol{x}$.
2. The *Parikh* or *frequency vector* of a string $\boldsymbol{x}[1..n]$ that contains α' distinct letters is an integer vector $\phi_{\boldsymbol{x}} = \phi_{\boldsymbol{x}}[1..\alpha']$, where $\phi_{\boldsymbol{x}}[h]$ counts the number of occurrences in $\boldsymbol{x}$ of the hth letter [P66]. Express the complexity of Algorithm C in terms of n and the Parikh vector. Use this expression as a basis for commenting on the behaviour of Algorithm C in cases where α' is large.
3. In this section the term "sequence" has been used frequently, but without formal definition. Observing that a sequence can be thought of as a kind of string on an indexed alphabet (Section 4.1), provide the missing definitions for sequence, subsequence, and perhaps other related terms.
4. Show that at most $2n$ sequence indices are required in order to handle refinement of all sequences at level ℓ into subsequences at level $\ell+1$. Can you sharpen this upper bound?
5. Solve equations (12.2) and (12.3) to show that $t_n \in O(n \log n)$. Then explain how the same method enables us to use (12.4) to show that $t_n \in \Omega(n \log n)$.

   Hint: Observe that if for some nonnegative integer r, $2^r < n < 2^{r+1}$, the result will hold for n if we prove it for 2^{r+1}.
6. Suppose that (12.2) is generalized to

   $$t_n \le k_2 f_n + 2t_{\lceil n/2 \rceil},\ f_n \in \Omega(n^2),$$

   for every $n \ge 1$, while (12.3) continues to hold. Show that

   $$t_n \in O(f_n).$$

   Remark: We shall make use of this result in Section 13.3 to establish the complexity of Schmidt's Algorithm for approximate repeats.
7. Show that condition (b) of Lemma 12.1.5 is equivalent to $i \le p + \text{LP}[p+1]$.
8. Assuming that the arrays LP and LS have been precomputed for given strings $\boldsymbol{u}[1..n_1]$ and $\boldsymbol{v}[1..n_2]$, specify a $\Theta(n_2)$-time algorithm to compute all the newright squares in $\boldsymbol{uv}$.
9. Prove the following analogue of Lemma 12.1.5:

   Let $\boldsymbol{u}[1..n_1]$ and $\boldsymbol{v}[1..n_2]$ be given strings, and let $p < n_1$ and i be positive integers such that $n_1 - 2p < i - 1 < n_1 - p$. Then there is a square of length $2p$ in $\boldsymbol{uv}$ beginning at position i of $\boldsymbol{u}$ if and only if

   $$(n_1 - p) - \text{LS}'[n_1 - p] \le i - 1 \le (n_1 - 2p) + \text{LP}'\big[(n_1 - p) + 1\big].$$

10. Assuming that the arrays LP$'$ and LS$'$ have been precomputed for given strings $\boldsymbol{u}[1..n_1]$ and $\boldsymbol{v}[1..n_2]$, specify a $\Theta(n_1)$-time algorithm to compute all the newleft squares in $\boldsymbol{uv}$.
11. Making use of linear-time routines to calculate the arrays and to compute the newright and newleft squares, specify the procedure C3 that computes all the squares of class (C3) in $\boldsymbol{uv}$, where $\boldsymbol{u} = \boldsymbol{u}[1..n_1]$, $\boldsymbol{v} = \boldsymbol{v}[1..n_2]$.
12. Give a formal proof of (12.5).
13. Prove that Algorithm 12.1.3 does in fact maintain the value of $k \in 2..i-1$ that maximizes $k + \text{LP}[k]$.
14. Analogous to calcLP and calcLP$'$, specify algorithms calcLS and calcLS$'$ that compute LS and LS$'$, respectively.

12.2 Runs

we'll kick the l out of the world and cuddle
up with the avenues and byways of the word ...

— A. R. AMMONS (1926–2001), *Garbage*

In this section we consider two approaches to the repetitions problem that reduce output and processing time by computing runs (i, p^*, r^*, t) rather than repetitions (i, p^*, r^*). Both approaches make use of the s-factorization whose calculation using suffix trees was described in Section 6.3. Furthermore, both algorithms make use of the arrays LS, LS$'$, LP, LP$'$ introduced in Subsection 12.1.2 to implement Algorithm ML. In fact, the second algorithm, Algorithm KK, is essentially an extension to all runs of the first, Algorithm Mn, that computes only ***leftmost*** runs (and as we shall see, perhaps a few others) — that is, only the leftmost occurrence of every distinct run in $\boldsymbol{x}$.

Since both Algorithms Mn and KK depend on the use of suffix trees, they will require $O(n \log n)$ time in the case of an unknown ordered alphabet; on the other hand, for a small known ordered alphabet, both algorithms can be thought of as requiring only $\Theta(n)$ time. As remarked in [KK00], the main difficulty with these algorithms is "the memory occupied by a data structure needed for computing the s-factorization" — in other words, the space (thus indirectly the time) required by suffix tree construction.

In order that Algorithms Mn and KK should execute in time linear in the string length, the size of the output must be linear. As we shall see, this follows for Algorithm Mn from the fact that the computation itself that identifies leftmost runs must be linear; it follows also from the later result [FS98], quoted below, that the number of distinct (hence leftmost) squares is linear. A much more difficult matter is the proof that the total number of runs, leftmost or otherwise, is linear:

Theorem 12.2.1 *Let $\rho(n)$ be the maximum number of runs that can occur in any string of length n on any alphabet. Then there exist positive constants k_1 and k_2 independent of n such that for every integer $n \geq 1$,*

$$\rho(n) \leq k_1 n - k_2 \sqrt{n} \log_2 n. \tag{12.11}$$

Proof See [KK00]. □

We do not include here the proof of this landmark result because it is very long and technical. Although Theorem 12.2.1 constitutes an important advance in the study of periodicity in strings, it is also an indicator of how much more remains to be learned. The proof of (12.11) does not permit any estimate to be made of the size of k_1 and k_2: k_1 could equal 10^9 and k_2 could be 10^{-9}. At the same time, exhaustive computer experiments up to $n = 60$ [KK98] suggest that in general:

(1) $\rho(n) < n$;

(2) $0 \leq \rho(n+1) - \rho(n) \leq 2$;

(3) there exists a cube-free string on $\{a, b\}$ that contains $\rho(n)$ runs.

It seems likely that all of these very fundamental results are true, but nobody seems to have any idea how to prove them!

Instead of $\rho(n)$, a recent paper [FS98] has considered $\sigma(n)$, the maximum number of distinct squares that can occur in any string of length n, and shown that $\sigma(n) \leq 2n - 8$ for every $n \geq 5$. As suggested in [KK00], there may be some connection between $\rho(n)$ and $\sigma(n)$ that could help to specify more precise bounds on ρ. Further, in [KK99], a stronger result than Theorem 12.2.1 has been proved using similar methods: it is shown that the sum of the exponents of all the runs in any string of length n is also linear in n.

12.2.1 Leftmost Runs — Main

The word, even the most contradictory word,
preserves contact — it is silence that isolates.

— Thomas MANN (1875–1955), *The Magic Mountain*

Crochemore [C84] seems to have been the first to realize that the s-factorization [LZ76] employed in [ZL77] for the purpose of string compression (Section 6.3) could also be used for the efficient computation of repetitions. In this subsection we describe Algorithm Mn [M89] that uses s-factorization together with techniques developed in [ML84] (Subsection 12.1.2) to compute the leftmost occurrence of each distinct run in a given string $\boldsymbol{x}$. The basis of this algorithm is the following theorem:

Theorem 12.2.2 *Suppose $\boldsymbol{x}[1..n]$ is a string with s-factorization $\boldsymbol{w_1 w_2} \cdots \boldsymbol{w_k}$, and let $\boldsymbol{r}$ denote the leftmost occurrence of a run $\boldsymbol{x}[i..j]$ in $\boldsymbol{x}$ whose final position j lies within $\boldsymbol{w_h}$ for some $h \leq k$. Then*

(a) *the start position i of $\boldsymbol{r}$ lies within $\boldsymbol{w_{h'}}$ for some $h' < h$;*

(b) *if we let $\boldsymbol{r} = \boldsymbol{r_L u_R}$ for some prefix $\boldsymbol{r_R}$ of $\boldsymbol{w_h}$,*

$$|\boldsymbol{r_L}| < 2|\boldsymbol{w_{h-1}}| + |\boldsymbol{w_h}|.$$

Proof Suppose that (a) does not hold. Therefore the run $\boldsymbol{r}$ is a substring of $\boldsymbol{w_h}$, and so $|\boldsymbol{w_h}| \geq 2$. But then by the definition of s-factorization, there must exist another occurrence of $\boldsymbol{w_h}$ in $\boldsymbol{x}$ to the left of the factor $\boldsymbol{w_h}$. Hence there also exists on the left another occurrence of the run, which can accordingly not be leftmost, in contradiction to the hypothesis of the theorem. Thus (a) must hold.

By virtue of (a), we may suppose that $\boldsymbol{r} = \boldsymbol{r_L u_R}$ for some nonempty $\boldsymbol{r_L}$ and some nonempty prefix $\boldsymbol{r_R}$ of $\boldsymbol{w_h}$. Suppose (b) does not hold, so that $|\boldsymbol{r_L}| \geq 2|\boldsymbol{w_{h-1}}|+|\boldsymbol{w_h}|$. Then at least half of $\boldsymbol{r}$ occurs to the left of the factor $\boldsymbol{w_{h-1}}$. Consider the suffix $\boldsymbol{v} = \boldsymbol{w_{h-1} u_R}$ of $\boldsymbol{r}$ as shown in Figure 12.4.

Since $\boldsymbol{v}$ is a suffix of a run $\boldsymbol{r}$ with period p^* and exponent $r^* \geq 2$, and since $|\boldsymbol{v}|$ is at most half the length of $\boldsymbol{r}$, there exists another occurrence of $\boldsymbol{v}$ distance p^* to the left, as shown in Figure 12.4. But since $\boldsymbol{w_{h-1}}$ is a proper prefix of $\boldsymbol{v}$, and $\boldsymbol{v}$ is a repeating substring, $\boldsymbol{w_{h-1}}$ cannot be the maximum-length repeating substring required by the s-factorization, a contradiction. We conclude that (b) holds. □

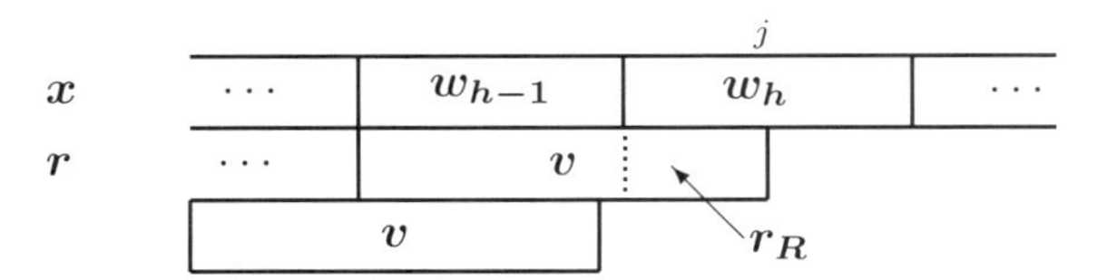

Figure 12.4. The periodic suffix $\boldsymbol{v}$ of a leftmost run $\boldsymbol{r}$

Actually, this unpretentious theorem provides us with a little more information than appears on the surface. Observe that the proof of Theorem 12.2.2(b) requires only that the run $\boldsymbol{r}$ terminate in $\boldsymbol{w_h}$ without being a substring of $\boldsymbol{w_h}$ — it need not be leftmost in the sense defined above. Thus we can divide all runs in $\boldsymbol{x}$ into two disjoint classes:

(I) those that are substrings of a single factor $\boldsymbol{w_h}$, $2 \leq h \leq k$, in the s-factorization of $\boldsymbol{x}$;

(II) those that terminate in some $\boldsymbol{w_h}$, $h \geq 2$, and begin in some preceding $\boldsymbol{w_{h'}}$, $h' < h$.

In this subsection we show how to compute all runs of Class II — that is, all leftmost runs together possibly with others. We will deal with Class I runs in the next subsection.

To gain some intuition about Class II runs, consider our ubiquitous example $\boldsymbol{f_6}$, whose s-factorization is

$$\boldsymbol{w_1 w_2} \cdots \boldsymbol{w_6} = a/b/a/aba/baaba/ab.$$

Then we can easily relate the Class II runs to their corresponding factors:

$$(aba)^2\ (\boldsymbol{w_1} - \boldsymbol{w_4}),\ \ a^2\ (\boldsymbol{w_3} - \boldsymbol{w_4}),\ \ (ab)^2 a\ (\boldsymbol{w_3} - \boldsymbol{w_4})$$
$$(abaab)^2 a\ (\boldsymbol{w_1} - \boldsymbol{w_5}),\ \ (aba)^2 ab\ (\boldsymbol{w_4} - \boldsymbol{w_6}).$$

Note that the Class I run aa in $\boldsymbol{w_5}$ is not included in this list.

Algorithm 12.2.1 (Mn)

– *Compute all the Class II runs in a string* $\boldsymbol{x} = \boldsymbol{x}[1..n]$

– *Execute Algorithm 6.3.1*
compute the s-factorization $\boldsymbol{x} = \boldsymbol{w_1 w_2} \cdots \boldsymbol{w_k}$

$j^* \leftarrow 0$;
for $h \leftarrow 2$ **to** k **do**
  – *Compute the maximum-length prefix* $\boldsymbol{r_L^*}$
  – *of a run ending in* $\boldsymbol{w_h}$ *(Theorem 12.2.2)*
  $j^* \leftarrow j^* + |\boldsymbol{w_{h-1}}|$
  $\ell \leftarrow \min\{j^*, 2|\boldsymbol{w_{h-1}}| + |\boldsymbol{w_h}|\}$
  $\boldsymbol{r_L^*} \leftarrow \boldsymbol{x}[j^* - \ell + 1..j^*]$
  – *Compute all the Class II runs ending in* $\boldsymbol{w_h}$
  calcruns$(\boldsymbol{r_L^*}, \boldsymbol{w_h})$

Observe that by virtue of Theorem 12.2.2(b), the prefix $\boldsymbol{r_L}$ of every Class II run is bounded in length by $2|\boldsymbol{w_{h-1}}| + |\boldsymbol{w_h}|$. As it turns out, this bound permits the techniques of Algorithm ML to be employed to compute all the Class II runs terminating in each factor $\boldsymbol{w_h}$, $h \geq 2$, in time exactly proportional to $2|\boldsymbol{w_{h-1} w_h}|$. Since

$$2n \leq \sum_{h=2}^{k} 2|\boldsymbol{w_{h-1} w_h}| < 4n, \tag{12.12}$$

it is immediately clear that all the Class II runs in $\boldsymbol{x}$ can be computed in $\Theta(n)$ time based on a previously computed s-factorization. Algorithm 12.2.1 describes the processing of Main's algorithm Mn, and we state the following theorem.

Theorem 12.2.3 *Given the s-factorization $\boldsymbol{w_1 w_2} \cdots \boldsymbol{w_k}$ of $\boldsymbol{x} = \boldsymbol{x}[1..n]$, and based on a linear-time implementation of routine* calcruns, *Algorithm 12.2.1 computes all Class II runs in $\boldsymbol{x}$ in $\Theta(n)$ time.* □

Here we make an important observation. As we shall see, each call to *calcruns*$(\boldsymbol{u}, \boldsymbol{v})$ results in the output of at most $|\boldsymbol{u}| + |\boldsymbol{v}|$ Class II runs. Then for the entire string $\boldsymbol{x}[1..n]$, the upper bound (12.12) applies also to the number of Class II runs, which we see cannot therefore exceed $4n$ — a more useful bound than that provided by Theorem 12.2.1, but also unfortunately more limited, since it does not include Class I runs.

In order to complete our description of Algorithm Mn, we need now to explain the operation of the routine *calcruns*$(\boldsymbol{u}, \boldsymbol{v})$ that computes all the runs in $\boldsymbol{uv}$ that are neither entirely in $\boldsymbol{u}$ nor entirely in $\boldsymbol{v}$. Let $\boldsymbol{u} = \boldsymbol{u}[1..n_1]$, $\boldsymbol{v} = \boldsymbol{v}[1..n_2]$. Recall that in Subsection 12.1.2 we distinguished between (and separately calculated) newright and newleft squares depending on whether or not the second occurrence of the generator of the square lay entirely within $\boldsymbol{v}$. Here we make a very similar distinction, dividing *calcruns* into two disjoint subroutines calcR and calcL, that compute "rightmax" and "leftmax" runs, respectively, defined as follows:

- ***rightmax:*** a Class II run in $\boldsymbol{uv}$ has a suffix in $\boldsymbol{v}$ of length at least p, where p is a period of the run;
- ***leftmax:*** a Class II run in $\boldsymbol{uv}$ has a nonempty suffix in $\boldsymbol{v}$ of length $\ell < p$.

In the example $\boldsymbol{f_6}$ given above, all the runs listed except $(aba)^2ab$ are rightmax runs; $(aba)^2ab$ is leftmax because it has period $p = 3$ but only a suffix ab of length 2 in $\boldsymbol{w_6}$.

The analogy with Subsection 12.1.2 continues. Corresponding to rightmax runs we define two arrays:

- **LP**$[2..n_2 + 1]$: for each position $i = 2, 3, \ldots, n_2$ in $\boldsymbol{v}$, LP$[i] = \text{lcp}(\boldsymbol{v}[i..n_2], \boldsymbol{v})$, while for $i = n_2 + 1$, LP$[i] = 0$;
- **LS**$[1..n_2]$: for each position $i = 1, 2, \ldots, n_2$ in $\boldsymbol{v}$, LS$[i] = \text{lcs}(\boldsymbol{uv}[1..i], \boldsymbol{u})$.

The definition of LP is identical to that given in Subsection 12.1.2; the definition of LS is very slightly changed to consider the suffix of $\boldsymbol{uv}[1..i]$ rather than of $\boldsymbol{v}[1..i]$.

Similarly, corresponding to leftmax runs, arrays LP$'$ and LS$'$ are defined, virtually identical to those used to locate newleft squares in Algorithm ML:

- **LP**$'[2..n_1]$: for each position $i = 2, 3, \ldots, n_1$ in $\boldsymbol{u}$, LP$'[i] = \text{lcp}(\boldsymbol{u}[i..n_1]\boldsymbol{v}, \boldsymbol{v})$;
- **LS**$'[1..n_1 - 1]$: for each position $i = 1, 2, \ldots, n_1 - 1$ in $\boldsymbol{u}$, LS$'[i] = \text{lcs}(\boldsymbol{u}[1..i], \boldsymbol{u})$.

In the remainder of this subsection, we confine ourselves to a description of the subroutine calcR and its arrays LP and LS only, leaving calcL together with LP$'$ and LS$'$ to Exercises 12.2.3 and 12.2.4. Furthermore, we omit also the algorithms that compute LP and LS, since they differ only in minor details from the algorithms calcLP and calcLS discussed in Subsection 12.1.2. We suppose then that each of the four arrays can be computed in time proportional to its length.

In order to understand the operation of calcR, we need a lemma that plays the role for Algorithm Mn that is played for Algorithm ML by Lemma 12.1.5:

Lemma 12.2.4 *Let $\boldsymbol{u}[1..n_1]$ and $\boldsymbol{v}[1..n_2]$ be given strings, and let $\boldsymbol{r} = \boldsymbol{r_L r_R}$ be a substring of $\boldsymbol{uv}$ such that $\boldsymbol{r_L}$ is a nonempty suffix of $\boldsymbol{u}$ and $\boldsymbol{r_R}$ is a nonempty prefix of $\boldsymbol{v}$. Suppose that $|\boldsymbol{r_R}| \geq p$ for some integer $p \in 1..\lfloor |\boldsymbol{r}|/2 \rfloor$. Then $\boldsymbol{r}$ has period p if and only if both of the following conditions hold:*

(a) $|\boldsymbol{r_L}| \leq \text{LS}[p]$;
(b) $|\boldsymbol{r_R}| \leq p + \text{LP}[p+1]$.

Proof In order that $\boldsymbol{r}$ have period p, it is necessary and sufficient that $\boldsymbol{r}$ have a border of length $r-p$, where $r = |\boldsymbol{r}|$. Observe that by hypothesis,

$$r - p \geq r - |\boldsymbol{r_R}| = |\boldsymbol{r_L}|.$$

Therefore, the border exists if and only if

(A) $\boldsymbol{r}\big[1..|\boldsymbol{r_L}|\big] = \boldsymbol{r}\big[p+1..p+|\boldsymbol{r_L}|\big]$;
(B) $\boldsymbol{r}\big[|\boldsymbol{r_L}|+1..r-p\big] = \boldsymbol{r}\big[p+|\boldsymbol{r_L}|+1..r\big]$.

Condition (A) is equivalent to the statement that $\boldsymbol{r}\big[p+1..p+|\boldsymbol{r_L}|\big]$ must be a suffix of $\boldsymbol{u}$. But $\boldsymbol{r}\big[p+1..p+|\boldsymbol{r_L}|\big]$ is itself a suffix of $\boldsymbol{uv}[1..p]$, and so (A) holds if and only if the length $|\boldsymbol{r_L}|$ of the suffix is at most $\text{LS}[p]$. Thus (A) is equivalent to (a).

Condition (B) is equivalent to the statement that $\boldsymbol{r}\big[p+|\boldsymbol{r_L}|+1..r\big]$ must be a prefix of $\boldsymbol{v}$. But $\boldsymbol{r}\big[p+|\boldsymbol{r_L}|+1..r\big]$ is itself a prefix (possibly empty) of $\boldsymbol{v}[p+1..n_2]$, and so (B) holds if and only if the length $r-p-|\boldsymbol{r_L}|$ of the prefix is at most $\text{LP}[p+1]$ — this is just the statement of (b). □

Observe that whenever the substring $\boldsymbol{r}$ described by this lemma exists, it can be written in the form

$$\boldsymbol{r} = \big(\boldsymbol{r}[1..p]\big)^k \boldsymbol{r}[1..p'], \tag{12.13}$$

for some integers $k \geq 2$, $p' \in 0..p-1$. Further, the lemma tells us, by virtue of its "if and only if" condition, that there exists a maximum-length such substring, $\boldsymbol{r}^* = \boldsymbol{r}_L^* \boldsymbol{r}_R^*$ say, with left part $\boldsymbol{r}_L^*$ satisfying

$$|\boldsymbol{r}_L^*| = \text{LS}[p] > 0 \tag{12.14}$$

and right part $\boldsymbol{r}_R^*$ satisfying

$$|\boldsymbol{r}_R^*| = p + \text{LP}[p+1]. \tag{12.15}$$

Recall from Definition 2.3.3 that a run takes the form (12.13), but that in addition it must be NE (nonextendible): it cannot be a proper substring of any other substring with the same period p. Since by Theorem 12.2.2 and our choice of $\boldsymbol{uv}$ we may suppose that every Class II run cannot extend either to the left or to the right beyond $\boldsymbol{uv}$, we see that the substring $\boldsymbol{r}^*$ defined by (12.14) and (12.15) is in fact both of maximum-length and NE within $\boldsymbol{x}$, therefore a run of $\boldsymbol{x}$. Furthermore, it is by definition a rightmax run.

Thus by first computing LP$[p+1]$ and LS$[p]$ corresponding to $\boldsymbol{u}$ and $\boldsymbol{v}$ for every $p \in 1..n_2$, we can identify every rightmax run $\boldsymbol{r}^*$ in $\boldsymbol{uv}$ by applying conditions (12.14) and (12.15) together with the condition $|\boldsymbol{r}^*| \geq 2p$. Since these three conditions are together equivalent to

$$\text{LS}[p] > 0 \ \text{ and } \ \text{LS}[p] + \text{LP}[p+1] \geq p,$$

we can express the processing required to compute all the rightmax runs in $\boldsymbol{uv}$ as shown in Algorithm 12.2.2.

Algorithm 12.2.2 (calcR)

– *Compute all the rightmax runs in* $\boldsymbol{uv} = \boldsymbol{u}[1..n_1]\boldsymbol{v}[1..n_2]$

compute LP$[2..n_2+1]$, LS$[1..n_2]$ for $\boldsymbol{uv}$

for $p \leftarrow 1$ **to** n_2 **do**
  if {LS$[p] > 0$ **and** LS$[p]$ + LP$[p+1] \geq p$ **then**
    $i \leftarrow n_1 - \text{LS}[p] + 1$ – *start of run in* $\boldsymbol{u}$
    $j \leftarrow p + \text{LP}[p+1]$ – *end of run in* $\boldsymbol{v}$
    output i, j

The **for** loop of Algorithm calcR requires $\Theta(n_2)$ time, and as shown in Exercise 12.2.2 the calculation of LP and LS requires $\Theta(n_1 + n_2)$ time: thus calcR$(\boldsymbol{u}, \boldsymbol{v})$ executes in time linear in $|\boldsymbol{uv}|$ and space linear in $|\boldsymbol{v}|$. Similarly, calcL$(\boldsymbol{u}, \boldsymbol{v})$ executes in time linear in $|\boldsymbol{uv}|$ and space linear in $|\boldsymbol{u}|$. Thus *calcruns*$(\boldsymbol{u}, \boldsymbol{v})$ executes in time and space $\Theta(|\boldsymbol{u}| + |\boldsymbol{v}|)$, Theorem 12.2.3 holds, and Algorithm Mn computes all Class II runs in $\boldsymbol{x}[1..n]$, given the s-factorization, in $\Theta(n)$ time.

It should be remarked that it is possible for Algorithm Mn to output the same Class II run more than once if the run has more than one period. Consider for example $\boldsymbol{x} = (ab)^3(aba)^4a$ with s-factorization

$$\boldsymbol{w_1 w_2 w_3 w_4} = a/b/ababab/abaabaabaa.$$

The rightmax run $(aba)^4a$ corresponding to factor $\boldsymbol{w_4}$ will be output twice, once for $p = 3$ (generator aba), again for $p = 6$ (generator $(aba)^2$). We see then that the period used to identify the run need not always be the minimum period, hence that the generator of the run may sometimes itself be a repetition. But even with this potential duplication of runs, the total number of outputs must still be less than $4n$.

We remark also that by requiring in Algorithm calcR only that

$$\text{LS}[p] + \text{LP}[p+1] \geq 1,$$

we output, in addition to all the runs, all the minimum-length nonprimitive substrings of period p satisfying $|\boldsymbol{r}_{\boldsymbol{L}}^*| \geq 1$, $|\boldsymbol{r}_{\boldsymbol{R}}^*| \geq p$. Still the number of outputs, therefore the total number of such substrings, must be less than $4n$.

As observed in [M89], a "further modification may allow the algorithm to find all [Class I runs] in a simple manner". We shall see in the next subsection that this can indeed be done, also in $\Theta(n)$ time.

12.2.2 All Runs — Kolpakov and Kucherov

For words divide and rend;
But silence is most noble till the end.

— Algernon Charles SWINBURNE (1837–1909), *Atalanta in Calydon*

In this subsection we briefly describe a straightforward extension [KK00] to Algorithm 12.2.1 that computes the Class I runs in a given string $\boldsymbol{x}[1..n]$ in $\Theta(n)$ time. Thus the original Algorithm Mn together with the extension computes all the runs in $\boldsymbol{x}$ in linear time, based on the prior computation of the s-factorization of $\boldsymbol{x}$, hence also of the suffix tree $T_{\boldsymbol{x}}$ of $\boldsymbol{x}$ (Section 6.3).

Algorithm KK represents each Class II run $\boldsymbol{r} = \boldsymbol{x}[i,j]$ found by Algorithm Mn as a pair (i,j) where i is the start position and j the end position of the run in $\boldsymbol{x}$. (In case more than one run (i,j) is found by Mn, the duplicates are eliminated, as we shall see shortly.) An array OCCURS$[1..n]$ is formed in which OCCURS$[i]$, $i = 1, 2, \ldots, n$, is a pointer to a linked list of positions

$$j_{i1}, j_{i2}, \ldots, j_{ir_i}$$

that are initially end positions of Class II runs starting at position i of $\boldsymbol{x}$. We maintain the invariant that

$$j_{i1} < j_{i2} < \ldots < j_{ir_i}. \tag{12.16}$$

If no run begins at position i, OCCURS$[i]$ will be `NULL`.

In order to compute OCCURS and its associated lists, a bucket sort is first performed on the pairs (i,j) that puts them in ascending order of j. OCCURS can then be updated by accessing the sorted pairs in ascending order and, for each value of i, adding the corresponding value of j at the end of the list pointed to by OCCURS$[i]$: this will guarantee that (12.16) holds. Since the number of Class II runs is $O(n)$, and since i and j are both integers in the range $1..n$, the bucket sort and the formation of OCCURS can both be performed in $\Theta(n)$ time. Of course duplicate pairs (i,j) can easily be eliminated during the sort process.

In order to make effective use of this data structure, we need to store two values for each factor $\boldsymbol{w_h}$, $1 \le h \le k$, in the s-factorization of $\boldsymbol{x}$: its start position i_h and the start position $i'_h < i_h$ of some previous copy of $\boldsymbol{w_h}$ in $\boldsymbol{x}$. In the case that $\boldsymbol{w_h}$ is the leftmost occurrence of some letter, no previous copy of $\boldsymbol{w_h}$ exists, and so we set $i'_h \leftarrow 0$. We may thus construct two arrays $I = I[1..k+1]$ (with $I[k+1] = n+1$) and $I' = I'[1..k]$ to store this information as a byproduct of the original computation of the s-factorization — the required values in I' are easily computed from the suffix tree.

Using these data structures, we can now imagine locating a copy of $\boldsymbol{w_h}$ (when it exists) at position $I'[h] < I[h]$, then transferring all the runs of suitable length from the copy to $\boldsymbol{w_h}$ itself. Since for each start position i the runs are available from OCCURS$[i]$ in increasing order of length, the transfer corresponding to each i can be implemented in time proportional to the number of runs transferred; since by Theorem 12.2.1 the total number of runs is $O(n)$, therefore the additional time required to store the Class I runs in the data structure is also $O(n)$. The details are given in Algorithm 12.2.3, and we are thus able to state our main result:

Algorithm 12.2.3 (KK)

– *Given the arrays $I[1..k+1]$ and $I'[1..k]$ that describe the s-factorization $\boldsymbol{w_1 w_2} \cdots \boldsymbol{w_k}$ of $\boldsymbol{x}[1..n]$, and given lists* OCCURS$[1..n]$ *that specify in increasing order of length all Class II runs that occur at each position i, this algorithm updates the lists to include* all *runs in $\boldsymbol{x}$*

```
for h ← 2 to k do
  if I'[h] > 0 then
    δ ← I[h]−I'[h]   – the offset of w_h from its copy
    for i ← I[h] to I[h+1]−1 do
      ∀ j ∈ list(OCCURS[i−δ])
          – A Class I run must begin and end in w_h
        if (j+δ)−i < I[h+1]−i then
          insert list(OCCURS[i]) ← j+δ
```

Theorem 12.2.5 *Given the s-factorization $\boldsymbol{w_1 w_2} \cdots \boldsymbol{w_k}$ of $\boldsymbol{x}[1..n]$, the array* OCCURS$[1..n]$ *and its associated lists together with the arrays $I[1..k+1]$ and $I'[1..k]$ can all be computed in $\Theta(n)$ time. Based on this data structure, Algorithm 12.2.3 requires $O(n)$ time to update* OCCURS *and its associated lists to include all the runs in $\boldsymbol{x}$.* □

There are several other algorithms that compute collections of repetitions in $\boldsymbol{x}[1..n]$ in $\Theta(n)$ time on an alphabet of size α. All of them make use of suffix trees and treat the $\log \alpha$ factor normally inherent in suffix tree construction as a constant. [K94] describes an algorithm that computes the shortest square at each position in $\boldsymbol{x}$. Inspired by the result in [FS98], mentioned earlier, that the maximum number of distinct squares in any string is $O(n)$, [GS98] computes all the distinct squares in $\boldsymbol{x}$. In [G97] an algorithm is described that computes all the distinct nonextendible (Definition 2.3.2) repeating substrings in $\boldsymbol{x}$: these substrings are the distinct generators, not only of repetitions, but also of repeats that may be split (nontandem) or overlapping (Definition 2.3.1). The calculation of all *pairs* of nonextendible repeating substrings that satisfy various constraints is described in [BLPS00].

Finally, we propose in Exercise 12.2.10 another approach to the computation of Class I runs — one whose complexity is not quite clear!

Exercises 12.2

1. Prove (12.12).
2. The definition of LP is identical to that used in Subsection 12.1.2, and therefore the routine that computes it is identical. The definition of LS is however slightly different. Write an algorithm calcLS that computes LS for given $\boldsymbol{u}$ and $\boldsymbol{v}$, and show that it requires $\Theta(|\boldsymbol{u}| + |\boldsymbol{v}|)$ time.
3. Write a $\Theta(|\boldsymbol{u}| + |\boldsymbol{v}|)$-time algorithm calcLP$'$ that computes LP$'$ for given $\boldsymbol{u}$ and $\boldsymbol{v}$.
4. State and prove the analogue of Lemma 12.2.4 for leftmax runs. Hence specify Algorithm calcL.
5. Suppose that the substring $\boldsymbol{x}[i..j]$ is a run of period p in $\boldsymbol{x}$. Prove that for every $i' \in i+p..j-p+1$, no substring of $\boldsymbol{x}$ beginning at i' is a run of period p. Show by example that the bounds on i' are sharp.
6. Suppose that the substring $\boldsymbol{x}[i..j]$ is a run of period p and also of period $p' \neq p$ in $\boldsymbol{x}$. Prove that $\gcd(p, p')$ is also a period of $\boldsymbol{x}[i..j]$.

   Hint: Don't forget about the Periodicity Lemma!
7. Explain how the array I' can be computed in linear time from $T_{\boldsymbol{x}}$ during the calculation of the s-factorization of $\boldsymbol{x}$.
8. Let h be the least integer such that a factor $\boldsymbol{w}_h$ in the s-factorization of $\boldsymbol{x}$ contains the end point of a run. Is every run in $\boldsymbol{w}_h$ necessarily of Class II?
9. Suppose that a factor $\boldsymbol{w}_h$ in the s-factorization of $\boldsymbol{x}$ begins at position i of $\boldsymbol{x}$ and that it is a copy of a substring of $\boldsymbol{x}$ beginning at position $i' < i$. Suppose further that $i - i' \leq |\boldsymbol{w}_h|/2$. Show then that $\boldsymbol{x}[i'..i+|\boldsymbol{w}_h|-1]$ is a run in $\boldsymbol{x}$.
10. Observe that many of the Class I runs that occur within any factor $\boldsymbol{w}_h$ of the s-factorization of $\boldsymbol{x}$ will themselves be Class II runs of the s-factorization of $\boldsymbol{w}_h$. Use this observation to design an algorithm that recursively calls Algorithm Mn and computes all the runs in $\boldsymbol{x}$. What can you say about this algorithm's complexity?
11. Let $\boldsymbol{z}_0 = \lambda_0^2$, and let $\boldsymbol{z}_i = \boldsymbol{z}_{i-1}\lambda_i\boldsymbol{z}_{i-1}$ for every $i = 1, 2, \ldots, n$, where the letters $\lambda_0, \lambda_1, \ldots, \lambda_{n-1}$ are pairwise distinct. Roman Kolpakov suspects that the string $\boldsymbol{z}_n$ represents a worst case for the recursive algorithm described in the preceding exercise. Based on the s-factorization of $\boldsymbol{z}_n$, determine the time complexity of the algorithm applied to $\boldsymbol{z}_n$.

Chapter 13

Extensions of Periodicity

A word to the wise is enough,
and many words won't fill a bushel.

— Benjamin FRANKLIN (1706–1790), *Poor Richard's Almanac*

In this chapter we discuss topics that are direct extensions of those arising out of the concept of periodicity. In general terms, the two main computational problems related to periodicity are as follows:

- compute all the periods of a given string $\boldsymbol{x} = \boldsymbol{x}[1..n]$;
- compute all the repetitions in $\boldsymbol{x}$.

The first of these problems is handled by Algorithm Border Array (Section 1.3), the first algorithm studied in this book: in time $\Theta(n)$ it computes the border array $\boldsymbol{\beta}[1..n]$ of $\boldsymbol{x}$ that specifies every border, hence every period, not only of $\boldsymbol{x}$, but of every prefix of $\boldsymbol{x}$. As discussed in Section 5.1, the border array corresponds to a tree structure called the border tree whose root node is labelled zero and whose other nodes are labelled by the positions $i \in 1..n$ in $\boldsymbol{x}$: the labels of the nodes on the path from the root of the tree to any node i (except for i itself) are precisely the lengths (in ascending order) of the borders of $\boldsymbol{x}[1..i]$.

We discover in Section 13.1 the rather remarkable fact that the quasiperiods of $\boldsymbol{x}$ (see Section 2.3 for definitions) can be computed using structures that are formally identical to those used to compute the periods:

- in time $\Theta(n)$ we can compute the ***cover array*** $\gamma[1..n]$ that specifies every cover (that is, every quasiperiod) of every prefix of $\boldsymbol{x}$;
- the cover array defines a ***cover tree*** with nodes labelled $0..n$ as for the border tree: the nonzero labels on the path from the root to any node i of the cover tree are precisely the lengths (in ascending order) of the covers of $\boldsymbol{x}[1..i]$.

Like the border array, the cover array is perhaps more properly thought of as an intrinsic pattern: every string has a cover array. We discuss it here, however, as a generic pattern because of its close connection to periodicity.

In Sections 13.2 and 13.3, we turn instead to natural extensions of the idea of a repetition: the calculation of repeats (Section 2.3), exact and approximate. We find in Section 13.2 that, like repetitions, nonextendible repeats can be computed in $O(n \log n)$ time and, like runs, be represented in $O(n)$ space. As a byproduct, we discover further that Crochemore's all-repetitions algorithm (Subsection 12.1.1) is ultimately a suffix tree construction algorithm, that a suffix tree can be economically represented as an array of integer pairs, and that the calculation of NE repeats is a natural prelude to the calculation of all NE coverable substrings (Problem 2.17—Section 2.3).

In Section 13.3 we consider the further extension of the all-repeats problem to approximate repeats, now making use rather of the ideas and techniques introduced in Chapters 9 and 10 to compute distance and to perform approximate pattern-matching. The computation of approximate repeats is of particular importance to computational biology, where it may for example be of interest to identify lengthy sequences of DNA that have been replicated elsewhere in the genome, but with minor alteration.

Finally, Section 13.4 investigates some of the difficulties that arise in the determination of the approximate generators of a string.

13.1 All Covers of a String — Algorithm LS

A man of words and not of deeds
Is like a garden full of weeds.

— *Nursery rhyme*

Recall from Section 2.3 that a repeat

$$M_{\boldsymbol{x},\boldsymbol{u}} = (p;\ i_1, i_2, \ldots, i_r) \tag{13.1}$$

is said to be a ***cover*** of $\boldsymbol{x}[i_1..i_r+p-1]$ if and only if $r \geq 2$, $p = |\boldsymbol{u}|$,

$$\boldsymbol{u} = \boldsymbol{x}[i_1..i_1+p-1] = \boldsymbol{x}[i_2..i_2+p-1] = \cdots = \boldsymbol{x}[i_r..i_r+p-1],$$

and each ***gap*** $g_j = i_{j+1} - i_j$, $j = 1, 2, \ldots, r-1$, lies in the range $1..p$. In this section we consider the problem of whether or not, for a given string $\boldsymbol{x} = \boldsymbol{x}[1..n]$, there exists a substring $\boldsymbol{u}$ such that (13.1) is a cover for the special case $i_1 = 1$ and $i_r + p - 1 = n$. When it is not necessary to specify the r locations at which $\boldsymbol{u}$ occurs, we shall thus say informally that $\boldsymbol{u}$ is a cover of $\boldsymbol{x}$ or that $\boldsymbol{u}$ ***covers*** $\boldsymbol{x}$. For example, given $\boldsymbol{x} = abaababaaba$, we would say that both $\boldsymbol{u} = abaaba$ and $\boldsymbol{u}' = aba$ are covers of $\boldsymbol{x}$. Note that because in (13.1) the value of r must be at least 2, $\boldsymbol{x}$ is never a cover of itself.

Supposing then that $\boldsymbol{u}$ is a cover of $\boldsymbol{x}$, we state the following elementary facts, leaving the proofs to Exercise 13.1.1:

(F1) $0 < |\boldsymbol{u}| < |\boldsymbol{x}|$.

(F2) $\boldsymbol{u}$ is a border of $\boldsymbol{x}$.

(F3) Let $\boldsymbol{u}'$ be a string such that $|\boldsymbol{u}'| < |\boldsymbol{u}|$. Then $\boldsymbol{u}'$ is a cover of $\boldsymbol{u}$ if and only if $\boldsymbol{u}'$ is a cover of $\boldsymbol{x}$.

(F4) Corresponding to every gap g_j, $j = 1, 2, \ldots, r-1$, determined by (13.1), $\boldsymbol{u}$ has a border of length $p - g_j$.

In the 1990s several $\Theta(n)$-time algorithms were published to compute covers of $\boldsymbol{x}$: in [AFI91] only the shortest cover of $\boldsymbol{x}$ was computed, then in [B92] the shortest cover of every prefix, then later in [MS94, MS95] every cover of $\boldsymbol{x}$ but not of any of its proper prefixes, finally in [LS02] every cover of every prefix of $\boldsymbol{x}$. We describe here the last of these algorithms, whose performance is made possible by the introduction of the ***cover array*** $\boldsymbol{\gamma} = \boldsymbol{\gamma}[1..n]$ of $\boldsymbol{x}$, defined as follows:

for every $i \in 1..n$, $\boldsymbol{\gamma}[i]$ is the length of the longest cover of $\boldsymbol{x}[1..i]$; zero if no cover exists.

	1	2	3	4	5	6	7	8	9	10	11	12	13	14	15	16	17	18	19	20	21	22	23
$x =$	a	b	a	a	b	a	b	a	a	b	a	a	b	a	b	a	a	b	a	b	a	b	a
$\beta =$	0	0	1	1	2	3	2	3	4	5	6	4	5	6	7	8	9	10	11	7	8	2	3
$\gamma =$	0	0	0	0	0	3	0	3	0	5	6	0	5	6	0	8	9	10	11	0	8	0	3

It is an immediate consequence of Facts (F1) and (F3) that the lengths of the covers of $\boldsymbol{x}[1..i]$ are exactly the elements of the descending sequence

$$\langle \boldsymbol{\gamma}[i], \boldsymbol{\gamma}^2[i], \cdots, \boldsymbol{\gamma}^{k-1}[i] \rangle, \tag{13.2}$$

where for every $h \in 2..k$, $\boldsymbol{\gamma}^{\boldsymbol{h}}[i] = \boldsymbol{\gamma}\big[\boldsymbol{\gamma}^{\boldsymbol{h-1}}[i]\big]$ and k is the least positive integer such that $\boldsymbol{\gamma}^{\boldsymbol{k}}[i] = 0$. This is the analogue of Lemma 1.3.1 which tells us that $\boldsymbol{u} = \boldsymbol{x}[1..p]$ is a border of $\boldsymbol{x}[1..i]$ if and only if $p = \boldsymbol{\beta}^{\boldsymbol{h}}[i]$ for some integer $h \in 1..k$, where now k is the least positive integer such that $\boldsymbol{\beta}^{\boldsymbol{k}}[i] = 0$. These results provide the basis for representing all the borders (respectively, covers) of $\boldsymbol{x}$ using only n storage locations. As illustrated in Figure 13.1, the relationship between any position i and $\boldsymbol{\beta}[i]$ (respectively, $\boldsymbol{\gamma}[i]$) is a child-

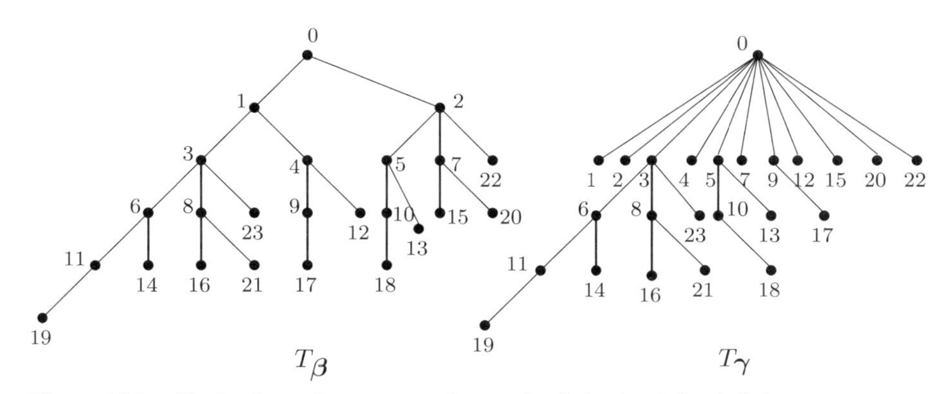

Figure 13.1. The border and cover trees of $\boldsymbol{x} = abaababaabaababaabababa$

parent relationship in the tree T_β (respectively, T_γ) that permits the elements of the set of borders (respectively, covers) of $\boldsymbol{x}[1..i]$ to be accessed simply by accessing ancestors of i in the tree.

The discussion to this point may have persuaded the reader that the cover array is a combinatorial object of some interest. In an effort to persuade a little further, we prove in the following subsection two theorems that place severe restrictions on the values allowable in a cover array, and in fact relate them to those in the border array. At least to the author, it is rather wonderful that the proof of one of these theorems depends on a not-so-difficult exercise that involves two topics seemingly quite unrelated to covers: Hamming distance and rotation of strings.

In the final subsection we go on to show how to compute the cover array in $\Theta(n)$ time.

Properties of the Cover Array

In the course of solving Exercise 2.2.6, the dedicated reader has already become satisfied that for any string $\boldsymbol{x}[1..n]$,

$$d_H\big(\boldsymbol{x}, R_j(\boldsymbol{x})\big) \neq 1, \qquad (13.3)$$

where d_H is the Hamming distance (Section 2.2) and $R_j(\boldsymbol{x})$ is the jth rotation of $\boldsymbol{x}$, $j \in 0..n-1$ (Section 1.4). We use this result to prove a lemma that is in its turn used to establish the first of two interesting properties of the cover array:

Lemma 13.1.1 *If $\boldsymbol{u}$ is a cover of $\boldsymbol{x}$ and $|\boldsymbol{u}| = p$, then every substring $\boldsymbol{v}$ of length p in $\boldsymbol{x}$ satisfies*

$$d_H(\boldsymbol{u}, \boldsymbol{v}) \neq 1.$$

Proof We consider any two distinct occurrences of $\boldsymbol{u}$ in $\boldsymbol{x} = \boldsymbol{x}[1..n]$ that are either adjacent or overlapping. Let the left occurrence be $\boldsymbol{u_1} = \boldsymbol{x}[i_1..i_1+p-1]$ and the right $\boldsymbol{u_2} = \boldsymbol{x}[i_2..i_2+p-1]$, where $1 \leq i_1 \leq n-p+1$ and $i_1 < i_2 \leq i_1+p$.

We show that the result holds for every $\boldsymbol{v}$ in $\boldsymbol{U} = \boldsymbol{x}[i_1..i_2+p-1]$, hence for $\boldsymbol{x}$ itself.

Observe first that if indeed $i_2 = i_1+p$, it follows that $\boldsymbol{U} = \boldsymbol{u^2}$, so that every substring $\boldsymbol{v}$ of length p in $\boldsymbol{U}$ is a rotation of $\boldsymbol{u}$. The result is then in this case an immediate consequence of (13.3).

Suppose therefore that $i_2 < i_1+p$, so that $\boldsymbol{u_1}$ and $\boldsymbol{u_2}$ overlap by some nonempty border, say $\boldsymbol{u'}$, of $\boldsymbol{u}$. Then, as shown in Figure 13.2, we may write $\boldsymbol{u} = \boldsymbol{u'u''}$ for some nonempty suffix $\boldsymbol{u''}$ of $\boldsymbol{u}$.

Let $\boldsymbol{u''} = \boldsymbol{w^k}$ for some largest integer $k \geq 1$ (if $\boldsymbol{u''}$ is not a repetition, $\boldsymbol{w} = \boldsymbol{u''}$ and $k = 1$). Then $\boldsymbol{U}$ has period $|\boldsymbol{w}|$, and so every substring $\boldsymbol{v''}$ of length $|\boldsymbol{u''}|$ in $\boldsymbol{U}$ also has period $|\boldsymbol{w}|$. Thus such a substring $\boldsymbol{v''}$ must satisfy one of the following two conditions:

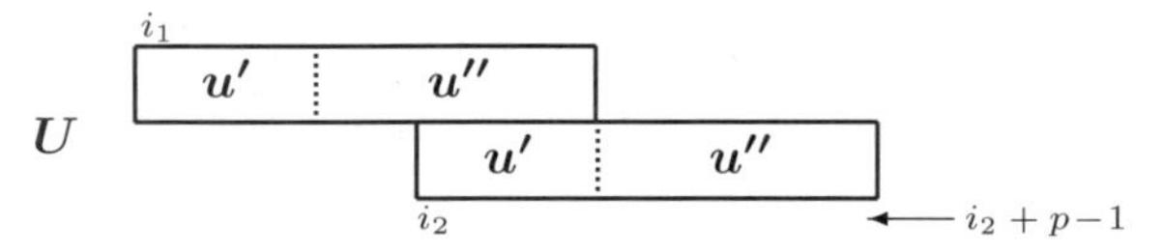

Figure 13.2. Overlapping occurrences of $\boldsymbol{u}$ in $\boldsymbol{U} = \boldsymbol{x}[i_1..i_2+p-1]$

- $\boldsymbol{v}''$ is a rotation of $\boldsymbol{u}''$ but not equal to $\boldsymbol{u}''$, and so $d_H(\boldsymbol{u}'', \boldsymbol{v}'') \geq 2$ by (13.3). Hence for every occurrence of $\boldsymbol{v} = \boldsymbol{v}'\boldsymbol{v}''$ of length p in $\boldsymbol{U}$, $d_H(\boldsymbol{u}, \boldsymbol{v}) \geq 2$.
- $\boldsymbol{v}'' = \boldsymbol{u}''$. In this case, since $\boldsymbol{v}''$ is a substring of $\boldsymbol{U}$ of period $|\boldsymbol{w}|$, and since we know that

$$\boldsymbol{u} = \boldsymbol{u}'\boldsymbol{u}'' = \boldsymbol{u}'\boldsymbol{v}''$$

occurs at least once in $\boldsymbol{U}$, therefore every occurrence of $\boldsymbol{v}''$ in $\boldsymbol{U}$ must be preceded by an occurrence of $\boldsymbol{u}'$ such that $d_H(\boldsymbol{u}, \boldsymbol{u}'\boldsymbol{v}'') = 0$.

We have shown that there exists no $\boldsymbol{v}$ in $\boldsymbol{U}$ for which $d_H(\boldsymbol{u}, \boldsymbol{v}) = 1$, as required. □

This result was first proved (very elegantly) in [Si00] using a variant of the Periodicity Lemma [BB99]; a weaker version was established (with painful determination) in [LS02]. With its help, we are able to establish the following fundamental property of the cover array $\boldsymbol{\gamma}[1..n]$:

Theorem 13.1.2 *For every integer $i \in 1..n-1$, if $\boldsymbol{\gamma}[i] \neq 0$, then either $\boldsymbol{\gamma}[i+1] > \boldsymbol{\gamma}[i]$ or $\boldsymbol{\gamma}[i+1] = 0$.*

Proof Let $p = \boldsymbol{\gamma}[i]$. Then $p > 0$ and $\boldsymbol{x}[1..p]$ is the longest cover of $\boldsymbol{x}[1..i]$. Suppose the theorem is false, so that there exists a positive integer $p' \leq p$ such that $\boldsymbol{x}[1..p']$ is the longest cover of $\boldsymbol{x}[1..i+1] = \boldsymbol{x}[1..i]\lambda$. Then (see Figure 13.3) it follows that $\boldsymbol{x}[1..p']$ is a cover of $\boldsymbol{x}[1..p]\lambda$.

If every occurrence of $\boldsymbol{x}[1..p]$ in $\boldsymbol{x}[1..i+1]$ were followed by λ, it would follow that $\boldsymbol{x}[1..p+1]$ would be a cover of $\boldsymbol{x}[1..i+1]$, contrary to the hypothesis that $\boldsymbol{x}[1..p']$ is the longest cover. Thus some occurrence of $\boldsymbol{x}[1..p]$ in $\boldsymbol{x}[1..i]$ must be followed by $\lambda' \neq \lambda$. This tells us two things:

- there are at least two distinct letters in $\boldsymbol{x}[1..i+1]$, hence also in $\boldsymbol{x}[1..p']$, so that $p' \geq 2$;
- corresponding to the cover $\boldsymbol{x}[1..p'] = \boldsymbol{x}[1..p'-1]\lambda$, there exists in $\boldsymbol{x}[1..i+1]$ a substring $\boldsymbol{u} = \boldsymbol{x}[1..p'-1]\lambda'$ such that

$$d_H\big(\boldsymbol{x}[1..p'], \boldsymbol{u}\big) = 1,$$

contrary to Lemma 13.1.1.

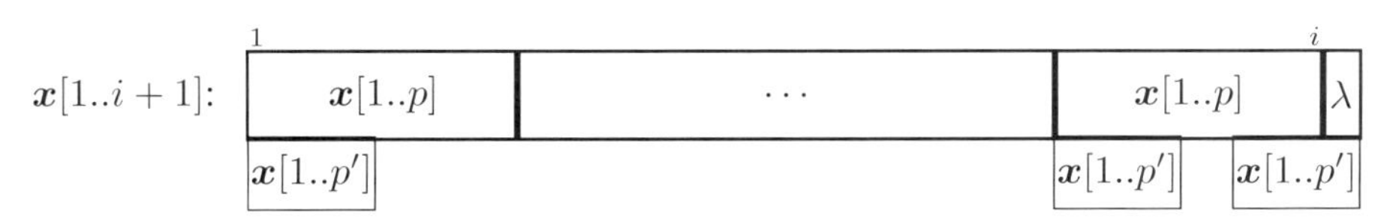

Figure 13.3. $\boldsymbol{x}[1..p']$ is a cover of $\boldsymbol{x}[1..p]\lambda$

We conclude that the assumption $0 < p' \le p$ is false, and the theorem is proved. □

The next result relates the values in the cover array $\gamma[1..n]$ to those in the border array $\beta[1..n]$: the value of $\gamma[i+1]$ can possibly be nonzero only if $\beta[i+1] = \beta[i] + 1$.

Theorem 13.1.3 *For every integer $i \in 1..n-1$, if $\beta[i+1] \le \beta[i]$, then $\gamma[i+1] = 0$.*

Proof It follows from Fact (F2) that if $\beta[i+1] = 0$, then $\gamma[i+1] = 0$. Hence we may assume without loss of generality that $\beta[i+1] > 0$. Therefore by hypothesis $\beta[i] > 0$.

Suppose now that the theorem is false, so that $\gamma[i+1] = p' > 0$. Then $x[1..p']$ is a cover of $x[1..i+1]$. Thus, setting $p = \beta[i]$, we have

$$0 < p' = \gamma[i+1] \le \beta[i+1] \le p.$$

If we let $x[i+1] = \lambda$, we see therefore that the cover

$$x[1..p'] = x[i-p'+2..i]\lambda$$

is a proper suffix of

$$x[i-p+1..i+1] = x[i-p+1..i]\lambda = x[1..p]\lambda = x[1..p+1]. \tag{13.4}$$

But since $\beta[i+1] \ne \beta[i]+1$, it cannot be true that $x[p+1] = \lambda$, in contradiction to (13.4). Therefore the assumption $\gamma[i+1] > 0$ must be false. □

This theorem suggests the idea of a ***staircase***; that is, a maximal sequence $S_{i,h}$ of $h \ge 1$ positions

$$i, i+1, \ldots, i-h+1$$

in the border array $\beta[1..n]$ such that $\beta[j+1] = \beta[j] + 1$ for every $j \in i..i+h-2$. Note that the maximality of the staircase implies that $\beta[i] \le \beta[i-1]$, for $i-1 \ge 1$, and $\beta[i+h-1] \ge \beta[i+h]$, for $i+h \le n$. For example, in $\beta = 0011232345645$, there are five staircases

$$S_{1,1},\ S_{2,2},\ S_{4,3},\ S_{7,5},\ S_{12,2}.$$

Theorem 13.1.3 tells us that $\gamma[i] = 0$ at the start position i of every staircase in the β array. We shall see later how the idea of a staircase is useful for the calculation of the cover array.

Another useful idea is that of a "live" vertex. If a string x is a prefix of a string y, we say that y is a ***right extension*** of x. If a prefix u of x can possibly be a cover of some right extension of x, then u is said to be ***live*** with respect to x; otherwise, u is said to be ***dead*** with respect to x. For example, if

$$x = abaab,$$

we see that $\boldsymbol{u} = aba$ is live with respect to $\boldsymbol{x}$, since aba covers the right extension $\boldsymbol{x}a$; on the other hand, $\boldsymbol{u} = a$ and $\boldsymbol{u} = ab$ can cover no right extension of $\boldsymbol{x}$, and so are dead with respect to $\boldsymbol{x}$. Observe that $\boldsymbol{x}$ is always live with respect to itself.

Since a prefix is identified by its length, and since we speak always of a single given string $\boldsymbol{x}$, we extend this terminology to positions in $\boldsymbol{x}$: we say that $j \in 1..i$ is live with respect to i if and only if $\boldsymbol{x}[1..j]$ can possibly be a cover of some right extension of $\boldsymbol{x}[1..i]$. Also, since positions in $\boldsymbol{x}$ correspond to nodes in the cover tree, we extend the terminology further to speak of live and dead nodes in $T_{\boldsymbol{\gamma}}$. Observe that if j is dead with respect to i, it is dead also with respect to $i+1$. Thus, as the string $\boldsymbol{x}$ is scanned from left to right — that is, as i increases — the number of dead positions is monotone nondecreasing. In strings, as in the world, it appears, once dead, dead forever.

Lemma 13.1.4 *If $j \in i-\boldsymbol{\beta}[i]..i$, then j is live with respect to i.*

Proof By hypothesis, the string $\boldsymbol{x} = \boldsymbol{x}[1..i]$ has minimum period $p^* = i - \boldsymbol{\beta}[i]$ and so may be written in the normal form

$$\boldsymbol{x} = \boldsymbol{x}[1..p^*]^{r^*}\boldsymbol{x}[1..p']$$

for integers $r^* = \lfloor i/p^* \rfloor$ and $p' = i \bmod p^*$. For every $j \in p^*..i$, observe that each prefix

$$\boldsymbol{x}[1..j] = \boldsymbol{x}[1..p^*]^{\lfloor j/p^* \rfloor}\boldsymbol{x}[1..j \bmod p^*]$$

is a cover of the right extension

$$\boldsymbol{x}\boldsymbol{x}[p'+1..p^*]\boldsymbol{x}[1..j \bmod p^*] = \boldsymbol{x}[1..p^*]^{r+1}\boldsymbol{x}[1..j \bmod p^*],$$

and so j is live with respect to i. □

As an example of this lemma, consider $\boldsymbol{x}[1..10] = abcaabcaab$ of length $i = 10$ and period $p^* = 4$. Then for every $j \in 4..10$, $\boldsymbol{x}[1..j]$ is a cover of the right extension

$$\boldsymbol{x}\boldsymbol{x}[3..4]\boldsymbol{x}[1..j \bmod 4] = (abca)^3\boldsymbol{x}[1..j \bmod 4].$$

Observe however that the sufficient condition of Lemma 13.1.4 is not necessary: $\boldsymbol{x}[1..3] = aba$ is a cover of $\boldsymbol{x}[1..8] = ababaaba$ of length $i = 8$ and period $p^* = 5$ even though $j = 3$ does *not* lie in the range 5..8. The next lemma provides a complete characterization of live positions:

Lemma 13.1.5 *With respect to every $i \in 1..n$, j is live if and only if one of the following two conditions holds:*

(a) $j \in i-\boldsymbol{\beta}[i]..i$;
(b) $\boldsymbol{x}[1..j]$ *is a cover of* $\boldsymbol{x}[1..j']$ *for some* $j' \in i-\boldsymbol{\beta}[i]..i$.

Proof To prove that conditions (a) and (b) are necessary, suppose first that j is live with respect to i. In view of Lemma 13.1.4, we need show only that (b) holds for $j < p^* = i - \beta[i]$. Suppose therefore that (b) is false: $x[1..j']$ is not a cover of $x[1..j']$ for any $j' \in p^*..i$. But then, since $j < p^*$, $x[1..j]$ cannot cover any extension of x, a contradiction. We conclude that if (a) does not hold, then (b) must hold in order that j should be live with respect to i.

The sufficiency of condition (a) has already been established in Lemma 13.1.4. Thus, in order to prove the sufficiency of (b), suppose that for some integer $j \in 1..p^* - 1$, $x[1..j]$ is a cover of some $x[1..j']$, $j' \in p^*..i$.

First suppose that $j' < i - p^*$. Then since $x[1..i]$ has period p^*, if $x[1..j]$ covers $x[1..j']$, it must also cover $x[p^* + 1..p^* + j']$, hence also $x[1..p^* + j']$. In fact, for every $j' < i - p^*$, $x[1..j]$ covers $x[1..p^* + j']$, and so without loss of generality we may assume that $j' \geq i - p^*$.

We can now show that the right extension

$$y = xx[i - p^* + 1..j']$$

of length $p^* + j' \geq i$ has a border $x[1..j']$ of length $j' \geq (p^* + j')/2$, and so is covered by $x[1..j]$. Then, by definition, j is live with respect to i, as required. We leave to Exercise 13.1.4 the details of the proof that

$$y = x[1..j']x[j' + 1..i]x[i - p^* + 1..j'] = x[1..p^*]x[1..j']. \tag{13.5}$$

□

As we shall see in the next subsection, it is important for the cover array calculation to be able to characterize dead nodes with respect to the beginning of a staircase:

Lemma 13.1.6 *Let i be a position that starts a staircase in a border array β. Then $j < i - \beta[i]$ is dead with respect to i if and only if j has no children in T_γ that are live with respect to i.*

Proof Suppose $j < i - \beta[i]$ is dead with respect to i. Then $x[1..j]$ cannot possibly cover any right extension of $x[1..i]$. If j has a child j' in T_γ that is live with respect to i, then $x[1..j']$ is a potential cover of $x[1..k]$, for some $k \geq i$. But since j' is a child of j, it must be true that $x[1..j]$ covers $x[1..j']$ and so by (F3) also $x[1..k]$, a contradiction. Thus necessity is proved.

To prove sufficiency, suppose that $j < i - \beta[i]$ has no child live with respect to i, but that j itself is live with respect to i. Then, by Lemma 13.1.5, $x[1..j]$ must cover some string $x[1..k]$, $i - \beta[i] \leq k \leq i$. Therefore j has a child k in T_γ that, again by Lemma 13.1.5, must be live with respect to i, a contradiction. We conclude that j is dead with respect to i. □

As a corollary of this result, we observe that if a node j is live with respect to a node i, then so is its parent $\gamma[j]$ in T_γ.

We state as a final lemma a simple property of $i - \beta[i]$ that is required in order to compute the cover array:

Lemma 13.1.7 *The function $i - \boldsymbol{\beta}[i]$ is invariant for every i in the same staircase and monotone nondecreasing in i; in particular, for any position $i > 1$ that starts a new staircase in $\boldsymbol{\beta}$,*

$$(i-1) - \boldsymbol{\beta}[i-1] < i - \boldsymbol{\beta}[i].$$

Proof An immediate consequence of the fact that $\boldsymbol{\beta}[i] \le \boldsymbol{\beta}[i-1]$ if and only if i starts a staircase. □

We conclude from Lemmas 13.1.5 and 13.1.7 that the set of dead nodes remains unchanged over all values of i in the same staircase; thus only at the start of a new staircase does it need to be recomputed (using Lemma 13.1.6).

Computing the Cover Array

Algorithm LS performs two main tasks as the string $\boldsymbol{x}$ and border array $\boldsymbol{\beta}$ are processed from left to right, for each $i = 1, 2, \ldots, n$:

- compute $\boldsymbol{\gamma}[i]$ and add i as a new child of $\boldsymbol{\gamma}[i]$ in the cover tree $T_{\boldsymbol{\gamma}}$ — recall that by definition $\boldsymbol{\gamma}[i]$ is just the parent of i in $T_{\boldsymbol{\gamma}}$;
- for every i that marks the start of a staircase in $\boldsymbol{\beta}$, compute the nodes in $T_{\boldsymbol{\gamma}}$ that are dead with respect to i.

In addition to $\boldsymbol{\beta}[1..n]$ and $\boldsymbol{\gamma}[1..n]$, the algorithm also maintains the following arrays:

- $dead[0..n-1]$: by Lemma 13.1.6, $dead[j] =$ TRUE if and only if, for the current value of i, $j < i - \boldsymbol{\beta}[i]$ and j has no children in $T_{\boldsymbol{\gamma}}$ that are live with respect to i (as we shall see, the root node 0 in $T_{\boldsymbol{\gamma}}$ is always live);
- *livechildren*$[0..n]$: *livechildren*$[i] = k$ if and only if node i in the cover tree $T_{\boldsymbol{\gamma}}$ has exactly k live children;
- *largestlive*$[0..n]$: *largestlive*$[i] = j$ if and only if j is the largest live ancestor of i in the cover tree $T_{\boldsymbol{\gamma}}$ (of course i is always live with respect to itself).

The algorithm is initialized by placing node 0 in the cover tree $T_{\boldsymbol{\gamma}}$, setting every position in *dead* to FALSE, every position in *livechildren* to 0, and *largestlive*$[i]$ to i for every $i \in 1..n$. Then for the current ith position in $\boldsymbol{x}$, $i = 1, 2, \ldots, n$, Algorithm LS has just three steps:

- Step 1 ensures that the *largestlive* array is kept up to date.
- Step 2 attaches i to its proper parent in $T_{\boldsymbol{\gamma}}$.
- Step 3 ensures that the *dead* array is updated at the start of each new staircase.

These steps are specified in more detail in Algorithm 13.1.1.

Algorithm 13.1.1 (LS)

```
– Compute the cover array γ of x[1..n]
for i ← 1 to n do
   – STEP 1: if β[i] in T_γ is dead, β[i] should have
   – the same largest live ancestor that its parent has
  if dead[β[i]] = TRUE then
    largestlive[β[i]] ← largestlive[γ[β[i]]]

   – STEP 2: Compute γ[i]
   – if β[i] is live, set γ[i] ← β[i]; otherwise, since every
   – cover of x[1..i] must cover x[1..β[i]], set γ[i] equal
   – to the largest live ancestor (possibly 0) of β[i]
  γ[i] ← largestlive[β[i]]
  livechildren[γ[i]] ← livechildren[γ[i]] + 1

   – STEP 3: Identify all the nodes in T_γ that have
   – become dead as a result of starting the new staircase
  if i > 1 and β[i] ≤ β[i−1] then
    c1 ← i−β[i]; c2 ← (i−1)−β[i−1]   – Lemma 13.1.7
    for j ← c1 downto c2   – Lemma 13.1.6
      if not dead[j] and livechildren[j] = 0 then
        dead[j] ← TRUE
        j' ← γ[j]
        livechildren[j'] ← livechildren[j']−1
         – Set all ancestors dead that have no live children
        while livechildren[j'] = 0 do
          dead[j'] ← TRUE
          j' ← γ[j']
          livechildren[j'] ← livechildren[j']−1
```

We claim that

Theorem 13.1.8 *Algorithm 13.1.1 correctly computes the cover array $\boldsymbol{\gamma}[1..n]$ of a given string $\boldsymbol{x}[1..n]$.*

Proof We consider the steps separately, beginning with Step 3.

Let $1 = i_1 < i_2 < \cdots < i_k \le n$ denote the positions in $\boldsymbol{\beta}$ at which staircases start. In Step 3 each position $j \in i_{h-1}..i_h - 1, h = 2, 3, \ldots, k-1$, is tested to determine whether or not it should be set dead with respect to i_{h+1} — by Lemma 13.1.5, each such position was live with respect to i_h. By Lemma 13.1.6, the parent $\boldsymbol{\gamma}[j]$ of any position j that is set dead must be inspected: $\boldsymbol{\gamma}[j]$ must previously have been live, but it will now be dead if and only if it has no live children. Thus, Step 3 must recursively examine the parent of *any* position that has been set dead, until an ancestor is found that has at least one live child, and that therefore remains alive. Note that since i_{h+1} is always live with respect to itself, there must by Lemma 13.1.6 exist a path containing only live nodes that leads from i_{h+1} to the root. Thus the root is always live, and Step 3 will always terminate at the first live node along the path from j to the root.

Note that the positions $j \in i_{h-1}..i_h - 1$ are considered in the reverse order $i_h - 1, i_h - 2, \ldots, i_{h-1}$ in order to take account of the possibility that j may be set dead and that moreover $\gamma[j] \in i_{h-1}..i_h - 1$ also. It is for this same reason that, in order to avoid redundant processing of the same path to the root, it is necessary to check in the **for** loop of Step 3 that **not** $dead[j]$: j could have been set dead because it is the parent of a larger node, already set dead, in $i_{h-1}..i_h - 1$.

Since $livechildren[\gamma[j]]$ is always decremented for every node j that is set dead, we conclude that Step 3 deals correctly with the task of setting nodes dead at the start of each new staircase.

Step 2 is a straightforward update of T_γ based on the current position i. Its correctness depends entirely on the correct update of $largestlive[\boldsymbol{\beta}[i]]$ in Step 1.

The assignment statement in Step 1 will be executed only when $\boldsymbol{\beta}[i]$ has been set dead during one of the previous executions of the **for** loop in Step 3. As noted in the proof of Step 3, if some j is to be set dead with respect to i_{h+1}, it must have been live with respect to i_h, hence by Lemma 13.1.5 live with respect to $i_{h+1} - 1$. Therefore $\boldsymbol{x}[1..j]$ would be a *potential* cover of $\boldsymbol{x}[1..i_{h+1}]$; thus it would cover some right extension of $\boldsymbol{x}[1..\boldsymbol{\beta}[i_{h+1}]]$, which is a suffix of $\boldsymbol{x}[1..i_{h+1}]$. Hence the following fact holds: at the time that any position j is set dead, it must be true that j is greater than $\boldsymbol{\beta}[i_{h+1}]$, where i_{h+1} is the staircase-starting position used in Step 3.

Since by Lemma 1.3.1 values in the $\boldsymbol{\beta}$ array can increase by at most one from a given position to the next, it follows that every such dead position j must later become a value in the $\boldsymbol{\beta}$ array — that is, $j = \boldsymbol{\beta}[i']$ for some $i' > i_{h+1}$ — in order for Step 1 to be executed. In particular, the values j must be processed in Step 1 in ascending order of magnitude; that is, in descending order in the cover tree T_γ. This means that Step 1 will pass correct values of the largest live ancestor from parent to child. □

Now consider the time required by Algorithm LS. Each of the steps except the **for** loop in Step 3 requires only constant time. Then LS requires $\Theta(n)$ time plus the total time used within the **for** loop. To estimate this total time, observe that the time required for *each* execution of the **for** loop is proportional to

$$\max\{(c_1 - c_2), \text{no. of nodes set dead}\}.$$

Since the sum of $c_1 - c_2$ over all staircases is at most $n-1$ and since each of the n nodes may be set dead at most once, it follows that the total time for the **for** loop over all executions is $O(2n)$. Hence

Theorem 13.1.9 *Algorithm 13.1.1 requires $\Theta(n)$ time and $\Theta(n)$ space for its execution.* □

Algorithm LS is an optimal **on-line** algorithm in the weak sense defined in Section 4.1: it computes γ, thus making available all the covers of every prefix of $\boldsymbol{x}$, in the least possible asymptotic time and space. Note that since the calculation of β is also on-line, Algorithm LS can easily be modified to simultaneously compute on-line both the border and the cover arrays.

The reader will find it instructive to follow the algorithm as it applies to the example given in Figure 13.1. Note particularly that, as a result of the fact that $\beta[22] = 2$, the following nodes are all set dead: $5, 6, 9, 10, 11, 13$–19.

Exercises 13.1

1. Prove Facts (F1)–(F4).
2. Based on Facts (F1)–(F3), show that $\boldsymbol{u} = \boldsymbol{x}[1..p]$ is a cover of $\boldsymbol{x}[1..i]$ if and only if $p = \gamma^h[i]$ for some integer $h \in 1..k-1$, where k is the least positive integer such that $\gamma^k[i] = 0$.
3. Strictly speaking, the cover array corresponds most closely, not to the border array, but rather to a ***period array*** $\pi = \pi[1..n]$ in which $\pi[i]$ is the largest period of $\boldsymbol{x}[1..i]$ if and only if $\boldsymbol{x}[1..i]$ is a repetition, zero otherwise. Without computing the border array, specify an algorithm that directly computes π in linear time. Then show that the period array can be represented by a ***period tree*** with properties analogous to those of the border tree and cover tree.
4. Using the facts that $\boldsymbol{x}$ has period p and that $j' \geq \max\{p, i-p\}$, prove (13.5).
5. Modify Algorithm LS so as to compute *both* β and γ on-line.

13.2 All Repeats — Algorithm FST

Men of few words are the best men.

— William SHAKESPEARE (1564–1616), *Henry V*

In this section we deal with the exact repeats problem in its most general form: in the language of Section 2.3, we compute all the NE (nonextendible) complete repeats in a given string $\boldsymbol{x}[1..n]$ in $O(n \log n)$ time, and we express these repeats in either an array or tree form that requires only $\Theta(n)$ space [FST02]. We describe Algorithm FST first in terms of suffix trees (Section 5.2), then go on to show how it can also be implemented in terms of suffix arrays (Subsection 5.3.2) at a considerable saving in storage space. The all-repeats algorithm is based upon a simple fact:

Lemma 13.2.1 *Let $\widehat{\boldsymbol{x}}$ be the reverse string $\boldsymbol{x}[n]\boldsymbol{x}[n-1]\cdots\boldsymbol{x}[1]$ of a given string $\boldsymbol{x}$. Then a repeat $M_{\boldsymbol{x},\boldsymbol{u}}$ is LE if and only if $M_{\widehat{\boldsymbol{x}},\widehat{\boldsymbol{u}}}$ is RE.*

Proof See Exercise 13.2.1. □

This result suggests a straightforward approach to computing all the NE repeats in $\boldsymbol{x}$:

- **Step 1** Compute all the NRE (non-RE) repeats both of $\boldsymbol{x}$ and of $\widehat{\boldsymbol{x}}$.
- **Step 2** Compare (somehow) the NRE repeats of $\boldsymbol{x}$ with those of $\widehat{\boldsymbol{x}}$ to identify repeats that are in both lists (by Lemma 13.2.1, the NE repeats).

Since there may be a great many NRE/NLE repeats, and since it is far from clear how they could be efficiently compared with each other, this simple methodology appears to have insuperable flaws: in particular, it could be very costly in terms of its use of both time and space. Nevertheless, it is essentially this strategy that is used, rendered attractive by two main observations:

- As remarked in Subsection 12.1.1, the NRE repeats are represented naturally by a suffix tree (and, as we shall see later, also by a suffix array). Thus Step 1 can be implemented by constructing suffix trees (arrays) for $\boldsymbol{x}$ and for $\widehat{\boldsymbol{x}}$.

  To see this, consider the suffix tree for our usual example $\boldsymbol{f_6}$, displayed in Figure 13.4 in a stripped-down form that is slightly modified from Figure 12.1. The leaf (square) nodes are labelled with the starting positions $i \in 1..13$ of the suffixes $\boldsymbol{f_6}[i..13]$, while each internal (round) node gives the lcp of the collection of suffixes identified by the leaf nodes below it. Thus the lcp values can also be thought of as specifying the lengths of the complete NRE repeats in the string, and a suffix

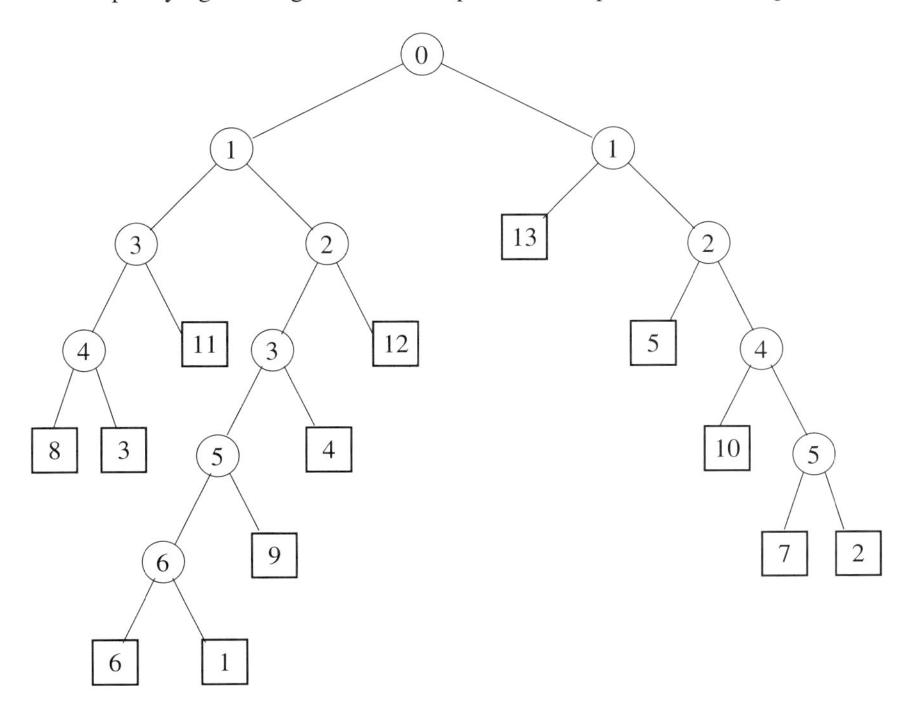

Figure 13.4. $T_{\boldsymbol{f_6}}$ for $\boldsymbol{f_6} = abaababaabaab$

tree can therefore just as well be referred to as an ***NRE tree***, whose round and square nodes can be called ***lcp nodes*** and ***position nodes***, respectively.

In our example, the substrings

$$\boldsymbol{u} = \boldsymbol{x}[9..13] = \boldsymbol{x}[6..10] = \boldsymbol{x}[1..5]$$

defined by the subtree rooted at the lcp node 5 constitute a complete NRE repeat

$$M^*_{\boldsymbol{x},\boldsymbol{u}} = (5;\ 9,6,1).$$

- As described below, Step 2 can be achieved by linear-time processing that matches the NRE trees (arrays) for $\boldsymbol{x}$ and $\widehat{\boldsymbol{x}}$ to produce a linear space ***NE tree*** (***NE array***) that describes all the NE repeats in $\boldsymbol{x}$.

In accordance with these remarks, we show in Subsection 13.2.1 how to compute the NE tree given NRE trees for $\boldsymbol{x}$ and $\widehat{\boldsymbol{x}}$, then in Subsection 13.2.2 how to compute the NE array given suffix arrays for $\boldsymbol{x}$ and $\widehat{\boldsymbol{x}}$.

13.2.1 Computing the NE Tree

After all, all he did was string together
a lot of old, well-known quotations.

— H. L. MENCKEN (1880–1956), *on Shakespeare*

We suppose that the suffix trees, say $T_{\boldsymbol{x}}$ and $T_{\widehat{\boldsymbol{x}}}$ of $\boldsymbol{x}$ and $\widehat{\boldsymbol{x}}$ respectively, have already been computed, thereby identifying complete NRE repeats in both $\boldsymbol{x}$ and $\widehat{\boldsymbol{x}}$. From Lemma 13.2.1 we know that the NRE repeats in $\widehat{\boldsymbol{x}}$ identify NLE repeats in $\boldsymbol{x}$. For brevity we refer to the generator $\boldsymbol{u}$ of an NRE (respectively, NLE, NE) repeat $M_{\boldsymbol{x},\boldsymbol{u}}$ as an NRE (respectively, NLE, NE) repeating substring, while asking the reader to bear in mind that the terms NRE, NLE, NE really relate to the repeat that is a collection of repeating substrings.

Suppose that some lcp node $\widehat{p} > 0$ in $T_{\widehat{\boldsymbol{x}}}$ has as child a position node $\widehat{\imath}$. Then in $\widehat{\boldsymbol{x}}$ there exists an NRE repeating substring

$$\widehat{\boldsymbol{u}} = \widehat{\boldsymbol{x}}[\widehat{\imath}..\widehat{\imath}+\widehat{p}-1]$$

that is the reverse of an NLE repeating substring

$$\boldsymbol{u} = \boldsymbol{x}\big[n-(\widehat{\imath}+\widehat{p}-1)+1..n-\widehat{\imath}+1\big]$$

in $\boldsymbol{x}$. Thus the assignment

$$i \leftarrow n-(\widehat{\imath}+\widehat{p}-2) \tag{13.6}$$

identifies one start position i of an NLE repeating substring $\boldsymbol{u}$ in $\boldsymbol{x}$ of length $\widehat{p}$. Observe that in fact i is also a start position of one of a set of NLE repeating substrings of every length $j \in 1..\widehat{p}$; thus if in $T_{\boldsymbol{x}}$ we find for some i the largest value of $\widehat{p}$ such that i is one of a collection of substrings of length $\widehat{p}$ that are both NRE and NLE, then every parent of $\widehat{p}$ in $T_{\boldsymbol{x}}$ will also identify collections of substrings that are both NRE and NLE.

Let us say that a substring $\boldsymbol{u}$ of $\boldsymbol{x}$ is a ***maximal NE repeating substring*** of $\boldsymbol{x}$ if and only if

- $\boldsymbol{u}$ occurs at least twice in $\boldsymbol{x}$;
- $\boldsymbol{u}$ is not a proper substring of any repeating substring of $\boldsymbol{x}$.

The following result, again easily proved, then tells us that a maximal NE repeating substring must be identifiable in both $T_{\boldsymbol{x}}$ and $T_{\widehat{\boldsymbol{x}}}$:

Lemma 13.2.2 *If $\boldsymbol{u} = \boldsymbol{x}[i..i+p-1]$ is a maximal* NE *repeating substring of $\boldsymbol{x}$, then the position node i occurs as a child of the lcp node p in $T_{\boldsymbol{x}}$, and $n-(i-p-2)$ occurs as a child of p in $T_{\widehat{\boldsymbol{x}}}$.* □

Proof See Exercise 13.2.4. □

Recall the observation made above that every lcp node in $T_{\boldsymbol{x}}$ that is an ancestor of an NE repeating substring must itself be the root of a subtree of $T_{\boldsymbol{x}}$ whose position nodes are the repeating substrings of a complete NE repeat. This lemma therefore provides us with a simple strategy for identifying all NE repeating substrings: just find the maximal ones (largest value of p), then locate all their ancestors in $T_{\boldsymbol{x}}$. Algorithm 13.2.1 provides an outline of how this can be accomplished.

Algorithm 13.2.1 (Compute the NE Tree)

– *Given the suffix trees $T_{\boldsymbol{x}}$ & $T_{\widehat{\boldsymbol{x}}}$, compute the* NE *tree of $\boldsymbol{x}[1..n]$*

(1) traverse $T_{\boldsymbol{x}}$ to create a table POINTER$[i]$ that for each position i in $\boldsymbol{x}$ points to the corresponding node in $T_{\boldsymbol{x}}$

(2) **for** every lcp node p in $T_{\boldsymbol{x}}$ **do**
  NE$[p] \leftarrow$ FALSE

(3) **for** every parent-child pair $(\widehat{p}, \widehat{\imath})$ in $T_{\widehat{\boldsymbol{x}}}$ **do**
  – *Here use* POINTER$[i]$*:*
  if $i = n-(\widehat{\imath}+\widehat{p}-2)$ is a child of lcp node $\widehat{p}$ in $T_{\boldsymbol{x}}$ **then**
   while not NE$[\widehat{p}]$ **do**
    NE$[\widehat{p}] \leftarrow$ TRUE
    if $\widehat{p} \neq 0$ **then**
     $\widehat{p} \leftarrow$ parent of $\widehat{p}$ in $T_{\boldsymbol{x}}$

(4) traverse $T_{\boldsymbol{x}}$ deleting every subtree rooted at an lcp node p for which NE$[p] =$ FALSE

The algorithm is expressed in terms of four steps. Step (1) is a traversal of $T_{\boldsymbol{x}}$ that sets up a table enabling each position node i in $T_{\boldsymbol{x}}$ to be accessed later in constant time. In Step (2) a Boolean variable corresponding to each lcp node p in $T_{\boldsymbol{x}}$ is initialized to FALSE, indicating that no terminal nodes in the subtree rooted at p have currently been identified as NE. Step (3) processes every parent-child pair $(\widehat{p}, \widehat{\imath})$ in $T_{\widehat{\boldsymbol{x}}}$, testing to determine whether or not the equivalent parent-child pair $\big(\widehat{p}, n-(\widehat{\imath}+\widehat{p}-2)\big)$ exists in $T_{\boldsymbol{x}}$; if so, then the lcp node $\widehat{p}$ and all its ancestors in $T_{\boldsymbol{x}}$ must be NE — accordingly, until an ancestor is found that is already NE, $\widehat{p}$ and its ancestors in $T_{\boldsymbol{x}}$ are identified as NE. A final step traverses $T_{\boldsymbol{x}}$ to eliminate all subtrees rooted at any node p for which NE$[p]$ = FALSE; the remaining tree, $T_{\boldsymbol{x}}^{\text{NE}}$, is the NE tree of $\boldsymbol{x}$.

Each of the steps (1), (2), (4) of Algorithm 13.2.1 is a traversal of an NRE tree that requires $O(1)$ time at each node, hence $O(n)$ time in total. Step (3) is another tree traversal, but with slightly more complex processing: since the **if** statement makes use of the POINTER array, it too requires only $O(1)$ time at each execution, hence also $O(n)$ time overall. The **while** loop in Step (3) causes the NE variable for at most n lcp nodes to be set TRUE, and tests the current setting of at most n lcp nodes for which NE is already TRUE — thus the total time consumed in the **while** loop is $O(n)$. We have established:

Theorem 13.2.3 *Given the suffix trees $T_{\boldsymbol{x}}$ and $T_{\widehat{\boldsymbol{x}}}$ for $\boldsymbol{x}[1..n]$, Algorithm 13.2.1 correctly computes the* NE *tree $T_{\boldsymbol{x}}^{\text{NE}}$ using $O(n)$ time and $\Theta(n)$ additional space.* □

Figure 13.5 displays the suffix tree $T_{\widehat{\boldsymbol{f_6}}}$, modified so that position nodes $\widehat{\imath}$ are replaced by $i = n-(\widehat{\imath}+\widehat{p}-2)$. Note that only positions 1, 6, 9 give rise to NLE repeats; thus the NE tree reduces to the subtree of $T_{\boldsymbol{f_6}}$ that corresponds to these positions — that is, the path from the root of $T_{\boldsymbol{f_6}}$ to lcp node 6 that corresponds to the substring *abaaba*. The complete NE

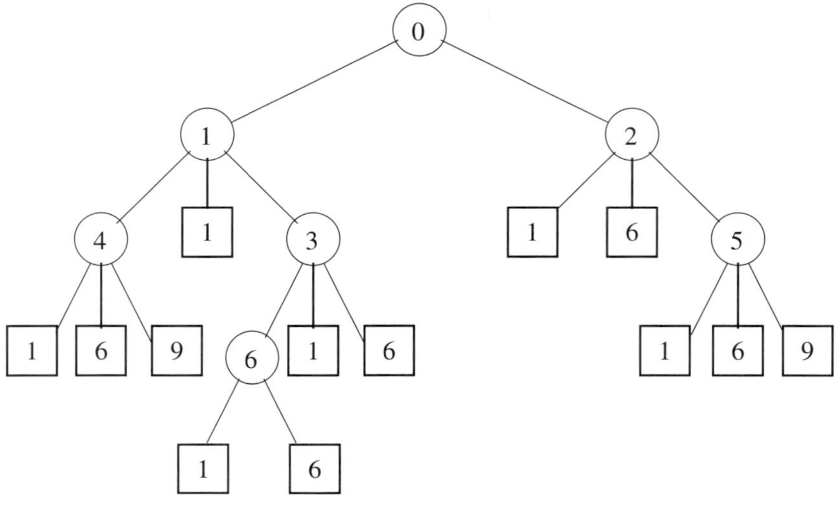

Figure 13.5. $T_{\widehat{\boldsymbol{f_6}}}$ for $\widehat{\boldsymbol{f_6}} = baabaababaaba$ with $i = n-(\widehat{\imath}+\widehat{p}-2)$

repeats of $\boldsymbol{f_6}$ are thus seen from Figure 13.4 to be all the occurrences of $a, ab, aba, abaab$ and $abaaba$. In general, the occurrences of NE repeats, whether all of them or selected according to various criteria, can be located and output in time proportional to their number by traversal of the NE tree. Such problems are also discussed in [BLPS00]. Another quite different approach to the computation of nonextendible repeats (though still making use of suffix trees) is described in [G97].

13.2.2 Computing the NE Array

The best way to keep one's word is not to give it.

— Napoleon BONAPARTE (1769–1821)

In Subsection 5.3.2 we defined a suffix array $\boldsymbol{\sigma_x} = \boldsymbol{\sigma}[1..n]$ of a given string $\boldsymbol{x}[1..n]$ to be a permutation of $1..n$ with the property that, for every $1 \le i < j \le n$,

$$\boldsymbol{x}\big[\boldsymbol{\sigma_x}[i]..n\big] < \boldsymbol{x}\big[\boldsymbol{\sigma_x}[j]..n\big].$$

Thus $\boldsymbol{\sigma_x}$ specifies the starting positions of the suffixes of $\boldsymbol{x}$ in ascending lexicographical order. For example, if

$$\begin{array}{rl} & \scriptstyle 1\ 2\ 3\ 4\ 5\ 6\ 7\ 8\ 9\ 10\ 11\ 12\ 13 \\ \boldsymbol{f_6} = & a\ b\ a\ a\ b\ a\ b\ a\ a\ b\ a\ a\ b, \end{array} \tag{13.7}$$

then

$$\boldsymbol{\sigma_{f_6}} = 11\ 8\ 3\ 12\ 9\ 6\ 1\ 4\ 13\ 10\ 7\ 2\ 5. \tag{13.8}$$

We saw also in Subsection 5.3.2 that an auxiliary array $\boldsymbol{\pi_x}$ could be defined giving lcp values for ranges of positions in $\boldsymbol{\sigma_x}$. Here we define a simpler array $\boldsymbol{\lambda_x} = \boldsymbol{\lambda}[1..n]$, where $\boldsymbol{\lambda}[1] = 0$ and for every $i \in 2..n$,

$$\boldsymbol{\lambda}[i] = \text{lcp}\Big(\boldsymbol{x}\big[\boldsymbol{\sigma}[i-1]..n\big], \boldsymbol{x}\big[\boldsymbol{\sigma}[i]..n\big]\Big],$$

which we write in the abbreviated form

$$\boldsymbol{\lambda}[i] = \text{lcp}\big(\boldsymbol{\sigma}[i-1], \boldsymbol{\sigma}[i]\big). \tag{13.9}$$

Thus, for our example,

$$\boldsymbol{\lambda_{f_6}} = 0341256301452, \tag{13.10}$$

reflecting the fact that $\text{lcp}(11, 8) = 3$, $\text{lcp}(8, 3) = 4$, and so on.

Just like the more complex array $\boldsymbol{\pi_x}$, the array $\boldsymbol{\lambda_x}$ can be computed in a straightforward way as a byproduct of suffix array computation. Recall from Subsection 5.3.2 that suffix arrays can be computed directly [MM93] without reference to a suffix tree.

We provide here an outline of a version of the algorithm that avoids the use of suffix trees completely: the entire calculation is carried out based on suffix arrays (both $\boldsymbol{\sigma}_{\boldsymbol{x}}$ and $\boldsymbol{\lambda}_{\boldsymbol{x}}$), and its end-product is not an NE tree, but rather an NE array. Since the suffix array of $\boldsymbol{x}[1..n]$ requires only $2n$ computer words for its storage, this revision yields significant saving in space.

For every $j \in 1..n$, let $p_j = \boldsymbol{\lambda}[j]$. It then follows from (13.9) that for every $j \in 2..n$,

if $p_j > p_{j-1}$ **then**
  p_j is a descendant of p_{j-1} in the suffix tree
elsif $p_j < p_{j-1}$ **then**
  p_j is an ancestor of p_{j-1} in the suffix tree
else
  p_j and p_{j-1} identify the same node in the suffix tree

Since the suffix array contains the same lcp information that is provided by the suffix tree, these relationships imply that the suffix array must be decomposable into subarrays, each of which specifies nodes that lie on a single path from the root to a terminal node in the suffix tree. For example, the lcp values in the array (13.10) can be separated into three subarrays

$$0, 3, 4, \underline{1} \; / \; \underline{1}, 2, 5, 6, 3, \underline{0} \; / \; \underline{0}, 1, 4, 5, 2$$

corresponding to the three main paths in the suffix tree of Figure 13.4. Here the repeated lcp values $\underline{0}$ and $\underline{1}$ identify the root of a subtree at which a new path begins. We therefore call these nodes ***branch nodes***, since they mark the points at which a new path diverges from a preceding one.

These ideas can be made more precise as follows:

Definition 13.2.4 *An* ***I-run*** $I_{j,h}$ *in* $\boldsymbol{\lambda}_{\boldsymbol{x}}$ *is a sequence of* $h \geq 2$ *lcp values* $p_j, p_{j+1}, \ldots, p_{j+h-1}$, *not all equal, such that*

(a) *either* $j = 1$ *or else* $p_{j-1} > p_j$;

(b) $p_j \leq p_{j+1} \leq \cdots \leq p_{j+h-1}$;

(c) *either* $j+h-1 = n$ *or else* $p_{j+h-1} > p_{j+h}$.

Thus an I-run is a maximum-length sequence of nondecreasing lcp values in $\boldsymbol{\lambda}_{\boldsymbol{x}}$; similarly, a ***D-run*** $D_{j,h}$ is a maximum-length sequence of nonincreasing lcp values. Using these definitions, we can now characterize paths in the suffix tree in terms of runs in $\boldsymbol{\lambda}_{\boldsymbol{x}}$:

Lemma 13.2.5 *Every sequence* $I_{j,h}D_{j+h-1,h'}$ *or* $I_{j,h}$, *where* $j+h-1 = n$, *in* $\boldsymbol{\lambda}_{\boldsymbol{x}}$ *identifies nodes that lie on a single path in the suffix tree.*

Proof Since $p_1 = 0$, it follows from Definition 13.2.4(a)–(b) that either there are no runs in $\boldsymbol{\lambda}_{\boldsymbol{x}}$ ($p_j = 0$ for every $j \in 1..n$) or else the sequence of runs begins with an I-run $I_{1,h}$. Observe that Definition 13.2.4(c) ensures that every I-run is followed by a D-run or else by the end of the array; similarly, every D-run must be followed by an I-run or else by the end of the array.

Now consider an I-run $I_{j,h}$. Since $p_{j+(t-1)} \leq p_{j+t}$ for every $t \in 1..h-1$, it follows that $I_{j,h}$ determines a sequence of descendants of p_j that lie on a single path in the suffix tree. If in fact $i+h-1=n$, then $I_{j,h}$ is in itself the final path identified in $\boldsymbol{\lambda_x}$. If not, however, then $I_{j,h}$ is followed by a D-run $D_{j+h-1,h'}$, where $p_{(j+h-1)+(t-1)} \leq p_{(j+h-1)+t}$ for every $t \in 1..h'-1$. These lcp values determine a sequence of ancestors of p_{j+h-1}, all on the same path from the root already identified by $I_{j,h}$. The subsequent I-run $I_{j+h+h'-2,h''}$, if it exists, of course specifies descendants of the branch node $p_{j+h+h'-2}$ that lie on a path distinct from the path specified by $I_{j,h}$. □

We see then that the ID-runs in $\boldsymbol{\lambda_x}$ essentially decompose the internal (lcp) nodes of the suffix array into paths. Adjacent pairs of runs in $\boldsymbol{\lambda_x}$ identify two paths in the tree that have exactly one node (a branch node) in common. Observe also that for each ID-run, the minimum node in the corresponding path must be either the first node of the I-run or the last node of the D-run.

Since a branch node is *both* the last node of a D-run *and* the first node of an I-run, the values of pairs of adjacent branch nodes in $\boldsymbol{\lambda_x}$ therefore determine the way in which two paths in the suffix tree are related. If we include the first lcp value ($p_1 = 0$) in $\boldsymbol{\lambda_x}$ as a branch node, and denote consecutive branch nodes by b_1 and b_2, respectively, we see that there are only three possible relationships between the corresponding paths, as shown in Figure 13.6. Notice in particular that in each of these three cases, future paths can be added only as descendants of b_2 on the current path or as descendants of ancestors of b_2 in the tree. No further change can be made to the path rooted at b_2 that was specified by the previous path containing b_1.

Finally, we remark that the regular structure of the suffix array enables us to determine the parent in the corresponding suffix tree $T_{\boldsymbol{x}}$ of each position node $\sigma[j]$, $j = 1, 2, \ldots, n$. We make two important observations, whose proof is left as Exercise 13.2.12:

(O1) The parent in $T_{\boldsymbol{x}}$ of $\sigma[j]$ is $\max\{\lambda[j], \lambda[j+1]\}$, where we take $\lambda[n+1] = 0$.

(O2) (Due to Andrew Francis.) If $\lambda[j]$ is a branch node and $j < n$, $\sigma[j]$ has parent $\lambda[j+1]$ in $T_{\boldsymbol{x}}$; otherwise, $\sigma[j]$ has parent $\lambda[j]$.

Note that these observations provide an alternate means of identifying branch nodes in $\lambda_{\boldsymbol{x}}$.

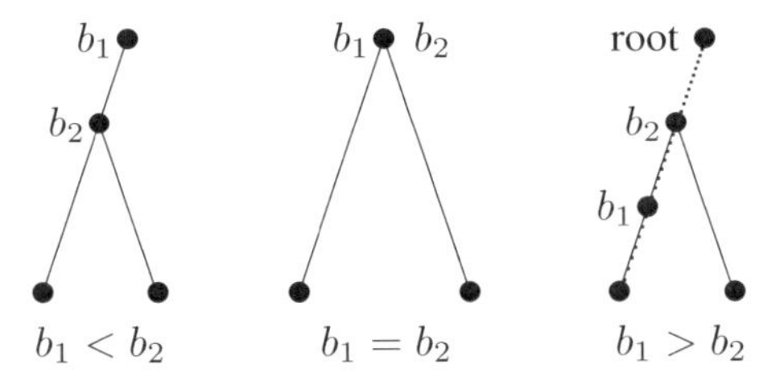

Figure 13.6. Branch nodes b_1 and b_2 of adjacent paths (ID-runs)

This discussion puts us in a position to describe an approach to the calculation of NE repeats that is based on suffix arrays only. As before we denote the position and lcp arrays for $\boldsymbol{x}$ by $\boldsymbol{\sigma_x}$ and $\boldsymbol{\lambda_x}$, respectively; for $\widehat{\boldsymbol{x}}$ the corresponding arrays are $\widehat{\boldsymbol{\sigma_x}}$ and $\widehat{\boldsymbol{\lambda_x}}$. We also make use of a LOC array that gives for each position node $\widehat{\imath}$ in $T_{\widehat{\boldsymbol{x}}}$ its location in $\widehat{\boldsymbol{\sigma_x}}$; this enables us to find any position in $\widehat{\boldsymbol{\sigma_x}}$ in constant time. A Boolean vector NE$[1..n]$ specifies for each $j \in 1..n$ whether (TRUE) or not (FALSE) the lcp node $\boldsymbol{\lambda}[j]$ is also NLE. Finally, in order to identify the beginning and end of ID paths in $\boldsymbol{\lambda_x}$, we use the ID patterns (Lemma 13.2.5) to compute a BRANCH array that includes, in addition to the branch nodes defined above, also the first and last positions in $\boldsymbol{\lambda_x}$. An outline of the processing, more complex than the tree-based calculation but not necessarily more time-consuming, is displayed as Algorithm 13.2.2.

Algorithm 13.2.2 (Compute the NE Array)

- *Given the suffix (NRE) arrays of $\boldsymbol{x} = \boldsymbol{x}[1..n]$ and $\widehat{\boldsymbol{x}}$, compute the NE array of $\boldsymbol{x}$*

(1) **for** $j \leftarrow 1$ **to** n **do**
  LOC$\left[\widehat{\boldsymbol{\sigma}}[j]\right] \leftarrow j$; NE$[j] \leftarrow$ FALSE

(2) compute BRANCH$[1..b^*]$, where for $b \in 1..b^*$, BRANCH$[b]$
  is the position of the bth branch node in $\boldsymbol{\lambda_x}$
  (include positions $j = 1$ and $j = n$ as branch nodes)

(3) $b \leftarrow 1$; $p_0 \leftarrow 0$
  for $j \leftarrow 1$ **to** n **do**
    determine j' such that $(p, i) = \left(\boldsymbol{\lambda}[j'], \boldsymbol{\sigma}[j]\right)$ is a
      parent-child pair in the suffix tree of $\boldsymbol{x}$
    if $p > 0$ **then**
      if $p < p_0$ **then**
        NE$[j'] \leftarrow$ TRUE; $p_0 \leftarrow p$
      else
        $\widehat{\imath} \leftarrow n-(i+p-2)$; $\widehat{\jmath} \leftarrow$ LOC$[\widehat{\imath}]$
        determine $\widehat{j'}$ such that $(\widehat{p}, \widehat{\imath}) = \left(\widehat{\boldsymbol{\lambda}}[\widehat{j'}], \widehat{\boldsymbol{\sigma}}[\widehat{\jmath}]\right)$ is a
          parent-child pair in the suffix tree of $\widehat{\boldsymbol{x}}$
        if $p = \widehat{p}$ **then**
          NE$[j'] \leftarrow$ TRUE; $p_0 \leftarrow p$

      if $j =$ BRANCH$[b+1]$ **then**
        determine $j^* \in$ BRANCH$[b]$..BRANCH$[b+1]$ such that
          $p^* = \boldsymbol{\lambda}[j^*]$ is the maximum lcp
          for which NE$[j^*] =$ TRUE
        for every $j' \in$ BRANCH$[b]$..BRANCH$[b+1]$ **do**
          if $\boldsymbol{\lambda}[j'] \leq p^*$ **then**

$\qquad\qquad \text{NE}[j'] \leftarrow \text{TRUE}$
$\qquad b \leftarrow b+1$

(4) process the paths identified by BRANCH in reverse order, setting NE true for ancestors of p^* as in Step (3)

Step (1) of the algorithm is a straightforward **for** loop that initializes the LOC array and the bit vector NE. Step (2) computes the BRANCH array based on Lemma 13.2.5 and a linear scan of $\boldsymbol{\lambda}[1..n]$. Step (3) is another **for** loop that first, using Observation (O1), identifies all the matches of (p,i) pairs in the NRE arrays for $\boldsymbol{x}$ and $\widehat{\boldsymbol{x}}$, then for each path ensures that every ancestor of a matched (p,i) pair within the path is NE. Each of the two sections of Step (3) requires $\Theta(n)$ time. In Step (4) the processing of the second section of Step (3) is repeated, but in reverse order to ensure that all ancestors of NE positions are properly set to be themselves NE. Finally the NE array is determined by all those positions j in $\boldsymbol{\lambda}_{\boldsymbol{x}}$ for which $\text{NE}[j] = \text{TRUE}$. Hence

Theorem 13.2.6 *Given the suffix (NRE) arrays corresponding to $\boldsymbol{x}[1..n]$ and $\widehat{\boldsymbol{x}}$, Algorithm 13.2.2 correctly computes the* NE *array using $\Theta(n)$ time and $\Theta(n)$ additional space.* □

We remark that the NE array can if desired be used to compute the corresponding NE tree in linear time.

Since the NE tree (or its corresponding NE array) specifies *all* the NE repeats in $\boldsymbol{x}$, it can be used in a natural way to identify NE tandem repeats (repetitions) and NE coverable substrings (Problem 2.17—Section 2.3) as well. Recall that the leaf nodes of the subtree rooted at any lcp node p in the NE tree are exactly the NE repeats corresponding to p. If these nodes were radix-sorted into a sequence

$$I = <i_1, i_2, \ldots, i_r>,$$

$i_1 < i_2 < \cdots < i_r$, then a single pass of I would easily determine both the maximum-length tandem repeats and and the maximum-length coverable substrings corresponding to p. In the first case the gap between adjacent entries in I must be exactly equal to p, in the second case at most p. For every choice of p, this process can be carried out using $O(n)$ space and time.

More challenging however is the task of somehow reporting *all* the NE coverable substrings in $\boldsymbol{x}$. There exist three algorithms for the related problem of computing all NRE coverable substrings (using the NRE tree rather than the NE tree). The first published algorithm [AE93] performs complex operations on the NRE (suffix) tree and requires $O(n \log^2 n)$ time, while the second algorithm [BP00] applies even more complex techniques to the NRE tree in order to reduce the time requirement to $O(n \log n)$. A third algorithm [IM01] directly processes the sequences defined by the Crochemore decomposition, also executing in $O(n \log n)$ time.

The world awaits a "simple" $O(n \log n)$-time algorithm that computes the NE coverable substrings of $\boldsymbol{x}$, preferably with modest space requirements.

Exercises 13.2

1. Prove Lemma 13.2.1.
2. The definition of the doubly-linked list maintained by Algorithm FST inevitably recalls the definition of string given in Section 1.1. If the doubly-linked list is indeed a string, then what alphabet is it defined on?
3. Show that in an NRE tree no node, except possibly the root, can have only a single child. Specify the conditions under which the root does in fact have a single child.
4. Prove Lemma 13.2.2.
5. Compute the NRE array of $\widehat{\boldsymbol{f}_6}$, then use it to verify that the NRE tree of Figure 13.5 is correct. Then use the NRE trees of Figures 13.4 and 13.5 to compute $T_{\boldsymbol{f}_6}^{\text{NE}}$ according to Algorithm 13.2.1.
6. The string $\boldsymbol{g} = abaababaababa$ differs only in its last two positions from $\boldsymbol{f}_6$. Compute the NRE arrays of $\boldsymbol{g}$ and $\widehat{\boldsymbol{g}}$, then the NRE trees, then $T_{\boldsymbol{g}}^{\text{NE}}$. Reflect upon the striking differences between these structures and those of the preceding exercise.
7. Observe that for each i computed from $n-(\widehat{\imath}+\widehat{p}-2)$ in $T_{\widehat{\boldsymbol{x}}}$, there is exactly one possible parent node in $T_{\boldsymbol{x}}$. Explain therefore how Algorithm 13.2.1 could possibly be improved by employing a radix sort.
8. Specify an algorithm that, given an NRE array of length n, computes a corresponding string $\boldsymbol{x}[1..n]$ on an indexed alphabet $1, 2, \ldots, \alpha$.
9. Give a precise definition of a D-run in the style of Definition 13.2.4.
10. Describe an algorithm that computes the NE tree of $\boldsymbol{x}$ from the NE array.
11. Describe an algorithm that computes both the NE tandem repeats and the NE coverable substrings corresponding to a given lcp node p in the NE tree.
12. Prove Observations (O1) and (O2).

13.3 k-Approximate Repeats — Schmidt

I hate false words, and seek with care, difficulty, and moroseness, those that fit the thing.

— Walter Savage LANDOR (1775–1864), *Imaginary Conversations*

In this section we present an outline of an algorithm that efficiently computes approximate repeats in the general case that is of particular interest to computational biologists: a general scoring matrix (Section 2.2) is used in the computation of distance between strings. The algorithm, due to Schmidt [S98], is a complicated one with a number of rather messy technicalities, just the kind of algorithm that we would normally strive to avoid. We include Algorithm S, however, at least in *précis* form, because it provides an efficient, indeed perhaps even asymptotically optimal, solution to a particularly important string-processing problem. Moreover, Algorithm S extends and combines several algorithmic ideas encountered earlier in this book in a way that provides new insight into their meaning and effectiveness:

- Algorithm S is based on the usual dynamic-programming approach to the cost array $\boldsymbol{c}$ (Section 9.1), but reinterpreted now so as to be applicable to repeating substrings. Part of this reinterpretation involves computation of maximum "scores" rather than minimum distance; thus the cost array $\boldsymbol{c}$ in which distance is minimized becomes a ***benefit array*** $\boldsymbol{b}$ in which scores are maximized.
- The idea of a dependency graph, introduced in Section 9.5 as the basis of the string-distance Algorithm UM, reappears here as a ***grid graph***, whose "highest-scoring paths" are the main focus of Algorithm S.
- The idea of nonextendibility, introduced in Section 2.3, then applied in Section 12.2 to exact repetitions and in Section 13.2 to exact repeats, is reformulated for approximate repeats as ***local optimality***. Indeed, it appears that the original definition of local optimality [ES83] actually predates that of nonextendibility [M89].
- The recursive divide-and-conquer approach employed in Algorithm ML (Subsection 12.1.2) to compute squares in terms of their centre points is extended to approximate tandem squares in grid graphs.

In addition to extensions of these familiar ideas, Algorithm S also introduces some interesting new ones: we simply cannot avoid discussing it!

The algorithm does not in fact compute repeats in the sense that we have defined; rather, it computes the set of all "locally optimal" (nonextendible) approximate nonoverlapping squares that satisfy a given threshold k. As we shall see shortly, overlapping squares are excluded because their inclusion could result in important tandem squares being missed. The calculation of squares rather than repeats of arbitrary exponent relates to a remark made at the end of Section 2.3 about the nontransitivity of k-approximate matching, hence the inherent ambiguity in k-approximate repetitions: in a substring $\boldsymbol{u_1 u_2 u_3}$, it is quite possible that

$$d(\boldsymbol{u_1}, \boldsymbol{u_2}) \leq k,\ d(\boldsymbol{u_2}, \boldsymbol{u_3}) \leq k,$$

so that both $\boldsymbol{u_1 u_2}$ and $\boldsymbol{u_2 u_3}$ are k-approximate squares, whereas

$$d(\boldsymbol{u_1}, \boldsymbol{u_3}) > k,$$

implying that $\boldsymbol{u_1 u_3}$ is not. Thus the idea of a k-approximate repetition or repeat is not easy to define in a useful way: in a loose sense, $\boldsymbol{u_1 u_2 u_3}$ could be a k-approximate repetition while $\boldsymbol{u_1 u_3}$ was not even a k-approximate square, while in a stricter sense the designation of $\boldsymbol{u_1 u_2 \cdots u_m}$ as a k-approximate repetition would require $\binom{m}{2}$ possible k-approximate squares (both tandem and split) to be verified. Indeed, as also remarked in Section 2.3, the generators $\boldsymbol{u_1}$ and $\boldsymbol{u_2}$ of a k-approximate square $\boldsymbol{u_1 u_2}$ do not even need to be of the same length.

Hence for the computation of approximate repetitions, there appears to be no hope of making use of any analogue of Crochemore's (i,p,r) encoding of exact repetitions (Subsection 12.1.1) that reduces the number of outputs to $O(n \log n)$, still less of making use of the (i,p,r,t) encoding of runs (Section 12.2) that reduces output size to $O(n)$. From the beginning, therefore, the computation of approximate repetitions or repeats seems to require the output of squares, hence of $\Omega(n^2)$ squares in the worst case, as described in Section 2.3 for $\boldsymbol{x} = a^n$. This observation is confirmed by the theoretical result [H88] that the string distance problem over arbitrary scoring matrices has an $\Omega(n^2)$ lower time bound.

As we shall see in the next section, essentially the same difficulties arise in the definition and computation of the approximate generators of a string $\boldsymbol{x}$.

Symmetry Restored

Recall that the cost array (or so-called dynamic-programming matrix) $\boldsymbol{c}$ was first introduced in Section 9.1, where we considered the problem of computing the distance between $\boldsymbol{x_1}[1..n_1]$ and $\boldsymbol{x_2}[1..n_2]$. The calculation proceeded by computing, for every $i \in 0..n_1$ and every $j \in 0..n_2$, the distance from $\boldsymbol{x_1}[1..i]$ to $\boldsymbol{x_2}[1..j]$; thus the initial values in row 0 and column 0 of $\boldsymbol{c}$ were computed in symmetric ways, giving respectively the distance from ε to $\boldsymbol{x_2}[j]$ and from $\boldsymbol{x_1}[i]$ to ε. Using these initial values, the other entries in array $\boldsymbol{c}$ could then be filled in using dynamic-programming as specified in Lemma 9.1.1. Thus at each step for each string we considered a *prefix*, and so it was always true that

$$\boldsymbol{c}[i,j] = d\big(\boldsymbol{x_1}[1..i], \boldsymbol{x_2}[1..j]\big), \tag{13.11}$$

a relation in which $\boldsymbol{x_1}$ and $\boldsymbol{x_2}$ again play symmetric roles. For this problem the final result could always be found in $\boldsymbol{c}[n_1, n_2]$.

This agreeable symmetry was lost however in Chapter 10 when the cost array $\boldsymbol{c}$ was used to find k-approximate patterns $\boldsymbol{p}[1..m]$ in a given string $\boldsymbol{x}[1..n]$. Now, for each position $i \in 1..n$ in $\boldsymbol{x}$, we looked for the nonempty *substring* $\boldsymbol{x}[i'..i]$, $i' \in 1..i$, that provided the best k-approximate match with the *prefix* $\boldsymbol{p}[1..j]$, $j \in 1..m$. As explained in Section 10.1, the new problem required that the column entries be initialized in a different way, a situation that we dealt with by eliminating column 0 altogether and making a slight change in the initialization of column 1. Thus $\boldsymbol{c}$ became $\boldsymbol{c}[0..n, 1..m]$, but, as it turned out, the same dynamic-programming approach could still be used for every $i \in 1..n$ and $j \in 2..m$, yielding the asymmetric relation

$$\boldsymbol{c}[i,j] = \min_{1 \le i' \le i} d\big(\boldsymbol{x}[i'..i], \boldsymbol{p}[1..j]\big). \tag{13.12}$$

The k-approximate matches of $\boldsymbol{p}$ with $\boldsymbol{x}$ were in this case identified by the positions i in column m of $\boldsymbol{c}$ such that $\boldsymbol{c}[i,m] \le k$.

A further extension of the problem occurs when we attempt to determine, for each $i \in 1..n_1$ and each $j \in 1..n_2$, matches between nonempty substrings $\boldsymbol{x_1}[i'..i]$ and $\boldsymbol{x_2}[j'..j]$ of both given strings $\boldsymbol{x_1}[1..n_1]$ and $\boldsymbol{x_2}[1..n_2]$ that satisfy a given threshold k. Then symmetry is restored between row 1 and column 1 of $\boldsymbol{c} = \boldsymbol{c}[1..n_1, 1..n_2]$, where now equivalent initializations take place, as well as for each entry $\boldsymbol{c}[i,j]$, where the computation must somehow yield

$$\boldsymbol{c}[i,j] = \min_{\substack{1 \le i' \le i \\ 1 \le j' \le j}} d\big(\boldsymbol{x_1}[i'..i], \boldsymbol{x_2}[j'..j]\big). \tag{13.13}$$

Here the matches of interest are signalled by values $\boldsymbol{c}[i,j] \leq k$.

As we shall see later, the cost array $\boldsymbol{c}$ whose entries are specified by (13.13) will provide the basis for the calculation of k-approximate squares when $\boldsymbol{x} = \boldsymbol{x_1} = \boldsymbol{x_2}$, but with a benefit array $\boldsymbol{b}$ replacing $\boldsymbol{c}$. In this case, for $i \neq j$, $\boldsymbol{b}[i,j] \geq k$ will identify a square formed by two distinct substrings

$$\boldsymbol{x}[i'..i] \overset{(k)}{=} \boldsymbol{x}[j'..j].$$

However, in order to be able to perform an efficient calculation, we first need to show that the analogue of (13.13) can be computed using the dynamic-programming approach of Lemma 9.1.1. We deal with this question in the next subsection.

The Benefit Array b

It is natural first to consider implementing (13.13) using exactly the same dynamic-programming calculation specified in Lemma 9.1.1:

$$\begin{aligned} \boldsymbol{c}[i,j] \leftarrow \min \Big\{ & \boldsymbol{c}[i-1,j] + d(\boldsymbol{x_1}[i], \varepsilon), \\ & \boldsymbol{c}[i,j-1] + d(\varepsilon, \boldsymbol{x_2}[j]), \\ & \boldsymbol{c}[i-1,j-1] + d(\boldsymbol{x_1}[i], \boldsymbol{x_2}[j]) \Big\}, \end{aligned} \tag{13.14}$$

applicable now for every $i \in 2..n_1$, $j \in 2..n_2$. But we immediately encounter difficulties that become evident if, as shown in Table 13.1, we apply (13.13) to our former example $\boldsymbol{x_1} = rests$, $\boldsymbol{x_2} = stress$ using edit distance d_E.

First we find that the only possible values in the array are unhelpful ones, 0 and 1 — according to (13.13), the former is the minimum that occurs whenever there exist positions $i' \leq i$ and $j' \leq j$ such that

$$\boldsymbol{x_1}[i'..i] = \boldsymbol{x_2}[j'..j],$$

while the latter occurs whenever $\boldsymbol{x_1}[i] \neq \boldsymbol{x_2}[j]$. Thus (13.13) really only tells us whether or not positions $\boldsymbol{x_1}[i]$ and $\boldsymbol{x_2}[j]$ match, while giving no information about any preceding range of positions in the two strings over which a partial match may have occurred. But a second difficulty is even more serious: the dynamic-programming approach seems no

	j	**1**	**2**	**3**	**4**	**5**	**6**
i		s	t	r	e	s	s
1	r	1	1	0	1	1	1
2	e	1	1	1	0	1	1
3	s	0	1	1	1	0	0
4	t	1	0	1	1	1	1
5	s	0	1	1	1	0	0

Table 13.1: A cost array for $\boldsymbol{x_1} = rests$ and $\boldsymbol{x_2} = stress$ using d_E and (13.13)

longer to work! For example, in Table 13.1 we cannot compute $\boldsymbol{c}[2,2] = 1$ or $\boldsymbol{c}[5,5] = 0$ using (13.14).

The standard approach used to circumvent these difficulties [SW81] is to make use of what we shall call a benefit array $\boldsymbol{b}$ in which maximum benefit rather than minimum cost is of interest. At the same time, distances $d(\lambda, \mu)$ between letters are replaced by ***scores*** $S(\lambda, \mu)$ that will be positive in desirable circumstances (for example, $\lambda = \mu$) and negative (or at least less positive) in undesirable circumstances (for example, $\lambda \neq \mu$). We suppose that these scores appear as entries in a ***scoring matrix*** S similar to the matrix W introduced in Section 2.2 except that now numerically larger entries usually identify matches rather than mismatches. The use of this matrix of course requires that the alphabet be indexed, as defined in Section 4.1.

Suppose that for our example a scoring matrix S is given as follows:

$$S = \begin{array}{cccccc} & \varepsilon & e & r & s & t \\ \varepsilon & 0 & -1 & -1 & -1 & -1 \\ e & -1 & 2 & -1 & -1 & -1 \\ r & -1 & -1 & 2 & -1 & -1 \\ s & -1 & -1 & -1 & 2 & -1 \\ t & -1 & -1 & -1 & -1 & 2 \end{array} \tag{13.15}$$

Thus in this simple case a penalty (negative benefit) of -1 is assessed for any mismatch, deletion, or insertion, while a benefit of $+2$ is awarded for any match except as usual for the empty string: $S(\varepsilon, \varepsilon) = 0$. More generally we remark that it may occur, for example in biological applications, that some scores $S(\lambda, \mu)$ are positive even when $\lambda \neq \mu$.

Just as distance between letters was extended to distance between strings, so also here we use scores between letters to define scores between strings. We suppose still that $\boldsymbol{x_1} = \boldsymbol{x_1}[1..n_1]$ is to be transformed into $\boldsymbol{x_2} = \boldsymbol{x_2}[1..n_2]$ using edit operations on single letters: insertion, deletion, substitution. We suppose also that these operations are irreversible (it is not allowed to substitute $a \rightarrow b$, then resubstitute $b \rightarrow a$, even if by doing so we increase the total score). Thus the transformation of $\boldsymbol{x_1}$ into $\boldsymbol{x_2}$ can be represented as a sequence of edit operations in which no backtracking occurs, exactly the same model that was used previously for distance.

Of course, as before, there may be more than one sequence of edit operations that achieves the transformation of $\boldsymbol{x_1}$ into $\boldsymbol{x_2}$. For each such sequence we can compute the sum of all the scores of the edit operations as given by the scoring matrix. Then for strings $\boldsymbol{x_1}$ and $\boldsymbol{x_2}$ we define the ***score***, written $S(\boldsymbol{x_1}, \boldsymbol{x_2})$, to be the maximum sum attained over all sequences of irreversible edit operations that transform $\boldsymbol{x_1}$ into $\boldsymbol{x_2}$.

We are now in a position to introduce the analogue of (13.13), the ***benefit array*** $\boldsymbol{b} = \boldsymbol{b}[1..n_1, 1..n_2]$, in which

$$\boldsymbol{b}[i,j] = \max_{\substack{1 \le i' \le i \\ 1 \le j' \le j}} S\big(\boldsymbol{x_1}[i'..i], \boldsymbol{x_2}[j'..j]\big), \tag{13.16}$$

for every $i \in 1..n_1$, $j \in 1..n_2$. This equation describes the problem; that is, it specifies the elements of the benefit array that we wish to compute using a dynamic-programming algorithm that is an analogue of (13.14):

$$\begin{aligned} \boldsymbol{b}[i,j] \leftarrow \max \Big\{ 0, & \boldsymbol{b}[i-1,j] + S(\boldsymbol{x_1}[i], \varepsilon), \\ & \boldsymbol{b}[i,j-1] + S(\varepsilon, \boldsymbol{x_2}[j]), \\ & \boldsymbol{b}[i-1,j-1] + S(\boldsymbol{x_1}[i], \boldsymbol{x_2}[j]) \Big\}, \end{aligned} \tag{13.17}$$

for every $i \in 2..n_1$, $j \in 2..n_2$. For $j = 1$ we initialize using an analogue of (10.2)–(10.3). We suppose for convenience that there exists a single value $\boldsymbol{b}[0,1] = 0$, so that for every $i = 1, 2, \ldots, n_1$, we can compute

$$\boldsymbol{b}[i,1] \leftarrow \max \Big\{ 0, \boldsymbol{b}[i-1,1] + S(\boldsymbol{x_1}[i], \varepsilon), S(\boldsymbol{x_1}[i], \boldsymbol{x_2}[1]) \Big\}. \tag{13.18}$$

Similarly, for $i = 1$, we suppose $\boldsymbol{b}[1,0] = 0$ and initialize with

$$\boldsymbol{b}[1,j] \leftarrow \max \Big\{ 0, \boldsymbol{b}[1,j-1] + S(\varepsilon, \boldsymbol{x_2}[j]), S(\boldsymbol{x_1}[1], \boldsymbol{x_2}[j]) \Big\}, \tag{13.19}$$

for $j = 1, 2, \ldots, n_2$. We shall now be interested in entries $\boldsymbol{b}[i,j] \geq k > 0$ — in other words, those that provide benefit at least equal to some strictly positive threshold k.

Observe particularly that in (13.17)–(13.19) the values of $\boldsymbol{b}[i,j]$ are constrained to be nonnegative. The reason for this is that in (13.16) we seek always to identify i' and j' that maximize a positive score; thus in the dynamic-programming calculation (13.17), preceding values

$$\boldsymbol{b}[i-1,j], \boldsymbol{b}[i,j-1], \boldsymbol{b}[i-1,j-1]$$

would not be of interest if they were negative, because we could always get a better result by treating the negative values as if they were zero and choosing $i = i'$ or $j = j'$, as the case may be. Thus storing only nonnegative values in $\boldsymbol{b}$ ensures that (13.17) executes in accordance with (13.16), and so avoids the difficulties described above that arise in the use of (13.14).

Table 13.2 exhibits a benefit array for our example strings, in which local maxima $\boldsymbol{b}[4,2]$, $\boldsymbol{b}[3,5]$ and $\boldsymbol{b}[5,6]$ are highlighted: these values identify exact matches of st and res, as well as the approximate match of $rests$ with $ress$. The significance of these maxima will become clear later when we discuss local optimality.

	j	**1**	**2**	**3**	**4**	**5**	**6**
i		s	t	r	e	s	s
1	r	0	0	2	1	0	0
2	e	0	0	1	4	3	2
3	s	2	1	0	3	**6**	5
4	t	1	**4**	3	2	5	5
5	s	2	3	3	2	4	**7**

Table 13.2: A benefit array for $\boldsymbol{x_1} = rests$ and $\boldsymbol{x_2} = stress$ using S and (13.16)

The Grid Graph

In Section 9.5 we introduced the dependency graph, a directed graph whose vertices were labelled with the positions $[i, j]$, $i \in 1..n_1$, $j \in 1..n_2$, of the cost array $\boldsymbol{c}$ and whose only arcs were from those preceding positions

$$[i-1, j], [i, j-1], [i-1, j-1]$$

whose values had in fact contributed to the value $\boldsymbol{c}[i, j]$. This structure was made precise in (9.8) and illustrated for $\boldsymbol{x_1} = rests$, $\boldsymbol{x_2} = stress$ in Figure 9.2.

Here we make use of essentially the same structure, now called a grid graph and redefined for the benefit array $\boldsymbol{b}$. Given two strings $\boldsymbol{x_1}$, $\boldsymbol{x_2}$ and an associated scoring matrix S, the ***grid graph*** $G = G(\boldsymbol{x_1}, \boldsymbol{x_2}, S)$ is a directed acyclic graph (V, E) whose vertices V are labelled with positions $[i, j]$ in the benefit array $\boldsymbol{b}$, and where each vertex $[i, j]$ has inarcs of E if and only if $\boldsymbol{b}[i, j] > 0$ and the following rules, analogous to (9.8), are applied:

$$\begin{aligned}
([i-1, j], [i, j]) \in E \quad &\text{iff} \quad \boldsymbol{b}[i, j] - \boldsymbol{b}[i-1, j] = S(\boldsymbol{x_1}[i], \varepsilon); \\
([i, j-1], [i, j]) \in E \quad &\text{iff} \quad \boldsymbol{b}[i, j] - \boldsymbol{b}[i, j-1] = S(\varepsilon, \boldsymbol{x_2}[j]); \\
([i-1, j-1], [i, j]) \in E \quad &\text{iff} \quad \boldsymbol{b}[i, j] - \boldsymbol{b}[i-1, j-1] = S(\boldsymbol{x_1}[i], \boldsymbol{x_2}[j]).
\end{aligned} \tag{13.20}$$

Thus, according to these rules, every arc in a grid graph lies on a path corresponding to a set of edit operations that yields a maximum score $\boldsymbol{b}[i, j]$. Observe that, since the vertices $[0, j]$, $[i, 0]$ and $[0, 0]$ do not exist, vertex $[1, 1]$ is a source (no inarcs); in general we see that

---

Lemma 13.3.1 *A vertex $[i, j]$ in a grid graph is a source if and only if the integers $i' = i$ and $j' = j$ yield the maximum benefit $\boldsymbol{b}[i, j]$ in (13.16).*

Proof See Exercise 13.3.4. □

Figure 13.7 shows the grid graph for our example $\boldsymbol{x_1} = rests$, $\boldsymbol{x_2} = stress$: there are altogether 10 source vertices, of which eight are actually isolated, a condition that could not occur in dependency graphs.

As illustrated by Figure 13.7, the main role of the grid graph G is to provide a structure that tells us *how* each score $\boldsymbol{b}[i, j]$ can be arrived at: once G has been constructed, the traversal of any path in G from vertex $[i', j']$ to vertex $[i, j]$ permits the quantity $\boldsymbol{b}[i, j]$ to be easily computed, just by summing the scores of all arcs traversed and adding this amount to $\boldsymbol{b}[i', j']$. We call this sum of arc scores the ***path score*** from $[i', j']$ to $[i, j]$ and denote it by $S_{ij}^{i'j'}$, where of course

$$S_{ij}^{i'j'} = \boldsymbol{b}[i, j] - \boldsymbol{b}[i', j'].$$

Hence for every choice of $[i, j]$, it is natural to be interested in the source vertices $[i', j']$ that start paths leading to $[i, j]$ — as the following result shows, for fixed i and j, these source vertices maximize $S_{ij}^{i'j'}$:

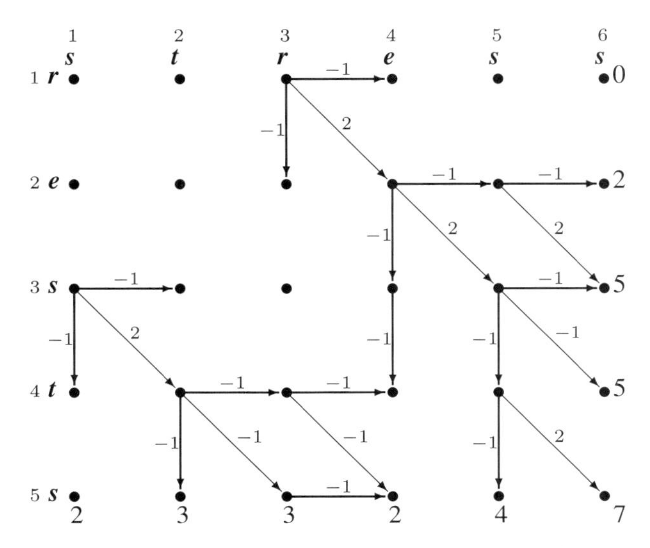

Figure 13.7. The grid graph $G = G(rests, stress, S)$

Lemma 13.3.2 *Let P denote a path in a grid graph G from a source vertex $[i', j']$ to some chosen vertex $[i, j]$. Then over all the vertices of P, $S_{ij}^{i'j'}$ is the unique maximum path score.*

Proof See Exercise 13.3.4. □

In fact, based on (13.16) and this result, we find that we can reach a very satisfying conclusion: source vertices $[i', j']$ identify integers i' and j' such that $\boldsymbol{x_1}[i'..i]$ and $\boldsymbol{x_2}[j'..j]$ yield a maximum score

$$S(\boldsymbol{x_1}[i'..i], \boldsymbol{x_2}[j'..j]).$$

Thus computation of highest-scoring paths (paths of highest path score) in the grid graph $G(\boldsymbol{x_1}, \boldsymbol{x_2}, S)$ is a natural approach to solving the problem stated by (13.16).

Observe in the example grid graph of Figure 13.7 that the vertex $[1, 3]$ is the source of the highest-scoring path leading to each of the vertices $[i, 6], i \in 2..5$; in view of Lemma 13.3.2, this tells us that the suffix $\boldsymbol{x_2}[3..6] = ress$ of $\boldsymbol{x_2}$ provides the best match with each of $\boldsymbol{x_1}[1..i]$, one that is furthermore the best match with the entire string $\boldsymbol{x_1}[1..i]$ itself. Note on the other hand that vertex $[4, 4]$ of G has two sources $[1, 3]$ and $[3, 1]$, reflecting the fact that a best match of suffixes of $\boldsymbol{x_1}[1..4] = rest$ and $\boldsymbol{x_2}[1..4] = stre$ can be formed in two ways, either by matching

$$\boldsymbol{x_1}[1..4] \text{ with } \boldsymbol{x_2}[3..4] = re$$

or by matching

$$\boldsymbol{x}[3..4] = st \text{ with } \boldsymbol{x_2}[1..4].$$

In view of Lemma 13.3.2, it is tempting to assume that maximum path scores are also necessarily attained by sinks $[i, j]$ as well as by sources $[i', j']$. But this is not true: in Figure 13.7 the vertex $[3, 5]$ lies on the path from $[1, 3]$ to the sink $[4, 6]$, but nevertheless $\boldsymbol{b}[4, 6] = 5 < 6 = \boldsymbol{b}[3, 5]$.

Generally we will not be interested in all high-scoring paths that exceed the given threshold k, but only a subset of them. As we have just seen, there are high-scoring paths leading both to $[3, 5]$ and to $[4, 6]$ in Figure 13.7, but $[3, 5]$ is likely to be more interesting. Similarly, vertex $[4, 2]$ will likely be more interesting than $[5, 3]$. In the next subsection we begin to deal with the important question of how to identify the most interesting high-scoring paths in grid graphs.

Local Optimality

We have seen in our discussion of exact repeats in strings (Section 13.2) that the idea of nonextendibility plays a critical role in focussing attention on the information that really needs to be output, and so acts to greatly reduce volume of output. Here, for approximate pattern-matching, the idea of local optimality plays a similar role. Both nonextendibility and local optimality depend upon extension of certain properties as far as possible both to the left (case (b) in the following definition) and to the right (case (c)).

Definition 13.3.3 [ES83] *A vertex pair* $([i', j'], [i, j])$ *in a grid graph* G *is said to be* ***locally optimal*** *if and only if the following three conditions are satisfied:*

(a) $S(\boldsymbol{x_1}[i'..i], \boldsymbol{x_2}[j'..j]) > 0$;

(b) *for every* $I \in 1..i$, $J \in 1..j$,

$$S(\boldsymbol{x_1}[i'..i], \boldsymbol{x_2}[j'..j]) \geq S(\boldsymbol{x_1}[I..i], \boldsymbol{x_2}[J..j]) > 0;$$

(c) *for every* $I \in i'..n_1$, $J \in j'..n_2$,

$$S(\boldsymbol{x_1}[i'..i], \boldsymbol{x_2}[j'..j]) \geq S(\boldsymbol{x_1}[i'..I], \boldsymbol{x_2}[j'..J]).$$

We see from the definition (13.16) of the benefit array that condition (b) will be satisfied by listing all the vertices $[i', j']$ that correspond to the maximum $\boldsymbol{b}[i, j]$ computed for each vertex $[i, j]$. Condition (c) can be satisfied by computing the benefit array $\widehat{\boldsymbol{b}}$ for the reversed strings $\widehat{\boldsymbol{x_1}}$ and $\widehat{\boldsymbol{x_2}}$, then listing all the vertices $[i', j']$ that correspond to each maximum value $\widehat{\boldsymbol{b}}[i, j]$. Matching the two lists of vertices, while at the same time taking appropriate account of transformations of positions from $\widehat{\boldsymbol{x_1}}$ to $\boldsymbol{x_1}$ and $\widehat{\boldsymbol{x_2}}$ to $\boldsymbol{x_2}$, identifies a set of vertex pairs that satisfy both conditions.

In terms of grid graphs, this processing amounts to identifying source vertices in G corresponding to each vertex $[i, j]$, as described above, then doing the same thing for the "reverse" grid graph $\widehat{G}$. The requirement in condition (a) that the score for a locally optimal pair be strictly positive essentially avoids the reporting of isolated vertices. Both to the reader's relief and to ours, we skip the details.

	j	**1**	**2**	**3**	**4**	**5**	**6**
i		s	s	e	r	t	s
1	s	2	2	1	0	0	2
2	t	1	1	0	0	2	1
3	s	2	3	2	1	1	**4**
4	e	1	2	5	4	3	3
5	r	0	1	4	**7**	6	5

Table 13.3: The reverse benefit array $\widehat{\boldsymbol{b}}$ for $\widehat{\boldsymbol{x_1}} = stser$ and $\widehat{\boldsymbol{x_2}} = sserts$

It is instructive however to look at what happens when local optimality is applied to our example.

Table 13.3 shows the benefit array $\widehat{\boldsymbol{b}}$ for $stser$ and $sserts$ with the local maxima highlighted for $\widehat{\boldsymbol{b}}[3,6]$ and $\widehat{\boldsymbol{b}}[5,4]$ that correspond to an exact match of ts and a good match of $stser$ with $sser$. Referring back to Table 13.2, we see that these matches correspond precisely to the exact match of st and the good match of $rests$ with $ress$ that were highlighted in $\boldsymbol{b}$. Thus these matches would be reported as locally optimal by the procedure just described, either with output

$$([1,3],[5,6]) \text{ and } ([3,1],[4,2])$$

identifying pairs of vertices in G, or alternatively by

$$([1,1],[5,4]) \text{ and } ([2,5],[3,6])$$

identifying pairs of vertices in $\widehat{G}$.

Note however that the exact match of ser, also highlighted in Table 13.2, will *not* be reported since $\widehat{\boldsymbol{b}}[3,2]$ does not correspond to a source node in the reverse grid graph $\widehat{G}$, and hence does not satisfy the local optimality requirement. In our example this is not a serious matter; in fact, since ser is a suffix of both $stser$ and $sser$, a reported match, we can claim that therefore ser should not have been reported since it is "left-extendible". However, examples can be constructed (see [S98]) in which "good" or "interesting" matches are missed because at least one end of the matched strings fails to correspond to a source node in G or $\widehat{G}$.

Thus the definition given here of local optimality does not quite provide for approximate matching all that nonextendibility provides for exact matching; still, it is the most useful definition we have.

A $\Theta(n^3)$ Algorithm

In the preceding three subsections, we have introduced in its full generality the problem (13.16) of computing approximate matches, especially locally optimal ones, between the substrings of two given strings $\boldsymbol{x_1}$ and $\boldsymbol{x_2}$. We turn now to the special case of this problem, mentioned earlier, where

$$\boldsymbol{x} = \boldsymbol{x_1} = \boldsymbol{x_2}.$$

Thus the approximate matches that are found will now be substrings of the *same* string $\boldsymbol{x}$, and so will become approximate squares.

The special case permits certain simplifications. In particular, as a consequence of the symmetry inherent in matching $\boldsymbol{x}$ against $\boldsymbol{x}$, only the upper triangular portion of the benefit matrix needs to be computed. Also, since we are not interested in matches of $\boldsymbol{x}[i'..i]$ with $\boldsymbol{x}[i'..i]$, the diagonal entries in $\boldsymbol{b}$ can all be initialized to zero.

With these simple amendments, it thus seems that our task is complete: given $\boldsymbol{x} = \boldsymbol{x}[1..n]$, we compute the corresponding upper triangular benefit array $\boldsymbol{b}$ in $\Theta(n^2)$ time using simplified versions of (13.17)–(13.19), then compute $\widehat{\boldsymbol{b}}$ using the same methodology applied to $\widehat{\boldsymbol{x}}$, and finally identify the locally optimal pairs that specify the approximate squares using straightforward housekeeping techniques. The total time requirement will be $\Theta(n^2)$ — what is there to stop us?

Alas, what stops us is the fact, remarked upon in the preceding subsection, that the local optimality criterion is too restrictive: there may be approximate squares in $\boldsymbol{x}$ that are not reported because at least one entry in the vertex pair $\big([i', j'], [i, j]\big)$ that defines such a square does not correspond to a source in G or $\widehat{G}$. [S98] gives an example of this phenomenon, in which various split or overlapping squares are located and output, while a probably more significant tandem square is missed entirely.

In order to deal with this difficulty, Algorithm S simply eliminates the calculation of overlapping squares altogether so as to ensure that all tandem and split squares are reported. Of course, in the case of exact matching, every overlapping square of a substring $\boldsymbol{u}$ necessarily implies the existence of a tandem square of a proper prefix of $\boldsymbol{u}$, so that in a sense little information is lost if overlapping squares are not computed. Similarly, for approximate matching, overlapping squares generally imply shorter tandem squares.

Given $\boldsymbol{x}[1..n]$, Algorithm S seeks therefore, at each position $t \in 1..n-1$, to determine values $i \in 1..t$ and $j \in t+1..n$ that maximize the scores

$$S(\boldsymbol{x}[i..t], \boldsymbol{x}[t+1..j]). \tag{13.21}$$

In terms of the grid graph G, this amounts to determining all highest-scoring paths from vertices $[i, t+1]$ in column $t+1$ of G to vertices $[t, j]$ in row t. A straightforward approach to this problem would compute

$$\max_{\substack{1 \le i \le t \\ t+1 \le j \le n}} S(\boldsymbol{x}[i..t], \boldsymbol{x}[t+1..j]), \tag{13.22}$$

requiring $\Theta(n^2)$ time for $\Theta(n)$ values of t, hence $\Theta(n^3)$ time overall to determine all the k-approximate tandem squares in $\boldsymbol{x}$.

Of course, in order to determine the more interesting subset of these squares that consists only of the locally optimal ones, the reverse problem $\widehat{G}$ would again need to be formulated and solved, as described above, so that highest-scoring paths from source vertices of G to source vertices of $\widehat{G}$ could be identified. Thus, finally, the locally optimal k-approximate tandem squares in $\boldsymbol{x}$ could be computed in $\Theta(n^3)$ time.

Algorithm S finds a way to reduce this time requirement to $\Theta(n^2 \log n)$, using techniques that are sketched below. Essentially, Algorithm S models the problem as $n-1$ subproblems that are special cases of (13.16), comparing strings $\boldsymbol{x}_1^t$ and $\boldsymbol{x}_2^t$, $t = 1, 2, \ldots, n-1$, where

$$\boldsymbol{x}_1^t = \boldsymbol{x}[1..t], \ \ \boldsymbol{x}_2^t = \boldsymbol{x}_2^t[1..n-t] = \boldsymbol{x}[t+1..n]. \tag{13.23}$$

The tth subproblem P^t can be thought of as giving rise to a benefit array

$$\boldsymbol{b^t} = \boldsymbol{b^t}[1..t, 1..n-t]$$

and a corresponding grid graph G^t in which the maximum scores

$$\max_{\substack{1 \le i \le t \\ 1 \le j \le n-t}} S\big(\boldsymbol{x_1^t}[i..t], \boldsymbol{x_2^t}[1..n-t]\big), \tag{13.24}$$

equivalent to those given in (13.22), are computed. Algorithm S recasts each of these problems in terms of arrays that have special properties, enabling their elements to be first calculated and then accessed quickly.

In fact, in [S98] it is shown how the approach can be extended, without sacrificing space or time complexity, to compute not only tandem, but also split, locally optimal k-approximate squares — here however we shall content ourselves with an outline of the tandem squares algorithm.

Every DIST Array is Monge

The grid graph G^t that corresponds to the maximized scores (13.24) will contain highest-scoring paths from source vertices $[i, 1]$ in the first column to sink vertices $[t, j]$ in the final row, where $i \in 1..t$, $j \in 1..n-t$. The idea was introduced in [AP88, AALM90] of defining DIST arrays

$$\boldsymbol{d^t} = \boldsymbol{d^t}[1..t, 1..n-t],$$

where $\boldsymbol{d^t}[i, j]$ is the score of the highest scoring path from $[i, 1]$ to $[t, j]$ in G^t, $t = 1, 2, \ldots, n-1$, and is therefore exactly the quantity specified by (13.24); if no such path exists, it is supposed now that $\boldsymbol{d^t}[i, j] = -\infty$. Each DIST array then, once computed, contains all the information required to solve subproblem P^t as defined in (13.24), while collectively the $n-1$ DIST arrays therefore provide the basis for computing all the tandem squares in $\boldsymbol{x}$ as defined in (13.21) and (13.22).

As it turns out, DIST arrays satisfy a special property that reduces the total time required for their calculation, and that furthermore permits the computed elements to be retrieved efficiently.

Definition 13.3.4 *A matrix $M[1..n_1, 1..n_2]$ is said to be* ***Monge*** *if and only if exactly one of the following conditions is satisfied:*

(a) $M[i, j] + M[i+1, j+1] \ge M[i, j+1] + M[i+1, j]$, *for every* $i \in 1..n_1-1$, $j \in 1..n_2-1$;

(b) $M[i, j] + M[i+1, j+1] \le M[i, j+1] + M[i+1, j]$, *for every* $i \in 1..n_1-1$, $j \in 1..n_2-1$.

Lemma 13.3.5 *Every DIST array $\boldsymbol{d^t}$ is Monge.*

Proof We show that $\boldsymbol{d^t}$ satisfies Definition 13.3.4(b). Suppose that for some $i < t$ and $j < n-t$, $\boldsymbol{d^t}[i,j] > 0$, the score of a highest-scoring path $\pi_{i,j}$ from $[i,1]$ to $[t,j]$ in G^t. If moreover $\boldsymbol{d^t}[i+1,j+1] > 0$, the score of a highest-scoring path $\pi_{i+1,j+1}$ from $[i+1,1]$ to $[t,j+1]$ in G^t, then as shown in Figure 13.8, the two paths must intersect at some vertex $[i',j']$ of G^t.

We find therefore that there must also exist (not necessarily highest-scoring) paths $\pi_{i,j+1}$ from $[i,1]$ to $[t,j+1]$ and $\pi_{i+1,j}$ from $[i+1,1]$ to $[t,j]$, both formed from subpaths that pass through $[i',j']$. Denoting by $|\pi|$ the score of a path π, we see that

$$|\pi_{i,j}| + |\pi_{i+1,j+1}| = |\pi_{i,j+1}| + |\pi_{i+1,j}|,$$

where the paths on the LHS are by definition highest-scoring paths, while the paths on the RHS may possibly not be highest-scoring. The result then follows in the case that both $\boldsymbol{d^t}[i,j]$ and $\boldsymbol{d^t}[i+1,j+1]$ are positive.

If one or both of $\boldsymbol{d^t}[i,j]$, $\boldsymbol{d^t}[i+1,j+1]$ is nonpositive, hence by definition equal to $-\infty$, it must again be true that

$$\boldsymbol{d^t}[i,j] + \boldsymbol{d^t}[i+1,j+1] = -\infty \leq \boldsymbol{d^t}[i,j+1] + \boldsymbol{d^t}[i+1,j].$$

Thus the lemma is proved. □

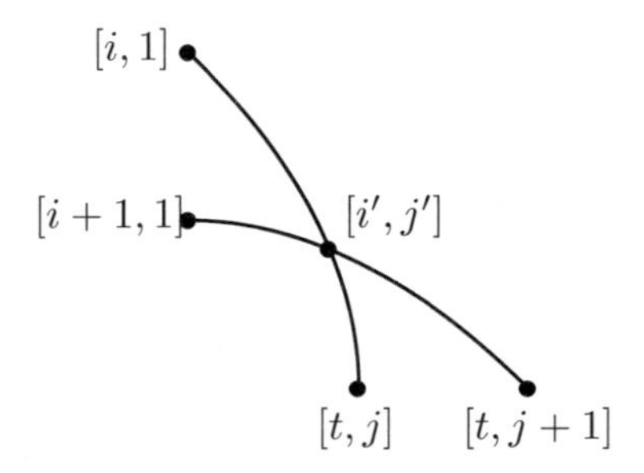

Figure 13.8. Intersecting paths in a grid graph G^t

More important for the subsequent analysis is a similar result that connects the values in the array $\boldsymbol{d^t}$ with those in $\boldsymbol{d^{t+1}}$, essentially by showing that Definition 13.3.4(a) is satisfied:

Lemma 13.3.6 *For every $t \in 1..n-2$, $i \in 1..t$, $j \in 1..n-t$,*

$$\boldsymbol{d^t}[i,j] + \boldsymbol{d^{t+1}}[i+1,j] \geq \boldsymbol{d^t}[i+1,j] + \boldsymbol{d^{t+1}}[i,j].$$

Proof Based on the same considerations as those used to prove Lemma 13.3.5. □

It is at this point that we part company with the detailed development and description of Algorithm S. Lemma 13.3.6 really shows that we can define arrays $X_j[t, i]$, $j = 1, 2, \ldots, n-t$, that are Monge arrays, and that relate the values of elements of $\boldsymbol{d}^{t+1}$ to those of $\boldsymbol{d}^t$. It therefore becomes possible to compute the elements of the arrays X_j, which are also elements of the arrays $\boldsymbol{d}^t$ and $\boldsymbol{d}^{t+1}$, in an efficient iterative fashion, as well as to access them efficiently once they have been computed. The techniques used for carrying out these calculations require the construction and use of a collection of binary trees that "represent" the arrays $\boldsymbol{d}^t$; the explanation and proof-of-correctness of these techniques is lengthy, complex and technical, an endeavour that we therefore regard as beyond the scope of this book. We do however state formally the two main results, given here in a slightly less general form than in [S98]. The first theorem deals with the original computation of the matrices $\boldsymbol{d}^t$ themselves.

Theorem 13.3.7 *A representation of the DIST matrices* $\mathbf{d}^t$, $t = 1, 2, \ldots, n-1$, *can be computed in* $O(n^2 \log n)$ *time using* $O(n^2 \log n)$ *space. Any individual element in any* $\mathbf{d}^t$ *can thereafter be retrieved in* $O(\log t)$ *time, while retrieval of all the elements in any column of any* $\mathbf{d}^t$ *requires only* $O(1)$ *time per element.* □

The second theorem deals with the subsequent computation of the row and column maxima in the $\boldsymbol{d}^t$ matrices; these maxima are required by Algorithm S for the calculation of highest-scoring paths.

Theorem 13.3.8 *All the row and column maxima in the DIST matrices* $\mathbf{d}^t$, $t = 1, 2, \ldots, n-1$, *can be computed in* $O(n^2 \log n)$ *time using* $O(n^2)$ *space.* □

Main and Lorentz Revisited

Once again avoiding pernicious details, we return in this section to the original formulation (13.22) of the approximate tandem squares problem, in order to describe how the interval-splitting technique used by Algorithm ML to specify exact squares (Subsection 12.1.2) finds application here also in Algorithm S for the specification of approximate squares.

Recall that Algorithm ML finds squares in any substring $\boldsymbol{u}$ of $\boldsymbol{x}$ by first splitting $\boldsymbol{u}$ into a left part $\boldsymbol{u_1} = \boldsymbol{u}[1..\lfloor n/2 \rfloor]$ and a right part $\boldsymbol{u_2} = \boldsymbol{u}[\lfloor n/2 \rfloor + 1..n]$, computing immediately the squares that overlap $\boldsymbol{u_1}$ and $\boldsymbol{u_2}$, then dealing with squares local to either $\boldsymbol{u_1}$ or $\boldsymbol{u_2}$ as recursive subproblems. Recall also from (13.21) that Algorithm S determines highest-scoring paths from vertices in column $t+1$ of a grid graph G to vertices in row t. Recall further that since our problem is reflexive — we are comparing substrings of $\boldsymbol{x}$ with substrings of $\boldsymbol{x}$ —, it is therefore necessary to consider only the upper triangular part of both the benefit array $\boldsymbol{b}$ and the grid graph G. It thus turns out, as shown in Figure 13.9, that the ML strategy can be applied to split G into subgraphs based on the location of the highest-scoring paths — those that cross the "mid-point" are computed right away, while those that do not become recursive subproblems:

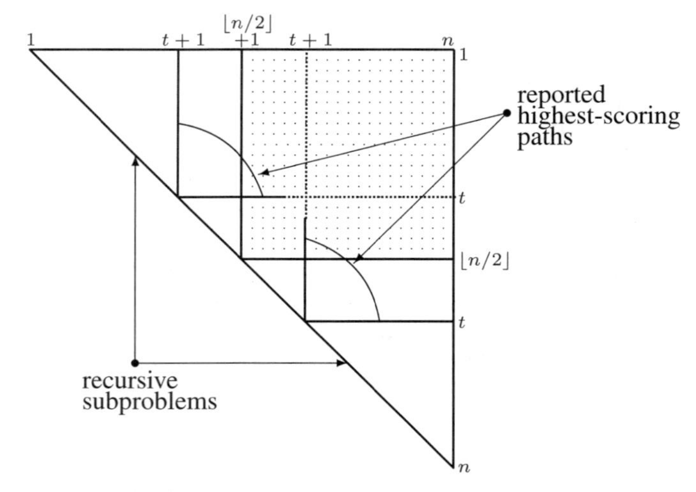

Figure 13.9. Recursive ML splitting of the grid graph

(a) Approximate squares corresponding to highest-scoring paths of score at least k that cross either column $\lfloor n/2 \rfloor + 1$ or row $\lfloor n/2 \rfloor$ are computed and reported immediately.

(b) Highest-scoring paths that are confined either to the upper left triangle or to the lower right triangle are handled as recursive subproblems.

(c) Highest-scoring paths that are confined to the near-square defined by

$$0 \leq i \leq \lfloor n/2 \rfloor, \ \lfloor n/2 \rfloor + 1 \leq j \leq n$$

cannot be solutions, since there is no value of t in this region that satisfies the requirement of the problem. Thus this portion of G can be discarded.

As indicated by Theorems 13.3.7 and 13.3.8, the highest-scoring paths of type (a) can be computed in $O(n^2 \log n)$ time. Thus, denoting by t_n the maximum time required by Algorithm S to compute the approximate tandem squares of any string $\boldsymbol{x}[1..n]$, (b) and (c) together tell us that t_n can be determined by solving the recurrence relation

$$t_n \leq k_2 n^2 \log n + 2t_{\lceil n/2 \rceil}, \tag{13.25}$$

where k_2 is a positive constant. Since of course there exists a positive constant k_1 such that $t_1 \leq k_1$, this is the exact analogue of the recurrence problem (12.2) for Algorithm ML. In fact, the truly devoted reader will recall solving a somewhat more general version of (13.25) in Exercise 12.1.6, in order to discover that

$$t_n \in O(n^2 \log n),$$

the time complexity of Algorithm S. Details of the algorithm, including aspects not touched upon here, will be found in [S98]. Here, based on Theorem 13.3.7 and the preceding analysis, we state formally the final result:

Theorem 13.3.9 *Algorithm S computes the locally optimal k-approximate nonoverlapping squares in a given string $\boldsymbol{x}[1..n]$ in $O(n^2 \log n)$ time and space.* □

Discussion

In this section we have endeavoured to present an intelligible outline of Algorithm S without becoming too enmeshed in its technicalities. We have selected this algorithm for discussion because it is the fastest one available for a problem of considerable practical interest, and because the techniques that it employs constitute a kind of stringological *smorgasbord*: Algorithm S is both practical and instructive. A previous algorithm [HM93] to compute all (not just locally optimal) nonoverlapping approximate squares required $O(n^2 \log^2 n)$ time and $O(n^2 \log n)$ space; the space requirement was reduced to $O(n^2)$ in [B94]. An algorithm to compute all "good" k-approximate tandem repeats in $O\big(kn \log k \log(n/k)\big)$ time appears in [LS93], but its result holds only for edit distance.

One of the most attractive features of Algorithm S is its exploitation of the local optimality criterion to restrict its output to approximate squares that are most likely to be of interest. We have noted however that this criterion poses problems not shared by the nonextendibility criterion used for exact repeats. It may be that a redefinition of local optimality could assist in the search for a more satisfactory algorithm for the approximate squares problem.

Exercises 13.3

1. Using edit distance, construct an example of a string $\boldsymbol{x} = \boldsymbol{u_1 u_2 u_3}$ such that $\boldsymbol{u_1 u_2}$ and $\boldsymbol{u_2 u_3}$ are 1-approximate squares while $\boldsymbol{u_1 u_3}$ is not. Then generalize your construction to work for arbitrary $k \geq 1$. Can you arrange that each of the strings is of a different length?
2. Suppose that reversible edit operations were allowed. Construct an example for which $S(\boldsymbol{x_1}, \boldsymbol{x_2})$ becomes arbitrarily large.
3. In this section a benefit array of nonnegative values has been described, based on positive scores for matches and generally negative scores for mismatches. Show that a cost array of nonpositive values would work just as well, and outline the required definitions.
4. Prove Lemmas 13.3.1 and 13.3.2.
5. The examples given in this section have sources $[i', j']$ of paths only for $i' = 1$ or $j' = 1$. Give examples to show that sources need not be either in the first row or in the first column, and that in fact they can occur arbitrarily far into the grid graph.
6. Draw the grid graph $\widehat{G}$ corresponding to $\widehat{\boldsymbol{x_1}} = stser$ and $\widehat{\boldsymbol{x_2}} = sserts$, then use it to identify all the locally optimal pairs in G corresponding to *rests* and *stress*.
7. Show that for exact matching every overlapping square of a substring $\boldsymbol{u}$ determines a tandem square of a prefix of $\boldsymbol{u}$. Can you construct an example for approximate matching in which this is not true?

8. Formulate the reverse problem of (13.22) and explain how it could be used to compute the locally-optimal k-approximate tandem squares in $\Theta(n^3)$ time.
9. Prove Lemma 13.3.6.

13.4 k-Approximate Periods — SIPS

Like a kite
Cut from the string,
Lightly the soul of my youth
Has taken flight.

— Ishikawa TAKUBOKU (1885–1912), *Song of My Youth*

The first algorithm described in this book computed the border array of a given string $\boldsymbol{x}$, thus determining in linear time every period, including the shortest one, of every prefix of $\boldsymbol{x}$. In this concluding section of the book, we have come almost full circle: we try to compute *approximate* periods of $\boldsymbol{x}$ — leading, as we shall see, to very much more difficult problems [SIPS01], especially when it is required that the computed items include one that is in some sense, like the shortest exact period, the "best" or most interesting. In fact, we shall see that the problems considered in this section are really at the frontier of knowledge: we can state several interesting ones, but even the simplest of them can at present only be partially solved.

First however we need to clarify the context, a process whose first step requires an answer to the following question:

> What use is it to report an "approximate period" — presumably the length of an "approximate generator" — when an approxi-mate generator is unlikely to be identified by its length?

Consider for example the simple string

$$\boldsymbol{x} = abacbc. \tag{13.26}$$

Suppose that under edit distance d_E we are asked to find a 1-approximate generator of $\boldsymbol{x}$ — presumably, a string $\boldsymbol{u}$ such that for some decomposition of $\boldsymbol{x}$, say

$$\boldsymbol{x} = \boldsymbol{x_1}\boldsymbol{x_2}\cdots\boldsymbol{x_r},\ r \geq 1,$$

it is true that $d_E(\boldsymbol{u}, \boldsymbol{x_j}) \leq 1$ for every $j \in 1..r$. But for our example string, and indeed for every string, surely the "best" solution to this problem is to choose $\boldsymbol{u} = \boldsymbol{x}$ and $r = 1$, so that $d_E(\boldsymbol{u}, \boldsymbol{x}) = d_E(\boldsymbol{u}, \boldsymbol{x_1}) = 0$. And if this solution is excluded, then again for our example string as for every string, we can choose $\boldsymbol{u} = \lambda$ for any letter λ, so that $d_E(\boldsymbol{u}, \boldsymbol{x}[i]) = d_E(\lambda, \boldsymbol{x}[i]) = 1$ for every position i in $\boldsymbol{x}$. Extrapolating a little, we discover that, at least under edit distance, the k-approximate generators of any string $\boldsymbol{x}$ will always include $\boldsymbol{x}$ itself as well as all strings $\boldsymbol{u}$ of length at most k — scarcely a helpful or "interesting" result.

Returning to our example (13.26), observe further that we can also choose $\boldsymbol{u} = ac$ and decompose

$$\boldsymbol{x} = \boldsymbol{x_1 x_2 x_3} = (ab)(ac)(bc),$$

so that $d_E(\boldsymbol{u}, \boldsymbol{x_j}) = d_E(ac, \boldsymbol{x_j}) = 1$ for every $j \in 1..3$. Thus ac is a 1-approximate generator, but not a prefix, of (13.26). Or, if we change the example somewhat, to

$$\boldsymbol{y} = abbcad,$$

we find that, while there is no substring $\boldsymbol{u}$ of $\boldsymbol{y}$ of length 2 that is a 1-approximate generator of $\boldsymbol{y}$, nevertheless we can again choose the string $\boldsymbol{u} = ac$, now not even a substring of $\boldsymbol{y}$, so that according to the decomposition

$$\boldsymbol{y} = \boldsymbol{y_1 y_2 y_3} = (ab)(bc)(ad),$$

$d_E(\boldsymbol{u}, \boldsymbol{y_j}) = d_E(ac, \boldsymbol{y_j}) = 1$ for every $j \in 1..3$. Thus ac is also a 1-approximate generator of $\boldsymbol{y}$.

In both of these examples, knowing the length of a nontrivial generator gives us little information — what is really required is the generator itself. Accordingly, our first step in stating the k-approximate periods problem is to eliminate any reference to periods! In this section we shall be dealing with the approximate *generators* of a given string $\boldsymbol{x}$.

We saw in the preceding section that in order to make sense of the idea of an approximate repeat, it was necessary to think carefully about the way in which "distance" was defined; we shall find here also that an appropriate definition of distance is a critical precondition for dealing with approximate generators. We shall throughout this section make use of ***weighted distance*** d_W, a generalization of edit distance that uses a scoring matrix W of nonnegative values as defined in Section 2.3 — according to the values used in W, d_W can be specialized to the other common forms of distance: d_E, d_L and d_H. Following [SIPS01], we introduce a useful new idea:

Definition 13.4.1 *Given any two nonempty strings $\boldsymbol{x}$ and $\boldsymbol{y}$ on an alphabet A, and given a distance function d_W defined on the letters of A, the quotient*

$$\delta_W(\boldsymbol{x}, \boldsymbol{y}) = \frac{2d_W(\boldsymbol{x}, \boldsymbol{y})}{|\boldsymbol{x}| + |\boldsymbol{y}|}$$

*is called the **relative distance** between $\boldsymbol{x}$ and $\boldsymbol{y}$.*

Thus the relative distance between two strings is just the weighted distance (or ***absolute distance***) divided by the average length of the strings. It is an entertaining exercise to show that δ_W retains the properties of a metric usually satisfied by d_W; see Exercise 13.4.1. More importantly, we observe that the use of relative distance rather than weighted distance greatly changes the results obtained when distances between candidate periods $\boldsymbol{u}$ and substrings $\boldsymbol{x}_j$ of $\boldsymbol{x}$ are computed. In our previous example (13.26), we find that for strings of length 1, the distances are unchanged:

$$\begin{aligned} \delta_E(\lambda, \mu) = d_E(\lambda, \mu) &= 1, \quad \lambda \neq \mu; \\ &= 0, \quad \textit{otherwise}. \end{aligned} \tag{13.27}$$

However, the distances for some strings of length 2 are decreased:

$$\delta_E(ac, ab) = \delta_E(ac, bc) = \delta_E(ac, ad) = 1/2.$$

Thus, if we were looking for a generator based on relative distance, we would be more likely to choose a string such as ac that achieves consistently lower scores, or a lower average score, than could be achieved by a single letter. Hence this example raises the question of what criteria should be used for the selection of generators. Note further that the choice $\boldsymbol{u} = \boldsymbol{x}$ still yields the result

$$\delta_E(\boldsymbol{u}, \boldsymbol{x}) = 0,$$

and so we will also need to apply a criterion that limits the length of $\boldsymbol{u}$.

In an effort to satisfy these constraints, we define an approximate generator as follows, slightly modified from [SIPS01]:

Definition 13.4.2 *Given a real number $k \geq 0$ and an integer $\rho \geq 1$, a string $\boldsymbol{u}$ is said to be a (ρ, k)-**approximate generator** of a string $\boldsymbol{x}$ if and only if there exists a decomposition*

$$\boldsymbol{x} = \boldsymbol{x_1 x_2} \cdots \boldsymbol{x_r}, \ r \geq \rho,$$

into nonempty factors $\boldsymbol{x_j}$, $j \in 1..r$, such that

$$\delta_W(\boldsymbol{u}, \boldsymbol{x_j}) \leq k$$

for every $j \in 1..r-1$, and

$$\delta_W(\boldsymbol{u'}, \boldsymbol{x_r}) \leq k$$

for some nonempty prefix $\boldsymbol{u'}$ of $\boldsymbol{u}$.

In its insistence that $\boldsymbol{u}$ should be sufficiently close to *every* factor $\boldsymbol{x_j}$, this definition seems to capture the property of uniformity that an exact generator satisfies. Also as in the exact case, it allows $\boldsymbol{x}$ to have a suffix that matches approximately with a prefix of $\boldsymbol{u}$. To get some intuition for the effect that the choice of the parameter ρ has on the determination of the generator, consider the example

$$\boldsymbol{x} = abcacaabaacaa.$$

Table 13.4 shows the approximate generators compatible with various values of the parameters ρ and k: in this case, in order to get "interesting" approximate generators, we need to choose $\rho \geq 3$, but it is not clear that any benefit is gained by setting $\rho = 4$, since doing so excludes the possibly interesting generator $abcaca$. On the other hand, there may be cases — for instance, where $\boldsymbol{x}$ has a long exact or approximate generator repeated several times — in which a larger value of ρ would be desirable. Observe further that in this example it is also not clear that, whatever the value of ρ, the generators corresponding to the minimum relative distance k are necessarily the most interesting!

Based on Definition 13.4.2, it is straightforward to define the idea of a "minimal" approximate generator:

Definition 13.4.3 *A (ρ, k)-approximate generator $\boldsymbol{u}$ of a string $\boldsymbol{x}$ is said to be **ρ-minimal** if and only if there exists no (ρ, k')-approximate generator of $\boldsymbol{x}$, $k' < k$.*

Observe that in Table 13.4 $abcaca$ and $abaaca$ are 3-minimal, while aba is 4-minimal.

We are now in a position to state three problems related to (ρ, k)-approximate generators [SIPS01], which we list in order of increasing difficulty:

(P1) Given nonempty strings $\boldsymbol{x}$ and $\boldsymbol{u}$, an integer ρ, and a relative distance function δ_W, find the minimum value k such that $\boldsymbol{u}$ is a (ρ, k)-approximate generator of $\boldsymbol{x}$.

(P2) Given a nonempty string $\boldsymbol{x}$, an integer ρ, and a relative distance function δ_W, find a substring $\boldsymbol{u}$ of $\boldsymbol{x}$ (if it exists) that is a ρ-minimal generator of $\boldsymbol{x}$.

(P3) Given a nonempty string $\boldsymbol{x}$, an integer ρ, and a relative distance function δ_W, find an arbitrary string $\boldsymbol{u}$ (if it exists) that is a ρ-minimal generator of $\boldsymbol{x}$.

In the next three subsections, we briefly discuss these three problems, only to discover that no solution currently exists for any of them! However, we shall see that for (P1) and

ρ	k	**Generators found**
1	0	$\boldsymbol{x}$, $abcacaabaaca$
2	0	$abcacaabaaca$
3	1/3	aba, $abcaca$, $abaaca$
3	1/6	$abcaca$, $abaaca$
4	1/3	aba
4	1/6	$\emptyset$

Table 13.4: (ρ, k)-approximate generators for $\boldsymbol{x} = abcacaabaacaa$

(P2) partial solutions are given by straightforward applications of string distance algorithms discussed in Chapter 9.

(P1) — Compute Minimum k for x and u

For uniformity of presentation, we have stated this problem for a relative distance function δ_W, but since both $\boldsymbol{x}$ and $\boldsymbol{u}$ are specified, it could just as well have been stated for an absolute distance function d_W.

The algorithm SIPS1 described here solves (P1) only for *some* value of ρ, not a specified one; that is, it computes a minimum value of k such that, for some ρ, $\boldsymbol{u}$ is a (ρ, k)-approximate generator of $\boldsymbol{x}$. SIPS1 breaks down into two stages, the first a repeated application of string distance calculations, the second an iterated calculation of the minimum threshold k.

In the first stage, n standard cost arrays (Section 9.1)

$$\boldsymbol{c}^{(i)}[0..n-i+1, 0..m], \quad i = 1, 2, \ldots, n \tag{13.28}$$

are computed corresponding to string distances $\delta_W(\boldsymbol{x}[i..n], \boldsymbol{u})$, $i = 1, 2, \ldots, n$. This calculation enables the distances $\delta_W(\boldsymbol{x}[i..h], \boldsymbol{u})$ from $\boldsymbol{x}[i..h]$ to $\boldsymbol{u}$ to be determined for every $h \in i..n-1$ — they will be available as the mth columns of the arrays $\boldsymbol{c}^{(i)}$, $i = 1, 2, \ldots, n-1$. Since the final occurrence of $\boldsymbol{u}$ in $\boldsymbol{x}$ need only be a nonempty prefix of $\boldsymbol{u}$, we can search the $(n-i+1)$th (last) row of each $\boldsymbol{c}^{(i)}$ for a minimum value, say $\text{minpref}(\boldsymbol{u}, i, n)$ — this will be the least distance from any prefix of $\boldsymbol{u}$ to $\boldsymbol{x}[i..n]$. Thus finally, using the last column and one element from the last row of each $\boldsymbol{c}^{(i)}$, we can form an upper triangular cost array $\boldsymbol{c}[1..n, 1..n]$, where for $i \in 1..n$,

$$\begin{aligned} \boldsymbol{c}[i,h] &= \delta_W(\boldsymbol{x}[i..h], \boldsymbol{u}), \quad \text{for } h \in i..n-1; \\ &= \text{minpref}(\boldsymbol{u}, i, n), \quad \text{for } h = n. \end{aligned}$$

The calculation of the $\boldsymbol{c}^{(i)}$ requires $\Theta(mn^2)$ time using Algorithm WF (Section 9.2), for example, while the calculation of $\boldsymbol{c}$ requires $\Theta(n^2)$ time — thus the overall time requirement for the first stage is $\Theta(mn^2)$.

We consider now the second stage. Suppose that $\boldsymbol{u}$ is an approximate generator of $\boldsymbol{x}[1..i]$: then $\boldsymbol{x}[1..i]$ breaks down into factors whose largest distance under δ_W from $\boldsymbol{u}$ is k_i, say. Then for any $h \in i+1..n-1$, we can determine a corresponding distance k_h by computing

$$\max\Big\{k_i, \delta_W(\boldsymbol{x}[i+1..h], \boldsymbol{u})\Big\} = \max\big\{k_i, \boldsymbol{c}[i+1, h]\big\},$$

while for $h = n$, we compute

$$\max\big\{k_i, \text{minpref}(\boldsymbol{u}, i, n)\big\} = \max\big\{k_i, \boldsymbol{c}[i+1, n]\big\}.$$

If moreover for every $i \in 1..h-1$, the factors of $\boldsymbol{x}[1..i]$ have already been selected in such a way as to minimize k_i over all selections of factors, then we see that computing

$$k_h = \min_{1 \le i < h} \Big\{ \max\big\{k_i, \boldsymbol{c}[i+1, h]\big\}\Big\} \tag{13.29}$$

must also yield a minimum distance k_h for $\boldsymbol{x}[1..h]$ over all selections of factors. This is a classical "dynamic-programming" approach that we can implement in $\Theta(n^2)$ time. Thus, over the two stages:

Theorem 13.4.4 *Given a general distance function δ_W and nonempty strings $\boldsymbol{x}$ and $\boldsymbol{u}$, Algorithm SIPS1 computes a minimum value of k such that, for some ρ, $\boldsymbol{u}$ is a (ρ, k)-approximate generator of $\boldsymbol{x}$. Its execution requires $\Theta(mn^2)$ time and $\Theta(n^2)$ space.* □

In view of the pronounced similarity between the first phase of SIPS1 and the calculation of benefit arrays $\boldsymbol{b^t}$ introduced as subproblems (13.24) of Algorithm S, it seems rather likely that DIST arrays can be used here also to reduce the $\Theta(mn^2)$ time requirement in the first phase to $O(mn \log n)$, hence the overall time to $\Theta(n^2)$. This intriguing possibility is left as a research project; see Exercise 13.4.4.

So far we have supposed that the distance function δ_W is quite general, based on an arbitrary scoring matrix. However, if instead edit distance δ_E or d_E is used, significant savings in time and space result. In view of (13.27), it is not necessary to compute edit distance between $\boldsymbol{u}$ and substrings of $\boldsymbol{x}$ of length greater than $2m$: such distances would necessarily exceed m and so could not possibly contribute to the computation of any minimum value k_i. Thus each of the n cost arrays (13.28) will have maximum size $2m \times m$, while the revised cost array $\boldsymbol{c}$ will have size $n \times 2m$. The time requirement for the first stage is correspondingly reduced to $\Theta(m^2 n)$, and in the second stage at most $2m$ positions in $\boldsymbol{c}$ need to be considered for each value $h = 1, 2, \ldots, n$, yielding a time requirement of $O(mn)$. Consequently using edit distance the overall time requirement of SIPS1 reduces to $\Theta(m^2 n)$, the space requirement to $\Theta(mn)$.

But a completely different approach (SIPS2) reduces the time even more. Recall that at the end of Section 9.5, reference was made to an incremental string distance algorithm [LMS98] that, given $d_E(\boldsymbol{x}, \boldsymbol{u})$, could compute either $d_E(\lambda\boldsymbol{x}, \boldsymbol{u})$ or $d_E(\boldsymbol{x}\lambda, \boldsymbol{u})$ for any letter λ in time $O(|\boldsymbol{x}| + |\boldsymbol{u}|)$. Suppose now that Algorithm LMS were applied to compute the edit distance between $\boldsymbol{u}$ and the following sequence of $n+1$ strings, each of length at most $2m$, and each formed from the preceding one by prepending a single letter:

$$\varepsilon, \boldsymbol{x}[n], \boldsymbol{x}[n-1..n], \ldots, \boldsymbol{x}[n-2m+1..n], \boldsymbol{x}[n-2m..n-1], \ldots, \boldsymbol{x}[1..2m].$$

These calculations make available all the distances between $\boldsymbol{u}$ and every substring of $\boldsymbol{x}$ of length at most $2m$. Since $|\boldsymbol{u}| = m$, the time required by LMS for each calculation will be $O(2m+m) = O(m)$, thus $O(mn)$ overall. The matrix $\boldsymbol{c}$ is formed as for SIPS1 in $\Theta(mn)$ time, then processed in the second stage in $\Theta(mn)$ time. Thus we have

Theorem 13.4.5 *Using edit distance and given nonempty strings $\boldsymbol{x}$ and $\boldsymbol{u}$, Algorithm SIPS2 computes a minimum value of k such that, for some ρ, $\boldsymbol{u}$ is a (ρ, k)-approximate generator of $\boldsymbol{x}$. Its execution requires $\Theta(mn)$ time and $\Theta(m^2)$ space.* □

(P2) — Compute ρ-Minimal Substring $\boldsymbol{u}$

One obvious way to identify a substring $\boldsymbol{u}$ of $\boldsymbol{x}$ that (for some ρ, but not a preselected one) is a ρ-minimal generator of $\boldsymbol{x}$ is to investigate all possible substrings of $\boldsymbol{x}$ using SIPS1 or SIPS2, then choose the substring $\boldsymbol{u}$ that as period gives rise to the minimum threshold value k. Recall from Exercise 1.2.10 that there are $O(n^2)$ distinct substrings in $\boldsymbol{x}$ of average length $O(n)$. Thus either SIPS1 (for general distance δ_W) or SIPS2 (for edit distance δ_E) would be executed $O(n^2)$ times using a generator whose average length $m \in O(n)$. Consequently,

Theorem 13.4.6 *For some unspecified value ρ, (P2) can be solved in $O(n^5)$ time using general distance δ_W, in $O(n^4)$ time using edit distance δ_E.* □

In [SIPS01] another method is given for the case of edit distance, but the time requirement is still $O(n^4)$.

(P3) — Compute Arbitrary ρ-Minimal $\boldsymbol{u}$

This problem is probably just the one that we would particularly like to be able to solve with reasonable efficiency — but it turns out, by a reduction from the shortest common supersequence problem, to be NP-complete! For details see [SIPS01].

Remark

In this section we have outlined algorithms that relate to a new topic area in stringology, an area in which the problems are not yet well defined, the solutions still less so. Over the course of 13 chapters, we have reached this point by a lengthy but nevertheless, in outline, fairly natural and understandable route. The fact remains, however, that the problems just considered are among the simplest of generalizations of the periodicity calculation that was the first algorithm studied in this book. The route was long, the distance covered short.

Exercises 13.4

1. Show that δ_W satisfies the conditions (2.2)–(2.5) of a metric if and only if d_W satisfies them.
2. Show that for edit distance, $0 \leq \delta_E(\boldsymbol{x}, \boldsymbol{y}) < 2$ in general, and that for $|\boldsymbol{x}| = |\boldsymbol{y}|$, $\delta_E(\boldsymbol{x}, \boldsymbol{y}) \leq 1$.
3. Explain how Algorithm SIPS1 achieves a space requirement of only $\Theta(n^2)$ rather than $\Theta(mn^2)$.
4. **Research Project:** Investigate the possibility of using DIST arrays (Section 13.3) to reduce the time of Algorithm SIPS1 to $\Theta(n^2)$.
5. Design a variant of Algorithm SIPS1 that solves Problem (P1) in $\Theta(n)$ time for Hamming distance d_H.

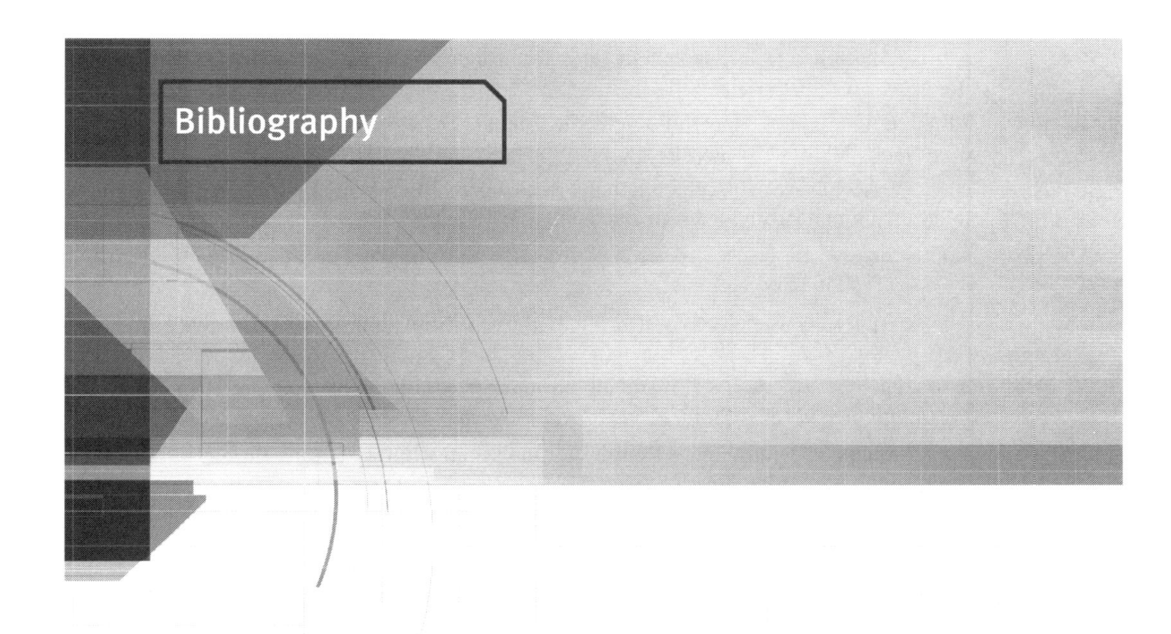

[A85] Alberto Apostolico, **The myriad virtues of subword trees**, *Combinatorial Algorithms on Words (NATO ASI Series F12)*, Alberto Apostolico & Zvi Galil (eds.), Springer-Verlag (1985) 85–96.

[A90] Alfred V. Aho, **Algorithms for finding patterns in strings**, *Handbook of Theoretical Computer Science*, J. van Leeuwen (ed.), Elsevier (1990) 255–300.

[A94] Jean-Paul Allouche, **Sur la complexité des suites infinies**, *Bull. Belg. Math. Soc. 1* (1994) 133–143.

[AALM90] Alberto Apostolico, Mikhail J. Atallah, Lawrence Larmore & Scott McFaddin, **Efficient parallel algorithms for string editing problems**, *SIAM J. Computing 19–5* (1990) 968–988.

[AARS94] Jean-Paul Allouche, Dan Astoorian, Jim Randall & Jeffrey Shallit, **Morphisms, square-free strings and the Tower of Hanoi puzzle**, *Amer. Math. Monthly 101–7* (1994) 651–658.

[AC75] Alfred V. Aho & Margaret J. Corasick, **Efficient string matching: an aid to bibliographic search**, *CACM 18–6* (1975) 333–340.

[AC91] Alberto Apostolico & Maxime Crochemore, **Optimal canonization of all substrings of a string**, *Information & Computation 95–1* (1991) 76–95.

[AD86] Lloyd Allison & Trevor I. Dix, **A bit string longest subsequence algorithm**, *IPL 23–6* (1986) 305–310.

[ADKF70] V. L. Arlazarov, E. A. Dinic, M. A. Kronrod & I. A. Faradžev, **On economical construction of the transitive closure of a directed graph**, *Soviet Math. Dokl. 11–5* (1970) 1209–1210.

[AE93] Alberto Apostolico & Andrzej Ehrenfeucht, **Efficient detection of quasiperiodicities in strings**, *TCS 119* (1993) 247–265.

[AFI91] Alberto Apostolico, Martin Farach & Costas S. Iliopoulos, **Optimal superprimitivity testing for strings**, *IPL 39* (1991) 17–20.

[AG86] Alberto Apostolico & Raffaele Giancarlo, **The Boyer-Moore-Galil string searching strategies revisited**, *SIAM J. Comput. 15–1* (1986) 98–105.

[AG87] Alberto Apostolico & C. Guerra, **The longest common subsequence problem revisited**, *Algorithmica 2* (1987) 315–336.

[AHU74] Alfred V. Aho, John E. Hopcroft & Jeffey D. Ullman, *The Design & Analysis of Computer Algorithms*, Addison-Wesley (1974).

[AHU76] Alfred V. Aho, Dan S. Hirschberg & Jeffrey D. Ullman, **Bounds on the complexity of the longest common subsequence problem**, *J. ACM 23–1* (1976) 1–12.

[ALS99] Arne Andersson, N. Jesper Larsson & Kurt Swanson, **Suffix trees on words**, *Algorithmica 23–3* (1999) 246–260.

[AP83] Alberto Apostolico & Franco P. Preparata, **Optimal off-line detection of repetitions in a string**, *TCS 22* (1983) 297–315.

[AP88] Alok Aggarwal & James Park, **Notes on searching in multidimensional monotone arrays**, *Proc. 29th Annual IEEE Symp. Foundations of Computer Science* (1988) 497–512.

[AR91] Pierre Arnoux & Gérard Rauzy, **Représentation géometriques de suites de complexité** $2n+1$, *Bull. Soc. Math. France 119* (1991) 199–215.

[ASU86] Alfred V. Aho, Ravi Sethi & Jeffrey D. Ullman, *Compilers: Principles, Techniques & Tools*, Addison-Wesley (1986).

[B63] C. H. Braunholtz, **Solution to Problem 5030: an infinite sequence of 3 symbols with no adjacent repeats**, *American Mathematical Monthly 70* (1963) 675–676.

[B80] Kellogg S. Booth, **Lexicographically least circular substrings**, *IPL 10–4/5* (1980) 240–242.

[B92] Dany Breslauer, **An on-line string superprimitivity test**, *IPL 44* (1992) 345–347.

[B94] Gary Benson, **A space efficient algorithm for finding the best non-overlapping alignment score**, *Proc. Fifth Annual Symp. Combinatorial Pattern Matching*, Maxime Crochemore & Dan Gusfield (eds.), Lecture Notes in Computer Science 807, Springer-Verlag (1994) 1–14.

[B96] Jean Berstel, **Recent results in Sturmian words**, *Developments in Language Theory*, World Scientific Press (1996) 13–24.

[BB99] Jean Berstel & Luc Boasson, **Partial words and a theorem of Fine and Wilf**, *TCS 218* (1999) 135–141.

[BBEH85] Anselm Blumer, Janet Blumer, David Haussler, Andrzej Ehrenfeucht, M. T. Chen & Joel Seiferas, **The smallest automaton recognizing the subwords of a text**, *TCS 40–1* (1985) 31–55.

[BEM79] Dwight R. Bean, Andrzej Ehrenfeucht & George F. McNulty, **Avoidable patterns in strings of symbols**, *Pacific J. Math. 85–2* (1979) 261–294.

[BF84] M. Boshernitzan & Aviezri S. Fraenkel, **A linear algorithm for nonhomogeneous spectra of numbers**, *J. Algs. 5* (1984) 187–198.

[BLPS00] Gerth S. Brodal, Rune B. Lyngsø, Christian N. S. Pedersen & Jens Stoye, **Finding maximal pairs with bounded gap**, *J. Discrete Algs. 1* (2000) 77–103.

[BM77] Robert S. Boyer & J. Strother Moore, **A fast string searching algorithm**, *CACM 20–10* (1977) 762–772.

[BP00] Gerth S. Brodal & Christian N. S. Pedersen, **Finding maximal quasiperiodicities in strings**, *Proc. Eleventh Annual Symp. Combinatorial Pattern Matching*, Raffaele Giancarlo & David Sankoff (eds.), Lecture Notes in Computer Science 1848, Springer-Verlag (2000) 397–411.

[BS93] Jean Berstel & Patrice Séébold, **A characterization of Sturmian morphisms**, *The Mathematical Foundations of Computer Science*, A. Borzyszkowski & S. Sokolowski (eds.), Springer-Verlag (1993) 281–290.

[BS97] Jon L. Bentley & Robert Sedgewick, **Fast algorithms for sorting and searching strings**, *Proc. Eighth Annual ACM-SIAM Symp. Discrete Algs.* (1997) 360–369.

[BY89a] Ricardo A. Baeza-Yates, **Algorithms for string searching: a survey**, *ACM SIGIR Forum 23–3/4* (1989) 34–58.

[BY89b] Ricardo A. Baeza-Yates, **String searching algorithms revisited**, *Algorithms & Data Structures*, Lecture Notes in Computer Science 382, Frank Dehne, Jörg-Rüdiger Sack & Nicola Santoro (eds.), Springer-Verlag (1989) 75–96.

[BYG92] Ricardo A. Baeza-Yates & Gaston H. Gonnet, **A new approach to text searching**, *CACM 35–10* (1992) 74–82.

[BYGR90] Ricardo A. Baeza-Yates, Gaston H. Gonnet & Mireille Régnier, **Analysis of Boyer-Moore-type string searching algorithms**, *Proc. First Annual ACM-SIAM Symp. Discrete Algs.* (1990) 328–343.

[BYN96] Ricardo A. Baeza-Yates & Gonzalo Navarro, **A faster algorithm for approximate string matching**, *Proc. Seventh Annual Symp. Combinatorial Pattern Matching*, Dan S. Hirschberg & Gene W. Myers (eds.), Lecture Notes in Computer Science 1075, Springer-Verlag (1996) 1–23.

[BYN97] Ricardo A. Baeza-Yates & Gonzalo Navarro, **Multiple approximate string matching**, *Proc. Fifth Annual Workshop on Algorithms & Data Structures*, F. Dehne *et al.* (eds.) (1997) 174–184.

[BYP92] Ricardo A. Baeza-Yates & Chris H. Perleberg, **Fast and practical approximate string matching**, *Proc. Third Annual Symp. Combinatorial Pattern Matching*, Alberto Apostolico, Maxime Crochemore, Zvi Galil & Udi Manber (eds.) Lecture Notes in Computer Science 644, Springer-Verlag (1992) 185–192.

[BYR92] Ricardo A. Baeza-Yates & Mireille Régnier, **Average running time of the Boyer-Moore-Horspool algorithm**, *TCS* 92 (1992) 19–31.

[C01] Maxime Crochemore, **Reducing space for index implementation**, *TCS*, to appear.

[C02] James D. Currie, **There are ternary circular square-free words of length n for $n \geq 18$**, *Electronic J. Combinatorics*, to appear.

[C81] Maxime Crochemore, **An optimal algorithm for computing all the repetitions in a word**, *IPL 12–5* (1981) 244–248.

[C84] Maxime Crochemore, **Linear searching for a square in a word**, *Bull. EATCS 24* (1984) 66–72.

[C86] Maxime Crochemore, **Transducers and repetitions**, *TCS 45–1* (1986) 63–86.

[C91] James D. Currie, **Which graphs allow infinite nonrepetitive walks?**, *Discrete Math. 87* (1991) 249–260.

[C94] Richard Cole, **Tight bounds on the complexity of the Boyer-Moore string matching algorithm**, *SIAM J. Comput. 23–5* (1994) 1075–1091.

[CCGJ94] Maxime Crochemore, Artur Czumaj, Leszek Gąsieniec, Stefan Jarominek, Thierry Lecroq, Wojciech Plandowski & Wojciech Rytter, **Speeding up two string-matching algorithms**, *Algorithmica 12* (1994) 247–267.

[CCGL99] Maxime Crochemore, Artur Czumaj, Leszek Gąsieniec, Thierry Lecroq, Wojciech Plandowski & Wojciech Rytter, **Fast practical multi-pattern matching**, *IPL 71–3/4* (1999) 107–113.

[CFL58] K. T. Chen, R. H. Fox & Roger C. Lyndon, **Free differential calculus IV**, *Ann. Math. 68–1* (1958) 81–95.

[CGG90] Livio Colussi, Zvi Galil & Raffaele Giancarlo, **On the exact complexity of string matching**, *Proc.* 31st *Annual IEEE Symp. on Foundations of Computer Science, Vol I* (1990) 135–143.

[CH92] Richard Cole & Ramesh Hariharan, **Tighter bounds on the exact complexity of string matching**, *Proc.* 33rd *Annual IEEE Symp. Foundations of Computer Science* (1992) 600–609.

[CH98] Richard Cole & Ramesh Hariharan, **Faster approximate string matching**, *Proc. Ninth Annual ACM-SIAM Symp. Discrete Algs.* (1998) 463–472.

[CHL01] Maxime Crochemore, Christophe Hancart & Thierry Lecroq, *Algorithmique du Texte*, Vuibert, Paris (2001).

[CHL02] Maxime Crochemore, Christophe Hancart & Thierry Lecroq, **A unifying look at the Apostolico-Giancarlo string matching algorithm**, *J. Discrete Algs. 2–1*, to appear.

[ChL90] William I. Chang & Eugene L. Lawler, **Approximate string matching in sublinear expected time**, *Proc.* 31st *Annual IEEE Symp. Foundations of Computer Science, vol. I* (1990) 116–124.

[ChL92] William I. Chang & Jordan Lampe, **Theoretical and empirical comparisons of approximate string matching algorithms**, *Proc. Third Annual Symp. Combinatorial Pattern Matching*, Alberto Apostolico, Maxime Crochemore, Zvi Galil & Udi Manber (eds.) Lecture Notes in Computer Science 644, Springer-Verlag (1992) 172–181.

[ChL94] William I. Chang & Eugene L. Lawler, **Sublinear approximate string matching and biological applications**, *Algorithmica 12–4/5* (1994) 327–344.

[ChL97] Christian Charras & Thierry Lecroq, *Exact String Matching Algorithms*, Laboratoire d'Informatique, Université de Rouen (1997):
`http://www-igm.univ-mlv.fr/~lecroq/string/index.html`

[CHPZ95] Richard Cole, Ramesh Hariharan, Michael S. Paterson & Uri Zwick, **Tighter lower bounds on the exact complexity of string matching**, *SIAM J. Comput. 24–1* (1995) 30–45.

[CIP01] Maxime Crochemore, Costas S. Iliopoulos & Yoan J. Pinzón, **Speeding up Hirschberg and Hunt-Szymanski LCS algorithms**, *Proc. Eighth IEEE Internat. Symp. String Processing & Information Retrieval*, IEEE Computer Science Press (2001) 59–67.

[CIPR01] Maxime Crochemore, Costas S. Iliopoulos, Yoan J. Pinzón & James F. Reid, **A fast and practical bit-vector algorithm for the longest common subsequence problem**, *IPL 80–6* (2001) 279–285.

[CIR98] Tim Crawford, Costas S. Iliopoulos & Rajeev Raman, **String matching techniques for musical similarity and melodic recognition**, *Computing in Musicology 11* (1998) 73–100.

[CL97] Maxime Crochemore & Thierry Lecroq, **Tight bounds on the complexity of the Apostolico-Giancarlo algorithm**, *IPL 63–4* (1997) 195–203.

[CMR99] M. Gabriella Castelli, Filippo Mignosi, Antonio Restivo, **Fine & Wilf's theorem for three periods and a generalization of Sturmian words**, *TCS 218* (1999) 83–94.

[Co91] Livio Colussi, **Correctness and efficiency of pattern matching algorithms**, *Information & Computation 95* (1991) 225–251.

[Co94] Livio Colussi, **Fastest pattern matching in strings**, *J. Algs. 16–2* (1994) 163–189.

[CP91] Maxime Crochemore & Dominique Perrin, **Two-way string matching**, *JACM 38–3* (1991) 651–675.

[CR94] Maxime Crochemore & Wojciech Rytter, *Text Algorithms*, Oxford University Press (1994) 412 pp.

[CS85] M. T. Chen & Joel Seiferas, **Efficient and elegant subword-tree construction**, *Combinatorial Algorithms on Words (NATO ASI Series F12)*, Alberto Apostolico & Zvi Galil (eds.), Springer-Verlag (1985) 97–107.

[CV97] Maxime Crochemore & Renaud Vérin, **Direct construction of compact directed acyclic word graphs**, *Proc. Eighth Annual Symp. Combinatorial Pattern Matching*, Alberto Apostolico & Jotun Hein (eds.), Lecture Notes in Computer Science 1264, Springer-Verlag (1997) 116–129.

[CW79] Beate Commentz-Walter, **A string matching algorithm fast on the average**, *Proc. Sixth Internat. Colloquium on Automata, Languages & Programming*, Hermann A. Maurer (ed.), Lecture Notes in Computer Science 71, Springer-Verlag (1979) 118–132.

[D65] Bálint Dömölki, **Algorithms for the recognition of properties of symbol strings** (Russian), *J. Computing Math. & Math. Physics 5–1* (1965).

[D68] Bálint Dömölki, **A universal computer system based on production rules**, *BIT 8* (1968) 262–275.

[D83] Jean-Pierre Duval, **Factorizing words over an ordered alphabet**, *J. Algs. 4* (1983) 363–381.

[DB86] G. Davies & S. Bowsher, **Algorithms for pattern matching**, *Software — Practice & Experience 16–6* (1986) 575–601.

[DLL02] Jean-Pierre Duval, Thierry Lecroq & Arnaud Lefebvre, **Border array on bounded alphabet**, *Proc. Prague Stringology Conf. '02* (2002).

[E61] Pál Erdős, **Some unsolved problems**, *Magyar Tud. Akad. Mat. Kutató Intézet Közl. 6* (1961) 221–254.

[EC96] Nadia El-Mabrouk & Maxime Crochemore, **Boyer-Moore strategy to efficient approximate string matching**, *Proc. Seventh Annual Symp. Combinatorial Pattern Matching*, Dan S. Hirschberg & Gene W. Myers (eds.), Lecture Notes in Computer Science 1075, Springer-Verlag (1996) 24–28.

[EH88] Andrzej Ehrenfeucht & David Haussler, **A new distance metric on strings computable in linear time**, *Discrete Appl. Math. 20* (1988) 191–203.

[ES83] Bruce W. Erickson & Peter H. Sellers, **Recognition of patterns in genetic sequences**, *Time Warps, String Edits & Macromolecules: The Theory & Practice of Sequence Comparison*, David Sankoff & Joseph B. Kruskal (eds.), Addison-Wesley (1983) 55–91.

[F97] Martin Farach, **Optimal suffix tree construction with large alphabets**, *Proc.* 38th *Annual IEEE Symp. Foundations of Computer Science* (1997) 137–143.

[FGLR02] František Franěk, Shudi Gao, Weilin Lu, Patrick J. Ryan, W. F. Smyth, Yu Sun & Lu Yang, **Verifying a Border Array in Linear Time**, *J. Combinatorial Math. & Combinatorial Comput. 42* (2002).

[FJLS02] František Franěk, Jiandong Jiang, Weilin Lu & W. F. Smyth, **Two-pattern strings**, *Proc. Thirteenth Annual Symposium on Combinatorial Pattern Matching*, Alberto Apostolico & Masayuki Takeda (eds.), Lecture Notes in Computer Science, Springer-Verlag (2002) 76–84.

[FKS00] František Franěk, Ayşe Karaman & W. F. Smyth, **Repetitions in Sturmian strings**, *TCS 249* (2000) 289–303.

[FM96] Martin Farach & S. Muthukrishnan, **Optimal logarithmic time randomized suffix tree construction**, *Proc.* 23rd *Internat. Colloquium Automata, Languages & Programming* (1996).

[FS98] Aviezri S. Fraenkel & R. Jamie Simpson, **How many squares can a string contain?**, *J. Combinatorial Theory (A) 82* (1998) 112–120.

[FS99] Aviezri S. Fraenkel & R. Jamie Simpson, **The exact number of squares in Fibonacci words**, *TCS 218–1* (1999) 83–94.

[FST02] František Franěk, W. F. Smyth & Yudong Tang, **Computing all repeats using suffix arrays**, *Proc. Thirteenth Australasian Workshop on Combinatorial Algorithms*, Elizabeth Billington, Diane Donovan & Abdollah Khodkar (eds.) (2002) 171–184.

[FSX02] František Franěk, W. F. Smyth & Xiangdong Xiao, **A note on Crochemore's repetitions algorithm — a fast space-efficient approach**, *Proc. Prague Stringology Conf. '02*, M. Balík & M. Simánek (eds.) (2002) 36–43.

[FW65] N. J. Fine & Herbert S. Wilf, **Uniqueness theorems for periodic functions**, *Proc. Amer. Math. Soc. 16* (1965) 109–114.

[G01] Shudi Gao, *New Properties of Borders & Covers of Strings*, M. Sc. thesis, Department of Computing & Software, McMaster University (2001).

[G79] Zvi Galil, **On improving the worst case running time of the Boyer-Moore string matching algorithm**, *CACM 22–9* (1979) 505–508.

[G97] Dan Gusfield, *Algorithms on Strings, Trees, & Sequences*, Cambridge University Press (1997) 534 pp.

[GBY91] Gaston H. Gonnet & Ricardo A. Baeza-Yates, *Handbook of Algorithms & Data Structures in Pascal and C*, 2nd edition, Addison-Wesley (1991).

[GG86] Zvi Galil & Raffaele Giancarlo, **Improved string matching with k mismatches**, *SIGACT News 17* (1986) 52–54.

[GG91] Zvi Galil & Raffaele Giancarlo, **On the exact complexity of string matching: lower bounds**, *SIAM J. Comput. 20–6* (1991) 1008–1020.

[GG92] Zvi Galil & Raffaele Giancarlo, **On the exact complexity of string matching: upper bounds**, *SIAM J. Comput. 21–3* (1992) 407–437.

[GK95] Robert Giegerich & Stefan Kurtz, **A comparison of imperative and purely functional suffix tree constructions**, *Science of Computer Programming 25–2/3* (1995) 187–218.

[GO80] Leo J. Guibas & Andrew M. Odlyzko, **A new proof of the linearity of the Boyer-Moore string searching algorithm**, *SIAM J. Comput. 9–4* (1980) 672–682.

[GO81a] Leo J. Guibas & Andrew M. Odlyzko, **Periods in strings**, *J. Combinatorial Theory (A) 30* (1981) 19–42.

[GO81b] Leo J. Guibas & Andrew M. Odlyzko, **String overlaps, pattern matching and nontransitive games**, *J. Combinatorial Theory (A) 30* (1981) 183–208.

[GP90] Zvi Galil & Kunsoo Park, **An improved algorithm for approximate string matching**, *SIAM J. Comput. 19–8* (1990) 989–999.

[GS83] Zvi Galil & Joel Seiferas, **Time-space optimal string matching**, *J. Computer & System Sci. 26–3* (1983) 280–294.

[GS98] Dan Gusfield & Jens Stoye, *Linear Time Algorithms for Finding & Representing all the Tandem Repeats in a String*, Tech. Rep. CSE-98–4, Computer Science Department, University of California, Davis (1998).

[H71] John E. Hopcroft, **An $n \log n$ algorithm for minimizing states in a finite automaton**, *Theory of Machines & Computations*, Z. Kohavi & A. Paz (eds.), Academic Press (1971).

[H73] Jean-Paul Haton, *Contribution à l'Analyse, Paramétrisation et la Reconnaissance Automatique de la Parole*, Doctoral thesis, Université de Nancy (1973).

[H75] Dan S. Hirschberg, **A linear space algorithm for computing maximal common subsequences**, *CACM 18–6* (1975) 341–343.

[H77] Dan S. Hirschberg, **Algorithms for the longest common subsequence problem**, *JACM 24–4* (1977) 664–675.

[H78] Dan S. Hirschberg, **An information-theoretic lower bound for the longest common subsequence problem**, *IPL 7–1* (1978) 40–42.

[H80] R. Nigel Horspool, **Practical fast searching in strings**, *Software — Practice & Experience 10–6* (1980) 501–506.

[H82] Richard Hamming, *Coding & Information Theory*, Prentice Hall (1982).

[H88] Xiaoqiu Huang, **A lower bound for the edit-distance problem under arbitrary cost function**, *IPL 27–6* (1988) 319–321.

[H93] Cristophe Hancart, *Analyse Exacte et en Moyenne d'Algorithmes de Recherche d'un Motif dans un Texte*, Ph. D. thesis, Université de Paris (1993) 96 pp.

[HC02] Maxime Crochemore & Jan Holub, **On the implementation of compact DAWGs**, *Proc. Seventh Internat. Conf. Implementation & Application of Automata* (2002) 293–298.

[HM93] Sampath K. Hannan & Gene W. Myers, **An algorithm for locating nonoverlapping regions of maximum alignment score**, *Proc. Fourth Annual Symp. Combinatorial Pattern Matching*, Alberto Apostolico, Maxime Crochemore, Zvi Galil & Udi Manber (eds.), Lecture Notes in Computer Science 684, Springer-Verlag (1993) 74–86.

[HS77] James W. Hunt & Thomas G. Szymanski, **A fast algorithm for computing longest common subsequences**, *CACM 20–5* (1977) 350–353.

[HS91] Andrew Hume & Daniel Sunday, **Fast string searching**, *Software — Practice & Experience 21–11* (1991) 1221–1248.

[HT84] Dov Harel & Robert Endre Tarjan, **Fast algorithms for finding nearest common ancestors**, *SIAM J. Computing 13–2* (1984) 338–355.

[HU79] John E. Hopcroft & Jeffrey D. Ullman, *Introduction to Automata Theory, Languages, & Computation*, Addison-Wesley (1979).

[IHST01] S. Inenaga, H. Hoshino, A. Shinohara, Masayuki Takeda, S. Arikawa, G. Mauri & G. Pavesi, **On-line construction of compact directed acyclic word graphs**, *Proc. Twelfth Annual Symp. Combinatorial Pattern Matching*, Amihood Amir & Gad M. Landau (eds.) Lecture Notes in Computer Science 2087, Springer-Verlag (2001) 169–180.

[IM01] Costas S. Iliopoulos & Laurent Mouchard, **Fast local covers**, preprint.

[IMS97] Costas S. Iliopoulos, Dennis Moore & W. F. Smyth, **A characterization of the squares in a Fibonacci string**, *TCS 172* (1997) 281–291.

[IP01] Costas S. Iliopoulos & Yoan J. Pinzón, **Recovering an LCS in $O(n^2/w)$ time and space**, *Proc. Twelfth Australasian Workshop on Combinatorial Algorithms*, Edi T. Baskoro (ed.), Institute of Technology Bandung (2001) 106–117.

[IS95] Costas S. Iliopoulos & W. F. Smyth, **A fast average case algorithm for Lyndon decomposition**, *Internat. J. Computer Math. 57–3/4* (1995) 15–31.

[J02] Christopher G. Jennings, *A Linear-Time Algorithm for Fast Exact Pattern Matching in Strings*, M. Sc. thesis, Department of Computing & Software, McMaster University (2002).

[JV92] Guy Jacobson & Kiem-Phong Vo, **Heaviest increasing/common subsequence problems**, *Proc. Third Annual Symp. Combinatorial Pattern Matching*, Alberto Apostolico, Maxime Crochemore, Zvi Galil & Udi Manber (eds.) Lecture Notes in Computer Science 644, Springer-Verlag (1992) 52–66.

[K72] Richard M. Karp, **Reducibility among combinatorial problems**, *Complexity of Computer Computations*, Raymond E. Miller & James W. Thatcher (eds.), IBM Research Symposia Series (1972) 85–103.

[K73a] Donald E. Knuth, *The Art of Computer Programming I: Fundamental Algorithms*, 2nd edition, Addison-Wesley (1973) 722 pp.

[K73b] Donald E. Knuth, *The Art of Computer Programming III: Sorting & Searching*, Addison-Wesley (1973) 634 pp.

[K76] Donald E. Knuth, **Big omichron, big omega, and big theta**, *ACM SIGACT News* (April 1976) 18–24.

[K92] Veikko Keränen, **Abelian squares are avoidable on 4 letters**, *Automata, Languages & Programming*, Lecture Notes in Computer Science 623, Springer-Verlag (1992) 41–52.

[K94] S. Rao Kosaraju, **Computation of squares in a string**, *Proc. Fifth Annual Symp. Combinatorial Pattern Matching*, Maxime Crochemore & Dan Gusfield (eds.), Lecture Notes in Computer Science 807, Springer-Verlag (1994) 146–150.

[K99] Stefan Kurtz, **Reducing the space requirement of suffix trees**, *Software — Practice & Experience 29–13* (1999) 1149–1171.

[KK00] Roman Kolpakov & Gregory Kucherov, **On maximal repetitions in words**, *J. Discrete Algs. 1* (2000) 159–186.

[KK98] Roman Kolpakov & Gregory Kucherov, *Maximal Repetitions in Words, or How to Find All Squares in Linear Time*, Rapport Interne LORIA 98-R-227, Laboratoire Lorrain de Recherche en Informatique et ses Applications, France (1998).

[KK99] Roman Kolpakov & Gregory Kucherov, *On the Sum of Exponents of Maximal Repetitions in a Word*, Rapport Interne LORIA 99-R-034, Laboratoire Lorrain de Recherche en Informatique et ses Applications, France (1999).

[KM84] G. Kowalski & A. Meltzer, **New multi-term high speed text search algorithms**, *Proc. First IEEE Internat. Conf. on Computer Applications* (1984) 514–522.

[KMP77] Donald E. Knuth, James H. Morris & Vaughan R. Pratt, **Fast pattern matching in strings**, *SIAM J. Comput. 6–2* (1977) 323–350.

[KP00] Sung-Ryul Kim & Kunsoo Park, **A dynamic edit distance table**, *Proc. Eleventh Annual Symp. Combinatorial Pattern Matching*, Raffaele Giancarlo & David Sankoff (eds.), Lecture Notes in Computer Science 1848, Springer-Verlag (2000) 60–68.

[KR87] Richard M. Karp & Michael O. Rabin, **Efficient randomized pattern-matching algorithms**, *IBM J. Res. Develop. 31–2* (1987) 249–260.

[L00] Thierry Lecroq, *New Experimental Results on Exact String-Matching*, Rapport LIFAR 2000.03, Université de Rouen (2000).

[L02] M. Lothaire (ed.), *Algebraic Combinatorics on Words*, Encyclopedia of Mathematics and its Applications, Vol.90, Cambridge University Press (2002). See also `http://www-igm.univ-mlv.fr/\ ~ \,berstel/`

[L57] John Leech, **A problem on strings of beads**, *Math. Gazette 41* (1957) 277–278.

[L66] V. I. Levenshtein, **Binary codes capable of correcting deletions, insertions and reversals**, *Cybernetics & Control Theory 10–8* (1966) 707–710.

[L83] M. Lothaire (ed.), *Combinatorics on Words*, Addison-Wesley (1983).

[L92] Thierry Lecroq, **A variation on the Boyer-Moore algorithm**, *TCS 92* (1992) 119–144.

[L95] Thierry Lecroq, **Experimental results on string matching algorithms**, *Software — Practice & Experience 25–7* (1995) 727–765.

[L96] Yin Li, *A Windows-Based String Algorithm Testing System*, Undergraduate Computer Science Project, Department of Comp. Sci. & Systems, McMaster University (1996).

[L97] M. Lothaire (ed.), *Combinatorics on Words*, 2nd edition, Cambridge University Press (1997).

[La96] N. Jesper Larsson, **Extended application of suffix trees to data compression**, *Proc. IEEE Data Compression Conference* (1996) 190–199.

[LM94] Aldo de Luca & Filippo Mignosi, **Some combinatorial properties of Sturmian words**, *TCS 136* (1994) 361–385.

[LMS98] Gad M. Landau, Gene W. Myers & Jeanette P. Schmidt, **Incremental string comparison**, *SIAM J. Comput. 27–2* (1998) 557–582.

[LP98] Harry R. Lewis & Christos H. Papadimitriou, *Elements of the Theory of Computation*, 2nd edition, Prentice-Hall (1998).

[LS02] Yin Li & W. F. Smyth, **Computing the cover array in linear time**, *Algorithmica 32–1* (2001) 95–106.

[LS62] Roger C. Lyndon & Marcel P. Schützenberger, **The equation $a^M = b^N c^P$ in a free group**, *Michigan Math. J. 9* (1962) 289–298.

[LS69] André Lentin & Marcel P. Schützenberger, **A combinatorial problem in the theory of free monoids**, *Combinatorial Mathematics & Its Applications*, R. C. Bose & T. A. Dowling (eds.), University of North Carolina Press (1969) 128–144.

[LS93] Gad M. Landau & Jeanette P. Schmidt, **An algorithm for approximate tandem repeats**, *Proc. Fourth Annual Symp. Combinatorial Pattern Matching*, Alberto Apostolico, Maxime Crochemore, Zvi Galil & Udi Manber (eds.), Lecture Notes in Computer Science 684, Springer-Verlag (1993) 120–133.

[LV85] Gad M. Landau & Uzi Vishkin, **Efficient string matching in the presence of errors**, *Proc.* 26th *Annual IEEE Symp. Foundations of Computer Science* (1985) 126–136.

[LV86a] Gad M. Landau & Uzi Vishkin, **Efficient string matching with k mismatches**, *TCS 43* (1986) 239–249.

[LV86b] Gad M. Landau & Uzi Vishkin, **Introducing efficient parallelism into approximate string matching and a new serial algorithm**, *Proc. Eighteenth ACM Symp. Theory of Comput.* (1986) 220–230.

[LV88] Gad M. Landau & Uzi Vishkin, **Fast string matching with k differences**, *J. Computer & System Sci. 37* (1988) 63–78.

[LV89] Gad M. Landau & Uzi Vishkin, **Fast parallel and serial approximate string matching**, *J. Algs. 10* (1989) 157–169.

[LW75] Roy Lowrance & Robert A. Wagner, **An extension of the string-to-string correction problem**, *JACM 22–2* (1975) 177–183.

[LZ76] Abraham Lempel & Jacob Ziv, **On the complexity of finite sequences**, *IEEE Trans. Information Theory 22* (1976) 75–81.

[M68] Donald R. Morrison, **PATRICIA — practical algorithm to retrieve information coded in alphanumeric**, *JACM 15–4* (1968) 514–534.

[M76] Edward M. McCreight, **A space-economical suffix tree construction algorithm**, *JACM 23–2* (1976) 262–272.

[M86] Gene W. Myers, **An $O(ND)$ difference algorithm and its variations**, *Algorithmica 1* (1986) 251–266.

[M88] Gene W. Myers, *A Four Russians Algorithm for Regular Expression Pattern Matching*, Technical Report 88–34, Dept. of Computer Science, University of Arizona (1988).

[M89] Michael G. Main, **Detecting leftmost maximal periodicities**, *Discrete Applied Maths. 25* (1989) 145–153.

[M99] Gene W. Myers, **A fast bit-vector algorithm for approximate string matching based on dynamic programming**, *JACM 46–3* (1999) 395–415.

[Mi89] Filippo Mignosi, **On the number of factors of Sturmian words**, *TCS 82* (1989) 221–242.

[Mi89a] Filippo Mignosi, **Infinite words with linear subword complexity**, *TCS 65* (1989) 221–242.

[ML79] Michael G. Main & Richard J. Lorentz, *An $O(n \log n)$ Algorithm for Recognizing Repetition*, Tech. Rep. CS-79–056, Computer Science Department, Washington State University (1979).

[ML84] Michael G. Main & Richard J. Lorentz, **An $O(n \log n)$ algorithm for finding all repetitions in a string**, *J. Algs. 5* (1984) 422–432.

[ML85] Michael G. Main & Richard J. Lorentz, **Linear time recognition of squarefree strings**, *Combinatorial Algorithms on Words (NATO ASI Series F12)*, Alberto Apostolico & Zvi Galil (eds.), Springer-Verlag (1985) 271–278.

[MM01] Panagiotis D. Michailidis & Konstantinos G. Margaritis, **On-line string matching algorithms: survey and experimental results**, *Internat. J. Computer Math. 76–4* (2001) 411–434.

[MM90] Udi Manber & Gene W. Myers, **Suffix arrays: a new method for on-line string searches**, *Proc. First Annual ACM-SIAM Symp. Discrete Algs.* (1990) 319–327.

[MM93] Udi Manber & Gene W. Myers, **Suffix arrays: a new method for on-line string searches**, *SIAM J. Comput. 22–5* (1993) 935–948.

[MM96] Robert Muth & Udi Manber, **Approximate multiple string search**, *Proc. Seventh Annual Symp. Combinatorial Pattern Matching*, Dan S. Hirschberg & Gene W. Myers (eds.), Lecture Notes in Computer Science 1075, Springer-Verlag (1996) 75–86.

[MP70] James H. Morris & Vaughan R. Pratt, *A Linear Pattern-Matching Algorithm*, Tech. Rep. 40, University of California, Berkeley (1970).

[MP80] William J. Masek & Michael S. Paterson, **A faster algorithm for computing string-edit distances**, *J. Computer & System Sci. 20–1* (1980) 18–31.

[MS93] Filippo Mignosi & Patrice Séébold, **Morphismes sturmiens et règles de Rauzy**, *J. Théorie de Nombres de Bordeaux 5* (1993) 221–233.

[MS94] Dennis Moore & W. F. Smyth, **An optimal algorithm to compute all the covers of a string**, *IPL 50–5* (1994) 239–246.

[MS95] Dennis Moore & W. F. Smyth, **A correction to: An optimal algorithm to compute all the covers of a string**, *IPL 54* (1995) 101–103.

[MSM99] Dennis Moore, W. F. Smyth & Dianne Miller, **Counting Distinct Strings**, *Algorithmica 13–1* (1999) 1–13.

[MY60] Robert F. McNaughton & H. Yamada, **Regular expressions and state graphs for automata**, *IRE Trans. Electronic Computers EC9–1* (1960) 39–47.

[N01] Benedek Nagy, private communication.

[Nav01] Gonzalo Navarro, **A guided tour to approximate string matching**, *ACM Computing Surveys 33–1* (2001) 31–88.

[NW70] Saul B. Needleman & Christian D. Wunsch, **A general method applicable to the search for similarities in the amino-acid principle of two proteins**, *J. Molecular Biology 48* (1970) 443–453.

[P66] Rohit J. Parikh, **On context-free languages**, *JACM 13–4* (1966) 570–581.

[P70] Peter A. B. Pleasants, **Non-repetitive sequences**, *Proc. Camb. Phil. Soc. 68* (1970) 267–274.

[R77] Ronald L. Rivest, **On the worst-case behaviour of string-searching algorithms**, *SIAM J. Comput. 6–4* (1977) 669–674.

[R80] Wojciech Rytter, **A correct preprocessing algorithm for Boyer-Moore string searching**, *SIAM J. Comput. 9–3* (1980) 509–512.

[R94] Günter Rote, **Sequences with subword complexity** $2n$, *J. Number Theory 46–2* (1994) 196–213.

[RCW73] Thomas A. Reichert, Donald N. Cohen & Andrew K. C. Wong, **An application of information theory to genetic mutations and the matching of polypeptide sequences**, *J. Theoretical Biology 42* (1973) 245–261.

[RS92] Patrick J. Ryan & W. F. Smyth, **An approach to asymptotic complexity**, *Mathematics & Computer Education 26–2* (1992) 135–146.

[S61] Craige Schensted, **Largest increasing and decreasing subsequences**, *Canadian J. Math. 13* (1961) 179–191.

[S72] David Sankoff, **Matching sequences under deletion-insertion constraints**, *Proc. USA National Acad. Sci. 69* (1972) 4–6.

[S76] Kenneth B. Stolarsky, **Beatty sequences, continued fractions, and certain shift operators**, *Canad. Math. Bull. 19–4* (1976) 473–481.

[S79] Yossi Shiloach, **A fast equivalence-checking algorithm for circular lists**, *IPL 8–5* (1979) 236–238.

[S81] Yossi Shiloach, **Fast canonization of circular strings**, *J. Algs. 2* (1981) 107–121.

[S82] G. de V. Smit, **A comparison of three string matching algorithms**, *Software — Practice & Experience 12* (1982) 57–66.

[S90] Daniel M. Sunday, **A very fast substring search algorithm**, *CACM 33–8* (1990) 132–142.

[S98] Jeanette P. Schmidt, **All highest scoring paths in weighted grid graphs and their application to finding all approximate repeats in strings**, *SIAM J. Comput. 27–4* (1998) 972–992.

[Sa98] Kunihiko Sadakane, **A fast algorithm for making suffix arrays and for Burrows-Wheeler transformation**, *Proc. IEEE Data Compression Conference* (1998) 129–138.

[SC71] Hiroaki Sakoe & Seibi Chiba, **A dynamic programming approach to continuous speech recognition**, *Proc. Internat. Congress of Acoustics*, Budapest (1971) paper 20C13.

[SG98] Jens Stoye & Dan Gusfield, **Simple and flexible detection of contiguous repeats using a suffix tree**, *Proc. Ninth Annual Symp. Combinatorial Pattern Matching*, Martin Farach-Colton (ed.), Lecture Notes in Computer Science 1448, Springer-Verlag (1998) 140–152.

[Si00] R. Jamie Simpson, **Covers and the periodicity lemma**, private communication.

[SIPS01] Jeong Seop Sim, Costas S. Iliopoulos, Kunsoo Park & W. F. Smyth, **Approximate periods of strings**, *TCS 262* (2001) 557–568.

[SK83] David Sankoff & Joseph B. Kruskal (eds.), *Time Warps, String Edits, & Macromolecules: the Theory & Practice of Sequence Comparison*, Addison-Wesley (1983).

[SM97] João Setubal & João Meidanis, *Introduction to Computational Molecular Biology*, PWS Publishing (1997) 296 pp.

[SP95] N. J. A. Sloane & Simon Plouffe, *The Encyclopedia of Integer Sequences*, Academic Press (1995). See also

`http://www.research.att.com/~njas/sequences/`

[St94] Graham A. Stephen, *String Searching Algorithms*, World Scientific Publishing (1994) 243 pp.

[SW81] Temple F. Smith & Michael S. Waterman, **Identification of common molecular subsequences**, *J. Molecular Biology 147* (1981) 195–197.

[T06] Axel Thue, **Über unendliche zeichenreihen**, *Norske Vid. Selsk. Skr. I. Mat. Nat. Kl. Christiana 7* (1906) 1–22.

[T55] J. R. R. Tolkien, *The Lord of the Rings, Part II: The Two Towers*, Houghton Mifflin (1955).

[T68] Ken Thompson, **Regular expression search algorithm**, *CACM 11–6* (1968) 419–422.

[TU93] Jorma Tarhio & Esko Ukkonen, **Approximate Boyer-Moore string matching**, *SIAM J. Comput. 22–2* (1993) 243–260.

[U85a] Esko Ukkonen, **Algorithms for approximate string matching**, *Information & Control 64* (1985) 100–118.

[U85b] Esko Ukkonen, **Finding approximate patterns in strings**, *J. Algs. 6* (1985) 132–137.

[U92a] Esko Ukkonen, **Constructing suffix trees on-line in linear time**, *Proc. IFIP 92*, vol. I (1992) 484–492.

[U92b] Esko Ukkonen, **Approximate string-matching with q-grams and maximal matches**, *TCS 92* (1992) 191–211.

[UW93] Esko Ukkonen & Derick Wood, **Approximate string matching with suffix automata**, *Algorithmica 10* (1993) 353–364.

[V68] T. K. Vintsyuk, **Speech discrimination by dynamic programming**, *Cybernetics 4–1* (1968) 52–57.

[VZ70] V. M. Velichko & N. G. Zagoruyko, **Automatic recognition of 200 words**, *Internat. J. Man-Machine Studies 2* (1970) 223–234.

[W73] Peter Weiner, **Linear pattern matching algorithms**, *Proc. 14th Annual IEEE Symp. Switching & Automata Theory* (1973) 1–11.

[W94] Alden H. Wright, **Approximate string matching using within-word parallelism**, *Software — Practice & Experience 24* (1994) 337–362.

[WF74] Robert A. Wagner & Michael J. Fischer, **The string-to-string correction problem**, *JACM 21–1* (1974) 168–173.

[WM92] Sun Wu & Udi Manber, **Fast text searching allowing errors**, *CACM 35–10* (1992) 83–91.

[WMM96] Sun Wu, Udi Manber & Gene W. Myers, **A subquadratic algorithm for approximate limited expression matching**, *Algorithmica 15* (1996) 50–67.

[Y79] Andrew Chi-Chih Yao, **The complexity of pattern matching for a random string**, *SIAM J. Comput. 8–3* (1979) 368–387.

[ZL77] Jacob Ziv & Abraham Lempel, **A universal algorithm for sequential data compression**, *IEEE Trans. Information Theory 23* (1977) 337–343.

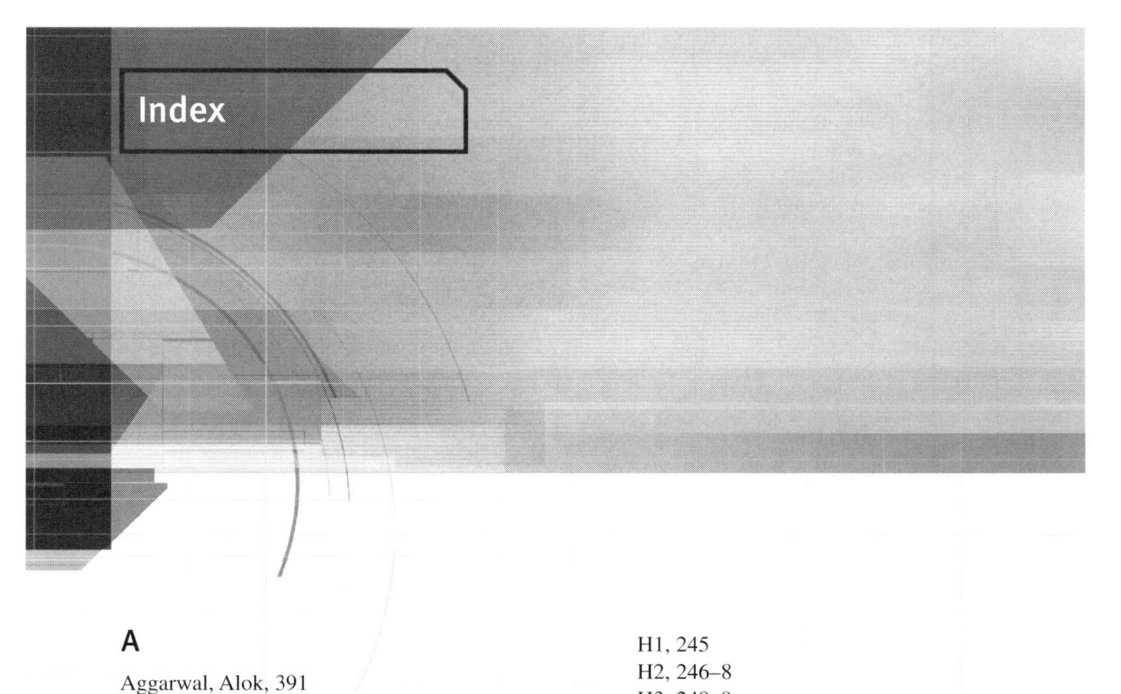

Index

A

B

C

G

H

I

J

K

L

M

N

O

P

Q

R

S

T

U

V

W

X

Y

Z